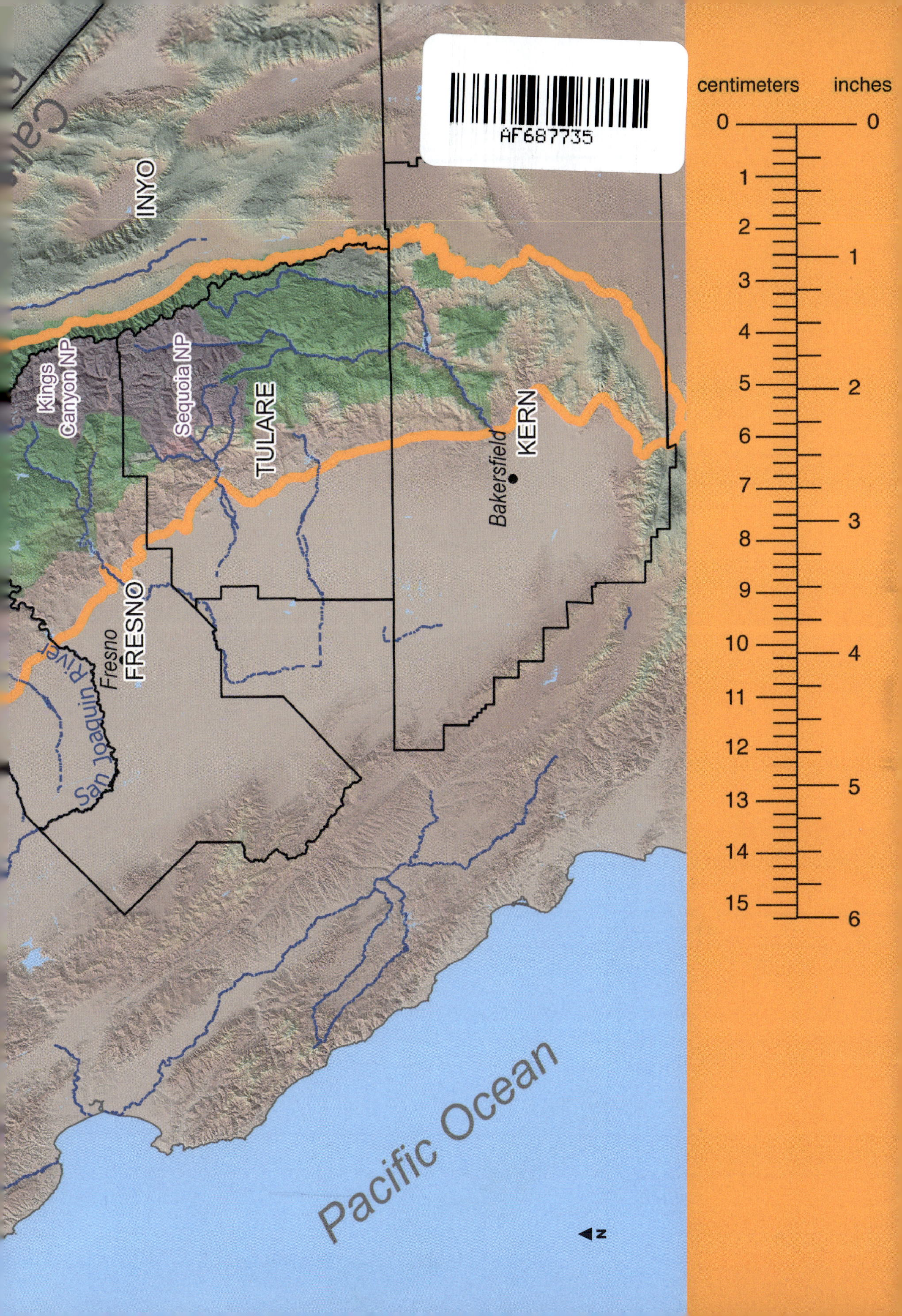
INYO
Kings Canyon NP
Sequoia NP
TULARE
KERN
Bakersfield
FRESNO
Fresno
San Joaquin River
Pacific Ocean
N
centimeters
inches
0 1 2 3 4 5 6 7 8 9 10 11 12 13 14 15
0 1 2 3 4 5 6

Flowering Plants of the Sierra Nevada

Flowering Plants of the Sierra Nevada

A Photographic Guide to over 1,000 Wildflowers

Joanna Clines
Stephen Sharnoff
Foreword by Peter H. Raven

HEYDAY, BERKELEY, CALIFORNIA
SIERRA COLLEGE PRESS, ROCKLIN, CALIFORNIA

Names: Clines, Joanna author | Sharnoff, Stephen, 1944- author
Title: The flowering plants of the Sierra Nevada : a photographic guide to over 1,000 wildflowers / Joanna Clines and Stephen Sharnoff.
Description: Berkeley, California : Heyday, 2025. | Includes bibliographical references and index.
Identifiers: LCCN 2025003476 (print) | LCCN 2025003477 (ebook) | ISBN 9781597146876 | ISBN 9781597146883 epub
Subjects: LCSH: Angiosperms--Sierra Nevada (Calif. and Nev.)--Identification | Angiosperms--Sierra Nevada (Calif. and Nev.)--Pictorial works | Wild flowers--Sierra Nevada (Calif. and Nev.)--Identification | Wild flowers--Sierra Nevada (Calif. and Nev.)--Pictorial works | Sierra Nevada (Calif. and Nev.)--Geography | LCGFT: Field guides
Classification: LCC QK142.7 .C5 2025 (print) | LCC QK142.7 (ebook) | DDC 582.1309794/4--dc23/eng/20250519
LC record available at https://lccn.loc.gov/2025003476
LC ebook record available at https://lccn.loc.gov/2025003477

Cover Art: Stephen Sharnoff
Cover Design: Victor Mingovits
Interior Design/Typesetting: Victor Mingovits
Published by Heyday in collaboration with Sierra College Press
P.O. Box 9145, Berkeley, California 94709
(510) 549-3564
heydaybooks.com
Printed in China by Imago
10 9 8 7 6 5 4 3 2 1

FSC
www.fsc.org
MIX
Paper | Supporting responsible forestry
FSC™ C005748

This book is dedicated to Sylvia Sharnoff, who conceived the idea for it in the summer of 1997. It is also dedicated to her father, Victor Duran, who spent years photographing and collecting plants in the Sierra Nevada and White Mountains, and to Peter Raven, who with enormous energy and enthusiasm supported the work of Sylvia and myself, first with our lichen projects, then with Sierra Nevada plants. And, finally, to my wife, Suzanne Guerlac, for her encouragement and support.

STEPHEN SHARNOFF

This book is dedicated to my mother, Jennifer Williamson, who has steadily supported me in this effort and in countless ways over the years. She is my rock. I also dedicate this book to the memory of Robert Winter, whose field biology course at Fresno City College changed my life entirely by opening my eyes to the world of biology, and to the memory of Howard Latimer, my professor, thesis advisor, mentor, and friend.

I am extremely grateful to the Western Mono people for the friendships, field trips, and mutual learning that we have endeavored together over the past thirty years. The days in the field with the Beecher sisters, Lou and Pat Beihn, Dirk Charley, Ron Goode, Gaylen and Judy Lee, Ruby Pomona, and many, many others have been deeply rewarding. Information presented about Native American uses of plants in the notes sections is documented in published books and articles about ethnobotany in California and beyond (cited in the bibliography). For central Sierra plants, my direct experience learning firsthand from the people themselves added a sparkle that exists to this day, and I look forward to more outings and projects in the future.

JOANNA CLINES

Contents

Foreword

My family has long had a special connection with the Sierra Nevada, memorably described as the "Range of Light." In the winter of 1846–47, my great-grandfather, then ten years old, and his parents survived a harrowing trek over the mountains to reach California. A century later, in the early spring of 1947, I was, at the age of ten, making my first botanical forays into the foothills of the Sierra. These led, beginning in 1950, to seven years of botanical exploration on Sierra Club outings in the High Sierra and a lifetime of interest in the rich and varied flora of the range.

But what is this "Range of Light," and how was it formed? The Sierra Nevada runs for some 400 miles north and south through eastern California, extending from the Cascades south to the Mojave Desert; it covers about a quarter of the state's land area. The mountains are about 80 miles wide at Lake Tahoe, where the range straddles the Nevada border, and 50 miles wide in the south. The Sierra Nevada separates the high desert basins to the east from the Great Valley of California to the west, rising steeply from the desert and sloping more gently down to the valley. Huge cliffs and slopes of amazing silvery granitic rocks dominate much of the Sierra, which led the early naturalist and conservationist John Muir to give it the lasting name "the Range of Light," a lovely and apt term that helped bring the mountains to the attention of a wider public. The Sierra Nevada was well known, of course, to the Indigenous peoples who occupied and tended the land in what is now California for over ten thousand years. This huge body of light-colored granitic rock was uplifted as a result of the tectonic movement of plates that spanned tens of millions of years. During more recent times, volcanic activity took place in the Sierra north of Yosemite, and its traces are still evident.

Worldwide earth cooling over the past 2.6 million years resulted in the widespread formation of glaciers in the Sierra. This, in turn, carved out the sharply defined canyons and other features of the range that so impressed John Muir and other visitors, in recent years numbering some four million people annually visiting Yosemite alone. This 2.6-million-year period is one in which ice sheets expanded in the north and glaciers formed in valleys and other suitable places further south. The Yosemite waterfalls resulted from side canyons with glacial ice draining into the main valley and merging with the enormous glacier that carved the U-shape of Yosemite Valley. These drainages were left high on the valley walls when the glaciers melted, creating immense drops for streams draining into Yosemite Valley.

Obviously, the presence of particular kinds of plants and animals in the Sierra Nevada is closely related to these geological events. With glaciers and their associated climate came cold winters and an influx of plants and animals from the north. Once established in the range, these organisms could often survive warmer climates by climbing to higher elevations; examples include the pika (*Ochotona princeps*), snow apollo butterflies (*Parnassius behrii*), or perhaps the Sierra Nevada sky pilot (*Polemonium eximium*), a distinctive wildflower which may have been derived from one of the other two dozen species of its genus as the climate warmed. Lower down on the east (desert) side are many kinds of plants that range north or west from the Mojave Desert, while on the west side is a rich assemblage of plants and animals ranging from the largely endemic (found nowhere else) to some shared with the Coast Ranges (e.g., manzanitas, gilias, and clarkias).

The geological history of California west of the Sierra Nevada is complex and partly explains plant diversity patterns in the Sierra. Before about two million years ago, what is now the Central Valley was occupied by what was essentially a bay, an arm of the Pacific Ocean, open to the south. A lake formed in the basin between about 758,000 and 665,000 years ago, and most of the valley may have been marshy when the lake was "dry." Thus, there seems to have been a barrier to migration of plant species between the Coast Ranges and the Sierra for a great deal of recent time. Closely related species of the same genera are often found on opposite sides of the valley, although there has been a great deal of migration around the northern end of the Sacramento Valley and long-distance dispersal between the two areas.

What does this book include? About half of the vascular plant species found in the US occur in California, some 5,500 species counting both native and introduced ones; about 3,500 of these occur in the Sierra Nevada. This book covers some 950 species of wildflowers throughout the Sierra Nevada, illustrated with beautiful images and sometimes markers to indicate the distinguishing features. It does not include trees; there are many tree books and apps available. The book mostly leaves out wind-pollinated species with reduced flowers, like grasses and sedges that are often difficult to identify in the field. Information about uses of plants and methods for managing the land by Indigenous peoples brings to life the connection between humans and plants. Using this book to identify the flowers seen along a trail in the Sierra will make the hike a special joy!

John Thomas Howell (1903–94), Curator of Botany at the California Academy of Sciences, San Francisco, and successor to Alice Eastwood (1859–1953) in that role, was the mentor for my first studies in botany, from age ten onward. In his later years at the Academy, in the 1950s and '60s, he was devoted to the idea of producing a book about the plants of the Sierra Nevada, collecting as many specimens as he could and keeping track of them. He was always an "encourager," knowing that the more people were involved in a project, large or small, the better and more complete the final product would be.

Tom would have been delighted with this book, and you, the reader, will be also as you review the beautiful images and useful tips for identifying these plants. I recommend it to you with great enthusiasm!

Peter H. Raven
President Emeritus of the
Missouri Botanical Garden

Acknowledgments

We gratefully acknowledge generous donations for the project from the Bechtel Foundation, Victoria and Dorn Dean, Lawrence R. Heckard Endowment Fund of the Jepson Herbarium, JiJi Foundation, Marisla Foundation, and Stanley Smith Horticultural Trust.

We are deeply grateful to Sierra College Press for their support as copublisher of this book. Thank you especially to Lynette Vrooman, Joe Medeiros, and Shawna Martinez.

Major financial support for the project was given by Peter Raven in memory of John Thomas Howell, his first mentor in botany, and by David and Meera Sharnoff in memory of Sylvia Sharnoff.

Other significant private donors include Averill Babson, Keith & Tsing Bardin, Mary Jane Brimhall, Megan Dalton, Arden & Lloyd Davidson, Barbara & Barry Deutsch, Mark Schwartz & Julie Feisst, Beatrice Fontana, Nancy Hair, Ella & Neal Krucoff, Glenn Lym, Nina McLellan, Gladys Sharnoff, and Jake Sigg.

The project was helped significantly in its earlier stages by Linda Vorobik and Elizabeth Wenk.

Staff at the Jepson Herbarium at UC Berkeley gave important help along the way, thanks to Jason Alexander, Bruce Baldwin, Andrew Doran, Kim Kersh, Brent Mishler, Richard Moe, Ana Penny, Staci Markos, John Strother, and Margriet Wetherwax.

Special thanks to Susan D'Alcamo for guiding the initial fundraising and providing much useful advice along the way.

Others who generously gave their time and assistance early on were Irene Acosta, Julie Carville, Phyllis Faber, Toni Fauver, Kathryn Ferrier, Linda Geiser, Annaliese Miller, Betsy Ringrose, Stew Winchester, and Rebecca Yahr.

Thanks also to Holly Forbes and Roger Raiche of the UC Botanical Garden, and to Steve Edwards, Burt Johnson, and Sue Rosenthal of the Tilden Botanical Garden.

Big thanks to Forest Service botanists and rangeland managers for help with collection permits, locations, and plant identification: Jennie Haas, Linnea Hanson, Lawrence Janeway, Kathleen Nelson, Dirk Rodriguez, Mark Stieler, Linda Tanner, Susie Urie, Kathy VanZuuk, Sue Weis, and Dave Weixelman. Also to National Park Service scientists Alison Colwell (now at UC Davis), David Graber, and US Geological Survey Plant Ecologist Peggy Moore.

We are grateful to the following people and organizations: Ihsan A. Al-Shehbaz, Kat Anderson, Mary Andrews, Bonnie Bladen and Ray LaClergue, Russ Cary, Lois Conner Bohna, Juri de Vos, Barbara Ertter, Peggy Fiedler, Naomi Fraga, Matt Guilliams, Dena Grossenbacher, Brett Hedrick, Sarah Hedrick, Leigh A. Johnson, Peter Hoch, Dean Kelch, Ronald B. Kelly, Susan Kephart, Ann Mendershausen, Kyle Merriam, Alan Moe, Nancy Morin, Michael Park, Craig Poole, James Reveal, Anne Russell, Mike Russler, Adam Schneider, Steve Schoenig, Jim Shevock, Sierra Foothill Conservancy, Michael Simpson, John Stebbins,

Thomas Stoughton, Dean Wm. Taylor, Genevieve Walden, Sue Weis, Margriet Wetherwax, Hugh A. Williamson, Tessa Williamson, and Chris Winchell.

Some of the photos in the book were taken by Sylvia Sharnoff and a large number by Joanna Clines. We wish to thank the following photographers who kindly donated their pictures: Barry Breckling, Debra Cook, Julie Kierstead, Neal Kramer, Steve Matson, Keir Morse, Aaron Schusteff, Michael Simpson, and Carol Witham. Please see the photo credits for a complete list.

Special thanks to Julie Kierstead for her careful manuscript review and editing assistance.

Finally, we wish to thank our wonderful team of editors at Heyday, Marthine Satris and Emmerich Anklam; Steven Scholl, who provided editorial guidance; Olivia Ngai, the copyeditor and proofreader; Archie Ferguson, Heyday's art director; Victor Mingovits, the designer; and the expert Heyday staff support from Emily Grossman and Suzanne Rivecca.

Abbreviations, Measurements, and Symbols

ABBREVIATIONS

ALL: In the distribution section of each species description, "All" means *occurs throughout the Sierra Nevada, within limits like elevation range and habitat type*; it does not mean that the species is guaranteed to occur in every county.
BLM: Bureau of Land Management
CAL-IPC: California Invasive Plant Council
CCH: Consortium of California Herbaria
CDFA: California Department of Food and Agriculture
CDFW: California Department of Fish and Wildlife
CNPS: California Native Plant Society
CRPR: California Rare Plant Rank. A rarity ranking assigned by the California Native Plant Society and botany experts across California.

- **1B:** Plants that are rare, threatened, or endangered in California and elsewhere. Most list 1B plants are California endemics, and a majority of them have declined in abundance over the last century.
- **2B:** Plants that are rare, threatened, or endangered in California but common elsewhere.
- **3:** Rare plants about which more information is needed. Often this is due to taxonomic uncertainty.
- **4:** Plants of limited distribution or infrequently encountered over a wider area. A watch list.

Threat rank, added to the CRPR rank to complete the rank (e.g. 1B.1, 4.3).

- Seriously threatened in California: over 80% of occurrences threatened, high degree of threat.
- Moderately threatened in California: 20%–80% of occurrences threatened, moderate degree of threat.
- Not very threatened in California: less than 20% of occurrences threatened, low degree of threat.

ID: identification
CA: California
SN: Sierra Nevada
SPP.: multiple species
SSP.: subspecies
SSPP.: multiple subspecies

Blue dicks and fiddlenecks in spring.

VAR.: variety
CM: centimeters
DM: decimeters
FT: feet
IN: inches
M: meters
MM: millimeters
LBS: pounds
OZ: ounces

MEASUREMENTS

(see inside front cover for meter-feet scale)
1 meter = 3.28 feet
100 meters = 328 feet
1000 meters = 3281 feet
1000 feet = 305 meters
(see inside back cover for centimeter-inches scale)

SYMBOLS

✿	Sierra Nevada endemic
°	degrees
±	more or less, plus or minus
=	equal to
+	more, plus
<	less than (5 < 10)
≤	less than or equal to
>	greater than (10 > 5)
≥	greater than or equal to

Peirson's paintbrush (*Castilleja peirsonii*) growing in the foreground at Little Lakes Valley, Mono County.

Introduction

The Sierra Nevada is the longest and tallest mountain range in the Lower 48 states, extending 400 miles along the eastern side of California and including a small part of western Nevada near Lake Tahoe (inside front cover map). Elevation varies from about 150 meters (500 feet) in the western foothills to above 4268 meters (14,000 feet) at the highest peaks of the crest, offering some of the most beautiful scenery in the United States. This elevation span combined with the influence of topography, latitude, geology, soils, and weather patterns results in a great diversity of growing conditions for plants. While the Sierra Nevada comprises less than a quarter of the land base of California, it contains nearly half of the vascular plant taxa (species, subspecies, and varieties) known to occur in the state.

The Sierra Nevada is known the world over for its scenery, geologic wonders, and year-round recreation opportunities. Wildflowers are a major attraction—many people flood to the Sierra in the spring and summer to see exquisite wildflower displays. The wildflowers featured in this book occur on and around spectacular background features that draw people to the Sierra Nevada. Some highlights are listed below.

- Mount Whitney in the southern Sierra is the tallest peak in the Lower 48 states at 4421 meters (14,505 feet). Towering above tree line, Mount Whitney's slopes feature alpine plants such as sky pilot, Sierra primrose, and Pacific hulsea.
- Giant sequoias are the world's most massive trees by volume. The Sierra Nevada is the only place on earth to see giant sequoias (*Sequoiadendron giganteum*) in their natural range. About seventy-three groves of giant sequoias are scattered from near Lake Tahoe to the southern Sierra, where the majority of groves occur. The oldest trees are over three thousand years old. Trees can reach over 300 feet in height, and the largest specimens are over 30 feet in diameter. The forest floor in sequoia groves has a rich mix of ferns and flowering plants; some common species are western azalea, bead lily, fairy bells, Sierra tiger lily, snowberry, and American vetch.

ABOVE: Alpine lakes seen from Puppet Pass, just south of French Canyon. **OPPOSITE:** Giant sequoia in Sequoia National Park.

- Yosemite Valley has nearly four million visitors each year, literally from all over the planet. The iconic rock profiles of Half Dome and El Capitan host a surprising number of flowering plants adapted to live on cliffs and in rock cracks. The diversity of flowers on and around these rocks is high enough to slow you down considerably if you stop to look at them along the trail (also a good excuse to take a rest!).
- Lake Tahoe is a natural freshwater lake lying at 1898 meters (6227 feet) average surface elevation in the northern Sierra. It is the second deepest lake in the US (after Crater Lake in Oregon), and one of the largest at 22 miles long by 12 miles wide. There are myriad trails in the Lake Tahoe basin for exploring the plant life. Wildflowers are abundant and varied.
- The Kings River canyon in the Sequoia National Forest is the second deepest in the US at 2500 meters (8200 feet), from river to rim.

This book focuses on flowering herbaceous plants (nonwoody annuals and perennials) and shrubs. Even for this relatively large guide, less than a third of 3,542 vascular plant taxa found in the Sierra Nevada could be included to keep the book to a reasonable size. Of these more than 3,500 taxa, about 13 percent are non-native. Our intent is to present a variety of plants people will see as they ramble the Sierra Nevada, including some non-natives. Taxa selected for inclusion meet at least one of the following criteria:

- Common native (and some non-native) species likely to be encountered within one or more vegetation types of the Sierra Nevada
- Native species characteristic of major Sierra Nevada ecosystems, especially showy and beautiful species that always catch the eye when encountered
- Some common or rapidly expanding invasive, non-native species that cause ecological damage and, if detected early in a new area, could be eradicated before they spread irreversibly
- Species endemic to the Sierra Nevada that are worth a special trip, even if they're not widespread
- Rare species growing in beautiful places that can be shown in a book for the general public without jeopardizing them
- Species pairs or groups that tend to be difficult to tell apart in the field but can be distinguished with the right diagnostic hints and photos

ABOVE: Basketmaker Lois Conner Bohna weaving a basket with deergrass (***Muhlenbergia rigens***), sedge or "white root" (***Carex*** **sp.**), bracken fern (***Pteridium aquilinum***), and redbud (***Cercis occidentalis***).
OPPOSITE: Lupines, sunflowers, and paintbrush, northern Sierra foothills.

We do not cover algae, fungi, lichens, mosses, ferns, or conifers (none of which are flowering plants). Of flowering plants, we leave out most grasslike plants (e.g., grasses, rushes, and sedges) and most wind-pollinated shrubs such as willows because they're more difficult to identify in the field, and because including them would result in an enormous book. Wind-pollinated plants tend not to have showy flowers, as there's no need to produce a colorful signal to attract insect pollinators if you're relying on the wind. Some trees are shown and named in the ecological zone descriptions, but trees are not included as entries in the Species Descriptions, as there are many books and apps covering the trees of the Sierra (exceptions are some species that can be either a shrub or a tree).

Native People and Flowering Plants

The Sierra Nevada has been inhabited by people for many thousands of years, long before Europeans arrived. Woven throughout many species descriptions in this book are notes on uses of plants by Native peoples as well as methods used to tend and manage plants (e.g., fire, digging, selective harvesting). Native Americans relied completely on native plants and animals for food, tools, weapons, utensils, shelter, medicine, clothing, art, and much more.

The significant role Sierra Nevada tribes played in land management over the millennia is becoming clearer and clearer. The Sierra Nevada was not a "pristine" wilderness prior to the arrival of Europeans. Complex and careful land management practices were integral to the daily lives of tribes. Burning, digging, tending, pruning, and replanting were among the practices applied at carefully selected times each year. The people employed purposeful, sustainable methods for ensuring the abundance of plants needed for basketry, food, medicine, cordage and nets, soap, living structures, weapons, traps for game animals, spiritual and ceremonial uses, and more.

Landscapes in the Sierra look significantly different from how they looked a few hundred years ago prior to the arrival of European people and European plants. This was brought home to me in the 1990s during an Indian Days event at the Auberry Library in Fresno County. Melba Beecher, an exceptionally knowledgeable Mono elder, brought four or five sandwich-sized baggies, each with about a cubic inch of tiny seeds. Each bag had the Mono name for the seeds written on it. Melba said the seeds had been passed on to her by her mother (at this time, Melba was in her eighties). I recognized chia seeds and figured out that another baggie had seeds of clarkias, and another of bushy blazing star, but I didn't recognize the others right away. They turned out to be seeds of at least one species of *Cryptantha* (Boraginaceae), several of which show up in great numbers after wildfires. It was an epiphany to learn that small seeds of these annual wildflowers were important food sources and that they had to be orders of magnitude more abundant in the past to have served this need. Even today, after wildfires, all of these plants have population explosions, and one can get a feel for how the world used to look. In addition, bulbs, corms, and tubers of plants with colorful flowers, like blue dicks, camas, Ithuriel's spear, mariposa lilies, some sanicles, and yampah, to name a few, were called Indian Potatoes. These plants were a mainstay of the diet of many California tribes and were harvested and tended to be a sustainable source of food. Tending included selective harvesting, digging to maintain soil texture, and burning. The larger bulbs of each species would be gathered, leaving smaller ones to grow larger for future years or replanting smaller bulblets that were attached to larger bulbs. To this day, bulb-forming wildflowers proliferate after wildfire, sometimes appearing in such abundance that entire slopes are covered with them (e.g., snake lilies and Mariposa lilies).

California Poppy (*Eschscholzia californica*).

Once most of the homelands of Sierra Nevada tribes became national forests, national parks, and private property, Native peoples could not continue burning as they had for millennia, and landscapes and gathering areas ceased being maintained by Traditional methods at the same scale. This, combined with fire suppression policies, past livestock grazing and logging practices, and the establishment of non-native annual grasses and forbs from Europe, changed the land immensely. Native peoples played a significant role in maintaining the landscape in an ecologically sustainable state, a previously underestimated factor in the dynamics of Sierran vegetation. All this is not to say that Traditional management is in the past—gathering areas are still tended, acorns are still gathered, cooked, and eaten, and more and more restoration projects using Traditional Ecological Knowledge are occurring. Incorporation of cultural burning into the burn programs of federal land management agencies is just beginning to be actualized in the SN.

Much of the information on ethnobotany and tribal land management in the notes of the species descriptions is from writings by Ron Goode, Gaylen Lee, Kat Anderson, and other sources listed in the bibliography. Complementing these published sources, I have been fortunate and honored to learn directly from the Western Mono People during countless field trips and restoration projects over the past thirty-plus years.

This is an exciting time of rapid positive change for Native Americans in California. There is growing interest in and respect for tribal land management practices and applying these techniques to heal degraded ecosystems. Scientific research on these practices in collaboration with tribes demonstrates benefits to the natural ecosystems of California, as well as increasing the traditional, healthier uses of native plants for Native peoples. On June 18, 2019, Governor Newsom issued Executive Order N-15-19 as a formal apology to tribes. One requirement of this order was to establish the California Truth & Healing Council, which has now been active for five years. In June 2024, the governor announced a commitment to return 2,800 acres of ancestral lands to the Shasta Indian Nation. On October 7, 2020, Executive Order N-82-20 was issued, committing to conserving 30 percent of California's native biodiversity by establishing the California Biodiversity Collaborative. Throughout the 2020 EO, tribes are acknowledged as integral players in accomplishing this goal: "Since time immemorial, California Native Americans have stewarded, managed and lived interdependently with the lands that now make up the State of California..." To learn more about the excitement and enthusiasm all of this is bringing about, see the resources section at the end of the book.

Structure of This Guide

Classification and arrangement mirror *The Jepson Manual, Second Edition*, and names reflect subsequent updates regularly made in the Jepson eFlora up to the time this book went to press. The Jepson eFlora is the accepted authority for plant taxonomy in California. The four major groups presented in the eFlora are arranged in the following order: Nymphaeales, Magnoliids, Eudicots, and Monocots (descriptions of each on page 8). Within each of the four groups, species are arranged alphabetically by the scientific name of the family, and within each family alphabetically by genus, then species, and subspecies or variety if applicable. In a few special cases, the alphabetical order of two species has been switched for best display.

Because the Jepson eFlora is available online at no cost (donations and memberships are always appreciated), with full descriptions of families and genera, we omitted family and genus descriptions to fit in more photos of plants with species descriptions. Additionally, we do not have dichotomous keys in this book, but you can use the eFlora online keys if you're inclined to dig deeper into how taxa are distinguished from each other.

Because this book covers about a thousand of the approximately 3,540 vascular plant taxa growing in the Sierra Nevada, you are bound to encounter wildflowers that are not in the book. Then what? We suggest that you photograph the plant and get closeups of flowers and leaves from different angles, fruits and seeds if present, leaf arrangement on the stem and at base, the entire plant, and the habitat. Take basic notes on location, associated plants, GPS coordinates, and elevation (a GPS-enabled smartphone records the latter two for you). Look into downloading a few helpful field guide apps that don't require a cell signal and that are either free or quite inexpensive. See the resources on page 585 for apps, books, and websites. iNaturalist is a citizen science website for posting photos of unknown taxa—members of the iNaturalist community can then suggest IDs. Also, get to know your way around Jepson eFlora and the Calflora website and mobile apps.

A flowery streambank in the subalpine, eastern Sierra.

Basics of Plant Identification

Let's start with how scientists name plants (and other living things). The scientific name is ideally a unique name that is internationally accepted as the *only* name for that plant—the name is called a binomial because it consists of both a genus and a species name. For example, the California poppy is in the genus *Eschscholzia*, and the species name (or epithet) is *californica* (meaning "of California"). The similar-looking tufted poppy is also in the genus *Eschscholzia*, but the species name is *caespitosa* (meaning "tufted"). Genera are grouped into families as species are grouped into genera; thus, *Eschscholzia* is placed, along with thirteen other genera in California, in the Poppy family, Papaveraceae.

Since the Vienna International Botanical Congress in 1905, plant family names must be based on a genus in that family plus the ending *-aceae*. However, nine family names that do not comply with this rule are still recognized as legitimate alternative names:

- The Pea family (Fabaceae) has long been known as Leguminosae.
- The Sunflower family (Asteraceae) also goes by Compositae, descriptive of the flower heads composed of disc flowers and ray flowers.
- The Carrot family (Apiaceae) also goes by Umbelliferae, descriptive of its umbrellalike inflorescences.
- The Mustard family (Brassicaceae) is also called Cruciferae, referring to the cross-shape of the four petals.
- The Mint family (Lamiaceae) is also called Labiatae, reflecting that most flowers have an upper and a lower lip.

Scientific names for all living things are based in Latin or Greek and are the same across the globe, no matter the country or the language spoken. The correct way to write a binomial is in italics, or underlined, with the genus name capitalized and the species name beginning with a lowercase letter (e.g., the binomial for humans is *Homo sapiens*). Because there can be a lack of consensus among experts in interpreting how genera, families, and species should be grouped and named, be aware that different books may not agree. For this book, we are following the Jepson eFlora, which may differ from names accepted in the *Flora of North America* or in iNaturalist, for example.

Some species are divided further into varieties or subspecies (subspecies is technically a higher rank than variety, but in reality one or the other is used; e.g., there are no plants in this book that have a subspecies as well as a variety). An example is given by the common spring wildflower baby blue eyes, *Nemophila menziesii,* which is represented by two varieties in the Sierra Nevada. Var. *menziesii* is the most common, the familiar large, sky-blue flowers with white centers. Less common is var. *integrifolia,* with smaller flowers that are often white or light blue have fewer seeds and fewer lobes on the leaves.

This book focuses on angiosperms, which are broadly defined as vascular flowering plants with seeds that form within a protective structure called the ovary. Simply put, the ovary houses ovules, which, after fertilization via a pollen grain conveying sperm to an egg, mature into seeds, while the ovary wall matures into the fruit wall. In contrast, gymnosperms (conifers and relatives) are vascular plants with seeds, ovules, and pollen, but ovules are not contained within an ovary, and there are no flowers.

Until recently, flowering plants were divided into two classes named after the number of cotyledons, or "seed leaves" (leaves that absorb stored nutrients and sustain the germinating seedling until the true leaves develop and begin photosynthesizing), contained within the embryo: The monocotyledons (monocots) have embryos with one cotyledon, and the dicotyledons (dicots) have embryos with two cotyledons. While the monocots remain intact as a group, the dicots are now broken into four major groups, or clades.

The group formerly called dicots now consists of:

1. **NYMPHAEALES**—aquatic plants with two cotyledons, flowers with parts in multiples of three or spirally arranged, and pollen with only one pore through which the pollen tube grows during fertilization
1. **MAGNOLIIDS**—terrestrial plants with two cotyledons, flowers with parts in multiples of three or spirally arranged, and pollen with only one pore through which the pollen tube grows during fertilization
1. **CERATOPHYLLALES**—hornworts (aquatic plants not included in this book; e.g., coon's tail/hornwort)
1. **EUDICOTS**—meaning "true dicots," with two cotyledons, flower parts in fours and fives, and pollen with three pores (most of the species in this book)

Monocots, the remaining group of plants, have one seed leaf (cotyledon), flower parts in multiples of three, and mostly leaves with parallel veins. See the table on page 9 for a summary of field-visible differences between eudicots and monocots.

Plants in this book are arranged alphabetically by their scientific names because the scientific name is the only name unique to that taxon (species; subspecies or variety) worldwide. This helps with learning to recognize families and genera, and we avoid two major problems of identifying by common names:

1. **WHEN THE SAME COMMON NAME IS USED FOR MORE THAN ONE SPECIES.** Let's take the common name "bluebells." In California this generally refers to members of the genus *Mertensia*, in the Boraginaceae (Borage family). Referring to a wildflower as bluebells in California can also mean a member of the genus *Phacelia*, in the Hydrophyllaceae (Waterleaf family), but in England a bluebell refers to *Hyacinthoides*, a member of the Hyacinth family.
2. **WHEN MANY COMMON NAMES ARE USED FOR ONE SPECIES.** For example, *Nuphar polysepala* is known by at least four common names: Rocky Mountain pond-lily, western spatterdock, cow-lily, yellow pond-lily, and western pond-lily.

Our use of scientific names follows that of the Jepson eFlora, and keeping up with name changes is a challenge. With the availability of molecular tools to study the genetics and evolutionary relationships of plants, name changes have been coming at us faster and faster! This will continue, and it reflects a wonderful thing—that we know vastly more about the evolution of plants and how taxa are related to each other; but it can be frustrating to get used to a scientific name and have it change, sometimes several times. Our use of common names generally follows the Jepson eFlora (if given), Calflora, or the USDA Natural Resources Conservation Service PLANTS database.

Character	Eudicot	Monocot
Arrangement of major leaf veins	Usually netlike—main veins pinnate like a feather or palmate like the fingers on a hand	Main veins usually parallel
Number of parts (petals, sepals, pistils, stamens) per flower	Often in multiples of four or five, rarely in multiples of three, sometimes indeterminate (numerous petals or stamens that are not always the same exact number for each flower)	Usually in multiples of three; typically three petals and three sepals, or six similar-appearing, petal-like "tepals"
Number of cotyledons (seed leaves) in embryo		
Examples		

Table of the major differences visible in the field between eudicots and monocots.

Parts of Flowering Plants Used in Identification

This section provides some basic concepts that help form the basis for plant identification, but even better, it builds a foundation to enhance the sense of wonder about flowering plants that you already feel if you are holding this book in the first place. Understanding the function of the stigma as the place where a pollen grain lands and grows a pollen tube that conveys the sperm nuclei to an egg within the ovary (often a considerable distance) allows you to marvel at this process of nature long after you've used the number of petals and their measurements to place a name on your mystery plant. The glossary on page 575 provides definitions of botanical terms used in the introduction and species descriptions.

FLOWER PARTS: A "typical" flower has the basic parts shown below. **PETALS** are modified leaves, generally serving to signal pollinators with color, patterns, shape, or fragrance that there is a nectar and/or pollen reward at this location. Petals may or may not actually be present; over evolutionary time some lineages have lost or modified petals. In some plants, petals are absent, but the sepals evolved to function as petals (e.g., California flannelbush). All the petals of a given flower are collectively called the **COROLLA**. In a flower with petals fused at the base, the term **COROLLA LOBE** is used rather than petal. **SEPALS** are the (usually) green modified leaves surrounding the petals (think of the green parts of a rosebud). Sepals serve to protect the developing bud, but their function has also evolved in a variety of ways in different lineages. Sometimes the sepals are petaloid (meaning "petal-like"), providing a visual signal for pollinators as mentioned above. In other plants, the sepals function in dispersal; e.g., in some *Lewisia* species they become papery "sails," which propel the ripe seeds in the wind and scatter them across granitic outcrops. All the sepals of a flower are collectively called the **CALYX**. In species with sepals fused at the base, the term **CALYX LOBE** is used rather than sepal.

The position of the **OVARY** in relation to the petals and sepals is a key feature in recognizing plant families. An ovary is **SUPERIOR** if it appears to be above the point of attachment of the petals and sepals, and it is **INFERIOR** if it appears positioned below the sepals and petals. The ovary contains **OVULES**, which contain **EGGS**. The eggs receive sperm when a pollen grain lands on the **STIGMA** and grows a pollen tube that conveys two sperm nuclei through the **STYLE**, one of which joins with an egg. The fertilized eggs ultimately become seeds if all goes well (sufficient water, light, and nutrients are available, the plant escapes herbivory, etc.). The ovary, style, and stigma are collectively called the **PISTIL.** The **STAMENS** consist of a **FILAMENT** and a pollen-producing **ANTHER**.

LEAVES AND STEMS: Below is a diagram of a "typical" leaf with the basic terminology used to describe leaves and their attachment to the stem. A common feature of leaves that helps with plant identification are whether they are **OPPOSITE** each other on the stem, **ALTERNATE, WHORLED,** and/or in a basal **ROSETTE.** Leaves can be **SIMPLE** or **COMPOUND** (made up of leaflets), and have an endless variety of shapes and margin types, from **ENTIRE** (smooth-margined) to **TOOTHED** or **LOBED**. The pattern of the veins is also key. Monocots usually have parallel veins. For eudicots, if they are arranged in a featherlike pattern, the venation is said to be **PINNATE.** If they are arranged more like fingers radiating from the palm of a hand, they are said to be **PALMATE.** The angle formed by the stem and the **PETIOLE** (leaf stalk) is called the **AXIL**.

The following diagrams illustrate some common inflorescence types and terms used throughout the book, as well as showing features of three common families that have unique terminology (Apiaceae, Asteraceae, and Fabaceae).

A note about conventions: Because descriptions of families and genera are available online via the Jepson eFlora, we decided to include more photos of plants and leave out descriptive text preceding each family and genus.

Measurements given in the species descriptions are for length unless otherwise indicated; thus, we do not write "long or tall" after the measurement as a rule. Measurements are given in metric units, as is standard in the science of botany; there is a ruler with both metric units and inches on the inside back cover. Carrying an inexpensive 15-centimeter (6-inch) plastic ruler is highly recommended, along with a hand lens with 10x or higher magnification. I find it helpful to memorize that 1 inch = 25.4 millimeters, or 2.54 centimeters.

We provide elevation ranges in meters (m), as in the Jepson eFlora. This can be tricky as many of us are used to feet as a standard of measurement. A quick calculation from meters to feet using a smartphone calculator is helpful: Multiply meters by 3.28 to get feet (and divide feet by 3.28 to get meters). See the inside front cover for a scale showing both meters and feet.

Keep in mind when using the elevation ranges and blooming times that plants bloom earlier at lower elevations and at lower latitudes.

Parts of a "typical" flower with a superior ovary.

Geophysical Features of the Sierra Nevada

The boundary of the Sierra Nevada for the purposes of this book is the area shown in the map on the inner cover. It matches the Sierra Nevada regions used by the Jepson Flora Project. The Sierra is bounded to the west by the Great Valley and to the east by the base of the eastern slope where it adjoins the Great Basin at approximately 6000 feet in elevation. The northern limit is in Plumas County around Lake Almanor at the southern end of the volcanic Cascade Range (approximately 40.1 degrees latitude), and the southern limit is the vicinity of Tehachapi Pass in Kern County (approximately 35.1 degrees latitude). The distance from Lake Almanor to Tehachapi Pass is about 400 miles along a northwest-southeast trending axis. The width of the range varies 60–80 miles.

The average gradient of the western slope of the Sierra Nevada varies 2–6 degrees and extends 50–65 miles from the Great Valley eastward to the crest, while the east slope can be as steep as 25 degrees where it abruptly drops to the flatter valleys of the Great Basin. (For reference, 45 degrees is a 100 percent slope).

The Sierra crest rises from north to south, achieving its highest point at Mount Whitney (4419 meters, 14,494 feet). South from there, Walker and Tehachapi Passes retain a lower crest but mark the southern terminus of the range.

Although the elevation of the crest is lower in the north, higher latitude results in continued alpine areas at lower elevations than are typical for the southern Sierra. The Tahoe Basin is an anomaly for the Sierra in having large, flatter expanses of land east of the crest. Mammoth Pass at 2861 meters (9386 feet) is the only point from south of Mount Whitney to the north of Yosemite National Park where the Sierra Crest drops below tree line. This low-lying section of the crest with subalpine and montane vegetation and no alpine areas is a barrier to alpine plants moving north–south along the crest and a pathway for lower-elevation species to move east–west.

The western slope is dissected by twelve major river drainages. From north to south, they are: Feather, Yuba, American, Cosumnes, Mokelumne, Stanislaus, Tuolumne, Merced, San Joaquin, Kings, Kaweah, and Kern. The rivers of the north and central Sierra run generally east–west and are shorter, with somewhat shallower gorges. The Kings and San Joaquin Rivers have drainages with many forks that wind in deep canyons through the highest Sierra and then merge into giant rivers in the Great Valley. The Kern River is the only major Sierra Nevada river to run north–south (along a fault line). On the east slope, the Truckee, Carson, Walker, and Owens Rivers drain eastward into the Great Basin. Sadly, water from the Owens River is now diverted to Los Angeles.

River canyons are characterized by cooler microclimates and often contain the lowermost locations for plant species that usually occur at higher elevations. For example, in the Merced River Canyon downstream of Yosemite National Park, one can see ponderosa pines and incense cedars growing in foothill chaparral much lower than usual due to the cooler, moister conditions found along the river.

BASIC PARTS OF A TYPICAL FLOWER

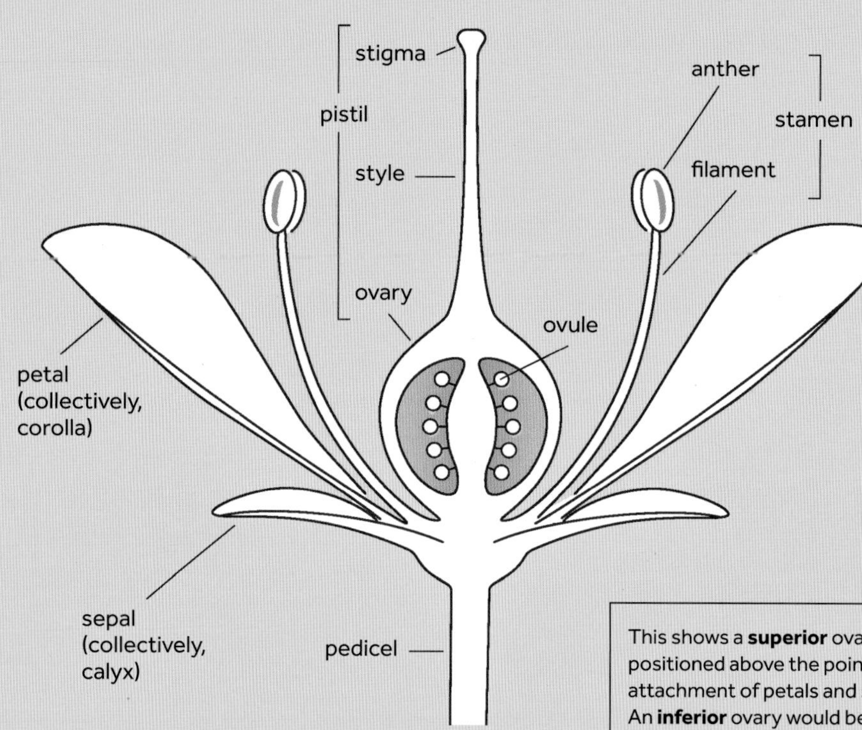

This shows a **superior** ovary, positioned above the point of attachment of petals and sepals. An **inferior** ovary would be positioned below the attachment point of petals and sepals.

SIMPLE AND COMPOUND LEAVES

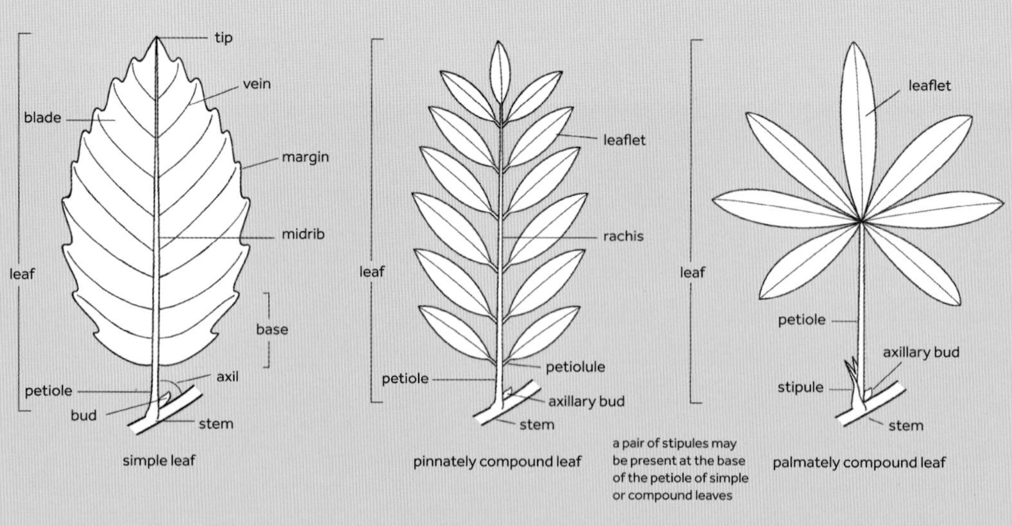

LEAF ARRANGEMENTS AND ORIENTATIONS

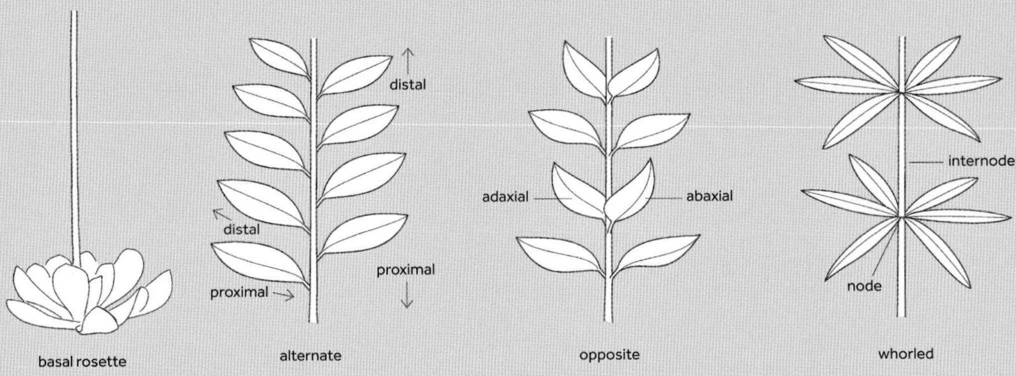

LEAF SHAPES

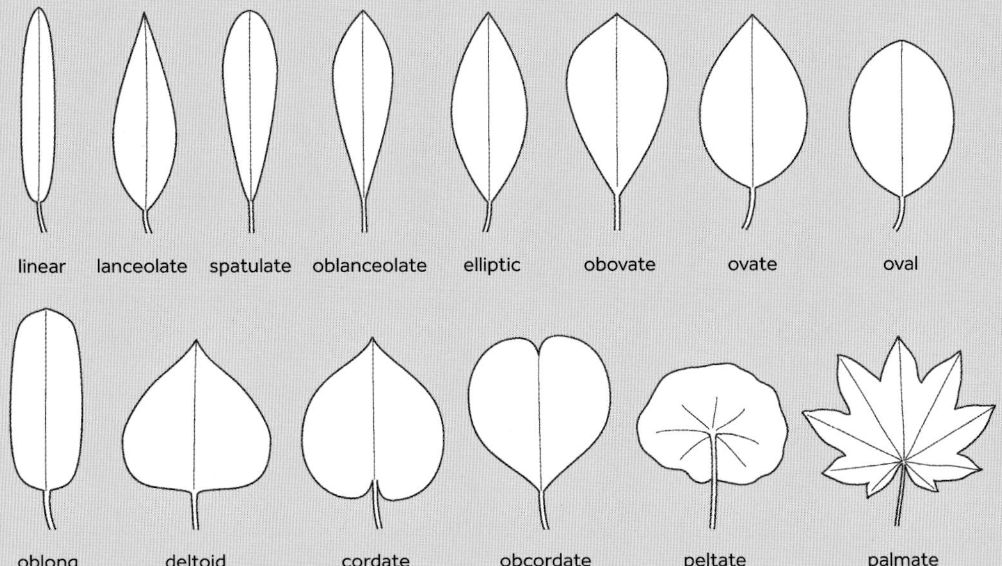

LEAF MARGINS

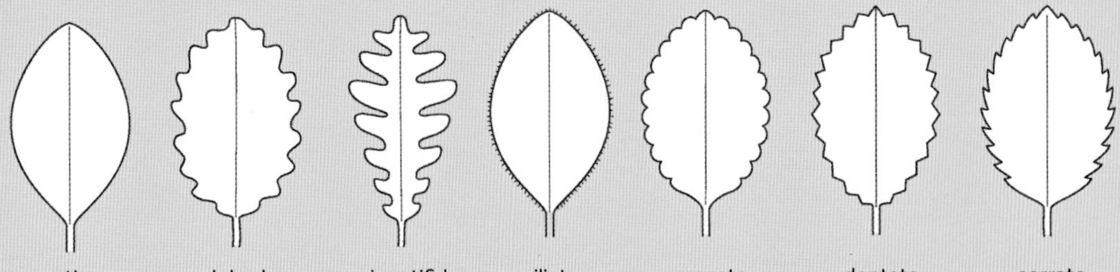

INFLORESCENCE TYPES

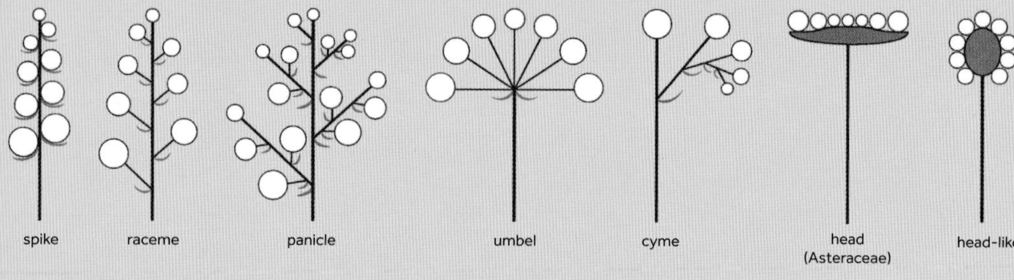

APIACEAE

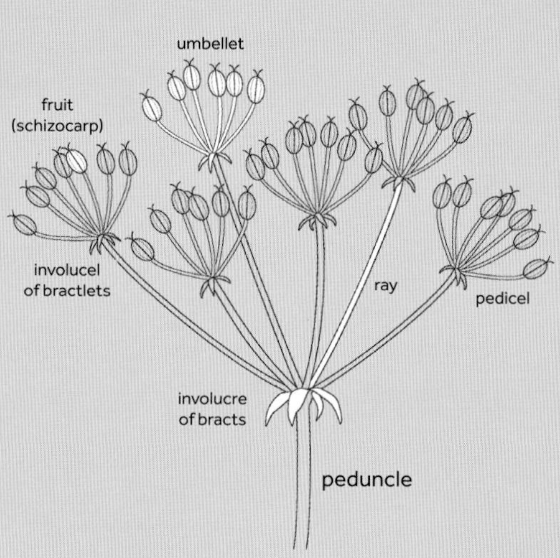

Compound Umbel

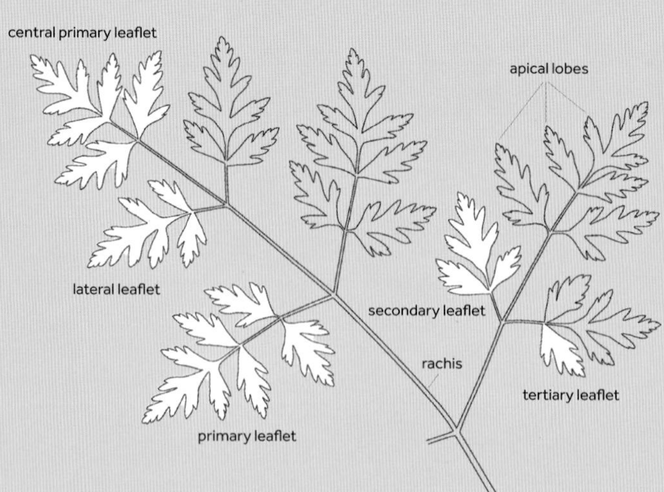

Typical Apiacaeae Leaf
(ternately compound)

ASTERACEAE

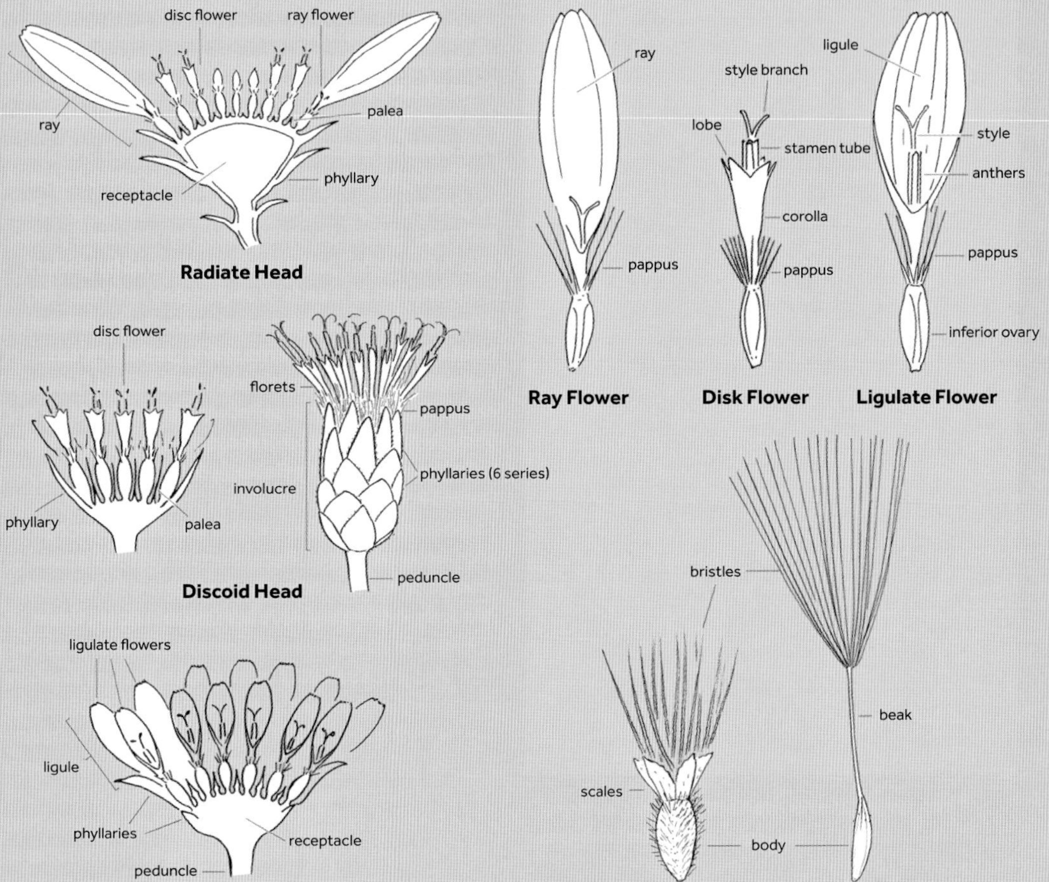

FABACEAE

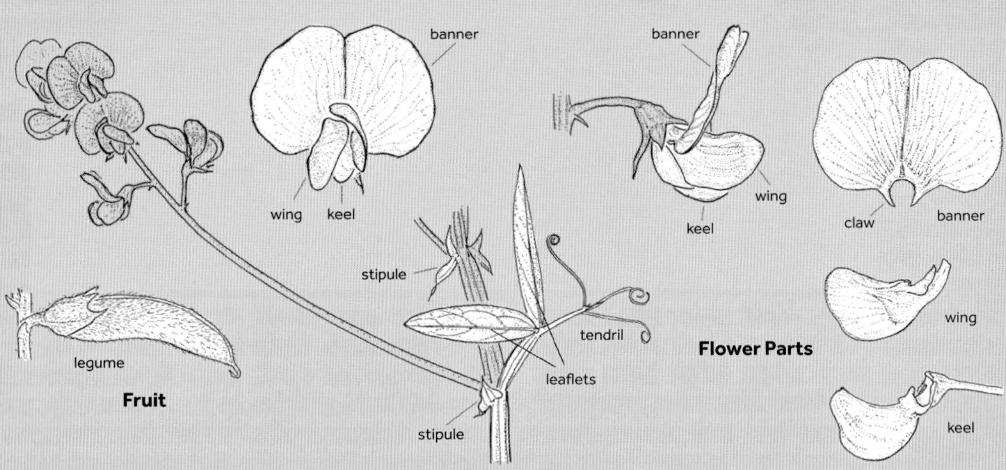

ASTERACEAE

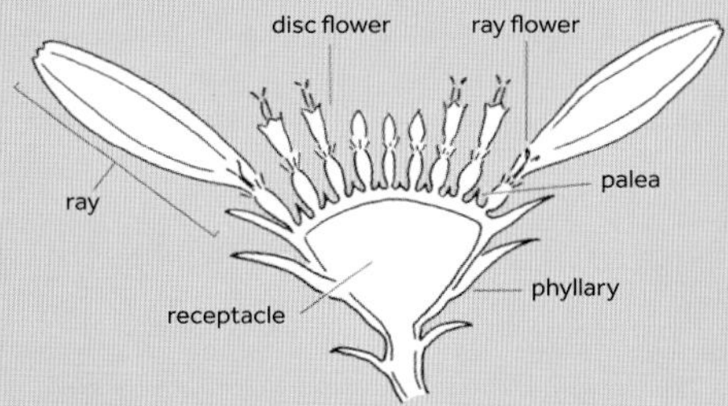

Radiate Head

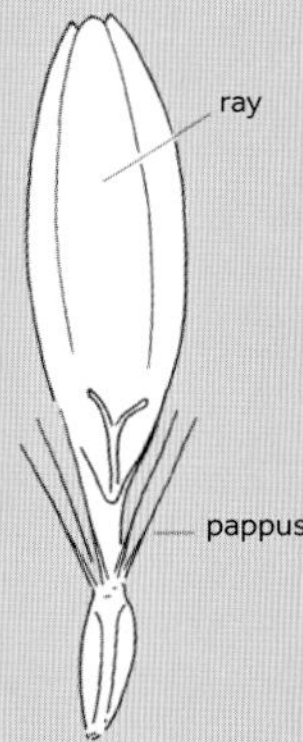

Ray Flower

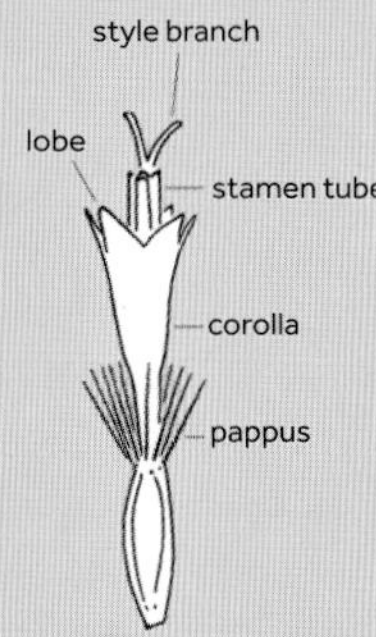

Disk Flower

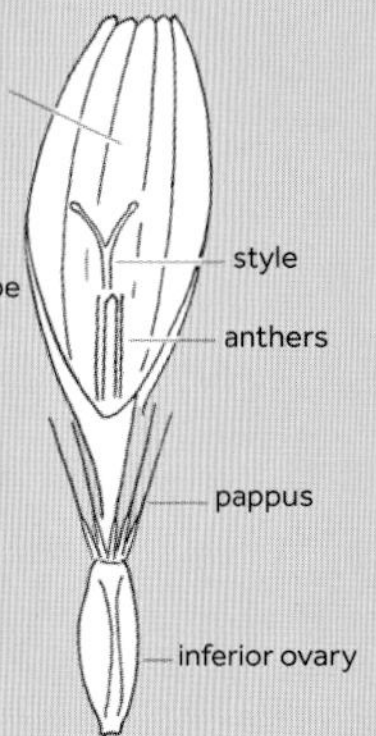

Ligulate Flower

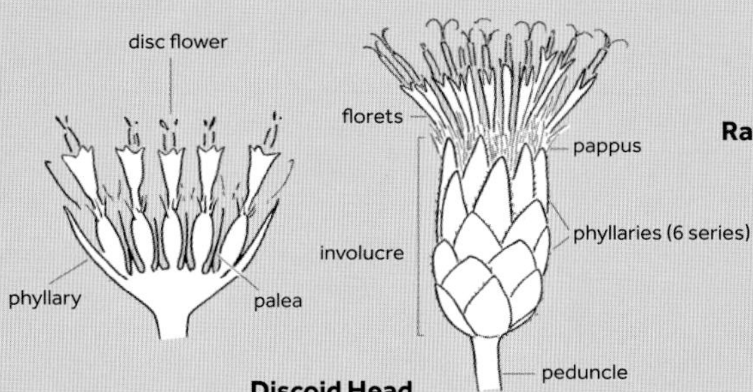

Discoid Head

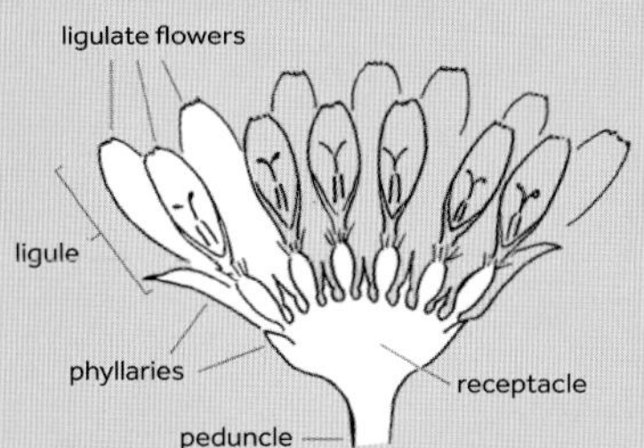

Ligulate Head

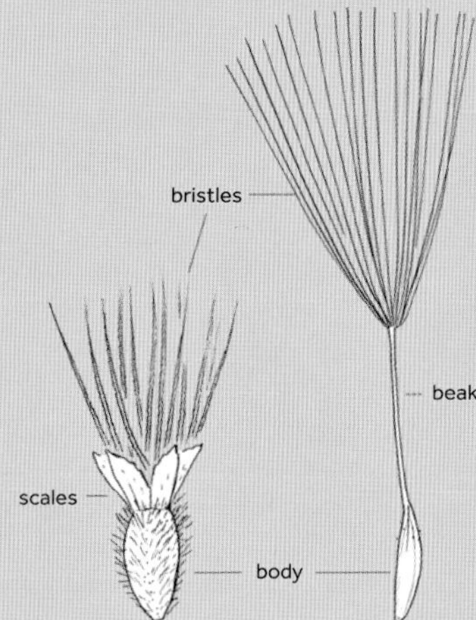

Achenes

FABACEAE

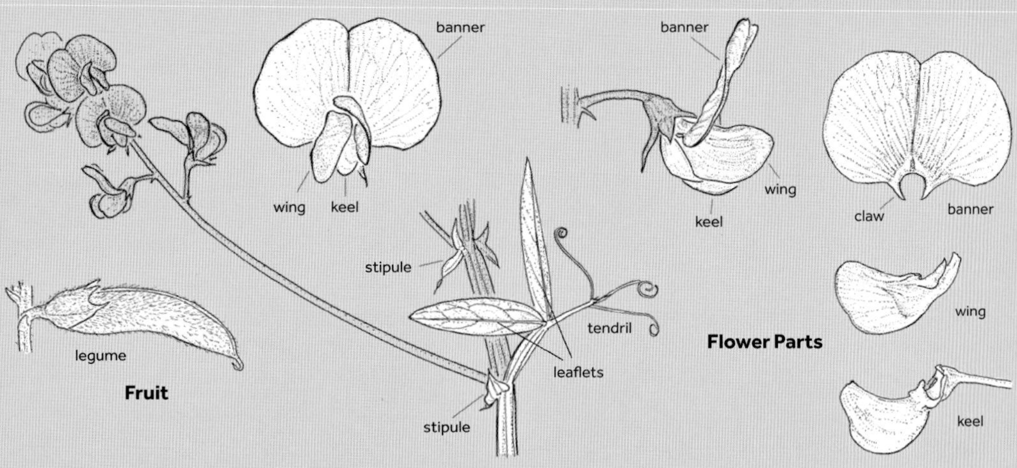

Fruit

Flower Parts

Geology and Soils

Simply put, the Sierra Nevada is a huge batholith formed by a series of granitic deposits. This core of granitic rock began to form as subduction occurred along the western edge of the continent over two hundred million years ago. The granitic rock represents the roots of an old volcanic chain similar to the Andes today. The granitic intrusions, which range about 80–125 million years old, metamorphosed (via heat and pressure) older sedimentary and volcanic rocks associated with ancient oceans. Uplift and erosion of the former top layers ultimately exposed the granitic batholith we know today as the Sierra Nevada. There are scattered "roof pendants" throughout the Sierra where the older metamorphosed rocks have not eroded away, and these sites often have a unique mix of plants, some of them rare, specialized to nongranitic conditions. Some roof pendants are enormous; for example, the Boyden Cave marble roof pendant in the southern Sierra is 5 miles long, 1500 feet thick, and has several rare plants and some magnificent caves.

The steep eastern escarpment and gentle western slopes are a reflection of the range front faulting on the east side and tilting of the Sierran block. Beginning around thirty million years ago, the northern Sierra from around the Stanislaus River northward was subject to extensive volcanic eruptions, the evidence of which can still be seen in vast areas of volcanic rock and soils.

The high country of the Sierra Nevada was sculpted by glaciers, beginning about 1.8 million years ago. The last major glaciation ended about ten thousand years ago. A few higher peaks were not covered by glaciers and remained as islands of unglaciated land retaining different soil characteristics and sometimes different plant life compared to surrounding glaciated areas. Examples of these "sky islands" are the Dana Plateau east of Yosemite National Park and Table Mountain in Sequoia National Park. For more details, a good book is *California's Amazing Geology, Second Edition*, by Donald R. Prothero.

Soils are formed from rock through weathering and biological agents. Because much of the Sierra is granitic, there is an abundance of thin, rocky soil with relatively low fertility. Although less abundant than

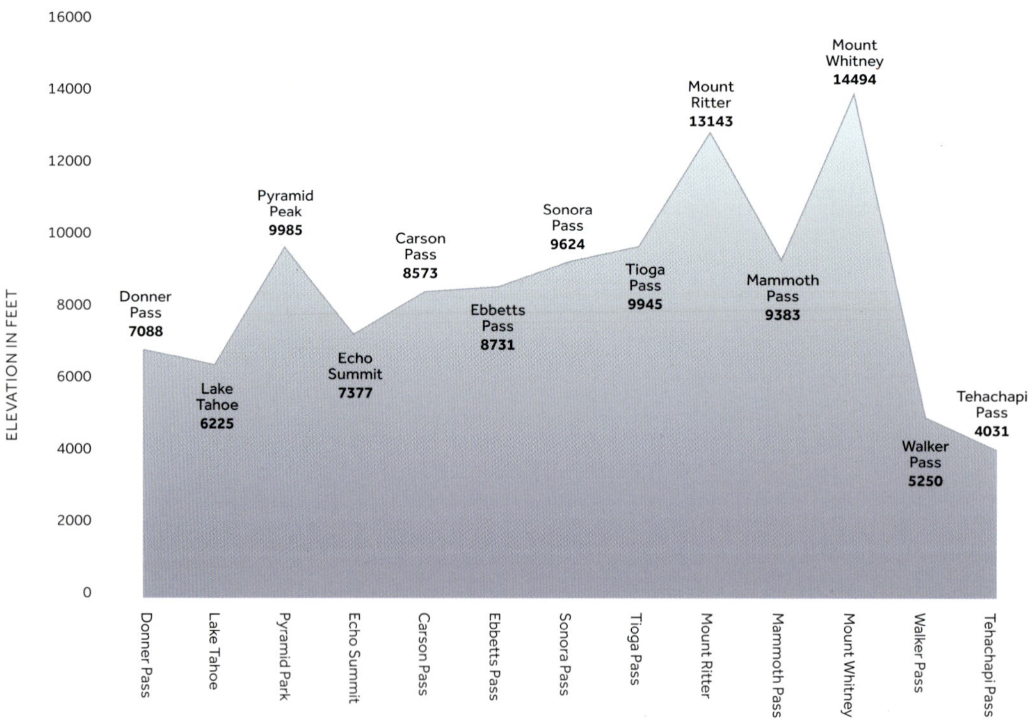

Longitudinal profile of the Sierra Nevada.

granitic rocks, the presence of volcanic, metamorphic, and sedimentary rock types adds to the variety of substrates that make the Sierra such a diverse place for plant life. Soils derived from different rocks obviously have different characteristics with regard to plant growth, and some plant species are specifically adapted to a particular soil type. A well-known example is the suite of rare species found on soils derived from serpentine, a type of metamorphic rock high in magnesium and iron.

Climate

Sierra Nevada climate is generally characterized as Mediterranean for the western slope below the subalpine zone, experiencing cold, high winter moisture, and mostly hot and dry summers. These mountains receive the full brunt of prevailing westerly winds; and as moisture-laden clouds from Pacific storms ascend the western slope, they cool and drop water in the form of rain or snow. Generally, the higher the elevation, the higher the amount of precipitation and the lower the temperature: On average, for each 91 meters (300 feet) gained in elevation, there is a 1°F decrease in average temperature; and a 2- to 4-inch increase in precipitation up to 1525–1830 meters (5000–6000 feet) in the central Sierra. Above this elevation, most precipitation falls as snow. By the time storm clouds reach the Sierra crest, they are depleted of moisture, creating a gigantic rain shadow that extends several hundred miles to the east. The vegetation of the eastern Sierra is more desertlike as a result, aside from riparian areas and wetlands.

Because the Sierra Nevada is a relatively long mountain range, the latitude differences from north to south cause climatic conditions to range from wetter and colder in the north to more desertlike in the south. Shifts in vegetation reflect this pattern—species characteristic of northern moist forests such as Douglas-fir (*Pseudotsuga menziesii*) drop out in Fresno County and are not found southward; desert species such as chaparral yucca (*Hesperoyucca whipplei*) in the southern Sierra extend only as far north as the Kings River drainage.

Chaparral yucca in the Kern River Canyon, the northern limit for this species.

The western slope and crest have historically accumulated massive winter snowpacks, providing ample water (from snowmelt and runoff) to sustain plant life after temperatures have warmed and precipitation is scarce or absent for the summer. This runoff feeds rivers and streams all year long, except in exceptionally dry years. Long after the low-elevation foothill bloom is over, the snow melts off at higher elevations and conditions become suitable for plants to emerge and flower. This progression occurs through August or September (and even sometimes October), mid to late summer being ideal for seeing high-elevation alpine areas in full glorious bloom.

Evidence of climate change in the Sierra Nevada is clear: Seven of the largest glaciers have retreated 30–70 percent in the last hundred years. Peak runoff from the melting snowpack is occurring earlier in the year, and when runoff shifts to winter months, there is less water melting in the spring and summer. The result is that perennial mountain streams that historically flowed all year are now drying up by midsummer. Significant shifts in plant distribution and reductions in range of native plant species are predicted by the year 2100; there is already evidence that young subalpine conifers such as limber pines are now found at higher elevations where conditions are cooler. Monitoring alpine plants to detect climate change–related shifts occurs annually through the Global Observation Research Initiative in Alpine Environments program (gloria.ac.at/scope/background).

Ecological Zones

Variation in elevation, latitude, precipitation, temperature, geology and soils, slope, light availability, fire frequency, and position of the snowline are some primary determinants of vegetation types of the Sierra Nevada, forming rough north–south zones.

The approach to vegetation classification taken for this book is a simple one, emphasizing broad zones recognizable to a layperson by the types and arrangements of the dominant plant species. Although the book features flowering herbs and shrubs, it is customary to define vegetation by the dominant woody species—in many cases, trees. The zones defined here are foothill, montane, subalpine, alpine, and east slope sagebrush/pinyon-juniper. Each zone contains many different plant communities. The term "plant community" is used to mean a group of plant species characteristically found growing together, usually defined by the trees or shrubs, although there are far more species of herbaceous plant species in most plant communities. For example, in the lower montane zone on the west slope one finds mixed conifer forest, a forest type typically dominated by ponderosa pine (*Pinus ponderosa*), white fir (*Abies concolor*), sugar pine (*Pinus lambertiana*), incense cedar (*Calocedrus decurrens*), black oak (*Quercus kelloggii*), and, in the northern Sierra, Douglas-fir (*Pseudotsuga menziesii*). These same four or five tree species are repeatedly found growing together within a given elevation band, and along with their shrub and herb associates can be understood as a "plant community." For a more detailed, fine-grained approach to the mosaic of vegetation in the Sierra Nevada, see *A Manual of California Vegetation*, by John O. Sawyer, Todd Keeler-Wolf, and Julie M. Evens.

The typical elevation range of each major vegetation zone changes with latitude. A zone that is found at one elevation in the northern part of the range is found at a higher elevation in the south. This reflects the similar effect on climate and plant life of going northward in latitude and going upward in elevation; moving 300 miles to the north is roughly equivalent to moving 304 meters (1000 feet) upslope. Keep in mind that these zones intergrade with each other and that there is enormous variability within each one due to microhabitats resulting from topographic features like river valleys, local rain shadows, etc.

ABOVE: **Vernal pools are natural depressions that accumulate water in winter. As water evaporates in spring, the pools are bordered by colorful wildflowers that emerge at specific stages of evaporation. The white meadowfoam bordering this pool is one of the first to bloom. Sacramento County.** **OPPOSITE:** **Gently rolling foothills of the southern west slope in early spring, with blue oak leafing out at right.**

Foothill Zone

- Lower reaches of west slope up to:
    - North SN: 912 m (3000 ft)
    - Central SN: 1216 m (4000 ft)
    - South SN: 1520 m (5000 ft)
- Precipitation falls mostly as rain, 8–40 inches annually
- Summers very hot and dry
- Average maximum temperature: 75–95°F
- Average minimum temperature: 29–42°F
- Growing season: 150–300 days
- A high proportion of non-native species (e.g., 23% in the foothill zone of Yosemite National Park)

TOP LEFT: Popcorn flowers and fiddlenecks, southern foothills in early spring. **TOP RIGHT:** Tufted poppies and purple owl's clover. **BOTTOM:** Chamise chaparral—lower to mid-slopes dominated by chamise; scattered gray pine visible in lower half of photo. Mariposa County. **OPPOSITE:** A flowery hillside along the Hite Cove Trail, Merced River Canyon.

Lower Montane Zone

- On west slope roughly between:
    - North SN: 274–1672 m (900–5500 ft)
    - South SN: 1094–1915 m (3600–6300 ft)
- Precipitation falls as both rain and snow, 25–80 inches annually
- Summers warm, summer rain less frequent than at higher elevations; winters cool to cold
- Average maximum temperature: 80–93°F
- Average minimum temperature: 22–34°F
- Growing season: 120–210 days
- A smaller proportion of species are non-native than in the foothills (e.g., 13% in Yosemite National Park)
- Dominant trees are ponderosa pine (*Pinus ponderosa*), sugar pine (*P. lambertiana*), white fir (*Abies concolor*), incense cedar (*Calocedrus decurrens*), Douglas-fir (*Pseudotsuga menziesii*, Fresno County north), canyon live oak (*Quercus chrysolepis*), black oak (*Q. kelloggii*), and valley oak (*Q. lobata*)

TOP: Mixed conifer forest, 2016; showing mortality of ponderosa pines from drought and bark beetles. Mariposa County. **BOTTOM:** A montane meadow, west slope. **OPPOSITE:** Nevada Falls from near Glacier Point, Yosemite National Park.

Upper Montane Zone

Occurs on both slopes and along the crest in areas where the Sierran crest drops to lower elevations below the subalpine and alpine zones.

- On the west slope roughly between:
    - North SN: 1672–2280 m (5500–7500 ft)
    - Central SN: 1976–2432 m (6500–8000 ft)
    - South SN: 2432–3040 m (8000–10,000 ft)
    - Higher on east slope
- Average maximum temperature: 73–85° F
- Average minimum temperature: 16–26° F
- Growing season: 120–195 days
- Precipitation falls mostly as snow, 35–65 inches annually
- Summers cool, some summer thundershowers; winters cold
- A smaller proportion of non-native species than the two previous zones (e.g., 5% in Yosemite National Park)
- Dominant trees are Jeffrey pine (*Pinus jeffreyi*), sugar pine (*P. lambertiana*), red fir (*Abies magnifica* var. *magnifica*), Douglas-fir (*Pseudotsuga menziesii*)

Red fir forest.

Subalpine Zone

- On the west slope roughly between:
  - North SN: 2188–2781 m (7200–9150 ft)
  - South SN: 2644–3343 m (8700–11,000 ft)
  - Higher on east slope
- Precipitation falls mostly as snow, 30–50 inches annually
- Summers cool, summer thundershowers; winters cold
- Average summer temperature 50°F, frost possible any month of the year.
- Growing season: 49–63 days
- Few non-native species (e.g., 1% in Yosemite National Park)
- Dominant trees are western white pine (*Pinus monticola*), lodgepole pine (*P. contorta* ssp. *murrayana*), whitebark pine (*P. albicaulis*), mountain hemlock (*Tsuga mertensiana*)

TOP LEFT: A subalpine meadow in Little Lakes Valley. **TOP CENTER:** Lupines, corn lily, paintbrush, and goldenrod in the subalpine zone. **TOP RIGHT:** Mounds of short-hair sedge (*Carex filifolia*) just east of Tioga Pass. **BOTTOM:** Stands of whitebark pine (*Pinus albicaulis*) on a rocky slope at timberline.

Alpine Zone

- Along crest and on both sides of the crest above tree line:
    - North SN: ≥ 2736 m (9000 ft)
    - South SN: ≥ 3343 m (11,000 ft)
- Less precipitation than subalpine, mostly as snow, though summer thundershowers common
- Summer maximum temperature rarely exceeds 80°F, frost nightly
- Growing season: 40–60 days
- Non-native species few to none
- Above timberline; dominant vegetation varies but consists of shrubs, mostly perennial herbaceous plants, and some annuals

ABOVE: An alpine rock garden with buckwheat and penstemon. **OPPOSITE:** An alpine fell field on the Dana Plateau, Yosemite National Park.

Sagebrush/Pinyon-Juniper Zone

- Mostly east slope:
    - North SN: 1280–1707 m (4200–5600 ft)
    - South SN: 1830–2134 m (6000–7000 ft)
- Precipitation falls mostly as snow, 10–30 inches annually
- Summers dry, hot; winters cold
- Summer average maximum temperature: 82–89°F
- Winter average minimum temperature: 10–20°F
- Growing season: 60–150 days
- Dominant vegetation are pinyon pine (*Pinus monophylla*), Sierra juniper (*Juniperus grandis*), mountain sagebrush (*Artemisia tridentata* ssp. *vaseyana*), bitterbrush (*Purshia tridentata*), and a diversity of rabbitbrushes (*Ericameria* spp.)

TOP LEFT: Jeffrey pine in sagebrush scrub. **TOP RIGHT:** A stream bordered by willows in sagebrush, east of Sonora Pass. **BOTTOM:** View from near Monitor Pass, looking eastward toward the White Mountains. In the foreground is green ephedra (*Ephedra viridis*), a gymnosperm, related to conifers. **OPPOSITE:** Mount Whitney from the east side, sagebrush in the foreground.

Yosemite lewisia (*Lewisia disepala*) blooming in March on a granitic dome near Bass Lake, Madera County.

Special Habitats

Some of the most interesting habitats for Sierran flowering plants at any elevation are sites of extreme conditions: super dry or super wet, rocky or riparian areas (streamsides and meadows), and harsh alpine areas above tree line. In addition, burned areas, especially in foothill chaparral and lower ponderosa pine forests, contain a unique set of species adapted to fire and only seen after a burn when their seeds are cued to germinate by heat or smoke.

Rock Outcrops and Cliffs

Rocky outcrops and the gravelly or sandy soil associated with them are desertlike in that they're exposed to relentless sun; soils are quickly drained of water and are exposed to the drying effect of wind. Often classified as "barren" on vegetation maps, these areas are far from barren botanically. While moisture is still available, either from rain in the foothills or melting snow at higher elevations, a diverse suite of specially adapted wildflowers emerges in these locations. Many Sierra Nevada endemics are in this group, as well as numerous common and beautiful species. The habitat section in the species descriptions tells if that particular plant is a plant of rocky or gravelly sites. Plants may grow primarily in cracks and soil crevices of vertical cliffs. Some plants are even more specialized and are found on only one type of rock, such as serpentine or limestone. Granitic outcrops and cliffs are abundant in the Sierra, and to spice things up, look for volcanic sites or areas where metamorphosed rocks persist.

Riparian Areas and Meadows

Sites with moist or wet soil along streams, in meadows and springs, and around high-country lakes support a strikingly different flora from adjacent uplands, and this is often where the highest diversity of wildflowers is found. Riparian plant communities dominated by water-loving shrubs and trees such as willows (*Salix* spp.), alders (*Alnus* spp.), cottonwoods (*Populus* spp.), and Oregon ash (*Fraxinus latifolia*) exist from low to high elevations along streams. Areas of high water tables and few to no trees support open, sunny wetlands called meadows. There is often a strip of meadow vegetation along streams and around lakes in the montane, subalpine, and alpine zones.

Meadows are found from the montane through alpine zones, and although they comprise less than 5 percent of the Sierra Nevada land base, they

A lake surrounded by meadow in the subalpine zone near Yosemite National Park.

contain most of the plant species diversity and attract people, livestock, and wildlife (and mosquitos) disproportionately. Most Sierra Nevada meadows are about ten thousand years old and, contrary to common lore, are not filled-in lakes that are on their way to becoming forests. Fascinating studies of washed-out meadows reveal alternating layers of organic matter, mineral deposits (sand and rocks), and volcanic dust from ancient explosions. These strata show few to no forest trees within the past ten thousand years, meaning that these sites have been meadows since the last ice age. Some meadows have petrified giant sequoia trees (*Sequoiadendron giganteum*) at the base of strata that have been carbon-dated as being about ten thousand years old.

Sierran meadows may contain areas where organic material accumulates called fens. If you are walking through a meadow and suddenly find yourself bouncing slightly or sinking into mire, you are probably in a fen. Fens are a type of peatland, areas of deep organic soils where accumulation of plant matter exceeds decomposition. They are important habitats for rare plants and mosses and play a major role in maintaining the hydrologic integrity of meadows.

Although there are plenty of colorful native wildflowers to see in meadows, these occur within a matrix of perennial grasslike plants in the Sedge (Cyperaceae), Rush (Juncaceae), and Grass (Poaceae) families. Members of the Sedge family (e.g., sedges, *Carex* spp.; spikerushes, *Eleocharis* spp.; and bulrushes, *Scirpus* spp.) are especially important because they form a protective layer of sod made up of exceptionally strong roots and rhizomes that hold the vegetation and soil in place even during periods of intense runoff at peak snowmelt. Bryophytes (mosses) are abundant in most meadows and contribute significantly to sustaining the hydrological integrity of these wetlands. As mentioned above, these groups lack showy colorful flowers and are challenging to identify in the field, thus they are not included in this book.

ABOVE: Mixed flowers in a damp mid-elevation montane meadow on the westside. **OPPOSITE:** Looking from the upper subalpine into the alpine, with paintbrush in the foreground. Middle Fork of Bishop Creek.

Many meadows suffered severe damage during the mid to late 1800s when they were subject to heavy grazing and trampling pressure from sheep and cattle. In addition, disruption of meadow hydrology was caused by harmful logging practices, mining, and roads. Due to such activities, large erosion gullies are still visible in some meadows today. The encroachment of lodgepole pine (*Pinus contorta* ssp. *murrayana*) into meadows with lowered water tables due to gullies is a problem because it causes loss of wetland plant species diversity, reduces wildlife habitat quality, and decreases livestock forage provided by the meadows. Meadow restoration is occurring across the Sierra Nevada, and collaborative efforts among land management agencies, Native Americans, universities, and nonprofit organizations are succeeding in repairing erosion gullies and returning water tables to more natural levels. The degree to which meadows can heal is remarkable, although in some cases the practices of yesterday (and sometimes today) remain all too visible.

ABOVE: Flowers of Shuteye Peak fawn lily. OPPOSITE: Lupines flowering after a burn, Madera County.

Burned Areas

Fire is and has always been an integral factor in Sierra Nevada plant communities. Scientists and land managers became aware of the ecological importance of fire in many Western ecosystems decades ago, in the 1950s and '60s. Native Americans managed ecosystems for millennia, burning to replenish the land, working with the natural role of fire. Managers began to reexamine policies of complete fire suppression and implement the idea of restoring the role of fire either by allowing natural fires to burn when safe or by actively lighting prescribed burns. In the prime timber-producing montane zone, land managers are striving in various ways to undo over a century of fire suppression and past logging practices, which have left forests unnaturally dense and with extreme fuel loads, and thus more likely to burn catastrophically when ignited, especially with climate change causing higher temperatures, drier fuels, and longer fire seasons. Montane-zone forests historically experienced frequent fire, as often as every eight to ten years in ponderosa pine and mixed conifer forests, as indicated by the study of tree ring scars. Natural fires were mostly ground fires that did not kill the larger trees while maintaining lower fuel loads and a sparser understory than is typically seen today, intermixed with patches of moderate to high fire intensity.

Newly burned areas at any elevation are interesting to botanize, but chaparral within the west-slope foothill zone is arguably the most fascinating. Foothill chaparral is comprised of plant species that evolved a wide variety of responses to fire; shrubs may sprout back, produce copious seedlings, or both. A suite of native annual plant species exists primarily as a bank of seeds lying dormant in the soil between fires; these often emerge en masse the first year after a fire. Whispering bells (*Emmenanthe penduliflora*) and golden ear drops (*Ehrendorferia chrysantha*) are examples of such species (pages 291 and 376). Seeds of specialized chaparral shrubs and herbs may germinate in response to heat, smoke, or rainwater containing compounds leached from charred wood. A good source of information about chaparral ecosystems is www.californiachaparral.com/; although the site focuses on Southern and Central California chaparral, the primary concepts hold true for Sierra Nevada foothill-zone chaparral.

Rare Plants

The Sierra Nevada is exceptionally floristically rich, containing more native plant taxa than any other area of its size in North America north of Mexico. Of the approximately 3,500 or so vascular plant taxa in the Sierra Nevada, about 405 are endemic to the range, meaning that they occur nowhere else on earth. A study of rarity and endemism by river basin (James R. Shevock, "Status of Rare and Endemic Plants") shows that the southern Sierra (Fresno, Inyo, Kern, and Tulare Counties) has the highest concentration

ABOVE: **Aftermath of the Aspen Burn, Sierra National Forest.** **OPPOSITE:** **In the middle of the Jose Basin Burn area, *Lupinus citrinus* var. *citrinus* flowers almost cover the exposed granite.**

of rare and endemic taxa, probably because this area contains the highest, steepest, and driest areas within the range. Across the range, areas of serpentinite and carbonate rock and the soils derived from them are also characterized by a high number of rare and endemic species.

It may seem that in this day and age, botanists would have discovered and named all the species in as highly populated a state as California. Not so, and new species are still being discovered; for example, during the eighteen years from 1968 to 1986, 220 new vascular plant taxa were described in California, sixty-five of which occur in the Sierra Nevada. Not surprisingly, most of these newly named species are rare (or they would have been discovered long ago!) and are usually added to agency rare plant lists to ensure that special management to ensure their long-term viability can be incorporated into planning of projects. The Shuteye Peak fawn lily, *Erythronium pluriflorum*, was discovered in 1988 and described as a new species soon thereafter, though it grows along a four-wheel-drive route to a lookout that gets plenty of visitors each summer. Although restricted to the Chiquito Ridge area of the Sierra National Forest in Madera County, there are millions of plants, and populations appear to be thriving, based on the presence of all age classes, from seedlings to larger, mature plants, and observations of native bees working the flowers during peak bloom.

Fortunately, many of the rare plants in the Sierra Nevada are on public lands where there are laws and policies aimed at conserving them. Some species are in rugged areas where direct threats are unlikely. Those rare plants occurring nearest to human habitation and on private lands are the most at risk. Development (building and paving that causes outright loss of populations and habitat) and alteration of habitat by invasive non-native plants are two of the most immediate threats at this time for many rare plant species; however, climate change is also looming as a serious threat. The California Department of Fish and Wildlife (CDFW) has a program devoted to conserving rare plants and tracking conservation trends through the California Natural Diversity Database. The CDFW's rare plant program is described on their website. Working hand-in-hand with CDFW is the California Native Plant Society (CNPS), a science-based nonprofit advocacy group that has worked successfully to protect and conserve rare plants in California for decades. CNPS maintains an inventory of rare plants, assigns status rankings to each species (California Rare Plant Rank, or CRPR), and maintains an online version accessible to the public (search for the rare plant inventory). Federal agencies, including the US Forest Service, National Park Service, and Bureau of Land Management, also have programs for conserving and enhancing rare plant populations and work in collaboration with CDFW and CNPS.

Invasive Non-native Plants

Of all the issues relevant to sustainability of Sierran ecosystems over the next few decades (human population expansion, development, climate change, pollution), the topic of invasive non-native plants deserves special mention in a book about Sierra Nevada flowering plants. Non-native plants began to arrive in California as early as the 1600s, mostly from Europe's Mediterranean region where they were perfectly suited to the similar climate found in California. The coastal areas were the first to be affected, with vast acreages taken over by non-native annual grasses, annual mustards, brooms, and more. As California's infrastructure of railways, roads, and towns has grown, along with a huge increase in the human population, invasive weeds have followed throughout much of the state.

Approximately 1,100 introduced plant species now live in the wild in California, which is about 20 percent of all species found in California. Of these 1,100 species, only a couple hundred are considered so harmful to California's ecology and economy that they are listed as Noxious Weeds by the California Department of Food and Agriculture, or as damaging wildland weeds by the California Invasive Plant Council. The rest of the introduced species are either so well established or pervasive that they will never be eliminated (e.g., many of the annual grasses in the genera *Avena*, *Bromus*, *Hordeum*, and *Festuca*), but a network of native plants persists and coexists, or are restricted to disturbed areas and not so aggressive that they spread into and devastate native plant communities.

The worst of the non-natives, however, cause serious problems by forming monocultures where complex native ecosystems formerly existed and, in some cases, damaging the land irreversibly. This is partly because these species arrive in California without the suite of predators, pathogens, and competitors that keep them in check in their countries of origin. Specific effects these weeds can have include increasing erosion and runoff, reducing available forage for wildlife and livestock, altering pollinator communities, making recreation areas unusable for people, reducing water availability for wildlife and native plants in drier areas, poisoning wildlife, and changing fire frequency in such a way as to favor more invasive plants and fewer native plants. A well-known invader, yellow starthistle, is a spiny disaster when it forms monocultures by replacing native plants completely, which leads to the land losing the diversity of native flora and fauna formerly present and making it painful for people and other animals to walk.

Remarkably, the high elevations of the Sierra Nevada are still largely free of invasive weed infestations. In Stephen Botti's 2001 *Flora of Yosemite National Park*, he estimated that in the central Sierra, 23 percent of plant species were non-native in the foothill zone, 13 percent in the lower montane zone, 5 percent in the upper montane zone, 3 percent in the subalpine zone, and 0 percent in the alpine zone. This is probably roughly representative of the central Sierra today. The Sierra Nevada has its share of invasive weed problems, some examples including yellow starthistle, Italian thistle, musk thistle, several species of knapweed, and ornamental brooms like Scotch and French brooms. Some of the worst weeds are featured in the book to help raise awareness because prevention and early detection are crucial to keeping the vast "clean" areas from becoming overrun by invasive weeds. The highest elevations of the Sierra Nevada will likely become more vulnerable to weed invasions as climate change progresses and as human populations grow. Devastating weeds, such as spotted knapweed and Canada thistle, already occupy millions of acres of similar habitat in the Rocky Mountains and are fully capable of thriving in the upper montane and subalpine zones given time. Detecting new outbreaks and promptly controlling them is essential.

And Finally, a Couple of Notes of Caution

To avoid harming populations of rare plants inadvertently or needlessly impacting populations of more common plants, it is best not to pick flowers just for fun, although collecting of professional-

Dense yellow starthistle infestation near Oakhurst.

quality vouchers for herbaria is important to expand our knowledge of species distributions and ecology. If you're a serious amateur or professional botanist interested in collecting scientific specimens, or aspire to be one, be sure to learn basic methods of collecting scientifically valuable herbarium specimens and obtain the required permits, especially for public lands (national forests, national parks, Bureau of Land Management lands, state parks, etc.). The process of obtaining your permit will put you in touch with knowledgeable and helpful experts in your region of interest and could lead to some valuable citizen science opportunities that help conserve native plants.

To avoid harming yourself, it is wise to approach the subject of edibility with extreme caution. Some harmless and edible plants have poisonous "look-alikes" that may be indistinguishable to the casual observer. For the most part, we do not provide information on edibility, though we do note some Native American uses of plants, which may include references to use as food. This is for interest and education, not to encourage foraging. Toxicity is only mentioned for species that are notorious for being highly poisonous, not for all toxic plants.

Species Descriptions

Species are organized following the arrangement in *The Jepson Manual, Second Edition*. Nymphaeales and Magnoliids are first, followed by Eudicots, then Monocots. Within each of the four groups, species are alphabetical by scientific name of family, genus, then species (and subspecies or variety, if applicable).

NYMPHAEALES

The Nymphaeales were formerly included with Eudicots in the Dicots. All Nymphaeales are aquatic perennials, with two cotyledons, floating leaf blades, bisexual flowers borne singly, perianth parts in threes or spirally arranged, and pollen with one aperture through which the pollen tube grows. Two of these families are represented by native species in the SN: Cabombaceae and Nymphaeaceae. Watershield, *Brasenia schreberi*, is the only SN member of Cabombaceae (Watershield family) and is not included in the book because it's quite rare in the SN. Locations and photos can be seen on CalPhotos and Calflora.

stamens
(hiding petals)
stigmatic disks
sepals

Nuphar polysepala

Western Pond-Lily

Nymphaeaceae (Waterlily family)

Aquatic perennial herb with floating leaves arising directly from stout, underwater rhizomes.

LEAVES: Petioles long; blades pinnately veined, 1–4 dm, cordate-oblong or ovate. Basal lobes rounded, at least ⅓ the length of midrib.

FLOWERS: Borne singly, arising from axils, 5–10 cm wide, globose, yellow. Sepals 6–12, petal-like; petals numerous, inconspicuous and hidden by the numerous petal-like stamens. Ovary is superior and has a disk-shaped top (stigmatic disk) with radiating lines of stigmatic tissue.

FRUITS: Fruits 4–6 cm x 3.5–6 cm wide, cylindric or ovoid, yellowish to green; strongly ribbed and constricted below stigmatic disk. Air chambers develop between carpels as fruits mature, and carpels break apart into floating units that carry the mostly nonfloating seeds across the water surface. Seeds 3.5–5 mm.

SIERRA DISTRIBUTION: Both slopes, Fresno and Mono Counties north.

ELEVATION: 1000–2900 m

HABITAT: Ponds, lakes, and slow-moving streams; montane and subalpine zones.

FLOWERING TIME: Apr–Sep

NOTES: Leaves, rhizomes, and roots have abundant air chambers that convey gases throughout the plant—oxygen travels from leaves to rhizomes that are buried in oxygen-poor mud via these chambers, and carbon dioxide and other gases travel up to be released by the leaves. Plants are used for food and medicine by Native Americans; seeds are roasted and eaten whole similar to popped corn kernels, or ground into flour. Leaves and rhizomes are used to treat skin boils and swelling, and an infusion made from rhizomes makes a gargle to ease sore throats. Pollinators include reed beetles (*Donacia* spp.) and various flies.

MAGNOLIIDS

Magnoliids were also formerly included with Eudicots in the former Dicot group. Three of the seven families in this group are featured in this book. These are terrestrial plants with two cotyledons, bisexual flowers with perianth parts in threes or spirally arranged, and pollen with one aperture through which the pollen tube grows. Most contain aromatic "ethereal oils," which are secondary (specialized) metabolites manufactured by the plant for survival (e.g., defense against herbivores or attracting pollinators).

Aristolochia californica

California Pipevine

Aristolochiaceae (Pipevine family)

Woody, twining vine with stems < 5 m; deciduous, herbage soft-hairy.

LEAVES: Alternate, 3–15 cm, ovate-cordate to sagittate.

FLOWERS: Solitary in axils, sepals fused into a U-shaped tube 2–4 cm, light brown or green with purple veins. Calyx lobes 3, red-brown. Petals absent.

FRUITS: Winged capsule 5–6 cm x 1–3 cm, cylindric or ovoid, 6-chambered. Seeds 1–1.2 cm, triangular.

SIERRA DISTRIBUTION: West slope, Fresno County north, more common from El Dorado County north.

ELEVATION: < 700 m

HABITAT: Streamsides in chaparral and foothill woodland; foothill zone.

FLOWERING TIME: Jan–May

NOTES: The leaves are the sole food source for larvae of the hairy pipevine swallowtail (*Battus philenor* ssp. *hirsuta*), which is a CA endemic, like its host. The striking black-and-orange caterpillars accumulate toxic aristolochic acid from the leaves, making themselves unpalatable to predators as larvae, then as butterflies. Butterflies feed on nectar from a variety of plants, including thistles, phlox, yerba santa, lupines, and California buckeye.

Asarum hartwegii

Hartweg's Wild Ginger

Aristolochiaceae (Pipevine family)

Perennial herb, arising from deep, vertical rhizomes (no aerial stems), growing in clumps rather than mats. Leaves and roots are aromatic.

LEAVES: Petioles 9–21 cm, hairy; blades 5.3–10 cm x 7–14 cm wide, cordate, usually variegated white along major veins. Blades appressed-hairy below, glabrous above or with sparse hairs along veins, margins fringed with hairs that curve toward leaf tip. Leaves emit a gingery fragrance when crushed.

FLOWERS: On erect pedicels 1–2.3 cm, close to the ground and facing upward. Sepals 3, petal-like (petals absent), forming a false calyx tube at base (sepal bases not actually fused). The tube is reddish and shaggy-hairy outside, with inner surface less shaggy, usually vertically red-striped with vertical bands of white hairs. Sepals brown or maroon, with long tips 1.2–2.7 cm spreading away from the tube during flowering. Ovary inferior, styles fused into a column; stamens 12.

FRUITS: Capsule spherical, fleshy, crowned by persistent calyces.

SIERRA DISTRIBUTION: West slope, Tulare County north.

ELEVATION: 400–2200 m

HABITAT: Shady moist to dry areas in conifer forest; montane zone.

FLOWERING TIME: Apr–Jul

NOTES: The white on major leaf veins varies from obvious to nearly nonexistent. Plants with little to no white along the major veins can be distinguished from *A. lemmonii* by their clumped habit (contrasted with continuous mats or patches in *A. lemmonii*), erect flowers, much longer calyx lobes, and drier habitat preference. Note that periodic posts of *A. caudatum*, creeping wild ginger, pop up for the Sierra Nevada, but to date these have turned out to be *A. hartwegii* with exceptionally long and narrow sepal tips (length of sepal lobes/tips varies considerably in *A. hartwegii*). *A. caudatum* is a plant of northwestern CA and northward, with all-green leaves and much longer (2–9 cm), ± linear sepal tips.

Asarum lemmonii

Lemmon's Wild Ginger

Aristolochiaceae (Pipevine family)

Perennial herb, spreading via shallow horizontal rhizomes to form loose mats (no aerial stems).

LEAVES: Petioles 6–26 cm, hairy; blades 4–6.5 cm x 7–10.5 cm, widely cordate, solid green without white veins. Hairiness of blades is similar to *A. hartwegii* except that marginal hairs may be perpendicular to the margin, eyelash-like. A gingery fragrance is released upon crushing.

FLOWERS: Borne singly on reflexed pedicels that are 2–3.5 cm long, flowers nodding. The 3 sepals form a false calyx tube (false because sepals are not actually fused into a tube). The calyx tube is cylindric, somewhat reddish on outer surface, inner surface white, without stripes. Sepal tips 0.8–1.5 cm, strongly reflexed, pressed against calyx tube, red.

FRUITS: Capsule round, fleshy, capped by calyces that persist after flowering.

SIERRA DISTRIBUTION: West slope, Tulare County north.

ELEVATION: 1000–1900 m

HABITAT: Moist to wet areas, often along streams in conifer forest; montane zone.

FLOWERING TIME: May–Jul

NOTES: Compare with *A. hartwegii*. *A. lemmonii* grows near water, is matted rather than clumped; leaves are solid green; and the short, red, folded-back sepals look quite different. Fondly called "hot-lips" by some Sierran botanists. Despite the common name "wild ginger," *Asarum* is not closely related to the common culinary ginger, *Zingiber officinale*, a monocot cultivated for its rhizome. Aristolochic acid is present in tissues of *Asarum*, and due to liver toxicity and carcinogenic qualities, eating any part of the plant is ill advised.

Calycanthus occidentalis

Spicebush, Sweet Shrub

Calycanthaceae (Sweet-Shrub family)

Deciduous shrub 1–3+ m, with a pleasant, spicy aroma (crush leaves to release fragrance).

LEAVES: Opposite, simple; petioles 3–10 mm. Blades 5–15 cm, ovate to lance-oblong, hairless or somewhat hairy below, scabrous above.

FLOWERS: Solitary at tips of branches. Perianth parts alike, no differentiation into sepals vs. petals (in this case called tepals), 2–6 cm, maroon, linear to ovate-elliptic, spirally arranged on a hollow ovoid receptacle. Pistils on inner walls of receptacle, with threadlike styles that extend to the outside of receptacle from 2-ovuled ovary; stamens 10–15, at top of receptacle. Inner tepals, stamens, and staminodes have succulent, white food bodies at their tips.

FRUITS: Achenes numerous, embedded on inner surface of woody receptacle, which is ovoid or slightly bell-shaped at maturity, 2–4 cm x 1–2 cm. Outer surface of fruits marked with scars left by perianth parts.

SIERRA DISTRIBUTION: West slope, Kern County north, patchy.

ELEVATION: < 1500 m

HABITAT: Moist, shaded to sunny sites, especially along rivers and streams; foothill and montane zones.

FLOWERING TIME: Mar–Aug

NOTES: Early in the flowering cycle, the inner perianth parts are curved toward the center to form a chamber with a small opening at the top. Pollination occurs when small beetles, attracted by the fragrance emitted through the opening, enter the chamber. Beetles feed on "food bodies" borne on staminodes, stamens, and inner perianth parts, and possibly on pollen (no nectar is produced). Due to downward-pointing hairs on the inner perianth parts, beetles are trapped for up to several days in the chamber, during which time pollen is released, coating their bodies. They exit when perianth parts move outward as flowering concludes, often going to a different spicebush plant, effecting pollination when they enter the chamber bearing pollen during the time when the stigma is receptive. Native peoples use spicebush branches for arrows, and shrubs were burned or pruned every 2–3 years to encourage straight, strong shoots.

Anemopsis californica

Yerba Mansa

Saururaceae (Lizard's-Tail family)

Perennial herb, stems 8–80 cm; hollow, hairless or hairy. From thick, woody rhizomes, forming clonal patches.

LEAVES: Basal large and conspicuous, petioles 2–40 cm; blades 5–60 cm, elliptic to oblong. Stem leaves smaller and of 2 kinds: sessile primary leaves with bases clasping stem, and secondary leaves emerging from axils of primary leaves.

FLOWERS: Conical spikes 1–4 cm, of 75–100 flowers, fragrant. Spikes subtended by showy, white or reddish petal-like bracts 5–35 mm. Flowers bisexual and without petals, each flower subtended by a small, white bract 3.5–6 mm.

FRUITS: Capsule brown, 5–7 mm. Seeds 18–40.

SIERRA DISTRIBUTION: Both slopes, most SN locations are in Kern County, a few in Inyo County.

ELEVATION: < 2000 m

HABITAT: Moist or wet saline or alkaline areas, often near hot springs; foothill and montane zones.

FLOWERING TIME: Jun–Sep

NOTES: Native Americans in CA and beyond have used yerba mansa for generations for a wide variety of medicinal purposes (e.g., wound treatment; flus and colds; pain and inflammation; and urinary, reproductive, and digestive tract ailments). Both belowground and aboveground parts of the plants are used. *Anemopsis* occurs only in western North American and northern Mexico. *A. californica* is the only species in the genus.

EUDICOTS

Eudicots are a large group of approximately 190,000 flowering plants worldwide (75 percent of angiosperms). This group is demonstrated through molecular data to be monophyletic (sharing a common evolutionary history, derived from a common ancestor). Eudicots are distinguished by having two cotyledons (seed leaves), flower parts generally in fours or fives, and leaves with pinnate or palmate venation (or pinnately or palmately divided compound leaves). Pollen grains have three or more apertures (areas with differentiated tissue through which pollen tubes exit to deliver sperm nuclei to eggs; apertures may also function to regulate humidity, allowing the pollen grain to contract or expand).

Rhus aromatica

Skunk Bush, Sourberry

Anacardiaceae (Sumac family)

Deciduous shrub 0.5–2.5 m, often forming large patches.
LEAVES: Alternate, petioles 5–15 mm; blades compound with 3 leaflets; upper surfaces dull; central leaflet/lobe 10–35 mm, tapered to petiole (no petiolule), lateral 5–18 mm.
FLOWERS: Panicles terminal on short twigs, open to dense. Flowers sessile, appearing before leaves. Sepals yellow-green to reddish; petals yellow. Styles 3, stamens 5.
FRUITS: Drupe 5–8 mm diameter, rusty red or orange; sticky.
SIERRA DISTRIBUTION: West slope.
ELEVATION: < 2200 m
HABITAT: Upland openings, burned areas; foothill and lower montane zones.
FLOWERING TIME: Mar–May
NOTES: Formerly *R. trilobata*. The Western Mono people (and other tribes) use sourberry stems extensively as a foundation for cooking baskets, winnowing trays, seed beaters, and carrying baskets. Although sourberry is a fairly common shrub, good gathering patches have become scarce; plants need fire or coppicing to produce straight, strong, supple shoots for baskets, and many sourberry stands today have old, brittle stems with short internodes. The tart berries are highly prized for eating and are also used to make a fizzy drink.

Toxicodendron diversilobum

Western Poison Oak

Anacardiaceae (Sumac family)

Deciduous shrub 0.5–4 m, or vine < 25 m with gray to reddish-brown twigs; from an extensive rhizome system.
LEAVES: Alternate, compound, leaflets 3 or sometimes 5, with shiny upper surfaces. Petioles 1–10 cm; leaflet bases truncate to rounded, tips obtuse to rounded. Terminal leaflets 1–13 cm, with a distinct stalk (arrow in photo); lateral leaflets 1–7 cm. Leaflet margin and size vary enormously, from smooth-edged to highly lobed or jagged (hence the species name *diversilobum*, which derives from "diverse," referring to the extreme variability in leaflet margins + "lobe," as the lobes are the features that vary.
FLOWERS: Plants dioecious (photo shows male flowers). Inflorescences axillary; pedicels 2–8 mm; sepals 5, green; petals 5, yellow to whitish-green. Male flowers have 5 stamens; ovaries are vestigial or absent. Female flowers have superior ovaries, fused styles, 3 stigmas, and a disklike nectary at the base; stamens are vestigial.
FRUITS: Drupe < 6 mm, roundish, cream to tan, often black-striped.
SIERRA DISTRIBUTION: West slope.
ELEVATION: < 1650 m
HABITAT: Riparian areas and uplands; foothill and montane zones.
FLOWERING TIME: Apr–Jun
NOTES: The genus name is from the Greek words *toxikos* ("poison") and *dendron* ("tree"). Poison oak contains urushiol in all parts of the plant. Urushiol causes severe contact dermatitis in some people. The symptoms can take hours or days to appear and can develop in people who have not been afflicted in the past. Because this plant is so variable in appearance, it can be a challenge to recognize in all its morphs: from low shrubs with erect or sprawling stems to vines climbing up to 25 m on trees. Leaves are reddish as they emerge, then green, until they again turn red prior to dropping in the fall. Leafless winter stems can inflict dermatitis! Poison oak fruits provide food for many native mammals and birds.

Angelica breweri

Brewer's Angelica

Apiaceae (Carrot family)

Erect perennial herb 1–2 m; taprooted, herbage usually hairless, sometimes hairy.

LEAVES: Basal and alternate, compound; basal blades 15–35 cm, 2-pinnate or ternate (meaning leaflets themselves have leaflets), triangular in overall outline. Leaflets 6–10 cm, lanceolate, dentate to serrate.

FLOWERS: Compound umbels, with 20–50 unequal rays, each 2–10 cm, giving umbels an open look. Calyx minute or absent; petals 5, white (occasionally pink, red, or purple), and hairy.

FRUITS: 8–12 mm, flattened front to back, winged, oblong or ovate, at least somewhat hairy.

SIERRA DISTRIBUTION: West slope, Tulare County north; both slopes from Alpine County north.

ELEVATION: 800–2800 m

HABITAT: Moist sites, especially meadows and streamsides; montane zone.

FLOWERING TIME: Jun–Aug

NOTES: Two other tall *Angelica* spp. differ from *A. breweri* as follows: California angelica, *A. californica*, is found from El Dorado County north on dry slopes < 1500 m; stems are 1–2.5 m, and leaflets are 4–8 cm, sharply serrate; inflorescence rays are of varying lengths; petals and ovaries are hairless, fruits 6–7 mm. Call's angelica, *A. callii*, is a rare plant found in Fresno, Tulare, and Kern Counties, mostly south of the range of *A. breweri*, along streambanks in the montane zone 1000–2000 m. Stems are 1–2 m, and leaflets are 3–13 cm, sharply serrate; the inflorescence rays are 2.5–7 cm, all about the same length; petals and ovaries are hairy, and fruits are small (3.5–5 mm).

Angelica capitellata

Ranger's Buttons

Apiaceae (Carrot family)

Erect perennial herb 0.5–1.8 m; from tuberous taproots.
LEAVES: Basal and alternate, varying from bright green to grayish. Petioles 1–4 dm, inflated at base, partially sheathing stem, sheaths enlarged on cauline leaves and sometimes without leaf blades, or with greatly reduced leaf blades as in photo. Basal blades 1–4 dm, triangular to ovate in overall outline, 1–2 pinnate or ternate-pinnate; leaflets 1–12 cm, margins vary from simply toothed to pinnately lobed.
FLOWERS: Compound umbels, tomentose, peduncles 7–40 cm. Bracts absent, rays 4–18, 1.5–10 cm, ascending to reflexed. Secondary umbels generally lack pedicels, thus umbels are composed of densely grouped flowers, headlike. Bractlets at base of these round heads are linear and bristlelike. Calyx lobes absent; petals 5, obovate, white or sometimes pink. Styles slender; stamens 5.
FRUITS: 5–8 mm, wedge-shaped, tomentose.
SIERRA DISTRIBUTION: All.
ELEVATION: 900–3500 m
HABITAT: Moist sites, especially meadows and streamsides; montane through alpine zones.
FLOWERING TIME: Jul–Aug
NOTES: Formerly *Sphenosciadium capitellatum*. Toxic to livestock, but rarely eaten; also toxic to humans and pets. The tomentose rays (stalks of secondary umbels) distinguish ranger's buttons from other tall, white-flowered Apiaceae.

Angelica lineariloba

Poison Angelica

Apiaceae (Carrot family)

Tall, striking, erect perennial herb 1–1.5 m; hairless or scabrous.
LEAVES: Basal and alternate; blades < 35 cm, pinnately divided into long, slender segments 2–10 cm, smooth-margined. Although the leaves may appear to be pinnately compound, the segments are connected to each other by a thin strip of leaf tissue along the midrib, thus each segment is not entirely separate as in a compound leaf. Petioles are inflated and partially sheathing stem at base.
FLOWERS: Compound umbels, of 20–40 ± equal rays, 3–7 cm. Calyx minute or absent; petals 5, white to pinkish, hairless.
FRUITS: 10–13 mm, oblong or wedge-shaped, flattened front to back, winged on sides, hairless.
SIERRA DISTRIBUTION: Both slopes, Kern and Inyo Counties north to Tuolumne and Mono Counties.
ELEVATION: 1700–3300 m
HABITAT: Exposed, rocky slopes; montane and subalpine zones.
FLOWERING TIME: Jun–Aug
NOTES: A confirmed host plant (plants where eggs are laid and caterpillars feed) for the anise swallowtail butterfly (*Papilio zelicaon*). This is a beautiful black-and-yellow butterfly with a wingspan of 5–8 cm; adults obtain nectar from a variety of native plants (e.g., spreading dogbane, western wallflower, wax currant).

Anthriscus caucalis (non-native)

Bur-Chervil

Apiaceae (Carrot family)

Erect annual herb ranging in height from 45 cm to (rarely) 1 m; often growing in dense stands in shady sites.

LEAVES: Petioles 3–8 cm; blades 5–15 cm, pinnately dissected with overall outline ovate to triangular. Base of petiole forms a sheath on the stem (can look like stipules).

FLOWERS: Compound umbels with peduncles ranging in length 0–2 cm, rays 3–6, each 1–2.5 cm. No bracts at base of rays, but bractlets present at the base of outer flower clusters.

FRUITS: Elongate and cylindric, ovate or lanceolate; the beak at the tip is far shorter than the main body. Covered with hooked bristles.

SIERRA DISTRIBUTION: West slope.

ELEVATION: < 1500 m

HABITAT: Shady, disturbed sites; foothill and montane zones.

FLOWERING TIME: Apr–Jun

NOTES: *Anthriscus* foliage gives off an aniselike aroma when crushed, and it is especially pungent when being mowed. It flowers earlier and is more restricted to shady sites than the somewhat similar-looking *Torilis arvensis*; the bristly fruits are not as strongly Velcro-like as *Torilis* and are less injurious.

Conium maculatum (non-native)

Poison Hemlock

Apiaceae (Carrot family)

Hairless biennial herb 0.5–3 m; stem is conspicuously purple-dotted or streaked.

LEAVES: Petioles expanded at base; blades 1.5–3 dm, 2-pinnate, overall outline is ovate.

FLOWERS: Compound umbels of 10–20 rays; bracts and bractlets present; calyx absent; petals white.

FRUITS: Ovoid, slightly flattened side-to-side; ribs wavy.

SIERRA DISTRIBUTION: All, continually spreading to new locations.

ELEVATION: < 2300 m

HABITAT: Moist to wet, flooded areas; all zones except subalpine and alpines.

FLOWERING TIME: Apr–Jul

NOTES: Extremely toxic to humans and animals; this plant contains alkaloids that have caused many accidental human deaths and is notorious for use in capital punishment in ancient Greece, especially for Socrates. Livestock generally avoid it. Contact dermatitis is possible if handled without gloves. Poison hemlock is very invasive and outcompetes native plants once well established.

Cymopterus terebinthinus var. *californicus*

California Wavewing

Apiaceae (Carrot family)

Perennial herb 1.5–4.5 dm; woody at base, taprooted; hairless, stem short or absent.

LEAVES: Basal, bright green to gray-green; petioles 2–16 cm; blades 1.5–18 cm, pinnately or ternately dissected, oblong or ovate in overall outline; ultimate segments linear.

FLOWERS: Compound umbels with 3–24 unequal rays, each 0.5–8 cm, bracts absent but outer flower clusters have linear bractlets 2–6 mm. Pedicels 1–8 mm. Calyx lobes sometimes present; petals 5, yellow.

FRUITS: 5–10 mm wide, irregularly wavy wings wider than fruit body.

SIERRA DISTRIBUTION: All.

ELEVATION: 500–3600 m

HABITAT: Rocky and gravelly areas; montane and subalpine zones.

FLOWERING TIME: May–Jun

NOTES: Resembles a *Lomatium*, but none of the SN *Lomatium* spp. have wavy-winged fruits.

Daucus carota (non-native)

Carrot, Queen Anne's Lace

Apiaceae (Carrot family)

Erect biennial herb 1.5–12 dm; branched, hairy.

LEAVES: Basal and alternate; petioles 3–10 cm; blades 5–15 cm, pinnately dissected.

FLOWERS: Inflorescences of dense compound umbels, rays numerous, lengths varying 3–7.5 cm, subtended by pinnately lobed bracts 1–4.5 cm. Secondary umbels with bractlets at base; petals white, though central flower is often purple.

FRUITS: 3–4 mm, oblong or ovate, somewhat flattened front to back, widest at middle, ribs bristly.

SIERRA DISTRIBUTION: Patchy throughout the SN, most abundant from Calaveras to Butte Counties; could appear anywhere that conditions are suitable.

ELEVATION: < 1650 m

HABITAT: Disturbed sites; foothill and montane zones.

FLOWERING TIME: May–Sep

NOTES: The "nest" created by the inflorescence bracts surrounding the densely arranged flowers or fruits is diagnostic (inset). This garden plant has become widespread in wildlands and is highly invasive, though it is not ubiquitous throughout the SN at this time. Domestic carrots are named *Daucus carota* ssp. *sativus* and originated in central Asia. Breeding over millennia to maximize sweetness and minimize the woody core of the root resulted in the tasty and healthy root vegetable we know today.

Daucus pusillus

American Wild Carrot

Apiaceae (Carrot family)

Erect annual herb 0.3–9 dm; simple or with a few branches, hairy.
LEAVES: Petioles 4–15 cm; blades 3–10.5 cm, pinnately dissected, with segments 1–5 mm, linear, often bristly.
FLOWERS: Inflorescences of dense, compound umbels, surrounded by large, leaflike bracts 1–4.5 cm; secondary umbels have smaller bractlets at base; petals white. Flowers with or without calyx lobes, white petals.
FRUITS: 3–5 mm, oblong or ovate, somewhat flattened, bristles on ribs not hooked.
SIERRA DISTRIBUTION: West slope.
ELEVATION: < 1650 m
HABITAT: Open areas, often sandy or rocky; foothill and montane zones.
FLOWERING TIME: Apr–Jun
NOTES: Compare this native Velcro-like plant with weedy non-native annuals *Torilis arvensis* and *Anthriscus caucalis*. The "nest" created by the long, inflorescence bracts surrounding the densely arranged flowers or fruits distinguishes this plant, along with the less aggressive tendency to attach to clothing, fur, and fingertips.

Eryngium castrense

Great Valley Coyote-Thistle

Apiaceae (Carrot family)

Spiny, hairless perennial herb 2–6 dm; ascending to erect. The numerous branches arise about 2–5 cm up from base.
LEAVES: Basal in rosettes, cauline alternate; petioles very short or absent. Blades 10–30 cm, pinnately lobed with spine-tipped lobes.
FLOWERS: Inflorescences 8–15 mm, round or ovoid. Bract 1 per flower, 15–30 mm, linear, distinctly protruding throughout inflorescence, with each bract spiny below and on margins. Sepals 3–3.5 mm, margin translucent and spine-tipped; petals 5, oblanceolate, white or purplish. Styles 3.5–4.5 mm, longer than sepals.
FRUITS: 3–3.5 mm, narrowly elliptic, densely covered with unequal, lanceolate, papillate scales.
SIERRA DISTRIBUTION: West slope.
ELEVATION: < 900 m
HABITAT: Vernal pools, swales, intermittent streams; foothill zone.
FLOWERING TIME: Apr–Jul
NOTES: From Madera County south, the rare *E. spinosepalum* can be found in similar habitats < 1270 m. It differs by having long (5–20 mm) petioles and shorter (2.5–3 mm) fruits without papillate scales, and in that the bracts protruding from the inflorescence are not as long and noticeable as in *E. castrense*. ID of *Eryngium* to species is best done with the guidance of someone experienced with this genus. Because several taxa in the SN foothills and adjacent Great Valley are protected and they can be hard to field ID, be careful not to harm any *Eryngium* encountered.

Heracleum maximum

Common Cow Parsnip

Apiaceae (Carrot family)

Erect, robust perennial herb 1–3 m, with hollow stems covered with woolly hairs. Strongly scented.

LEAVES: Basal and alternate; basal leaves usually ternately compound but sometimes just deeply 3-lobed, roundish in overall outline. Petioles 1–4 dm, with wide sheaths at base; blades 2–5 dm wide, each leaflet or lobe itself deeply lobed and toothed.

FLOWERS: Compound umbels, rays 15–30 (sometimes up to 45), of unequal lengths 5–10 cm, woolly. Calyx lobes absent; corollas mostly radial; petals white, obovate. Marginal flowers have outer petals larger, 2-lobed (these flowers are bilateral).

FRUITS: 8–12 mm, obovate/obcordate, flattened front to back (flat side facing axis of ray), hairy, with ribs thin, winged on side margins, other ribs not winged.

SIERRA DISTRIBUTION: Both slopes, Tulare and Inyo Counties north.

ELEVATION: < 2900 m

HABITAT: Moist sites, especially meadows and streamsides; montane and subalpine zones.

FLOWERING TIME: Apr–Jul

NOTES: Formerly *H. lanatum*. One of the tallest native herbs in the Sierra, with enormous basal leaves. The invasive giant hogweed, *H. mantegazzianum*, native to Eurasia, is likely to show up in wet areas of the SN at some point. It differs by being truly giant rather than just very large: Stems are 2–5 m, with leaves up to 1.5 m wide, and 50–150 rays in the inflorescence, which is also much wider in diameter (50 cm). Stems are blotched with purple. Touching the sap of this weed can cause severe dermatitis; sap in your eyes can cause temporary or permanent blindness. Any sightings of this weed should be reported to the local county agriculture department, Cal-IPC, or your local CNPS chapter.

Ligusticum grayi

Gray's Lovage

Apiaceae (Carrot family)

Erect perennial herb 2–8 dm; hairless, from fibrous taproots, stem bases also fibrous.

LEAVES: Hairless; basal leaves with petioles 0.2–3 dm; blades compound, 10–25 cm, 2-pinnate, and oblong or triangular in overall outline. Leaflets are 1–4 cm, oblong or ovate, acute, themselves pinnately compound or deeply lobed. Stem leaves 0–2, usually 1, and much smaller than basal.

FLOWERS: Compound umbels, rays 5–18, 1.5–5 cm, unequal; bracts absent, pedicels 5–10 mm. Bractlets absent or inconspicuous. Calyx lobes minute; petals 5, white or occasionally pink.

FRUITS: 4–5 mm, oblong, ribs narrowly winged. Seeds concave.

SIERRA DISTRIBUTION: Both slopes, Tulare and Mono Counties north.

ELEVATION: 1000–3300 m

HABITAT: Moist to wet sites, in shaded forests or more open areas, sometimes rocky sites; montane, subalpine, and alpine zones.

FLOWERING TIME: Jun–Sep

NOTES: Gray's lovage is exceedingly common and variable. Another less common species occurs in the SN: *L. californicum* occurs sparingly from El Dorado County north on the west slope, < 2400 m; it is taller (6–12 dm) than *L. grayi*, has 2–3 stem leaves approximately equal to basal leaves, and umbels with 16–30 unequal rays, each 2–6 cm.

Lomatium dissectum

Fern-Leaved Lomatium

Apiaceae (Carrot family)

Tall, stout perennial herb 3–14 dm; glaucous, hairless to puberulent, from thick taproots.

LEAVES: Basal and cauline; petioles 3–30 cm; blades 15–35 cm wide, ternately-pinnately dissected, and triangular-ovate to round in overall outline. Ultimate leaf segments 2–22 mm x 0.5–2 mm wide, linear to oblong.

FLOWERS: Peduncles glaucous; compound umbels of 10–30 rays, each 3–10 cm, spreading, no bracts, but linear bractlets at base of each secondary umbel. Calyx lobes usually absent; petals 5, yellow or maroon.

FRUITS: 12–16 mm, hairless, wings smaller than fruit body; pedicels 1–3 mm, fruits often appearing sessile.

SIERRA DISTRIBUTION: All, sparingly distributed.

ELEVATION: 150–2000 m

HABITAT: Deep moist soils in wooded or shrub-dominated areas; foothill, montane, and sagebrush/pinyon-juniper zones.

FLOWERING TIME: May–Jul

NOTES: This can be a huge plant! Much taller than most other *Lomatium* spp. in the SN, and quite striking. The photo shows yellow flowers, but they are more often maroon. *Lomatium multifidum* was formerly *L. dissectum* var. *multifidum,* but is now at full species level. It is equally tall and very similar to *L. dissectum*, but peduncles are not glaucous, flowers are always yellow, and fruiting pedicels are 5–15 mm. Distribution is patchy throughout the SN, 600–3000 m.

Lomatium torreyi

Torrey's Lomatium, Sierra Biscuitroot

Apiaceae (Carrot family)

Perennial herb 1–3 dm; hairless to scabrous, base fibrous.

LEAVES: Clustered at base; petioles 2–5 cm, sheathed entire length; blades 2–15 cm, ternately-pinnately dissected, oblong or ovate in overall outline. Leaf segments 3–10 mm linear, acute.

FLOWERS: Compound umbels, of 3–9 ascending or spreading rays, each 1–4 cm. Bractlets absent, pedicels 1–10 mm. Petals 5, yellow. Ovary hairless.

FRUITS: 10–16 mm, narrowly oblong, hairless; wings narrow, < ½ the width of fruit body.

SIERRA DISTRIBUTION: Both slopes, Kern and Inyo Counties north to Placer County.

ELEVATION: 1100–3300 m

HABITAT: Rocky sites, cliffs in coniferous forest; montane, subalpine, and alpine zones.

FLOWERING TIME: May–Aug

NOTES: This extremely variable plant is found above the foothill zone of the west slope, and to the crest and in scattered locations on the east slope.

Lomatium utriculatum

Common Lomatium

Apiaceae (Carrot family)

Perennial herb 1–5 dm; hairless to densely puberulent, from slender taproots.

LEAVES: Basal and alternate; petioles 1.5–10 cm, bases widely sheathing. Stem leaves have conspicuously widened membranous sheaths; blades 5–16 cm, pinnate or ternately-pinnately dissected, oblong or ovate in overall outline; leaf segments 2–25 mm, linear.

FLOWERS: Compound umbels, with 5–20 ascending or spreading rays, lengths unequal, 2–12 cm. Bractlets 3–12, 3–16 mm; pedicels 2–9 mm. Calyx lobes sometimes present; petals 5, yellow. Ovary puberulent to hairless.

FRUITS: 5–15 mm, hairless, wings thin and wider than fruit body.

SIERRA DISTRIBUTION: West slope.

ELEVATION: < 1500 m

HABITAT: Grassy and rocky slopes and flats; foothill and montane zones.

FLOWERING TIME: Feb–May

NOTES: One of the most common *Lomatium* spp. of the foothill and lower montane zones, and also one of the most variable. Plants can be almost woolly, with foliage appearing grayish, or entirely hairless with shiny, green foliage. Look for the conspicuous expanded membranous sheath ("utricle") at leaf bas

Oenanthe sarmentosa

Water Parsley
Apiaceae (Carrot family)

Ascending to decumbent perennial herb 5–15 dm, rooting from lower nodes when stems rest on wet ground or in water.
LEAVES: Basal and alternate; petioles 2–35 cm; blades 7–30 cm, bipinnate; leaflets 1–6 cm, toothed to lobed.
FLOWERS: Compound umbels, with 10–20 rays, each 1.5–3 cm, usually without bracts; pedicels 2–6 mm; bractlets many, 4–5 mm, lanceolate. Calyx lobes 0.5–1 mm; petals white, outer flowers sometimes bilateral. Styles 2–3 mm.
FRUITS: 2.5–3.5 mm, oblong, ribs thickened, not winged.
SIERRA DISTRIBUTION: West slope, Placer County south (likely more common than records indicate).
ELEVATION: < 1800 m
HABITAT: Wet areas such as streams, ponds, marshes; foothill and montane zones.
FLOWERING TIME: Jun–Oct
NOTES: Often found growing in shallow water as well as next to streams, ponds, or lakes. Often forms a connected network of stems (rhizomes, stolons) growing from lower nodes where rooting occurred, these often in the water. Once I became familiar with this semiaquatic plant, I found it to be widespread and abundant in Madera County around lakes and along streams. I suspect it has been overlooked to date.

Oreonana clementis ✿

Pygmy Mountain-Parsley
Apiaceae (Carrot family)

Cushion-forming perennial herb 3–8 cm, grayish, hairy, from thick taproots. Flowers appear simultaneously with leaf blades.
LEAVES: Base of stems with sheaths ≥ 1, 0.5–2.0 cm, scarious, with no leaf blade attached. Upper leaves have petioles 2–4 cm; blades 1.5–3.5 cm, pinnately or ternately dissected, ovate in overall outline; leaf segments 1–3 mm.
FLOWERS: Compound, headlike umbels, peduncles 2–7 cm, bractlets fused, rays 5–15. Flowers bisexual or male; calyx lobes 0.5–1.5 mm, yellow; corollas white. Anthers purple.
FRUITS: 3–4 mm, hairless or hairy, not winged.
SIERRA DISTRIBUTION: Both slopes, Fresno, Inyo, and Tulare Counties.
ELEVATION: 1500–4000 m
HABITAT: Rocky ridges; montane, subalpine, and sagebrush zones.
FLOWERING TIME: Jul–Aug
NOTES: In Tulare County, the similar but much rarer purple mountain parsley, *O. purpurascens*, may be encountered. Its flowers emerge before the leaves, the umbels are larger (2.4–4.5 cm) and have 20–35 rays, and the calyx lobes are purple rather than yellow.

Osmorhiza berteroi

Mountain Sweet-Cicely

Apiaceae (Carrot family)

Branched, leafy perennial herb 3–12 dm. Hairless or finely hairy.
LEAVES: Basal and alternate; petioles 5–16 mm; blades 4–20 cm wide, 2-ternate, ovate in overall outline; leaflets 2–8 cm.
FLOWERS: Compound umbels, of 3–8 rays, each 2–12 cm; usually with no bractlets at base of secondary umbels. Pedicels 4–20 mm, corollas white. Styles < 1 mm.
FRUITS: 12–25 mm, narrowly cylindric with a tapered beak at top, bottom tapered at base into a tail 2–8.5 mm, ribs covered with bristly hairs.
SIERRA DISTRIBUTION: All.
ELEVATION: < 3100 m
HABITAT: Moist to dry sites; foothill, montane, subalpine, and sagebrush/pinyon-juniper zones.
FLOWERING TIME: Mar–May
NOTES: Formerly *Osmorhiza chilensis*. *O. berteroi* is common and can be distinguished from other SN *Osmorhiza* spp. by the white flowers and the very long tails on fruits.

Osmorhiza brachypoda

California Sweet-Cicely

Apiaceae (Carrot family)

Branched, leafy perennial herb 3–8 dm. Covered with fine hairs.
LEAVES: Basal and alternate, 2–3 ternate; petioles 5–20 cm; blades 1–2 dm, triangular-ovate to ovate in overall outline; leaflets 2–8 cm, ovate, toothed to deeply lobed.
FLOWERS: Compound umbel, of 2–5 rays, 3.5–12 cm, with 2–6 clearly visible bractlets, 2–10 mm; corollas greenish to yellowish.
FRUITS: 12–20 mm, tapered at base into a tail 1–4 mm, ribs bristly.
SIERRA DISTRIBUTION: West slope.
ELEVATION: 200–2000 m
HABITAT: Moist to dry sites; foothill and montane zones.
FLOWERING TIME: Mar–May
NOTES: *O. brachypoda* is distinguished from other SN *Osmorhiza* spp. by the obvious bractlets at the base of each secondary umbel. The seeds become needlelike, easily caught in clothing or animal fur.

Osmorhiza occidentalis
Western Sweet-Cicely
Apiaceae (Carrot family)

Branched, leafy perennial herb 4–12 dm; hairless to fine-hairy.
LEAVES: Basal and alternate, 2-pinnate; petioles 5–25 cm; blades 1–2 dm, oblong to ovate in overall outline. Leaflets 2–10 cm, lance-oblong to ovate, margins toothed, irregularly cut or lobed to various degrees.
FLOWERS: Compound umbels, of 5–12 rays, 3–8 cm, usually no bractlets at base of secondary umbels. Pedicels 3–8 mm, corollas yellow. Styles 0.8–1.4 mm.
FRUITS: 12–22 mm, not tapered at base, no basal tail-like projection, hairless.
SIERRA DISTRIBUTION: Both slopes, Tulare and Mono Counties north; more common north of Madera County.
ELEVATION: 350–3200 m
HABITAT: Moist to dry sites; foothill, montane, subalpine, and sagebrush/pinyon-juniper zones.
FLOWERING TIME: May–Jul
NOTES: *O. occidentalis* is distinguished from other *Osmorhiza* spp. in the SN by the hairless fruits without basal tails, and the twice pinnate leaves.

Oxypolis occidentalis
Western Cowbane
Apiaceae (Carrot family)

Erect perennial herb 6–15 dm; from clustered tubers, hairless; stems are simple or branched.
LEAVES: Basal and alternate; blades 12–30 cm, pinnately compound, overall outline oblong. Leaflets 5–13, 3.5–9.5 cm, lanceolate to ovate, margins scalloped or toothed (never entire).
FLOWERS: Compound umbels, of 12–24 unequal rays, 2–8.5 cm; bracts 0–1, 5–20 mm; bractlets beneath secondary umbels < 10 mm.
FRUITS: 5–6 mm, oblong or ovate, flattened front-to-back, appearing 5-ribbed with lateral ribs broadly winged.
SIERRA DISTRIBUTION: All.
ELEVATION: < 2700 m
HABITAT: Wet areas, fens, streamsides; montane and subalpine zones.
FLOWERING TIME: Jul–Aug
NOTES: Presumably toxic to livestock, as "bane" means poisonous and fatal, though I have found no literature confirming this. On the other hand, the highly toxic native water hemlock, *Cicuta douglasii*, which looks similar, occurs sporadically throughout the SN < 2800 m. *C. douglasii* has stems 1.5–3 m, with leaves 15–45 cm; its overall outline is lanceolate or linear, acute or acuminate, entire to serrate; fruits are ovoid to spheric, with low, wide, corky ribs (not winged). Do not eat!

Perideridia parishii ssp. *latifolia*

Parish's Yampah

Apiaceae (Carrot family)

Erect, hairless perennial herb 2.5–9 dm; from a single, fusiform tuberous root 1–2.5 cm long.

LEAVES: Basal with petioles 3–10 cm; blades 10–20 cm, 1-ternate or pinnate, with 1–3 pairs of leaflets, 3–15 cm. Stem leaves 1-ternate.

FLOWERS: Compound umbels, flat-topped to rounded, of 12–14 rays, 1–4 cm; bracts 0–2. Secondary umbels of 13–27 flowers, with 6–8 bractlets, about equal in length to pedicels; petals 1-veined. Styles 1–1.5 mm.

FRUITS: 1.8–3.5 mm, oblong or round. Rays remain about equal in fruit (rather than elongating to different lengths, like the similar *P. lemmonii*). Ribs threadlike, not winged. Oil tubes 3–4 per rib interval (high magnification required).

SIERRA DISTRIBUTION: All.

ELEVATION: 1200–3400 m

HABITAT: Moist to wet meadows and streamsides; montane, subalpine, alpine, and sagebrush/pinyon-juniper zones.

FLOWERING TIME: Jun–Aug

NOTES: Nearly identical to Lemmon's yampah, *P. lemmonii*, which occurs 750–2500 m from Tulare County north. Both form extensive stands of solid white flowers in meadows, sometimes growing together. *P. lemmonii*'s tuberous root is 1.5 cm; the secondary umbels have 20–35 flowers, with 7–10 lance-linear bractlets 1–2 mm (shorter than pedicels), rays becoming unequal in fruit, and 1 oil tube per fruit rib interval (high magnification required). The tuberous roots of both species were and still are a hugely important traditional starch source for Native Americans; roasted tubers can be stored for later use.

Sanicula bipinnata

Poison Sanicle

Apiaceae (Carrot family)

Perennial herb 12–60 cm; usually erect, with a swollen taproot that looks elongate rather than like the tubers of *S. tuberosa*. The herbage smells like cilantro.

LEAVES: Basal and cauline, ternate, then 1–2 pinnate, 3–10 cm, green.

FLOWERS: Heads dense, composed of 3–10 bisexual flowers and 4–6 staminate flowers; all corollas yellow, inconspicuous. Staminate flowers do not stick out beyond bisexual flowers.

FRUITS: 2–3 mm wide, ovate to round; short, curved prickles with bulbous bases.

SIERRA DISTRIBUTION: West slope.

ELEVATION: < 1000 m

HABITAT: Grasslands, openings; foothill and montane zones.

FLOWERING TIME: Apr–May

NOTES: The common name implies this plant is toxic, but there are no records of poisonings, and it is likely to be no more toxic than other *Sanicula* spp. Some CA tribes are recorded to use it for treating venomous snake bites. However, because the Carrot family has many highly toxic plants, it would be prudent not to eat any of them unless you are 100% sure what you're doing. Compare with *S. tuberosa*; the male flowers do not stick out beyond the bisexual flowers/fruits in *S. bipinnata* as they do in *S. tuberosa.*

Sanicula bipinnatifida

Purple Sanicle, Shoe Buttons

Apiaceae (Carrot family)

Erect perennial herb 12–60 cm.

LEAVES: Basal and cauline, 4–19 cm, appearing bipinnately compound but with a strip of often jagged leaf tissue along the midribs, hence bipinnatifid rather than bipinnate (*-fid* means split, divided into). Foliage sometimes purplish.

FLOWERS: Umbel of 3–5 rays; heads dense and composed of 8–10 bisexual flowers and 10–12 staminate flowers; all corollas purple or yellow; stamens noticeably long.

FRUITS: 3–6 mm, ovate to round, with curved prickles.

SIERRA DISTRIBUTION: West slope.

ELEVATION: < 1850 m

HABITAT: Open grasslands, rocky areas; foothill and lower montane zones.

FLOWERING TIME: Mar–May

NOTES: The yellow-flowered form seems more common in the Sierra foothills, though both occur, often directly next to each other. This species has also been called poison sanicle (see notes for *S. bipinnata*).

Sanicula crassicaulis

Pacific Sanicle

Apiaceae (Carrot family)

Stout perennial herb 24–120 cm; erect, taprooted, branched from mid to upper stem if branched.

LEAVES: Basal and alternate, 3–12 cm, overall shape is rounded, palmately lobed, margins with teeth ± 1 mm.

FLOWERS: Simple heads with about 5 bracts. Heads of 3–8 bisexual flowers and 3–5 staminate flowers. Calyx lobes 5, 0.5–0.7 mm, lanceolate, acute, fused at base; corollas yellow; petals 5, styles twice as long as calyx lobes; stamens 5.

FRUITS: 2–5 mm, roundish, with prickles.

SIERRA DISTRIBUTION: West slope.

ELEVATION: < 1500 m

HABITAT: Open grasslands, shaded slopes; foothill and montane zones.

FLOWERING TIME: Mar–May

NOTES: This plant tends to blend into the background vegetation, with its less showy flowers and little to no smell of cilantro. The species name is from *crassi* ("thick or fat") and *caulis* ("stem"). A study in Canada showed a strong preference for *S. crassicaulis* by a few species of mining bees (*Andrena* spp.). The bees gathered pollen for their broods in underground nests and presumably pollinated the flowers.

Sanicula tuberosa

Turkey-Pea

Apiaceae (Carrot family)

Perennial herb 5–80 cm; erect or somewhat lax, with underground tubers 5–15 mm wide. The herbage smells like cilantro.

LEAVES: Basal and cauline ternate, then pinnatifid or 1–2 pinnate, 2–13 cm, green to glaucous.

FLOWERS: Umbel of 3 rays; heads dense, composed of 3–5 bisexual flowers with no pedicels, and 4–10 staminate flowers with pedicels 2–7 mm. All corollas bright yellow.

FRUITS: 1.5–2 mm, ovate to round, with rounded bumps, no prickles. Fruiting heads distinctive due to the staminate flowers sticking out far past the fruit.

SIERRA DISTRIBUTION: West slope, all counties.

ELEVATION: < 2700 m

HABITAT: Open grasslands, rocky areas; foothill and montane zones.

FLOWERING TIME: Mar–Jul

NOTES: One of many native plants called "Indian potatoes." Miwok and Western Mono people (and other tribes) gather tubers in the spring, carefully leaving plenty for the future. The spheric tubers are boiled like potatoes, and from the enthusiastic response I've seen to finding them in the field, they must be delicious! These plants are especially abundant after wildfire, and cultural burning serves to maintain the abundance of the plants.

Tauschia hartwegii

Hartweg's Umbrellawort

Apiaceae (Carrot family)

Perennial herb 3–10 dm; herbage scabrous.

LEAVES: Basal with petioles 5–25 cm; blades 12–24 cm, 1–2 pinnately divided into ovate or oblong leaflets with toothed margins.

FLOWERS: Compound umbels, no bracts, but reflexed bractlets 5–12 mm are present at the base of each secondary umbel. Calyx lobes tiny; petals 5, yellow.

FRUITS: 4–7 mm, roundish, unwinged.

SIERRA DISTRIBUTION: West slope.

ELEVATION: < 1800 m

HABITAT: Dry, often rocky sites; foothill and montane zones.

FLOWERING TIME: Mar–May

NOTES: Three other *Tauschia* spp. occur in the SN: *T. howellii* is found only at 2 disjunct locations in Fresno and Sierra Counties (its main range in Siskiyou County). Plants are grayish, 5–8 cm, stems are shorter than leaves, and bractlets are longer than umbels and like the leaves. *T. parishii* occurs mostly on the east slope and also looks quite different from *T. hartwegii*; plants are 1–4 dm, glaucous, and leaflets are pinnately lobed and sharply toothed. *T. kelloggii* looks the most like *T. hartwegii*, occurring in the western foothills from Nevada County north < 1700 m; plants are 2–7 dm with petioles 5–15 cm, leaf blades 8–20 cm, and leaflets more markedly toothed or lobed than *T. hartwegii*. Bractlets are few and shorter than flowers and fruits; fruits are 3–5 mm.

Torilis arvensis (non-native)

Spreading Hedge Parsley

Apiaceae (Carrot family)

Erect, branched annual herb 3–10 dm.

LEAVES: Basal and cauline, ovate in overall outline; lower 2–3 leaves pinnately dissected, upper smaller and only 1-pinnate.

FLOWERS: Compound umbels open, with 0–2 bracts at base of umbel and several bractlets, 1–4 mm, at base of secondary umbels. Flowers slightly bilateral, petals white.

FRUITS: This is where the fun begins! Fruits 3–5 mm, oblong, with barbed prickles that form a Velcro-like bond with passing animals and also on shoes, socks, clothing, etc.

SIERRA DISTRIBUTION: West slope.

ELEVATION: < 1600 m (probably higher)

HABITAT: Disturbed areas, grassy openings; foothill and montane zones.

FLOWERING TIME: Apr–Jul

NOTES: This is truly one of the most annoying plants ever introduced into CA. *T. nodosa* is also found in the Sierra < 1800 m; flower heads are more condensed, and fruits are prickly only on outer margins.

Yabea microcarpa

California Hedge Parsley

Apiaceae (Carrot family)

Erect annual herb 3–40 cm; hairy.

LEAVES: Alternate; petioles 2.5–3.5 cm; blades 2–6 cm, with overall outline oblong to ovate, pinnately dissected into threadlike or linear segments 2–8 mm.

FLOWERS: Compound (sometimes simple) umbels of 1–9 rays, 1–8 cm, unequal in length; bracts 2–5, leaflike; bractlets 1–5, 1–10 mm, entire or occasionally lobed; pedicels 2–9, < 15 mm; flowers tiny, white, usually bilateral.

FRUITS: 2–7 mm, oblong, compressed side-to-side, ribbed, bristly.

SIERRA DISTRIBUTION: West slope.

ELEVATION: < 1500 m

HABITAT: Grasslands, chaparral, foothill woodland; foothill and montane zones.

FLOWERING TIME: Apr–Jun

NOTES: The prickles on the ribs of mature fruits have recurved tips, qualifying them as "Velcro plants" along with *Torilis arvensis* and *Daucus pusillus*, but they don't seem to attach as aggressively to passersby.

Apocynum cannabinum

Indian Hemp

Apocynaceae (Dogbane family)

Erect, stout perennial herb with stems 1.2 dm; branched near top, growing in patches from thick rhizomes, shrublike.

LEAVES: Opposite, petioles 2–7 mm or absent; blades 5–8 cm, ovate to lanceolate, ascending, base clasping stem or not, yellowish-green on both surfaces.

FLOWERS: Inflorescences terminal (and sometimes in upper leaf axils), not extending above foliage. Pedicels 2–3 mm, the 5 calyx lobes triangular, ± length of corolla tube; corollas 2.5–5 mm, cylindric to urn-shaped, the 5 corolla lobes erect or spreading, 1.5–2 mm, corollas white, or white tinged with yellow or green. There are 5 nectaries surrounding the ovaries.

FRUITS: Capsules in pairs, 6–9 cm, long, slender, pendulous.

SIERRA DISTRIBUTION: All.

ELEVATION: < 2450 m

HABITAT: Moist to wet areas along streams, many plant communities, all zones except alpine.

FLOWERING TIME: Apr–Oct

NOTES: Because of its strong stem fibers, Indian hemp is an important cordage plant for Native Americans. Stems are used to make string, rope, and nets. Plants can be managed with fire and judicious harvesting and tending to produce strong, long fibers. Stems are harvested in the fall. Abundance is reduced since European contact—because of its toxicity to livestock, campaigns to eliminate it were carried out in the past

Apocynum androsaemifolium

Bitter Dogbane, Spreading Dogbane

Apocynaceae (Dogbane family)

Perennial herb 1.6–3 dm; with extensively branched stems, often growing horizontally as a ground-cover. Sap milky.

LEAVES: Opposite, petioles 3–5 mm, blades 4–6 cm, ovate to round, drooping to spreading horizontally, glabrous to pubescent. Before leaves are shed in the fall, they contribute a beautiful array of yellow, orange, and reddish hues to the below-the-knee fall colors.

FLOWERS: Inflorescences extending above foliage, flowers terminal and in upper leaf axils. Pedicels 1.5–7 mm, the 5 calyx lobes usually broadly triangular, much shorter than corolla tubes. Corollas 4–8 mm, bell-shaped, white or pale pink with darker veins or reddish-purple; lobes 5, spreading or recurved. There are 5 nectaries surrounding the ovaries.

FRUITS: Long, cylindric capsule, 7–11 cm, in pairs; pendulous or erect.

SIERRA DISTRIBUTION: All, except Kern County.

ELEVATION: 500–3200 m

HABITAT: Dry forest floor, open rocky areas; montane, subalpine, and sagebrush/pinyon-juniper zones.

FLOWERING TIME: May–Oct

NOTES: Provides nectar for a great variety of insect visitors, including monarch butterflies. The milky sap (seen by breaking a leaf or piece of stem) contains mildly toxic cardiac glycosides; and Native Americans use powdered roots and extracts of roots or leaves to treat a variety of maladies, including headache, asthma, cardiac conditions, and intestinal worms. Other Native American uses for this plant include using the tough stem fibers to make thread, string, nets, bowstrings, and basketry materials.

Asclepias californica
California Milkweed
Apocynaceae (Dogbane family)

White-woolly, decumbent to ascending perennial herb with stems 10–90 cm; usually with many stems (up to 20), not rhizomatous. Sap milky.

LEAVES: Opposite; sessile or with short petioles, densely woolly, widely ovate; blades 5–18 cm.

FLOWERS: Umbels of 6–12 flowers; pedicels 15–40 mm, calyx lobes 5, hidden by reflexed corolla lobes. Corolla lobes pink to purple; hairless on inner surfaces; white-woolly on outer surfaces. Hoods purple, rounded, shorter than anther heads, with no horns visible.

FRUITS: Follicles ovoid, 5–12.5 cm x 2–3 cm wide, tomentose, longitudinally ridged. Seeds ± oval, 1 cm, margin narrowly winged, tuft of hairs 1.5–2.5 cm.

SIERRA DISTRIBUTION: West slope, Mariposa County south.

ELEVATION: 200–2350 m

HABITAT: Open areas in grasslands, rocky areas, chaparral, coniferous forest; foothill, montane, and sagebrush/pinyon-juniper zones.

FLOWERING TIME: Apr–Jul

NOTES: This is a gorgeous milkweed, with its striking pink to purple flowers contrasting with woolly, white foliage. Along with other milkweeds, it is an important cordage plant and has many other uses by Native Americans.

Asclepias cordifolia
Purple Milkweed, Heart-leaved Milkweed
Apocynaceae (Dogbane family)

Hairless, green to purplish perennial herb ascending, 25–100 cm; no rhizomes. Sap milky.

LEAVES: Opposite, sessile, blades 5–15 cm, cordate, bases clasping stem; green to purplish.

FLOWERS: Inflorescences terminal as well as in upper axils; many flowers in loose umbels. Corolla lobes purple and reflexed, concealing the pubescent calyx lobes when fully open. Hoods light purple, at about the level as the anther heads; no horns; upper edges of hood taper into an acute "toothlike cusp" facing anther heads.

FRUITS: Follicles 10–14 cm, hairless. Seeds 8–9 mm, ovate, hairs 3–3.5 cm, white.

SIERRA DISTRIBUTION: West slope, all; east slope, Alpine County north.

ELEVATION: < 2000 m

HABITAT: Open areas in grasslands, chaparral, and coniferous forest; foothill, montane, and sagebrush/pinyon-juniper zones.

FLOWERING TIME: May–Jul

NOTES: This striking milkweed is easy to identify—no other species is hairless, purplish, with clasping, heart-shaped opposite leaves. Highly valued for its tall, fibrous stems by the western Miwok, western Mono, and other tribes, it is also a host plant for monarch butterfly larvae. The larvae feed exclusively on milkweeds, ingesting toxic alkaloids and becoming unpalatable and toxic to predators. After metamorphosis, monarch butterflies feed on nectar from a variety of native plants. Planting of tropical milkweed, *Asclepias curassavica*, to benefit monarchs backfires in a number of ways, as explained here: xerces. org/blog/tropical-milkweed-a-no-grow.

Asclepias eriocarpa

Kotolo, Indian Milkweed

Apocynaceae (Dogbane family)

Erect perennial herb 30–100 cm; tomentose or puberulent, sometimes becoming hairless in age. Sap milky. Forms continuous patches via rhizomes.

LEAVES: Whorled, with 3–4 leaves per whorl, sometimes opposite or alternate (all 3 arrangements can be found on 1 stem). Petioles 4–15 mm; blades 8–20 cm x 2–8 cm wide; oblong to lanceolate or ovate.

FLOWERS: Inflorescences terminal or in upper axils, with 12–57 flowers; pedicels 15–50 mm, densely woolly, Calyx lobes 3–4 mm, tomentose, hidden by the reflexed corolla lobes. Corollas hairless above; cream, yellow, greenish, or pinkish; hairless below except for a woolly patch at the tip. Hoods light pink, slightly shorter than anther heads; horns level with or slightly exserted from hoods, curved toward anther heads.

FRUITS: Follicles 5–10 cm, usually erect, bright green; tomentose; fruiting pedicels reflexed. Seeds 7–9 cm, ovate, margin with a narrow wing, hairs 2.5–3 cm.

SIERRA DISTRIBUTION: West slope, except Plumas, Sierra, Amador, Calaveras, and Tuolumne Counties.

ELEVATION: 200–2250 m

HABITAT: Open, dry areas; foothill, montane, and sagebrush/pinyon-juniper zones.

FLOWERING TIME: May–Oct

NOTES: Whorled leaves (with opposite and/or alternate leaves) and short, pointed horns that barely exceed the hoods distinguish this species from other milkweeds with densely hairy foliage. Note the monarch butterfly caterpillar in the right center of the photo.

Asclepias fascicularis

Narrow-Leaf Milkweed

Apocynaceae (Dogbane family)

Bright green, erect perennial herb 50–150 cm, often with many stems. No rhizomes. Sap milky.

LEAVES: Whorled, 3–5 leaves per whorl; sessile or short-petioled; blades linear to narrowly elliptic; 4.5–13 cm x 0.2–1.8 cm wide, often folded lengthwise, hairless. Additional axillary clusters of smaller leaves may be present.

FLOWERS: Inflorescences of 10–37 flowers terminal and in upper axils; pedicels 9–14 mm; corolla lobes strongly reflexed, hairless, and greenish-pale yellow to pink or purplish, hiding calyx lobes when fully open. Hoods about the same height as anther heads, and horns longer than both and curved inward.

FRUITS: Follicles 6–9 cm x 0.5–1 cm wide, erect, slender, hairless. Seeds 6 mm, ovate, with winged margins and slightly bumpy surfaces; hairs 2.5–3 cm.

SIERRA DISTRIBUTION: All.

ELEVATION: < 2000 m

HABITAT: Open areas, dry areas, or sites with moisture like roadside ditches; foothill, montane, and sagebrush/pinyon-juniper, and montane zones.

FLOWERING TIME: May–Oct

NOTES: The only milkweed with narrow, bright green whorled leaves. Attracts abundant and diverse native butterflies, bees, and other insects.

Asclepias speciosa

Showy Milkweed

Apocynaceae (Dogbane family)

Erect perennial herb with stems 30–125 cm; rhizomatous, tomentose to puberulent. Sap milky.

LEAVES: Opposite; petioles 4–12 mm; blades 6–20 cm, elliptic or oval.

FLOWERS: Heads large, showy, globelike, with 3–34 flowers. Corolla lobes reflexed, pink-purple, hairy on the back and with smooth inner surfaces, concealing the 5 calyx lobes. Hoods long and petal-like, white to light pink at first then turning yellow, with horns shorter than hoods, curved inward onto anther heads.

FRUITS: Follicles 6–12 cm, erect on reflexed pedicels, surface smooth or roughened, tomentose; seeds 6–9 mm, margins winged, hairs 2.5–3 mm.

SIERRA DISTRIBUTION: Both slopes, Tulare and Inyo Counties north.

ELEVATION: < 2600 m

HABITAT: Open to partly shady areas; often where soil moisture is high; foothill, montane, and sagebrush/pinyon-juniper zones.

FLOWERING TIME: May–Sep

NOTES: This milkweed is especially showy due to the exceptionally long hoods, which look, at first glance, like 5 petals forming a star. An important cordage species for the Mono, or *Nim*, people of the central SN. Flowering plants attract a diversity of native insect pollinators, and caterpillars of Monarch butterflies feed on the leaves (photo).

Asclepias vestita

Woolly Milkweed

Apocynaceae (Dogbane family)

Densely woolly perennial herb; stems prostate, decumbent, or ascending; 50–100 cm; no rhizomes. Sap milky.

LEAVES: Opposite; petioles 3–10 mm; blades 8–20 cm, elliptic, ovate, or lanceolate.

FLOWERS: Inflorescences of 19–45 flowers at top of plant and nearly sessile in upper axils, thus directly adjacent to the stem. Pedicels 15–35 mm, calyx lobes elliptic, 5–6 mm, hidden by the reflexed 6–9 mm corolla lobes, corollas usually cream but sometimes purple; hoods yellowish-white, at about the same level as the anther heads, with horns only slightly exserted above level of hoods.

FRUITS: Follicles ovoid, 5–6.5 x 2–2.5 cm, longitudinally ridged, on reflexed pedicels. Seeds 10–12 mm x 7–10 mm, margin wing very narrow, hairs 2–2.5 cm. cm (largest seeds of CA milkweeds).

SIERRA DISTRIBUTION: West slope, Kern to Madera Counties.

ELEVATION: < 1350 m

HABITAT: Open areas; foothill zone.

FLOWERING TIME: Apr–Jul

NOTES: Sparsely scattered throughout its SN range, though possibly underreported. The nearly sessile inflorescences are diagnostic.

Aralia californica

Elk Clover, Spikenard

Araliaceae (Ginseng family)

Tall, erect, hairless perennial herb to 3 m. Sap milky.

LEAVES: Deciduous, alternate, petioles < 3 dm; blades huge, to 2 m, 1–3 pinnately compound; leaflets 15–30 cm, ovate to oblong, toothed.

FLOWERS: Clusters of umbels in a paniclelike array. Flowers < 5 mm diameter, sepals 5, tiny; petals 5, white or greenish.

FRUITS: Berry ± 5 mm, black.

SIERRA DISTRIBUTION: West slope, Tuolumne County north.

ELEVATION: < 2500 m

HABITAT: Moist shady areas in canyons and along streams; foothill and montane zones.

FLOWERING TIME: May–Oct

NOTES: Uncommon in the SN. The only member of the Ginseng family native to CA (true ginseng is *Panax* spp.).

Achillea millefolium

Common Yarrow

Asteraceae (Sunflower family)

Hairy to woolly rhizomatous perennial herb 10–200 cm. Aromatic.

LEAVES: Basal and alternate, sessile, with finely divided blades sometimes appearing cylindrical in overall outline rather than flattened (especially at higher elevations). Stem leaves generally clasp stem and are smaller toward top of stem.

FLOWERS: Heads radiate, in flat-topped clusters, phyllaries 4–9 mm. Ray flowers 3–8 per head, ray oval to roundish, 2.5–4 mm, white (to pink), disk flowers 15–40, white to pinkish, with bright yellow anthers.

FRUITS: Hairless, and without pappus.

SIERRA DISTRIBUTION: All.

ELEVATION: < 3650 m

HABITAT: Many and varied habitats; all zones.

FLOWERING TIME: Apr–Sep

NOTES: This herb is circumboreal (occurs throughout the northern hemisphere) and is also one of the most widespread native plants in CA. Native Americans across North America use yarrow for a wide variety of ailments—headaches, earaches, fever reduction, and rashes, to name just a few remedies.

Adenocaulon bicolor
Trail Plant
Asteraceae (Sunflower family)

Erect perennial herb 3–10 dm; woolly-hairy at base of stem, with stalked glandular hairs toward the top of the plant.

LEAVES: Triangular to ovate, at base of stem; petioles winged, blades 3–25 cm; hairless above and white-woolly below.

FLOWERS: Heads with phyllaries 1–2.5 mm, disk flowers (in center of head) staminate, corollas white. These are surrounded by pistillate (female) flowers, also white.

FRUITS: The pistillate flowers give rise to club-shaped fruits, 5–9 mm, covered with stalked glands.

SIERRA DISTRIBUTION: West slope, Tulare County north.

ELEVATION: 700–2000 m

HABITAT: Shady floor and riparian areas of coniferous and mixed hardwood forests; montane zone.

FLOWERING TIME: Jun–Oct

NOTES: The common name of this plant derives from the fact that as you walk through a dense patch of plants, some of the leaves turn over, resembling white arrows creating a "trail." The genus name refers to the stalked glands on the upper stem: *adeno* ("gland") + *caulon* ("stem"). The flowers have a subtle, foul smell that attracts pollinating flies.

Ageratina occidentalis
Western Snakeroot
Asteraceae (Sunflower family)

Perennial herb to subshrub 1.5–7 dm; rhizomatous; stem puberulent, sometimes purplish.

LEAVES: Lower leaves opposite, upper alternate; petioles 5–12 mm, blades 2.5–5 cm, triangular, toothed, gland-dotted below.

FLOWERS: Heads clustered, discoid; involucres 3–3.5 mm, phyllaries puberulent and glandular; corollas 10–60 per head, 5–6 mm, pink (sometimes bluish or light purple), style branches exserted.

FRUITS: 3–3.5 mm, with sessile glands; pappus of 5–40 minutely barbed bristles.

SIERRA DISTRIBUTION: All.

ELEVATION: 400–3400 m

HABITAT: Rocky areas, open slopes, montane chaparral; gravelly/sandy stream bars and shores; montane, subalpine, and sagebrush/pinyon-juniper zones.

FLOWERING TIME: Jun–Oct

NOTES: The only *Ageratina* in the SN.

Agoseris aurantiaca var. *aurantiaca*

Orange Agoseris

Asteraceae (Sunflower family)

Erect perennial herb 10–50 cm. Sap milky.
LEAVES: Basal 10–30 cm; petioles purplish; overall outline oblanceolate with acute to long-tapered tips, entire or with 2–4 pairs of spreading lobes. Surfaces hairless or sparsely woolly.
FLOWERS: Heads solitary, top of peduncle and base of involucres densely hairy; involucres 15–20 mm in flower, phyllaries green or pinkish-purple, often purple-spotted. Flowers 15–100 per head, ligulate, corolla tubes 7–9, ligules 5–10 mm, orange, turning purple in age.
FRUITS: Achenes 3–9 mm, ribbed; beak 5–10 mm; pappus bristles in 2–4 series, 9–15 mm.
SIERRA DISTRIBUTION: Both slopes, Tulare and Inyo Counties north to Nevada County.
ELEVATION: 1500–3500 m
HABITAT: Meadows, streamsides, forest or shrublands; montane through alpine zones.
FLOWERING TIME: Jun–Sep
NOTES: Distinctive "dandelion" of the high country with orange flower heads.

Agoseris grandiflora var. *grandiflora*

Grassland Agoseris

Asteraceae (Sunflower family)

Erect, grayish perennial herb with flowering stems 25–100 cm; stems are woolly-hairy at base and below flower head. Sap milky.
LEAVES: Basal 10–25 cm, white-gray hairy, entire or with 3–5 pairs of teeth or lobes (variable) angled toward tip of leaf; blades 10–35 mm wide (excluding lobes).
FLOWERS: Heads solitary, involucres 3–5 cm, phyllaries in 4–5 series, purple-edged; flowers ligulate, 150–500 per head, ligule 3–7 mm, heads closed during the day.
FRUITS: Achenes 3–7 mm, ribbed, with tips tapered to beaks; beaks 10–20 mm; pappus bristles 7–15 mm, white.
SIERRA DISTRIBUTION: West slope.
ELEVATION: 300–2000 m
HABITAT: Open to shady, dry sites; foothill and montane zones.
FLOWERING TIME: Apr–Jul
NOTES: Very common, but often overlooked unless seen in the morning when flower heads are open. Differs from *A. retrorsa* by having more variable leaves—from entire to slightly lobed to deeply lobed, with lobe tips generally pointing toward leaf tip. The beaks on fruits are shorter than in *A. retrorsa*, and the top of the achene tapers to the beak rather than being flattened abruptly.

var. heterophylla

Agoseris heterophylla var. *cryptopleura*

Mountain Dandelion

Asteraceae (Sunflower family)

Erect annual herb 5–60 cm; flower heads open well into the day. Sap milky.

LEAVES: Mostly basal, sometimes on lower stem too. Blades 1–25 cm, oblanceolate, entire or with 3–5 pairs of teeth or short lobes, hairy (sometimes hairless). Leaves extremely variable in shape and hairiness.

FLOWERS: Heads solitary, liguliflorous, with peduncles hairless or becoming hairless, usually woolly just below involucres. Involucres 5–15 mm; phyllaries green, sometimes purplish in center, glabrous or with long, soft, white hairs or translucent stalked glandular hairs with purple tips and purple or yellow cross walls (barely visible at 10x). Flowers 20–100+ per head, corolla tube 2–5 mm, ligule 10–15 mm, much exceeding phyllaries. Anthers 2–4 mm.

FRUITS: Fruiting involucres 10–25 mm, bodies 2–5 mm, often with distinct ribs; beak 5–11 mm; pappus 4–9 mm, of white bristles. Outermost achenes often different from innermost.

SIERRA DISTRIBUTION: West slope.

ELEVATION: < 2100 m

HABITAT: Open habitats; foothill and montane zones.

FLOWERING TIME: May–Jul

NOTES: *Heter-* means "different," and *-phyll* means "leaf," referring to the variation in leaf shape, margins, and hairiness. Can be difficult to distinguish from the coexisting and common var. *heterophylla* (inset), which is found in similar habitats on the west slope. Var. *heterophylla* has fewer (20–50) flowers per head, shorter (2–6 mm) ligules not exceeding phyllaries, and shorter (1–1.5 mm) anthers. In fruit, the 2 varieties are virtually indistinguishable. Var. *cryptopleura*, with larger and more numerous flowers per head, is strictly outcrossing, while var. *heterophylla* is strictly self-fertilized.

Agoseris monticola

Mountain Agoseris

Asteraceae (Sunflower family)

Perennial herb 5–25 cm; decumbent to prostrate, sometimes erect. Sap milky.

LEAVES: Basal; petioles usually not purple; blades 2–10 cm. The blades are oblanceolate in outline, tips may be blunt or long-tapered; margins are generally toothed or lobed with 2+ pairs, but sometimes entire. If lobed, there are 2+ pairs of lobes, each lobe linear to narrowly oblanceolate, the bottom ones pointing downward, upper ones pointing toward leaf tips, or lobes spreading. Most often the leaves are densely covered with hairs white or translucent yellow, but can be hairless or glaucous.

FLOWERS: Heads solitary, with top of peduncles and bases of involucres densely glandular-woolly. Involucres 10–20 mm; phyllaries usually purplish except for a thin margin (or all green), with dark midstripes. Corollas 10–40, ligule 5–10 mm.

FRUITS: 6–9 mm, narrowly spindle-shaped; beak 1–3 mm or absent; pappus 8–11 mm, in 2 series.

SIERRA DISTRIBUTION: Both slopes, Tulare and Inyo Counties north.

ELEVATION: 1500–3800 m

HABITAT: Meadows, rocky slopes, open forests; upper montane through alpine zones.

FLOWERING TIME: Jul–Aug

NOTES: Formerly *Agoseris glauca* var. *monticola*. Ridiculously variable, plus it hybridizes with *A. glauca*, *A. aurantiaca*, and *A. parviflora*.

Agoseris retrorsa

Spearleaf Agoseris

Asteraceae (Sunflower family)

Erect perennial herb with ≥ 1 leafless flowering stalks 10–65 cm. Sap milky.

LEAVES: Basal, petioles often purple; blades 10–30 cm, with lobes 7–9, pointing backward toward base.

FLOWERS: Heads are solitary, flowers ligulate; involucres 20–30 mm, woolly hairy at base; ligules 6–15 mm (≥ phyllaries).

FRUITS: Achenes 5–7 mm, ribbed, and truncate (flat-topped, not tapered to beak); beaks 15–25 mm; pappus white, bristles 15–20 mm.

SIERRA DISTRIBUTION: All.

ELEVATION: 900–2900 m

HABITAT: Dry, open to shady sites; foothill and montane zones.

FLOWERING TIME: Apr–Aug

NOTES: The species name refers to the retrorse, or downward-pointing, leaf lobes. Flower heads open in the morning and close by midday. Hairiness on the plant is variable, but never glandular.

Agoseris parviflora

Steppe Agoseris

Asteraceae (Sunflower family)

Perennial herb with leafless flowering stalks 5–25 (sometimes up to 45) cm; usually erect, but sometimes decumbent. Sap milky.

LEAVES: Basal, with petioles purple; blades 10–20 cm, lanceolate, hairless and glaucous or with hairs white, glandless. Leaves have 5–8 pairs of teeth or lobes angled backward toward leaf base, or not angled.

FLOWERS: Heads solitary, involucres 1–3 cm, phyllaries purplish. Flowers 30–100 per head, ligulate, yellow; corolla tube 6–15 mm, ligule 10–20 mm. Anthers 3–5 mm.

FRUITS: Achenes 5–9 mm, beak 3–10 mm (0.5–2x body size); pappus white, bristles 10–20 mm, ± 3 rows.

SIERRA DISTRIBUTION: Both slopes, Tulare and Inyo Counties north.

ELEVATION: 1400–3400 m

HABITAT: Dry, sandy soils, dry meadows, sagebrush steppes, pinyon-juniper woodlands; montane, subalpine, alpine, and sagebrush zones.

FLOWERING TIME: Apr–Aug

NOTES: Formerly *Agoseris glauca* var. *laciniata*. Mountain agoseris, *A. monticola*, looks somewhat similar and is common in the same range, 1500–3800 m. *A. monticola* (formerly *A. glauca* var. *monticola*) is extremely variable. It differs by being most often decumbent or prostrate (but sometimes erect); having leaves with 2+ pairs of lobes, with lower lobes angled toward base and upper angled toward tip or leaves entire; heads with 10–40 flowers; corolla tube 4–10 mm, ligule 5–10 mm, yellow; fruits bodies 6–9 mm, narrowly fusiform, with beaks 1–3 mm (sometimes absent), stout; and pappus bristles 8–11 mm, in ± 2 series. These 2 spp. hybridize in addition to being inherently variable, so don't get too frustrated if you don't feel 100% confident with your ID!

Anaphalis margaritacea

Pearly Everlasting

Asteraceae (Sunflower family)

Erect, bushy perennial herb 2–12 dm; rhizomatous, with white-woolly stems.

LEAVES: Alternate, 2–10 cm, linear, lanceolate, or oblanceolate, with bases somewhat clasping stem and decurrent. Leaves entire, with margins turned under; generally green and glabrous or nearly so above, white and woolly below.

FLOWERS: Heads are in tight groups of disk flowers, heads unisexual or primarily male or female. Involucres 5–7 mm, phyllary bracts many, in 8–12 series, white, paperlike. Female flowers have corollas yellow, narrowly tubular, no anthers. Male flowers yellow with styles truncate, anther tips ovate.

FRUITS: 0.5–1 mm, base stalked; covered with club-shaped hairs; pappus bristles deciduous.

SIERRA DISTRIBUTION: Both slopes, Tulare and Inyo Counties north.

ELEVATION: < 3200 m

HABITAT: Roadsides, rocky sites, meadow edges, coniferous forest; montane and subalpine zones.

FLOWERING TIME: Jul–Oct

NOTES: Herbage of *Anaphalis* has relatively little odor compared to other common members of the everlasting group (e.g., compare with *Pseudognaphalium*). Blooms last a long time and retain their shape after fruiting; they're used in both fresh and dried flower arrangements. Also differs from *Pseudognaphalium* in being rhizomatous.

Anisocoma acaulis

Scalebud

Asteraceae (Sunflower family)

Annual herb with a leafless "stem" that is actually the inflorescence stalk, < 20 cm. Sap milky.

LEAVES: Basal, pinnately lobed (and lobes are also lobed), greenish to reddish-purple, hairless to hairy.

FLOWERS: Heads 1 per stem, ligulate; paleae attached to receptacles among flowers. Phyllaries often red-dotted, margins papery, and outer phyllaries much shorter; flowers yellow or creamy, 5-toothed.

FRUITS: Achene; pappus bristles 10, clearly different lengths, white, plumose.

SIERRA DISTRIBUTION: Southern Sierra, Tulare and Kern Counties; eastern Sierra, Inyo and Mono Counties.

ELEVATION: < 2600 m

HABITAT: Sandy flats and slopes in fully exposed areas; foothill, montane, and sagebrush zones.

FLOWERING TIME: Mar–Jun

NOTES: Mostly a desert and Great Basin plant, scalebud reaches up into the Sierra in the south and east. Distinctive because of its red-dotted phyllaries and basal leaves, combined with the asymmetric pappus (with a hand lens you can see that bristles are much longer on one side). The species name means "without a stem" (*a-* is "without"; *caul-* is "stalk" or "stem").

Antennaria corymbosa

Meadow Pussytoes

Asteraceae (Sunflower family)

Matted perennial herb with stems 6–15 cm; stolons present, 1–10 cm, dioecious.

LEAVES: Basal 18–22 mm, linear-oblanceolate (somewhat spoon-shaped), 1-veined, gray-tomentose. Cauline leaves alternate, 8–13 mm, linear.

FLOWERS: Male and female flower heads are on separate plants, with 3–7 heads in cymelike arrangement; involucres 4–5.3 mm, densely hairy at base; phyllaries white with black-brown spots near base, wide with acute tips in pistillate (female) flower heads, or blunt tips in staminate (male) flower heads. Pistillate corollas 2.5–3.5 mm with pappus bristles numerous, soft, weakly barbed; staminate corollas 2–3.2 mm, pappus tips enlarged.

FRUITS: Achenes 0.5–1 mm, slightly papillate; pappus 3.5–4.5 mm.

SIERRA DISTRIBUTION: Both slopes, Sierra County south.

ELEVATION: 1900–3200 m

HABITAT: Meadows, moist sites near streams and seeps; montane through alpine zones.

FLOWERING TIME: Jun–Aug

NOTES: Distinguished from other pussytoes by the spoon-shaped basal leaves and white-tipped phyllaries that have a distinct black dot at the base of the scarious (papery) portion.

Antennaria media

Rocky Mountain Pussytoes

Asteraceae (Sunflower family)

Matted perennial herb with stems 5–13 cm; stolons numerous.

LEAVES: Basal 6–19 mm, linear to spoon-shaped, 1-veined, densely gray-tomentose; stem leaves alternate, 5–20 mm, smaller toward top of stem.

FLOWERS: Male and female flower heads are on separate plants, 2–7 heads in cymelike arrangement (remember that each head in the Sunflower family is an inflorescence). Involucres 4–8 mm, base woolly; phyllaries narrow and acute in pistillate flower heads, or wide and blunt in staminate flower heads. Phyllaries dark brown, olive-brown, or black toward tips. Pistillate corollas 3–4.5 mm with pappus 4–5.5 mm; staminate corollas 2.8–4.5 mm.

FRUITS: Achenes 0.6–1.6 mm, papillate or smooth; pappus 3.5–4.5 mm.

SIERRA DISTRIBUTION: Both slopes, Sierra County south to Tulare and Inyo Counties.

ELEVATION: 1800–3900 m

HABITAT: Meadows, ridges, snow basins; montane through alpine zones.

FLOWERING TIME: Jul–Aug

NOTES: Distinguished from other pussytoes by the olive-brown to black phyllary tips.

Antennaria rosea ssp. *rosea*

Rosy Pussytoes

Asteraceae (Sunflower family)

Matted perennial herb with stems 10–40 cm; not glandular, stolons 20–70 mm, decumbent (not ascending).

LEAVES: Basal 8–40 mm, wedge- to spoon-shaped, 1-veined, gray-tomentose; cauline leaves alternate, 8–36 mm.

FLOWERS: Only pistillate plants present, male plants rarely seen. Inflorescences of 6–20 heads, arranged cymelike; involucres 5–8 mm and bases hairy; phyllaries wide, acute, with tips rose, white, or yellowish. Pistillate corollas 3–4.5 mm.

FRUITS: Achenes 0.7–1.8 mm, papillate or smooth; pappus 4–6 mm.

SIERRA DISTRIBUTION: Both slopes, Tulare and Inyo Counties north.

ELEVATION: 1200–3700 m

HABITAT: Meadow edges, rocky sites, dry ridges; montane through alpine zones.

FLOWERING TIME: Jun–Aug

NOTES: The most wide-ranging *Antennaria* in North America, and probably the most taxonomically challenging. Male plants are uncommon. Female plants reproduce via apomixis, whereby a viable seed develops without fertilization—a fancy way of reproducing asexually.

Arnica cordifolia

Heartleaf Arnica

Asteraceae (Sunflower family)

Erect, glandular perennial herb 10–50 cm; rhizomatous, with 1 to several usually unbranched stems. Aromatic.

LEAVES: Basal withered by flowering, or present as nearby sterile rosettes (flowers were not produced that year). Stem leaves 2–4 pairs, opposite, smaller toward top of stem, and the lower petioled. Blades 3–11 cm, cordate (heart-shaped), and shallowly toothed.

FLOWERS: Heads 1–5, radiate; phyllaries long-white, hairy. Disk corollas soft-hairy with glandular hairs. Ray flowers 6–15, ray < 3 cm.

FRUITS: 5–10 mm, dark gray, sparsely to densely hairy. Pappus of white bristles with minute barbs.

SIERRA DISTRIBUTION: Both slopes.

ELEVATION: 850–3384 m

HABITAT: Montane and high-elevation meadows and coniferous forest, in moist to dry soil or rocky areas, usually in shaded sites; montane through alpine zones.

FLOWERING TIME: May–Aug

NOTES: One of the more distinctive *Arnicas* because of the heart-shaped leaves and long, white hairs at the base of the involucre. Stunted alpine plants may have ovate leaves.

Arnica longifolia

Spearleaf Arnica

Asteraceae (Sunflower family)

Perennial herb 30–60 cm; stems emerge in clusters from short rhizomes, sometimes forming large patches. Stem is scabrous to short-hairy, sometimes sticky, toward top.

LEAVES: Basal few, shorter than stem leaves, usually withered by flowering time. Stem leaves 5–7 pairs, opposite, sessile, smaller toward top of stem, lanceolate. Leaf blades 5–12 cm, with bottom stem leaves often fused at their bases around the stem.

FLOWERS: Heads 3–20, radiate; involucres 7–10 mm, ± bell-shaped; phyllaries lanceolate, acute, long-hairy, glandular. Ray flowers 6–15, ray 10–20 mm. Disk flowers 6–11; anthers yellow.

FRUITS: Achenes 3–7 mm, hairless to glandular-hairy or with simple hairs; pappus plumose or with short barbs, brown to brownish-yellow.

SIERRA DISTRIBUTION: Both slopes, Tulare and Inyo Counties north.

ELEVATION: 1500–3500 m

HABITAT: Moist to wet sites, often following late snowmelt; montane through alpine zones.

FLOWERING TIME: Jul–Sep

NOTES: Reproduction is primarily vegetative in CA.

Arnica mollis

Hairy Arnica

Asteraceae (Sunflower family)

Erect perennial herb 15–70 cm; stem single or several in a clump (from short rhizome), rarely branched; glandular-hairy at base, and in age sometimes becoming hairless.

LEAVES: Stem leaves 3–5 pairs, sessile, 4–20 cm, smaller toward top of stem. Leaf shape varies from lanceolate to ovate, as well as from oblanceolate to obovate, entire or with uneven teeth. Blades sparsely to moderately hairy, with hairs short to long and silky, and stalked glands sometimes also present.

FLOWERS: Heads 1–3 (up to 7), radiate; involucres 10–16 mm, hemispheric or bell-shaped; phyllaries lanceolate, with soft hairs. Ray flowers 10–22, ray 15–30 mm. Disk corollas have dense, soft hairs as well as stalked-glandular hairs.

FRUITS: Achenes 4–8 mm, with forked hairs and glands; pappus plumose, brown (or yellowish).

SIERRA DISTRIBUTION: Both slopes, Tulare and Inyo Counties north.

ELEVATION: 2500–3500 m

HABITAT: Moist/wet sites, meadows, streamsides; montane through alpine zones.

FLOWERING TIME: Jul–Sep

NOTES: Very common. A 10x hand lens helps with seeing the various hairs and glands on the herbage, disk flowers, and fruits.

Arnica nevadensis

Sierra Arnica

Asteraceae (Sunflower family)

Erect, unbranched, glandular-hairy perennial herb 10–50 cm; from long rhizomes.

LEAVES: Basal sometimes present at flowering. Stem leaves 2–3 pairs, opposite, smaller toward top of stem, with petioles short, usually winged. Blades 3–8 cm, elliptic to widely ovate, with a cordate or rounded base.

FLOWERS: Heads 1–3, radiate; involucres 10–18 mm; phyllaries stalked-glandular. Ray flowers 6–14, ray 1–2 cm. Disk corollas have hairs glandular; anthers yellow.

FRUITS: 6–9 mm, with short, forked hairs or glandular hairs; pappus white or light tan, short-barbed to plumose.

SIERRA DISTRIBUTION: Both slopes, Tulare and Inyo Counties north.

ELEVATION: 2000–3000 m

HABITAT: Meadows, rocky areas, coniferous forest; montane through alpine zones.

FLOWERING TIME: Jul–Aug

NOTES: Intergrades with *A. cordifolia* in the subalpine zone.

Arnica ovata

Sticky Leaf Arnica

Asteraceae (Sunflower family)

Glandular-hairy perennial herb with stem 10–50 cm; 1 to several stems and not branched; sometimes hairless in age.

LEAVES: Stem leaves 2–3 (sometimes 4) pairs, short-petioled (middle one largest); blades 4–8 cm.

FLOWERS: Heads 1–3, radiate; involucres 10–14 mm, obconic; phyllaries linear-lanceolate, with soft hairs or short-glandular. Ray flowers 8–16, ray 10–20 mm. Disk corollas soft hairy.

FRUITS: Achenes 5–7 mm, variously hairy, with short-stalked glands; pappus ± plumose, light tan to brown.

SIERRA DISTRIBUTION: Both slopes, Inyo and Fresno Counties north.

ELEVATION: 1800–3600 m

HABITAT: Wet rocky sites, moist forest floor, meadows, streamsides; montane through alpine zones.

FLOWERING TIME: Jul–Sep

NOTES: Reproduces asexually.

Arnica parryi

Nodding Arnica

Asteraceae (Sunflower family)

Erect perennial herb 1–6 dm; from short rhizome, stem usually unbranched, long-hairy at base.

LEAVES: Basal, sessile; cauline in 3–4 pairs, lowermost with winged petioles; blades 5–20 cm, longer than basal, lanceolate to ovate, entire; uppermost pairs smaller, sessile.

FLOWERS: Heads 3–9 per stem, generally radiate but sometimes lacking ray flowers, nodding in bud; involucres 15–20 mm, obconic; phyllaries linear-lanceolate and sparsely hairy. Ray flowers 5–10 (sometimes none), ray < 15 mm; disk flowers 20–50, soft-hairy with stalked glands.

FRUITS: 4–6 mm, cylindric, hairless to hairy or glandular; pappus brownish-yellow, plumose.

SIERRA DISTRIBUTION: Both slopes, Tulare and Inyo Counties north, patchy.

ELEVATION: 1200–3400 m

HABITAT: Open rocky areas, conifer forest, meadows; montane through alpine zones.

FLOWERING TIME: Jul–Aug

NOTES: Look for flower heads nodding in bud within a given population/site; this is a good distinguishing trait, along with the densely long-hairy stem bases.

Artemisia douglasiana

Mugwort

Asteraceae (Sunflower family)

Erect perennial herb 5–25 dm; rhizomatous, often forming extensive patches in moist areas. Strongly aromatic.

LEAVES: Alternate, 1–1.5 cm x 1–2 cm wide, entire, toothed or 3–5 times pinnately lobed; green with sparse, woolly hairs on upper surface, densely woolly below.

FLOWERS: Each head has female flowers 6–10 and bisexual flowers 6–25; corollas pale yellow, hairless or glandular.

FRUITS: Achenes 0.5–1 mm, hairless, no pappus.

SIERRA DISTRIBUTION: All.

ELEVATION: < 2200 m

HABITAT: Moist areas; foothill and montane zones.

FLOWERING TIME: May–Nov

NOTES: Common and enormously variable. Could be confused with *Artemisia ludoviciana*, which is on average a smaller (2–8 dm), more delicate plant, with entire to pinnately lobed leaves < 1 cm wide. Several sspp. of *A. ludoviciana* occur in the SN. If you're fascinated with *Artemisia*, it is highly recommended that you use a dissecting microscope and the Jepson eFlora key, with a heavy dose of patience. California Indians line granaries with mugwort to protect stored grain against insects, in addition to other uses.

Artemisia michauxiana

Lemon Sagewort

Asteraceae (Sunflower family)

Many-stemmed perennial herb 3–10 dm; from rhizome, stems unbranched, green. Strongly lemon-scented.

LEAVES: Alternate, 1–2 pinnately divided, 1.5–11 cm, in overall shape ovate to obovate. Upper leaf surface green, hairless, gland-dotted; lower surface white tomentose, sometimes hairless in age.

FLOWERS: Heads are in racemelike or paniclelike clusters, sessile or nearly so, < 5.5 mm wide, disciform, nodding. Phyllaries often purple, elliptic or ovate/obovate, and mostly hairless but dotted with yellow glands, margins scarious. Pistillate (outer) flowers 9–12; disk flowers 15–35, bisexual; corollas 1–1.5 mm, yellow, glandular.

FRUITS: 0.5–1.0 mm, hairless; pappus absent.

SIERRA DISTRIBUTION: Both slopes, Tulare and Inyo Counties north, sporadic.

ELEVATION: 3000–3500 m

HABITAT: Talus, scree, rocky areas; subalpine and alpine zones.

FLOWERING TIME: May–Aug

NOTES: Uncommon, and distinctive among *Artemisia* taxa for the lemon scent and green, hairless upper leaf surfaces. Can be confused with mountain mugwort, *Artemisia ludoviciana* ssp. *incompta*, which occurs < 3500 m in the same range but in openings in forests and shrublands, has densely tomentose phyllaries, and is sage-scented.

Baccharis salicifolia ssp. *salicifolia*

Mule Fat, Seep-Willow

Asteraceae (Sunflower family)

Erect or ascending shrub < 4 m; hairless or with tiny hairs, resinous, sticky.

LEAVES: Alternate, < 15 cm, < 3 cm wide, gland-dotted and sticky, often clustered.

FLOWERS: Plants produce either male or female flowers (dioecious). Inflorescences appear in terminal clusters in fall/summer-blooming plants, and in lateral clusters in spring. Heads of 50–150 female flowers and 17–48 male flowers; pappus 3–4 mm.

FRUITS: 0.8–1.3 mm, hairless with 5 ribs; pappus 3–6 mm.

SIERRA DISTRIBUTION: West slope (records in CCH for Butte, Tuolumne, Madera, Fresno, Tulare, and Kern Counties as of 2024), sporadic; east slope, Kern and Inyo Counties.

ELEVATION: < 2400 m

HABITAT: Moist to wet areas; foothill and montane zones.

FLOWERING TIME: Year-round

NOTES: The species name *salicifolia* refers to the fact that the leaves resemble the genus *Salix*, or willow. An important fall-blooming plant for bees and butterflies. Young shoots are used by Native Americans as "fire drills" to create the intense friction necessary to start a fire. *B. glutinosa* looks somewhat similar and occurs sparingly on the west slope in wet areas; it differs by forming colonies via rhizomes and in having glandular-hairy fruits; it also never exceeds 2 m.

Artemisia tridentata ssp. *vaseyana*

Mountain Sagebrush

Asteraceae (Sunflower family)

Gray-hairy, evergreen shrub 5–10 dm; leafy canopy is flat topped. Strongly aromatic.

LEAVES: Alternate, 1–3 cm, wedge-shaped, usually 3-toothed or lobed at tip (sometimes unlobed or 5-lobed), often in axillary clusters.

FLOWERS: Inflorescences < 30 cm, extending beyond main leafy canopy. Heads 1–3 mm diameter, discoid, erect, with 3–6 disk flowers. Phyllaries densely tomentose, with transparent margins.

FRUITS: Achenes 1–2 mm, glandular, pappus absent.

SIERRA DISTRIBUTION: All, more common on east slope.

ELEVATION: 1800–3000 m

HABITAT: Dry, open, sometimes rocky areas; montane, subalpine, and sagebrush/pinyon-juniper zones.

FLOWERING TIME: Aug–Oct

NOTES: There are 3 other sspp. of *Artemisia tridentata* in the SN, but ssp. *vaseyana* is the most common. Ssp. *tridentata* occurs throughout the SN but is absent from much of the west slope, and is up to 2 m tall, with inflorescences surrounded by leafy branches; ssp. *parishii* occurs sparsely in the southern and eastern Sierra and has drooping inflorescence branches; ssp. *wyomingensis* is < 4 dm, leaves < 1.2 cm, and is known from just a few east slope counties. Rothrock sagebrush, *A. rothrockii*, occurs at 2000–3100 m from Kern and Inyo Counties to Sierra County. It resembles big sagebrush but is shorter (2–5 dm), sticky-resinous, and strongly aromatic; it has bushier, paniclelike inflorescences, and larger (3–5 mm diameter) flower heads, with 10–16 disk flowers.

Balsamorhiza deltoidea

Deltoid Balsamroot

Asteraceae (Sunflower family)

Early blooming perennial herb 2–9 dm, with huge leaves; bright green, sticky due to dense, glandular hairs; herbage also has longer, nonglandular hairs.

LEAVES: Basal 2–6 dm, triangular, with base heart-shaped to somewhat arrow-shaped, margins usually toothed, bright green above and below. Stem leaves smaller, bractlike.

FLOWERS: Heads usually 1, sometimes 2, per stem; 6–10 cm diameter. Phyllaries densely glandular-hairy (not woolly); ray flowers 12–20, 2–3 cm.

FRUITS: 7–8 mm, hairless; no pappus.

SIERRA DISTRIBUTION: West slope, sporadic; east slope, Nevada and Sierra Counties.

ELEVATION: < 2510 m

HABITAT: Open grassy slopes, forest floor; foothill and montane zones.

FLOWERING TIME: Apr–Jun

NOTES: Compare with *B. sagittata*, which has leaves with woolly undersides and more densely hairy herbage in general; also, *B. sagittata* has a more limited distribution than *B. deltoidea*. If you're south of Tuolumne County, you're probably looking at *B. deltoidea*.

Balsamorhiza sagittata

Arrowleaf Balsamroot

Asteraceae (Sunflower family)

Woolly perennial herb 2–6 dm, with large, triangular leaves.

LEAVES: Basal 2–6 dm, triangular, with base heart-shaped to somewhat arrow-shaped, margins entire, densely woolly below. Stem leaves smaller, bractlike.

FLOWERS: Heads usually 1, sometimes 2, per stem; 6–10 cm diameter. Phyllaries densely white-woolly; ray flowers 13–21, 2–4 cm.

FRUITS: 7–8 mm, hairless; no pappus.

SIERRA DISTRIBUTION: West slope, Tuolumne County north; east slope, all counties.

ELEVATION: 1200–2800 m

HABITAT: Openings or partially shaded areas; sagebrush and foothill through subalpine zones.

FLOWERING TIME: May–Aug

NOTES: Compare with *B. deltoidea*, which has much less hairy leaves and phyllaries, and a wider distribution on the west slope. Grazed by many animals.

Bidens frondosa

Sticktight

Asteraceae (Sunflower family)

Erect annual herb 2–12 cm; taprooted, hairless; stem is square, simple to many-branched.

LEAVES: Opposite, pinnately compound, overall outline lance-ovate or deltate; petioles 1–4 cm. Leaflets 3–5, 3.5–6 cm, stalked, lanceolate to lance-ovate, and toothed.

FLOWERS: Heads usually borne singly, with outer phyllaries 5–8, leaflike, and ciliate. Heads discoid or occasionally have up to 3 ray flowers that are 2–3.5 mm, yellow. Disk flowers 20–60; corollas 2.5–3 mm, orange-yellow.

FRUITS: Achenes 6–10 mm, flattened, wedge-shaped, with margins barbed; pappus awns 2, 3–4.5 mm, barbed.

SIERRA DISTRIBUTION: Mostly west slope.

ELEVATION: < 2100 m

HABITAT: Moist sites, streamsides, mud flats, often in disturbed areas; foothill and montane zones.

FLOWERING TIME: Jun–Oct

NOTES: Increasingly common in the central Sierra and likely elsewhere, the plant has barbed achenes with their barbed awns that are exceptionally good at sticking tightly to passing creatures, hominids included.

Blennosperma nanum var. *nanum*

Common Stickyseed

Asteraceae (Sunflower family)

Erect annual herb 3–12 cm; taprooted, branched throughout or simple.

LEAVES: Basal and alternate; blade linear or with 2–15 pinnate, linear lobes.

FLOWERS: Heads have 20–60 disk flowers, with stigmas yellow, pollen white. Ray flowers 5–13, often purple below.

FRUITS: 2.5–4.5 mm, bumpy, mucilaginous when wet; no pappus.

SIERRA DISTRIBUTION: West slope, Tulare County north.

ELEVATION: < 1500 m

HABITAT: Open grassy and vernally moist sites; foothill and lower montane zones.

FLOWERING TIME: Jan–May

NOTES: Disk flowers are staminate and thus functionally male and do not produce fruits, while the ray flowers are functionally female and produce fruits.

Brickellia californica

California Brickellbush

Asteraceae (Sunflower family)

Bushy shrub 5–20 dm, branched extensively from near base; herbage is short-hairy, glandular, and often grayish.

LEAVES: Alternate (lowest sometimes opposite); petioles 5–60 mm; blades 1–10 cm x 1–9 cm wide. Leaves deltate to ovate, with margins crenate to serrate, bases cordate to truncate, surfaces puberulent, and hairless in age.

FLOWERS: Heads discoid, in paniclelike clusters, with peduncles 1–5 mm; involucres 7–12 mm, cylindric; phyllaries 21–35, in 5–6 series, often tinged purple, with margins scarious; flowers 8–18; corollas 5.5–8 mm, pale yellowish-green.

FRUITS: 2.5–3.5 mm, cylindric, ribs 10, puberulent; pappus bristles 24–30, white, and minutely barbed.

SIERRA DISTRIBUTION: Both slopes, all counties.

ELEVATION: < 2700 m

HABITAT: Dry slopes and rocky sites; below subalpine zone.

FLOWERING TIME: Jul–Dec

NOTES: Brickellbush tends to blend in with its rocky habitat. This is a hardy shrub, thriving in harsh, hot, dry sites. Flowers produce nectar and pollen that sustains native pollinators well into the fall months when few other flowers are out, including bees, butterflies, and moths.

Calycadenia mollis

Soft Western Rosinweed

Asteraceae (Sunflower family)

Erect annual herb 30–90 cm; stems with dense, soft hairs; herbage glandular; stems sometimes have a zigzag appearance.

LEAVES: Alternate, 2–8 cm, linear.

FLOWERS: Heads radiate; leaflike bracts below involucres 3–9 mm, flat, oblanceolate, and glandular throughout, with 3–6 tack-shaped glands on thickened margins; phyllaries 4–6 mm. Heads of 2–10 disk flowers and 1–4 ray flowers, yellow, white, pink, or maroon. Each ray expands into 3 lobes, the middle slightly smaller than the laterals. Phyllaries and paleae lack tack-shaped glands.

FRUITS: Ray flower achenes are bumpy-wrinkled, hairless; fruits of disk flowers are smooth, with pappus of 6–8 scales.

SIERRA DISTRIBUTION: West slope, Tulare to Tuolumne Counties.

ELEVATION: 100–1500 m

HABITAT: Open areas, rock outcrops, sandy or gravelly soil; foothill and montane zones.

FLOWERING TIME: Jun–Oct

NOTES: The tack-shaped glands are well worth a look with a magnifier (10x or higher). In *C. mollis* the tacks are only on margins of the peduncle bract, which doesn't preclude them being terminal.

Calycadenia multiglandulosa

Sticky Rosinweed

Asteraceae (Sunflower family)

Erect annual herb 10–70 cm; simple or with short branchlets; hairy, and glandular-hairy at top.

LEAVES: Alternate (sometimes opposite on lower stem), 3–8 cm, linear.

FLOWERS: Heads clustered in spikelike or roundish heads, with peduncle bracts 4–20 mm, linear, sometimes cylindrical at tip, and often reddish, with 2–6 tack-shaped glands. Phyllaries 4–8.5 mm, paleae 4.5–10 mm, both with red, tack-shaped glands of various sizes in addition to smaller glands throughout. Heads have 4–20 disk flowers and 2–6 ray flowers; ray 5–10 mm, white, cream, yellow, or pink, and with central lobe much smaller than laterals (laterals asymmetric).

FRUITS: Ray fruits smooth; disk fruits have ± 11 pappus scales (2–3 scales shorter than others).

SIERRA DISTRIBUTION: West slope.

ELEVATION: < 1200 m

HABITAT: Open grassy or rocky areas, roadsides, open meadows; foothill and montane zones.

FLOWERING TIME: May–Oct

NOTES: This species is extremely variable, and some forms with peduncle bracts that are cylindrical at tips resemble *C. spicata*, which grows in scattered locations < 1400 m from Kern to Butte Counties. *C. spicata* differs in having grayish, aromatic herbage, only 1 tack-shaped gland at the tip of the peduncle bract, spikelike inflorescences with 1–3 heads per node, and the 15–20 peduncle bracts 3–7 mm, obscuring heads pre-bloom.

Carduus pycnocephalus ssp. *pycnocephalus* (non-native, invasive)

Italian Thistle

Asteraceae (Sunflower family)

Erect annual herb 2–20 dm; stem is hairless to woolly, with a narrow strip of green, spiny-winged tissue extending the length of stem. Tends to form enormous, continuous stands that can add up to acres in no time.

LEAVES: Basal 10–15 cm, with 4–10 spiny lobes; cauline leaves alternate, smaller toward top of plant.

FLOWERS: Heads in clusters of 2–5, discoid, essentially sessile; involucres 1–2 cm wide, cylindric. Phyllaries linear, tomentose at base, margins green (not papery), tips sharp. Corollas 10–14 mm, pink or purple, with a long tube and 5 slender lobes.

FRUITS: Achenes 4–6 mm, 20-veined; pappus bristles many and flat.

SIERRA DISTRIBUTION: Ubiquitous on the entire west slope below 1500 m.

ELEVATION: < 1500 m

HABITAT: Disturbed areas, roadsides, understory, and openings; primarily foothill zone and increasingly montane zone.

FLOWERING TIME: Mar–Jun

NOTES: Italian thistle, native to the Mediterranean, has spread rapidly across countless acres in the western foothills since about the year 2000. Once this weed is entrenched over a large area, it is virtually impossible to eradicate, though it can be controlled with a variety of methods and is worth treating if maintenance is possible on a given piece of land. Extensive stands of Italian thistle are clearly damaging to native ecosystems, painful and annoying to people, and a scourge for wildlife and livestock. Slender flowered thistle, *C. tenuiflorus*, looks similar but has clusters of 5–25 heads, and the spiny-winged tissue running the length of stems is wider than in *C. pycnocephalus*.

Centaurea benedicta (non-native)

Blessed Thistle

Asteraceae (Sunflower family)

Bushy annual herb to 10 dm (usually shorter), with stems spreading or prostate; much branched, not winged, and with cobwebby hairs.

LEAVES: Cauline, sessile or with winged petioles extending down stem, 6–25 cm, pinnately lobed or toothed with spines at tips, resin-dotted, hairy when young and often hairless in age.

FLOWERS: Heads disciform, sessile; involucres 20–40 mm; phyllaries in tightly overlapping series, some with pinnately divided spines at tips. Corollas numerous, yellow; fertile disk flowers 19–24 mm, and sterile flowers linear, slender, 3-lobed. Flower heads are subtended and sometimes partially hidden by leaflike bracts positioned directly below involucres.

FRUITS: 8–11 mm, cylindric with a slight curve, ribs 20, and hairless; pappus awns in 2 series, the inner 2–5 mm and the outer 9–10 mm.

SIERRA DISTRIBUTION: West slope, scattered locations, more locations expected.

ELEVATION: < 1700 m

HABITAT: Disturbed areas; foothill and montane zones.

FLOWERING TIME: Apr–Aug

NOTES: Formerly *Cnicus benedictus*. Native to the Mediterranean region as well as parts of Asia and Africa. It is now introduced worldwide and can be invasive, similar to other *Centaurea* spp. Has been used as a remedy for a variety of ailments in Eurasia, and research is ongoing to determine which chemicals produced by the plant show promise as antibiotics and cancer treatments, among other things.

Centaurea cyanus (non-native)

Bachelor's Buttons, Cornflower

Asteraceae (Sunflower family)

Annual herb 2–10 dm, finely woolly. Stems branch toward top and are not winged (unlike many other *Centaurea* spp.).

LEAVES: Lower leaves 3–10 cm, sparsely to densely woolly, green or grayish, and usually entire; upper of similar length, linear or few-toothed.

FLOWERS: Heads radiant, often in open clusters; involucres 10–16 mm, ovoid or bell-shaped. Main phyllaries green or purplish, and appendages at tip of each phyllary have slender white, black, or brown teeth; phyllary margins fringed with slender teeth. Flowers 23–35 per head, blue, purple, pink, or white; sterile flowers largest and showiest, raylike; 5–8 lobed; fertile flowers 10–15 mm.

FRUITS: 4–5 mm, straw-colored or pale blue, puberulent; pappus of numerous, stiff, unequal bristles.

SIERRA DISTRIBUTION: West slope, patchy; will likely become more abundant.

ELEVATION: < 1500 m

HABITAT: Open, disturbed areas; foothill and montane zones.

FLOWERING TIME: Apr–Sep

NOTES: Native to southern Europe; has been spread widely in the SN via wildflower seed mixes, birds, and vehicles. Invasive to some degree and best not seeded where seeds could escape to wildlands.

Centaurea melitensis (non-native)

Tocalote, Maltese Starthistle

Asteraceae (Sunflower family)

Erect, bushy annual herb 1–10 dm; gray-green, and resin-dotted; stem is winged with leafy tissue.

LEAVES: Basal in rosettes, 5–12 cm, pinnately lobed with terminal lobe largest and usually rounded. Stem leaves alternate, sessile, 1–3 cm, decurrent, entire.

FLOWERS: Heads disciform, solitary at tips of branchlets; involucres 10–15 mm, ovoid; main phyllaries have spines 5–10 mm, purple-tipped, often downward-pointing. Corollas 10–12 mm, yellow, glandular.

FRUITS: 2.5 mm; pappus 2.5–3 mm, white.

SIERRA DISTRIBUTION: All.

ELEVATION: < 2200 m

HABITAT: Open, disturbed areas; foothill and montane zones.

FLOWERING TIME: Apr–Jul

NOTES: *C. melitensis* is distinguished from the similar-looking *C. solstitialis* by its shorter, purple-tipped, downward-pointing phyllary spines, as well as its earlier blooming time and the fact that last year's phyllary spines stay on the flower head rather than falling off as they do for *C. solstitialis*. Although yellow starthistle is a worse problem for native ecosystems, tocalote is also a problem weed. Tocalote has become ubiquitous in the Sierra foothills to the point where control is basically impossible, but if found in a new area that is still uninfested, removing the plants (bagging if flowers/seeds present) is worthwhile before they have a chance to spread.

Centaurea solstitialis (non-native, invasive)

Yellow Starthistle

Asteraceae (Sunflower family)

Erect, bushy annual herb 1–10 dm; grayish-green, and woolly; stem is winged with ruffled, leafy tissue.

LEAVES: Basal in rosettes, 5–15 cm, pinnately lobed to dissected; stem leaves alternate, 1–3 cm, sessile, decurrent, and entire.

FLOWERS: Heads disciform, solitary at tips of branchlets; involucres to 18 mm, ovoid; main phyllaries have stout, straight, tan-colored spines 10–25 mm. Corollas 13–20 mm, yellow, without glands.

FRUITS: 2–3 mm, of 2 types: outer is brown without pappus, and inner is mottled with pappus 2–4 mm, white.

SIERRA DISTRIBUTION: All, except Inyo County.

ELEVATION: < 200 m

HABITAT: Open, disturbed areas; foothill and montane zones.

FLOWERING TIME: May–Oct

NOTES: Originally from southern Europe, introduced as a contaminant in alfalfa seed in the mid-1800s, yellow starthistle (YST) is a huge problem in wildlands. Roots can extend 10 ft deep, robbing native plants of water and outcompeting them for nutrients and sunlight. YST has taken over millions of acres of CA. SN communities have been effective at keeping YST from spreading to the higher reaches of the mountain range, but constant vigilance will be forever needed. More information at Cal-IPC.

Centaurea stoebe ssp. *australis* (non-native, invasive)

Spotted Knapweed

Asteraceae (Sunflower family)

Erect biennial or perennial herb; grayish-green, thinly woolly.

LEAVES: Alternate, deeply pinnately lobed.

FLOWERS: Heads radiant, in open panicles; involucres 10–13 mm; phyllaries with darkened tips that form "spots." Corollas 12–25 mm, pink to purple (sometimes white).

FRUITS: 3–3.5 mm; pappus 1–2 mm, white.

SIERRA DISTRIBUTION: Both slopes, Madera and Inyo Counties north; expected in new sites.

ELEVATION: < 2600 m

HABITAT: Disturbed areas; foothill and montane zones, could occur in subalpine zone.

FLOWERING TIME: Jul–Sep

NOTES: Formerly *Centaurea maculosa*, then until recently *C. stoebe* ssp. *micranthos*; in late 2022, changed to *C. stoebe* ssp. *australis*. Spotted knapweed is a serious pest that infests millions of acres in the Rocky Mountains where growing conditions are similar to the SN, so a strong early detection/rapid response program is important, as new infestations of this weed are continually popping up.

Centromadia fitchii

Fitch's Spikeweed

Asteraceae (Sunflower family)

Annual herb < 5 dm; soft-hairy, stiffly branched, and spiky.
LEAVES: Alternate, linear to few-lobed, with long-stalked glandular hairs, spine-tipped.
FLOWERS: Involucres 5–10 mm; phyllaries spine-tipped. Disk flowers have anthers red, brown, or purple.
FRUITS: Disk fruits have pappus of disk fruits of 8–12 linear scales.
SIERRA DISTRIBUTION: West slope, Tulare to Butte Counties.
ELEVATION: < 1000 m
HABITAT: Open grassy areas, roadsides; foothill zone.
FLOWERING TIME: May–Nov
NOTES: Formerly *Hemizonia fitchii*.

Chaenactis alpigena

Sharsmith Pincushion

Asteraceae (Sunflower family)

Clumping or mat-forming perennial herb 2–7 cm; woolly at base, with basal leaves only.
LEAVES: Basal 1–2.5 cm, ± flat, variously lobed to ± entire.
FLOWERS: Head 1 per stem; outer phyllaries silky-hairy. Ray flowers absent; disk flowers 5.5–8 mm, pink to creamy white, radially symmetrical.
FRUITS: 5–8 mm; pappus scales 10–20, 5–8 mm.
SIERRA DISTRIBUTION: Tulare and Inyo Counties north to El Dorado County.
ELEVATION: 2200–3900 m
HABITAT: Sandy or gravelly soil, rocky areas, fellfields; subalpine and alpine zones.
FLOWERING TIME: Jul–Sep
NOTES: Check the outer phyllaries. If they have glandular hairs, the plant is probably *C. douglasii* var. *alpina*.

Chaenactis douglasii var. *douglasii*

Dusty Maidens

Asteraceae (Sunflower family)

Biennial or perennial herb < 40 cm; cobwebby-woolly at base, with leaves present on stem.

LEAVES: Basal and cauline 1.5–12 cm, woolly; cauline leaves 2-pinnate, not flat, glandular.

FLOWERS: Heads 2–25 per stem, with outer phyllaries glandular. Ray flowers absent, and disk flowers 5–8 mm, pink to creamy white, radially symmetrical.

FRUITS: 5–8 mm; pappus scales 10–20, 3–6 mm, in 3–4 series.

SIERRA DISTRIBUTION: Both slopes, all.

ELEVATION: 900–3500 m

HABITAT: Sandy or gravelly soil, rocky areas, fellfields; montane and subalpine zones.

FLOWERING TIME: May–Sep

NOTES: Alpine dusty maidens, *C. douglasii* var. *alpina*, occurs > 3000 m in Tuolumne, El Dorado, and Alpine Counties. It has only basal leaves and is a matted perennial < 10 cm. The 2 varieties intergrade where they overlap. Compare with *C. alpigena*, which has nonglandular outer phyllaries.

Chaenactis glabriuscula var. *glabriuscula*

Yellow Pincushion

Asteraceae (Sunflower family)

Annual herb 10–60 cm; cobwebby-woolly at base, and hairless with age.

LEAVES: Basal often withered by flowering time; stem leaves < 9 cm, 1–2 pinnately lobed, with lobes linear.

FLOWERS: Heads 2–20+ per stem; all are disk flowers, but outer flowers have outer lobes enlarged similar in appearance to a ray floret. Inner corollas 4–6 mm.

FRUITS: Inner fruits 3–5.5 mm; pappus of 4 scales, 2–4 mm (< corolla).

SIERRA DISTRIBUTION: West slope, all (no records between El Dorado and Butte Counties).

ELEVATION: 100–2300 m

HABITAT: Open, sandy soils; foothill and lower montane zones.

FLOWERING TIME: Feb–Jul

NOTES: Two other varieties occur in the Sierra: Var. *heterocarpha* occurs in scattered locations on serpentine or other metamorphic substrates on the west slope. The plant is < 40 cm, has 2–5 heads per stem, fruits with 7–8 scales in 2 series, and blooms Mar–Jun. *C. glabriuscula* var. *megacephala* is 15–40 cm, has 1–3 heads per stem, fruits with 4 wide scales in 1 series, and blooms Mar–Jun.

Chrysothamnus viscidiflorus ssp. *viscidiflorus*

Yellow Rabbitbrush

Asteraceae (Sunflower family)

Shrub 1–15 dm; stems hairless, green at first, then white with age.

LEAVES: Alternate, sessile, 1–7.5 cm, 0.5–10 mm wide, and twisted. Leaves are wavy-margined and hairless, margins ciliate.

FLOWERS: Heads discoid; in clusters, flat-topped or rounded. Disk flowers 4–14, style branches exserted.

FRUITS: 3–5 mm, obconic; pappus bristles ± equal to corolla length, white or tan.

SIERRA DISTRIBUTION: Mostly east slope, Kern County north.

ELEVATION: 900–4000 m

HABITAT: Dry, open areas; montane, subalpine, and sagebrush/pinyon-juniper zones.

FLOWERING TIME: Jul–Sep

NOTES: Among the yellow, rabbitbrush-like shrubs, this one is unique because it is entirely hairless except for the ciliate leaf margins, and the leaves are often twisted a full 360°. Two other sspp. occur in the eastern Sierra: Ssp. *lanceolatus* is found on the east slope from Mono County north (1200–2500 m), has hairy stems and narrower leaves (2–6 mm wide) that are hairy below; ssp. *puberulus* is also primarily on the east slope (1500–3000 m), has gray-green puberulent leaves and stems.

Cichorium intybus (non-native)

Chicory

Asteraceae (Sunflower family)

Perennial herb 4–20 dm; hairless except for bristly-hairy base. Sap milky.

LEAVES: Basal and alternate, 1–2 cm, the lower with winged petiole, pinnately toothed or lobed, and the upper much smaller, sessile, sometimes clasping.

FLOWERS: Heads of 10–25 showy, ligulate, blue flowers, with phyllaries in 2 series; anthers blue.

FRUITS: 1.5–2.5 mm; pappus scales < 0.5 mm.

SIERRA DISTRIBUTION: Both slopes, scattered locations.

ELEVATION: < 1500 m

HABITAT: Disturbed sites, roadsides; foothill and montane zones.

FLOWERING TIME: Apr–Oct

NOTES: Introduced from Europe and Asia. The leaves are used as greens, and the ground roasted root is the source of the coffee substitute chicory.

Cirsium andersonii

Anderson's Thistle

Asteraceae (Sunflower family)

Erect perennial herb 4–10 dm; stem simple or with ascending branches, hairless to tomentose, and often reddish or maroon; spreads horizontally by rootstocks, forming patches of rosettes and stems in varying stages of development.

LEAVES: Basal and alternate, with lower leaves 8–35 cm, and upper smaller and spinier. Upper leaf surface green and without prickles; lower surface gray-tomentose. Petioles spiny-winged and clasping at midstem, sometimes with spiny-winged tissue continuing down stem below node a short distance.

FLOWERS: Heads 1 to few per stem, discoid; involucres 3–5 cm x 2–4 cm wide, usually cylindric, subtended by 1 to several bracts, hairless or loosely tomentose; phyllaries in 6–8 series, outer green and inner red to purple. Corollas 30–45 mm, bright rose-purple; tube 10–20 mm; throat 10–16 mm, abruptly narrowed where it meets tube; corolla lobes 9–11 mm. Style tips 3.5–5 mm.

FRUITS: Achenes 6–7 mm; pappus 25–40 mm.

SIERRA DISTRIBUTION: Both slopes, all.

ELEVATION: 1100–3500 m

HABITAT: Partial shade to open areas; montane through alpine zones.

FLOWERING TIME: Jul–Sep

NOTES: An easy thistle to identify. Note the maroon stems and cylindrical involucres, along with the large expanse of smooth stem not occupied by spiny-winged leaf tissue. Also note that Anderson's thistle is found in patches of many rosettes and stems resulting from horizontal spread underground. Other native SN *Cirsium* spp. do not form semicontinuous patches (but Canada thistle, *C. arvense*, a serious noxious weed, does). Attracts hummingbirds and a variety of butterflies and native bees.

Cirsium arizonicum var. *arizonicum*

Arizona Thistle

Asteraceae (Sunflower family)

Perennial herb 2–15 dm; often with multiple stems from base, hairless to thinly tomentose, from woody caudex (no runners or rhizomes).

LEAVES: Basal and alternate, with lower leaves 1–2 dm, unlobed to deeply divided, main spines 2–15 mm, and tomentose (especially below). Middle and upper cauline leaves not significantly smaller, with petioles spiny-winged to stem, clasping stem or decurrent.

FLOWERS: Heads 1 to few in loose cymes, discoid; upper leaves closely subtend involucres. Involucres 3–4 cm x 1.5–2.5 cm wide, cylindric to ovoid or bell-shaped, and subtended by 1 to several bracts, hairless or loosely tomentose; phyllaries linear to lanceolate, entire, and the middle and outer are tipped with spines up to 20 mm. Corollas 24–34 mm, bright red or red pink; tube 7–12 mm; throat 2–5 mm; corolla lobes 13–18 mm. Style tips 2.5 mm.

FRUITS: Achenes 3.5–7 mm; pappus 17–28 mm.

SIERRA DISTRIBUTION: Both slopes, Inyo, Mono, and Tulare Counties.

ELEVATION: 2300–3400 m

HABITAT: Open areas; subalpine, alpine, and sagebrush zones.

FLOWERING TIME: Jul–Aug

NOTES: Compare with Anderson's thistle, *Cirsium andersonii*, which has shorter phyllary spines as a rule, and corolla lobes shorter than the corolla tube. Look for it in the eastern Sierra out of Lone Pine, e.g., Cottonwood Lakes, the trail to Whitney Portal, and also south of Mineral King in Tulare County.

Cirsium occidentale var. *californicum*

California Thistle

Asteraceae (Sunflower family)

Erect biennial herb 5–20 dm; stem simple, branched from middle upward, tomentose (looks gray-woolly), most of stem smooth, not occupied by spiny wings.

LEAVES: Basal and alternate, with upper surfaces green to grayish-green; lower surfaces gray (and more densely tomentose). Basal and lower leaves 1–4 dm; blades oblanceolate; petioles spiny-winged, lobed ≥ to middle, lobes roughly triangular; main spines 1–10 mm. Cauline leaves gradually smaller toward top of stem, clasping or short-decurrent, often spinier than basal, and entire or less deeply lobed than basal.

FLOWERS: Heads single or in loose clusters, discoid; involucres 1.5–5 cm x 1.5–5 cm wide, hairless or densely cobwebby-tomentose; middle phyllary tips tipped by spines up to 1 cm. Corollas 18–35 mm, white, purple, or rose; tube 8–18 mm, throat 5–7 mm, corolla lobes 5–10 mm. Style tips 4–5 mm.

FRUITS: Achenes 5–6 mm, hairless; pappus 15–30 mm, white.

SIERRA DISTRIBUTION: Both slopes, all (no records for Inyo County).

ELEVATION: < 2300 m

HABITAT: Open areas; foothill through montane zones.

FLOWERING TIME: Apr–Jul

NOTES: Note that at least some stem surface is free enough of prickly leaf tissue to allow touching (gingerly perhaps), in contrast to invasive non-native thistles like *C. vulgare* or *Carduus pycnocephalus*.

Cirsium occidentale var. *candidissimum*

Snowy Thistle

Asteraceae (Sunflower family)

Erect or bushy biennial herb 4–20 dm; with stem simple, branched from middle upward; herbage densely white-tomentose.

LEAVES: Basal and alternate, densely snowy white. Basal and lower leaves 1–4 dm; petioles spiny-winged, blades oblanceolate, lobed ≥ to middle, with lobes roughly triangular, and main spines 1–10 mm. Cauline leaves gradually smaller toward top of stem, clasping or short-decurrent, often spinier than basal, and entire or less deeply lobed than basal.

FLOWERS: Heads single or in loose clusters, discoid; involucres 1.5–5 cm x 2–6 cm wide, densely cobwebby-tomentose; middle phyllary tips 1.5–3 cm x 2–3 mm wide. Corollas 26–40 mm, usually red (sometimes white or pink).

FRUITS: Achenes 5–6 mm, hairless; pappus white, 15–30 mm.

SIERRA DISTRIBUTION: Both slopes, Alpine and Mono Counties north.

ELEVATION: < 1900 m

HABITAT: Openings, disturbed sites; mostly montane zone.

FLOWERING TIME: Apr–Sep

NOTES: This variety is so striking and different-looking from var. *occidentale* that we feature it as a separate entry. An absolutely gorgeous thistle!

Cirsium scariosum var. *americanum*

Dinnerplate Thistle

Asteraceae (Sunflower family)

Biennial or short-lived perennial herb; stem usually absent. Once the plant flowers and produces fruits, it dies.

LEAVES: 1–4 dm, with upper surfaces hairless to loosely tomentose, and lower surfaces densely tomentose; margins unlobed to deeply lobed, spines 2–7 mm.

FLOWERS: Heads discoid, crowded, sessile; involucres 1.5–3 cm; outer phyllary tip spine 1–12 mm. Corollas 20–40 mm, white to pink-tinged; tube 7–24 mm, throat 4–12 mm, corolla lobes 4–10 mm. Style tips 3.5–8 mm.

FRUITS: Achenes 3–6.5 mm; pappus 13–35 mm.

SIERRA DISTRIBUTION: East slope primarily and crest, Tulare and Inyo Counties north.

ELEVATION: 1600–3500 m

HABITAT: Meadows, forest openings; subalpine and alpine zones.

FLOWERING TIME: Jun–Aug

NOTES: Occasionally plants with stems up to 45 cm are observed; they may be growing with stemless plants. The similar-looking rosette thistle, *C. scariosum* var. *congdonii* (formerly *C. tioganum*) has lavender to purple flowers and blooms Jun–Aug at similar elevations in Fresno, Inyo, Mono, and Tulare Counties.

Corethrogyne filaginifolia

Common Sandaster

Asteraceae (Sunflower family)

Perennial herb to subshrub 10–100 cm; stems erect to decumbent, many from base, densely white-tomentose, and sometimes glandular.

LEAVES: Alternate, sessile or with winged petioles, often crowded toward top of stem. Blades 10–70 mm x 3–19 mm wide, smaller toward top of stem, linear to oblanceolate to ovate, entire or toothed, and hairy (whitish).

FLOWERS: Heads radiate, single or clustered; involucres 6–14 mm x 3–10 mm diameter, variously shaped; phyllaries graduated in 3–9 series, with tips often spreading or reflexed, hairy or glandular. Ray flowers 10–43, sterile (styles absent), rays white, pink, or purple. Disk flowers 12–120; corollas 4–8 mm, yellow.

FRUITS: 2–5 mm, cylindric to obconic, ribs 5–7; pappus 3–8 mm, bristles 35–65, in 1–2 series, barbed, brown to reddish.

SIERRA DISTRIBUTION: Both slopes, Placer County south.

ELEVATION: < 3000 m

HABITAT: Open sandy areas, roadsides; foothill through subalpine zones.

FLOWERING TIME: Jul–Nov

NOTES: An extremely variable species; at times, plants now included in *Corethrogyne filaginifolia* have been assigned to 3–7 separate spp. with various sspp. or var. From 1992 to 2000, plants in the SN were included in *Lessingia*.

Cirsium vulgare (non-native, invasive)

Bull Thistle

Asteraceae (Sunflower family)

Biennial herb 3–20 dm; stem is simple with branching in upper half or branched from near base, appearing bushy, with herbage loosely tomentose and often glandular-hairy.

LEAVES: Basal and alternate, sessile, petioles spiny-winged; lower leaves 10–40 cm, upper surface with slender, appressed bristlelike spines, sometimes tomentose when young, and lower surface densely tomentose with main veins raised prominently, veins glandular. Cauline leaves gradually become smaller toward top. Main spines ≤ 15 mm, long-decurrent with spiny-winged tissue.

FLOWERS: Heads 1 to many, clustered, discoid, with bractlike uppermost leaves just below heads. Involucres 3–4 cm x 2–4 cm wide, hemispheric or bell-shaped, and subtended by bracts 1 to several, hairless or loosely tomentose; phyllary tips linear to lance-linear, and tipped by spines 1–5 mm. Corollas 25–35 mm, purple; tube 18–25 mm, throat 5–6 mm, corolla lobes 5–7 mm. Style tips 3.5–6 mm.

FRUITS: Achene 3–4.5 mm, hairless; pappus 30–40 mm, white, generally detaching from achene at maturity.

SIERRA DISTRIBUTION: All.

ELEVATION: < 2500 m

HABITAT: Disturbed areas, especially immediately after disturbance such as fire, timber harvest/thinning, or fuels reduction; thrives in moist areas like meadows and streamsides, but grows in both dry and wet sites; all zones except alpine.

FLOWERING TIME: Jul–Sep

NOTES: The only thistle with small, appressed spines on upper leaf surfaces (best seen with a 10x lens). Spiny wings extend virtually the entire length of the stem, making it impossible to touch stems without being poked (compare with *C. occidentale* and *C. andersonii*, both of which have at least small stretches of stem where you could put your finger safely). Native to Eurasia, and now found in all 50 states. In drier sites, it often fades in abundance as natives recolonize, and is more ecologically damaging in moist and wet habitats.

Crepis acuminata

Longleaf Hawksbeard

Asteraceae (Sunflower family)

Perennial herb from a swollen woody caudex, with 1–5 erect stems 2–7 dm tall, thinly tomentose at least near the base, and branched from middle or above; sap milky.

LEAVES: Basal 12–40 cm, with base narrowed to winged petiole, blade deeply pinnately lobed into narrow lobes with acuminate tips, leaf tip long-acuminate, gray-green due to minutely tomentose surfaces.

FLOWERS: Heads liguliflorous, 30–70 (up to 100) heads arranged in flat-topped, compound clusters. Involucres 8–16 mm, cylindric to bell-shaped; outer phyllaries 1–2 mm, narrowly triangular; inner phyllaries 5–8, 8–12 mm, hairless and shiny, sometimes tomentose, not strongly keeled. Flowers 5–10 per head; corollas 10–18 mm.

FRUITS: 6–9 mm, cylindric, yellow or yellow-brown, ribs 12, tip not beaked; pappus white.

SIERRA DISTRIBUTION: All.

ELEVATION: 1000–3300 m

HABITAT: Rocky sites, open forests; montane, subalpine, and sagebrush zones.

FLOWERING TIME: May–Aug

NOTES: A similar-looking SN *Crepis*, naked-stemmed hawksbeard, *C. pleurocarpa*, is found < 2200 m on the east slope from Mono to El Dorado Counties and both slopes northward, often on serpentine. *C. pleurocarpa* has only 7–10 (up to 30) heads, 5–8 inner phyllaries with bases strongly keeled and swollen in age, and 5–10 flowers per head, with corollas 15–19 mm. Intermediate hawksbeard, *C. intermedia*, occurs from Tulare and Inyo Counties north < 3300 m, and has 10–60 heads, 7–10 inner phyllaries, and 7–12 flowers per head, with corollas 14–30 mm.

Crepis occidentalis ssp. *pumila*

Largeflower Hawksbeard

Asteraceae (Sunflower family)

Perennial herb from woody caudex, stems 1–4 dm; gray-tomentose but not glandular. Sap milky.

LEAVES: Basal and alternate; blades 10–20 cm, elliptic, deeply pinnately lobed. Lobes closely spaced, themselves toothed or lobed.

FLOWERS: Heads are liguliflorous, 5–15 heads in loose, flat-topped clusters; peduncles minutely stalked-glandular. Involucres 11–19 mm, cylindrical; outer phyllaries 2–6 mm, linear to narrowly deltate; inner phyllaries 8, 7–13 mm, woolly, lanceolate, and sometimes with green or black gland-tipped hairs; flowers 12–20 per head.

FRUITS: 6–10 mm, fusiform, ribs 10–18, tip not beaked, brown; pappus white.

SIERRA DISTRIBUTION: Mostly northern Sierra, uncommon.

ELEVATION: 800–1800 m

HABITAT: Open forests, rocky sites; montane and sagebrush zones.

FLOWERING TIME: Jun–Jul

NOTES: Two other sspp. occur in the SN: ssp. *conjuncta* and ssp. *occidentalis*. Because all sspp. intergrade extensively, plants are highly variable, and describing how the sspp. differ wouldn't be that helpful; thus, we recommend consulting the Jepson eFlora and Flora of North America eFlora for *Crepis*.

Dieteria canescens var. *canescens*

Hoary Aster

Asteraceae (Sunflower family)

Annual, biennial, or short-lived perennial herb 1–5 dm; erect or spreading, and branched toward top.

LEAVES: Alternate, 3–10 cm, entire or irregularly serrate to dentate, with teeth bristle-tipped.

FLOWERS: Head 6–12 mm wide; phyllaries in 5–10 series, tips acute, spreading to reflexed, and glandular. Ray and disk flowers are fertile.

FRUITS: 2.5–3.5 mm, slightly curved and flattened; pappus 6–8 mm, bristles numerous and unequal.

SIERRA DISTRIBUTION: Most counties > 2000 m.

ELEVATION: 2000–3000 m

HABITAT: Open areas; montane and subalpine zones.

FLOWERING TIME: Jul–Sep

NOTES: Formerly *Machaeranthera canescens* ssp. *canescens*. Also occurring from Mono and Tuolumne Counties north is the similar *D. canescens* var. *shastensis*, which grows 1500–3400 m and has phyllaries in 3–5 series and sterile ray flowers without apparent styles protruding.

Doellingeria breweri

Brewer's Aster

Asteraceae (Sunflower family)

Erect to ascending perennial herb 1–10 dm, often bushy; herbage glandular to tomentose.

LEAVES: Alternate, sessile, 2–5 cm, entire or toothed.

FLOWERS: Heads discoid, phyllaries ± equal, bases pale.

FRUITS: Pappus 6-10 mm, in 1 series.

SIERRA DISTRIBUTION: All.

ELEVATION: 1200–3660 m

HABITAT: Openings in conifer forest; montane and subalpine zones.

FLOWERING TIME: Jul–Sep

NOTES: Formerly *Aster breweri*, then *Eucephalus breweri*. Very common.

Encelia actoni

Acton's Brittlebush

Asteraceae (Sunflower family)

Shrub 0.5–1 m; many-branched from base, with young stems hairy and older stems with fissured bark.

LEAVES: Alternate, scattered on stems; petioles 6–12 mm; blades 25–40 mm, ovate to deltate, silvery hairy.

FLOWERS: Heads are solitary, with peduncles canescent; involucres 8–14 mm; phyllaries in 2–3 series. Ray flowers 14–25, sterile, rays 10–25 mm; disk flowers yellow, corollas 5–6 mm.

FRUITS: 5–7 mm, compressed, ciliate, with no pappus.

SIERRA DISTRIBUTION: Both slopes, Kern County, Kern River watershed.

ELEVATION: < 2100 m

HABITAT: Open areas, rocky slopes, roadsides; foothill and sagebrush zones.

FLOWERING TIME: Feb–Jul

NOTES: Named for the town of Acton in Los Angeles County.

Ericameria arborescens

Golden-Fleece

Asteraceae (Sunflower family)

Erect shrub to 5 m.

LEAVES: Alternate, 15–25 mm (< 2 mm wide), linear, ascending but recurved with age; gland-dotted.

FLOWERS: Heads arranged in cymelike clusters to 10 cm wide; involucres 4–6 mm, 3.5–4.5 mm wide; phyllaries in 3–4 series. Disk flowers 10–25, yellow; no ray flowers.

FRUITS: 2–3 mm, narrowly top-shaped, tan to brown, and ribs 5; pappus 3.5–4.5 mm, bristlelike and tan.

SIERRA DISTRIBUTION: West slope, Kern County north.

ELEVATION: < 1800 m

HABITAT: Dry areas; foothill chaparral and lower montane zones. Especially abundant after fire and other disturbance.

FLOWERING TIME: Aug–Oct

NOTES: The foliage is used by Native Americans of the central SN to line granaries for acorn storage.

Ericameria bloomeri

Bloomer's Goldenbush

Asteraceae (Sunflower family)

Low-growing shrub 2–6 dm; stems hairless (sometimes slightly woolly) and gland-dotted.

LEAVES: Alternate, sessile, 20–70 mm, linear to narrowly oblanceolate, and shiny; dotted with sunken glands.

FLOWERS: Heads in leafy involucres 8–11 mm, with phyllaries equal, and loosely arranged in 3–6 series; margins scarious and tips recurved. Ray flowers 1–5, rays 8–12 mm; disk flowers 4–14, corollas 7–11 mm, yellow.

FRUITS: 5.5–8 mm, 5-angled, brown, and hairless or hairy; pappus 6–9 mm, tan or reddish.

SIERRA DISTRIBUTION: Both slopes, Tulare and Inyo counties north.

ELEVATION: 900–3600 m

HABITAT: Sandy and gravelly openings in coniferous forest; montane, subalpine, alpine, and sagebrush zones.

FLOWERING TIME: Jul–Oct

NOTES: A spectacular contributor to fall wildflower color, forming bright yellow masses over large areas. *Ericameria* spp. are hosts for a variety of butterflies and moths.

Ericameria cuneata var. *cuneata*

Cliff Goldenbush

Asteraceae (Sunflower family)

Low-growing shrub < 5 dm, wider than tall, hairless, and gland-dotted.
LEAVES: Alternate, sessile, 3–14 mm x 2–9 mm wide, wedge-shaped, and shiny, with tips often notched.
FLOWERS: Heads mostly discoid and crowded at branch tips; disk flowers 12–33, sometimes with ray flowers < 7.
FRUITS: 2.5–3 mm, silky-hairy; pappus sparse, brown.
SIERRA DISTRIBUTION: Sierra County south.
ELEVATION: 600–2800 m
HABITAT: Granitic outcrops, cliffs; foothill and montane zones.
FLOWERING TIME: Sep–Nov
NOTES: An attractive "rock-garden" species that provides splashes of yellow against glossy green leaves well into the fall. In Kern County, var. *spathulata* co-occurs; it differs by having petioled, spoon-shaped leaves that are longer (12–25 mm) than in var. *cuneata* and having fewer (7–15) disk flowers, rarely with ray flowers.

Ericameria discoidea

Whitestem Goldenbush

Asteraceae (Sunflower family)

Low-growing shrub < 4 dm, with stems white-woolly.
LEAVES: Alternate, sessile, 10–35 mm, oblong-oblanceolate, and covered with stalked glands, with margins wavy or not.
FLOWERS: Heads generally 1 per branch, and disk flowers only. Styles yellow, conspicuously protruding.
FRUITS: 5–6 mm, brown, hairy; pappus tan.
SIERRA DISTRIBUTION: Tulare and Inyo Counties to Nevada County (probably further north, but no specimen records in CCH).
ELEVATION: 2300–3800 m
HABITAT: Rocky slopes and flats; subalpine and alpine zones.
FLOWERING TIME: Jul–Sep.
NOTES: The white, feltlike stems distinguish this high-elevation shrub.

Ericameria linearifolia

Interior Goldenbush

Asteraceae (Sunflower family)

Shrub 4–15 dm, with stems puberulent or hairless; herbage is resinous and gland-dotted (glands in pits).

LEAVES: Alternate, 10–55 mm, linear, hairy.

FLOWERS: Heads radiate and solitary on peduncles 20–70 mm, with hemispheric involucres 8–14 mm x 10–18 mm wide; phyllaries of equal length, in 2–3 series, acuminate, stalked-glandular, and green with ciliate, scabrous margins. Ray flowers 12–18, corollas 9–20 mm; disk flowers 16–60, corollas 6–10 mm.

FRUITS: 4–5 mm, somewhat flattened, ribs 2, tan to brown, with silky hairs; pappus 5.5–7 mm, silvery.

SIERRA DISTRIBUTION: Tulare, Kern, and Inyo Counties.

ELEVATION: < 2000 m

HABITAT: Dry slopes and valleys, rocky sites; foothill, pinyon-juniper, and sagebrush zones.

FLOWERING TIME: Mar–May

NOTES: This is a characteristic shrub of the Great Basin, desert, and Coast Ranges. In the southern SN, it is limited to Tulare and Kern Counties and scattered sites on the east slope in Inyo County < 1800 m (technically below the lower east slope cutoff for this book).

Ericameria nauseosa var. *speciosa*

Showy Rabbitbrush

Asteraceae (Sunflower family)

Shrub 2–10 dm, with stems loosely tomentose, hence often grayish.

LEAVES: Alternate, 30–70 mm, linear, usually loosely tomentose, green or grayish.

FLOWERS: Heads discoid, in dense, rounded to flat-topped arrays. Involucres 7–13.5 mm x 2–4 mm wide, obconic or nearly cylindric; phyllaries 10–31, arranged in 3–5 vertically ranked series (like roofing tiles) and ± woolly. Disk corollas 5, 9–13 mm, yellow; corolla lobes 1.1–2.1 mm.

FRUITS: 3–10 mm, tan, usually hairy; pappus 3–13 mm, whitish.

SIERRA DISTRIBUTION: Both slopes, all.

ELEVATION: 600–2300 m

HABITAT: Many dry habitats; all zones.

FLOWERING TIME: Aug–Oct

NOTES: *Ericameria nauseosa*, referred to as rubber rabbitbrush at the species level, comprises 8 varieties in CA and 21 in western North America. The species is used by Native Americans for chewing gum, tea, basketmaking, and a wide variety of medicinal treatments. Plants contain 1.5%–6.5% natural rubber, and commercial rubber production occurs at a small scale. Varieties of rubber rabbitbrush have been planted in the SN for erosion control within and outside their natural range, which can be confusing when you come upon a road lined with it but see none beyond the road's edge.

Erigeron algidus

Stalked Fleabane

Asteraceae (Sunflower family)

Perennial herb 2–20 cm; arising from short, thick fibrous caudexes or rhizomes; stems are unbranched, with loosely spreading hairs, glandular.

LEAVES: Basal 2–7 cm, petioled, oblanceolate or spoon-shaped, spreading hairy and glandular; cauline leaves much smaller, strigose, and sometimes glandular.

FLOWERS: Heads 1 per stem; involucres 5–8 mm x 8–16 mm diameter; phyllaries approximately equal, ± purple-black, glandular, with spreading hairs and tips spreading or reflexed. Ray flowers 50–125, rays 7–13 mm, lavender, pink, or white; ray flower coils when dry. Disk flowers numerous, with corollas funnel-shaped, yellow.

FRUITS: Achenes 1.9–2.1 mm, oblong, ribs 2; pappus bristles 12–16.

SIERRA DISTRIBUTION: Inyo and Tulare Counties north to El Dorado County.

ELEVATION: 2600–3700 m

HABITAT: Meadows, talus; alpine zone.

FLOWERING TIME: Jul–Aug

NOTES: Grows with and could be confused with *E. pygmaeus*.

Erigeron breweri var. *breweri*

Brewer's Fleabane

Asteraceae (Sunflower family)

Ascending to erect perennial herb 1.2–6.0 dm; from woody base, with stems and herbage often tinged purple (sometimes growing horizontally). Stems and leaves are covered with short-bristly, downward-pointing, nonglandular hairs 0.1–0.4 mm.

LEAVES: No basal leaves. Stem leaves 0.5–4 cm, linear or oblanceolate, with upper and lower evenly spaced and about the same size; hairs 0.1–0.3 mm, not glandular.

FLOWERS: Heads 1–5; involucres 4–6 mm, 8–15 mm diameter; phyllaries arranged in 3–5 series, with dense, glandular hairs. Ray flowers 12–45, rays 4–7 mm, white, pink, or lavender, coiled and bluish upon drying.

FRUITS: Fruit ribs 2; pappus bristles 22–46.

SIERRA DISTRIBUTION: Both slopes, Placer County south.

ELEVATION: 1200–3100 m

HABITAT: Dry, rocky areas; montane and subalpine zones.

FLOWERING TIME: Jun–Sep

NOTES: Other SN varieties not pictured: *Erigeron breweri* var. *covillei* is restricted to the sagebrush, chaparral, and juniper scrub of the eastern slope of the southern SN. It has phyllaries without glands. *E. breweri* var. *porphyreticus* also occurs in the eastern Sierra; it is a smaller (2–3 dm tall), wiry plant and has long, thick-based, nonglandular phyllary hairs (and may have a few glandular hairs mixed in).

Erigeron clokeyi var. *pinzliae*

Pinzl's Fleabane

Asteraceae (Sunflower family)

Perennial herb 5–20 cm; from stout taproots and simple caudexes. Stem is unbranched, ascending to erect, and sparsely minutely glandular as well as moderately covered with short, stiffly spreading to reflexed, glandless hairs.

LEAVES: Basal 2–8 cm, oblanceolate; cauline leaves reduced in size by midstem and not clasping; blades minutely rough-hairy or bristly.

FLOWERS: Heads 1 per stem; involucres 4–7 mm x 6–12 mm diameter; phyllaries green, minutely glandular and with short, stiff-spreading-reflexed hairs. Ray flowers 25–55, rays 6–11 mm, lavender (occasionally white or pink), reflexed. Disk flowers numerous, corollas funnel-shaped, yellow.

FRUITS: Achenes 1.8–2 mm, oblong, ribs 2; pappus bristles 13–22.

SIERRA DISTRIBUTION: East slope and crest, Inyo and Mono Counties.

ELEVATION: 2200–3400 m

HABITAT: Rocky slopes, crevices, talus; alpine and sagebrush zones.

FLOWERING TIME: Jun–Sep

NOTES: All *E. clokeyi* plants in the SN are var. *pinzliae*. The other variety, var. *clokeyi*, is a rare plant of the Charleston Mountains in Nevada.

Erigeron compositus

Cut-Leaf Fleabane

Asteraceae (Sunflower family)

Perennial herb 3–15 cm; from stout taproot and branched caudexes. Stem is unbranched, minutely glandular toward top, and sometimes sparsely bristly as well.

LEAVES: Basal 1–5 cm, oblanceolate or spoon-shaped, 2–3 times ternately divided, densely puberulent, and minutely glandular. Cauline leaves absent, or linear if present.

FLOWERS: Heads 1 per stem, may be disciform (as in photo) or radiate. Involucres 5–10 mm x 8–20 mm diameter; phyllaries approximately equal, with rough, spreading hairs, minutely glandular, and purple-tipped. When present, ray flowers 30–60, rays 6–12 mm, white, pink, or lavender; not coiling, or coiling weakly, when dry. Disk flowers numerous, corollas funnel-shaped, yellow.

FRUITS: Achenes 1.6–2.7 mm, oblong, ribs 2, pappus bristles 12–20.

SIERRA DISTRIBUTION: Both slopes, Tulare and Inyo counties north.

ELEVATION: 2000–4300 m

HABITAT: Rocky slopes, crevices, talus; alpine zone.

FLOWERING TIME: May–Sep

NOTES: Both forms, disciform (no ray flowers) and radiate, are encountered. The finely dissected leaves are helpful in identifying this plant.

Erigeron foliosus var. *hartwegii*

Hartweg's Fleabane

Asteraceae (Sunflower family)

Perennial herb with erect, leafy stems ranging in height 20–85 cm; taprooted, hairless to sparsely hairy.

LEAVES: No basal leaves. Stem leaves 2–6 cm x 1–2 (up to 4) mm wide, linear, mostly smooth-edged and threadlike, with middle and upper leaves longer than the internodes and not reduced in length toward top of stem. Leaves often oriented mostly along one side of the stem.

FLOWERS: Heads radiate; involucres 10–16 mm diameter; phyllaries in 3–5 graduated series (like roof shingles), linear. Ray flowers 26–50, rays 7–12 mm, lavender, surrounding disk flowers yellow.

FRUITS: Fruits of ray and disk flowers have double pappus, with 20–34 inner bristles and an outer row of inconspicuous bristles.

SIERRA DISTRIBUTION: West slope, Mariposa County north.

ELEVATION: 100–600 m

HABITAT: Dry, rocky areas, often along rivers, in chaparral and foothill woodland; foothill zone.

FLOWERING TIME: Apr–Jul

NOTES: Leafy fleabane, *E. foliosus* var. *foliosus*, occurs from Mariposa County south < 2900 m. Leaves are 2–4 cm, midvein is raised and with orange resin, and there are fewer (15–49) ray flowers, rays 6–11 mm. Its blooming time is May-Aug.

Erigeron glacialis var. *glacialis*

Subalpine Fleabane

Asteraceae (Sunflower family)

Perennial herb 8–45 cm, branched above; rhizomatous.

LEAVES: Basal 5–20 cm, oblanceolate, hairless or sparsely hairy. Stem leaves smaller above, ovate to lanceolate, clasping at base.

FLOWERS: Heads 1–4 (up to 8); peduncles minutely strigose with hairs somewhat appressed and slightly crinkled. Involucres 6–12 mm, 12–25 mm diameter; phyllaries in 2–3 series, densely covered with stalked glands, and with tips long-acuminate, spreading. Ray flowers 30–105, coiled when dry.

FRUITS: Achenes with 4–7 ridges, elongate; pappus bristles 20–30.

SIERRA DISTRIBUTION: Both slopes, Fresno and Inyo Counties north.

ELEVATION: 150–2100 m

HABITAT: Openings, forest floor, rocky areas, meadows; subalpine and alpine zones.

FLOWERING TIME: Jul–Sep

NOTES: Formerly *E. peregrinus* var. *callianthemus*. *E. glacialis* var. *glacialis* is abundant throughout its range. Another variety, var. *hirsutus*, has peduncles that are bristly to stiffly long-hairy. It occurs more sparingly with the same range.

Erigeron inornatus var. *inornatus*

Western Rayless Fleabane

Asteraceae (Sunflower family)

Perennial herb 30–90 cm; from woody taproots. Stem is branched near tips, long-ascending, and appressed-hairy or hairless toward base.

LEAVES: Cauline 1–6 mm, linear or narrowly oblanceolate to oblong, ± evenly sized and spaced, and ciliate. Proximal and middle leaves short-strigose or hairless.

FLOWERS: Heads 5 to many per stem, discoid; involucres 4.5–6.5 mm x 7–12 diameter; phyllaries in 3–5 clear series, hairless or minutely glandular; inner phyllaries 4.5–5.5 mm.

FRUITS: Achenes 1.6–2.7 mm, oblong, ribs 2; pappus bristles 12–20.

SIERRA DISTRIBUTION: Mostly northern SN, west slope, but a few scattered sites in Mono and Tulare Counties.

ELEVATION: 2000–4300 m

HABITAT: Dry, rocky or sandy sites; foothill and montane zones.

FLOWERING TIME: Jun–Sep

NOTES: In the southern Sierra, 1200–2200 m in Tulare and Fresno Counties, the rare *E. inornatus* var. *keilii*, Keil's daisy, can be found in foothill and montane habitats. Plants are shorter (40–65 cm), erect, short, stiff-hairy throughout, and herbage appears gray.

Erigeron miser

Starved Daisy

Asteraceae (Sunflower family)

Perennial herb 5–25 cm; many-stemmed from woody caudexes. Stems are decumbent to ascending, usually unbranched, with dense, long-spreading hairs.

LEAVES: Cauline 7–16 mm, narrowly oblanceolate, evenly spaced and equally sized, white-spreading-hairy, and minutely glandular.

FLOWERS: Heads 1–4 per stem, discoid; involucres 4–5 mm x 7–14 diameter; phyllaries in 3–5 clear series, densely minutely glandular; inner phyllaries 3.5–5 mm.

FRUITS: Achenes 2–2.5 mm, oblong, ribs 2; pappus bristles 18–25.

SIERRA DISTRIBUTION: Placer County north.

ELEVATION: 1800–2600 m

HABITAT: Talus, rock crevices; montane zone.

FLOWERING TIME: Jun–Sep

NOTES: CRPR 1B.3.

Erigeron pygmaeus

Pygmy Fleabane

Asteraceae (Sunflower family)

Tufted perennial herb 1–6 cm; from taproot and short-branched caudex. Stem is unbranched, with dense, tiny, glandular hairs mixed with short, stiff, nonglandular hairs.

LEAVES: Mostly basal, 6–35 mm, linear to narrowly oblanceolate, and ciliate at base; blades strigose and densely minutely glandular. Cauline leaves absent or much smaller.

FLOWERS: Heads 1 per stem; involucres 4–7 mm x 6–15 mm diameter; phyllaries approximately equal length, ± purple-black, stiff-hairy, and densely minutely glandular. Ray flowers 20–37, ray 4–7 mm, blue or purple, and not coiled when dry. Disk flowers numerous, with corollas funnel-shaped and yellow.

FRUITS: Achenes 2.3–2.8 mm, oblong, ribs 2; pappus bristles 15–25.

SIERRA DISTRIBUTION: Inyo and Tulare Counties north to El Dorado County.

ELEVATION: 2900–4100 m

HABITAT: Talus, rocky slopes; subalpine and alpine zones.

FLOWERING TIME: Jul–Aug

NOTES: Grows with and could be confused with *E. algidus*.

Eriophyllum confertiflorum var. *confertiflorum*

Golden Yarrow

Asteraceae (Sunflower family)

Subshrub or shrub 2–7 dm; herbage tomentose.

LEAVES: Alternate, 1–5 cm, obovate in overall outline, and deeply 3- to 5-lobed to 2-pinnate, with margins rolled under.

FLOWERS: Heads in dense clusters of 10–30, with peduncles 0–2 cm. Involucres 3–5 mm diameter, phyllaries 4–7. Ray flowers 4–6, rays 2–5 mm; disk flowers 10–35; corollas 2–4 mm, puberulent to glandular.

FRUITS: 2–3 mm; pappus scales 5–14, ± equal.

SIERRA DISTRIBUTION: West slope, Kern to Calaveras Counties.

ELEVATION: < 3000 m

HABITAT: Dry soil, rocky sites, burned areas; foothill and montane zones.

FLOWERING TIME: Apr–Aug

NOTES: In the central Sierra, plants with clusters of 3–10 heads, peduncles 0.5–2.5 cm, involucres 5–7 mm diameter, and 35–75 disk corollas 3.5–4 mm are *E. confertiflorum* var. *tanacetiflorum*. This variety is known from Calaveras County north on the west slope < 800 m, and may have originated from hybridization of *E. confertiflorum* x *E. lanatum*.

Eriophyllum lanatum var. *integrifolium*

Common Woolly Sunflower

Asteraceae (Sunflower family)

Perennial herb to multistemmed subshrub; taprooted.

LEAVES: Alternate, 3-lobed, 1–8 cm, wedge-shaped to ovate, and entire or with 3–5 lobes; both surfaces are woolly, and margins are flat (not rolled under).

FLOWERS: Heads radiate, 1 per peduncle; peduncles 3–10 cm. Involucres 6–8 mm; phyllaries usually 8 but varies 5–10, and are keeled and woolly. Ray flowers usually 8 (5–10), rays 6–10 mm, yellow; disk corollas ± 4 mm; tube glandular.

FRUITS: 3–4 mm, glandular or hairy; pappus scales absent or 6–12, 0–2 mm, translucent.

SIERRA DISTRIBUTION: Both slopes, all (northern Kern County, Greenhorn Mountains).

ELEVATION: 1400–4100 m

HABITAT: Dry, forested, or rocky/gravelly sites; all zones.

FLOWERING TIME: Jul–Aug

NOTES: Five varieties of this exceedingly variable polyploid species complex are found in the SN, and they intergrade with each other, sometimes making ID to variety practically impossible. The most common variety is var. *integrifolium*, which in itself is hugely variable, but you can rely on the number of ray flowers (usually 8, but can be 5–10), as all other varieties have 12–13. Var. *croceum* is found almost exclusively on the west slope of the SN < 2000 m (a few scattered locations are known north of the SN); it is often clonal, less woolly, and has ray flowers (rays 8–10 mm) that tend toward orange-yellow. Var. *achilleoides* has pinnately linear-lobed leaves and is less hairy, with margins rolled under and rays 6–9 mm; it occurs from Madera County north < 1300 m on the west slope. Var. *grandiflorum* (no photo) has rays 10–20 mm, the largest leaves of the varieties at 3–8 cm, is less densely hairy, has margins rolled under, and is found < 1700 m from Mariposa County north on the west slope. Var. *obovatum* (no photo) is found on the west slope 1300–2500 m from Madera County south; it has rays 6–7 mm, peduncles often swollen below heads, and leaves 1–5 cm, with both surfaces densely woolly.

Eriophyllum pringlei

Pringle's Woolly Sunflower

Asteraceae (Sunflower family)

Tufted annual herb 1–5 cm; herbage white woolly.
LEAVES: Alternate, 3–10 mm, 3-lobed, wedge-shaped in overall outline, and densely woolly, with margins rolled under.
FLOWERS: Heads are discoid, in ± sessile leafy clusters. Involucres 3–6 mm; phyllaries 6–8. Disk flowers 10–20; corollas ± 2 mm, minutely glandular; anther tips triangular.
FRUITS: 1.5–2 mm, strigose; pappus scales ± 1 mm.
SIERRA DISTRIBUTION: Both slopes, Kern, Tulare, Inyo, and Mono Counties.
ELEVATION: 300–2200 m
HABITAT: Open, sandy or gravelly sites; foothill and sagebrush zones.
FLOWERING TIME: Feb–Jul
NOTES: Mostly found in the Mojave Desert, this cute annual ascends to the east slope of the southern Sierra and can also be found on the west slope in Kern and Tulare Counties.

Eurybia integrifolia

Thickstem Aster

Asteraceae (Sunflower family)

Perennial herb 1.5–7 dm; rhizomatous, glandular-hairy toward top.
LEAVES: Basal leaves present at flowering, petioled, 3–22 mm; hairless or hairy. Cauline leaves alternate, smaller toward top of stem, clasping stem, lance-ovate to obovate, entire, and hairless or hairy.
FLOWERS: Heads in flat-topped clusters, phyllaries in 4–5 series; heads have 9–15 white ray flowers 8–13 mm, white (sometimes tinged pale pink or purple), and 30–70 disk flowers.
FRUITS: Achenes strigose; pappus 2.7–3 mm, tawny.
SIERRA DISTRIBUTION: Both slopes, Tulare and Mono Counties north.
ELEVATION: < 1830 m
HABITAT: Openings along roads, after fire or other disturbance, and often extremely abundant in recently burned areas; chaparral and montane zones.
FLOWERING TIME: Jul–Sep
NOTES: Formerly *Aster integrifolius*.

Eurybia radulina

Roughleaf Aster

Asteraceae (Sunflower family)

Perennial herb 1–7 dm, stem with long, wavy hairs; rhizomatous. Plants form extensive patches, especially after fire.

LEAVES: Cauline and alternate, sessile, 3–8 cm, ovate to obovate, and coarsely toothed. Bases wedge-shaped or clasping stem; strigose/scabrous above and scabrous below.

FLOWERS: Heads in flat-topped clusters; phyllaries in 4–5 series. Ray flowers 9–15, rays 8–13 mm, white (sometimes tinged pale pink or purple); disk flowers 30–70.

FRUITS: Achenes strigose, with pappus 2.7–3 mm, tawny.

SIERRA DISTRIBUTION: West slope, Madera County north.

ELEVATION: < 1830 m

HABITAT: Openings in conifer forest, often abundant in recently burned areas; chaparral and montane zones.

FLOWERING TIME: Jul–Sep

NOTES: Formerly *Aster radulinus*. Large stands of this handsome aster are sometimes found in full bloom on burned slopes where the plants were not apparent before the fire. It seems more likely these fully mature, blooming plants originated from rhizomes that were dormant in the soil rather than from seeds. More information on this phenomenon would be interesting. How long can the rhizomes live underground without emerging?

Euthamia occidentalis

Western Goldenrod

Asteraceae (Sunflower family)

Erect perennial herb < 2 m; rhizomatous, forming large patches, with stems sometimes ± white.

LEAVES: Alternate, sessile, < 10 cm x 6 mm wide, linear to lance-linear, entire, and resin-dotted. Leaves toward base deciduous.

FLOWERS: Heads radiate, ± sessile in dense and sometimes flat-topped clusters. Involucres 3–5 mm; phyllaries in 3–4 series. Flowers yellow; ray flowers 15–28, rays 1.5–2.5 mm; disk flowers 6–18, corollas 3–4 mm.

FRUITS: Achenes ± 1 mm. Pappus bristles 25–45, in 1–2 series, with inner pappus 2.5–3 mm; outer 1.5 mm, has fewer bristles.

SIERRA DISTRIBUTION: Both slopes, all.

ELEVATION: < 2300 m

HABITAT: Moist or wet areas, such as marshes, meadows, streamsides; foothill, montane, and sagebrush zones.

FLOWERING TIME: Jul–Nov

NOTES: Formerly *Solidago occidentalis*. Attracts a plethora of native insects in the fall when few other flowers are out.

Gnaphalium palustre

Western Marsh Cudweed

Asteraceae (Sunflower family)

Cottony-looking annual herb 3–15 cm; erect or with decumbent branches, and woolly-tomentose.

LEAVES: Alternate, 1.3–5 cm x 3–8 mm wide, spoon-shaped to oblanceolate; surfaces tomentose.

FLOWERS: Heads in dense terminal and axillary clusters, disciform; there are leaflike bracts below heads 4–12 mm x 1.5–4 mm wide. Involucres 2.5–4 mm; phyllaries in 3–5 series, whitish to tan to brown. Disk flowers 4–7, bisexual, and surrounded by 40–130 pistillate flowers.

FRUITS: 5–10 mm, hairless; pappus 7–10 mm, brown.

SIERRA DISTRIBUTION: Both slopes, all.

ELEVATION: < 3590 m

HABITAT: Moist or vernally moist sites; subalpine zone and below.

FLOWERING TIME: May–Oct

NOTES: Common.

Grindelia camporum

Great Valley Gum Plant

Asteraceae (Sunflower family)

Erect, many-branched perennial herb 6–25 dm.

LEAVES: Alternate, simple, 2–15 cm, toothed or entire. Primarily cauline at full bloom (basal leaves withered then), smaller distally. Blades clasping or narrowed at base, sessile, ovate or lanceolate, hairless, and resinous.

FLOWERS: Involucres 10–22 mm wide and resinous, oozing a white gummy substance; phyllaries in 5–7 series, coiled 180–360°, or outer one(s) may be spreading to reflexed. Ray flowers 25–39 (but sometimes none), rays 5–11 mm.

FRUITS: 2–5 mm, white to gold-brown on ridged top of fruit; pappus awns 2–6.

SIERRA DISTRIBUTION: All, sporadic.

ELEVATION: < 1400 m

HABITAT: Disturbed areas, sandy or saline valley soils; foothill, montane, and sagebrush/pinyon-juniper zones.

FLOWERING TIME: May-Nov

NOTES: Native American uses for this plant are many. Boiled leaves and flower heads are used by some tribes for healing symptoms of poison oak and for wounds and sores, and the Miwok have used the resinous budding flower heads to treat blood disorders.

Grindelia squarrosa var. *serrulata* (non-native)

Curly-Cupped Gum Plant

Asteraceae (Sunflower family)

Many-branched biennial herb 1–6 dm; decumbent to erect.
LEAVES: Alternate, simple, 1.5–7 cm, mostly seen only on stems at full bloom (basal leaves withered then). Leaf margins crenate (round-toothed), and each tooth has a yellowish bump near tip.
FLOWERS: Involucres 10–17 mm wide and resinous with white gummy exudate; phyllaries arranged in 5–6 series and coiled ± 360°. Ray flowers 24–36 (or none), rays 8–10 mm.
FRUITS: 2.3–3 mm, light brown or yellowish; pappus awns 2–3 on a truncate fruit.
SIERRA DISTRIBUTION: East slope more than west, from Kern and Inyo Counties north to Lassen County; however, grows at lower elevations on the western slope in the northern Sierra.
ELEVATION: 700–2300 m
HABITAT: Disturbed areas, roadsides, streamsides. Could be found in any zone except subalpine and alpine.
FLOWERING TIME: Jul–Sep
NOTES: Native to US, but not CA (Wyoming to New Mexico).

Hazardia whitneyi var. *whitneyi*

Whitney's Goldenbush

Asteraceae (Sunflower family)

Perennial herb or subshrub 2–5 dm; resinous, with stems leafy, and hairless to scabrous glandular.
LEAVES: Alternate, sessile, 25–50 mm, oblong to oblanceolate, and serrate.
FLOWERS: Heads radiate; involucres 8–12 mm wide, bell-shaped; phyllaries 5–12 mm, lance-linear, acute, and sometimes recurved, grading into leaves. Ray flowers 5–18, rays 5–8 mm, corollas 9–13 mm; disk corollas 15–30, corollas 8–10 mm.
FRUITS: 5–10 mm, hairless; pappus 7–10 mm, brown.
SIERRA DISTRIBUTION: Mostly west slope, but no records for Placer, El Dorado, Amador, or Calaveras. East slope, Mono and Alpine Counties.
ELEVATION: 1200–3500 m
HABITAT: Open conifer forest; montane and subalpine zones.
FLOWERING TIME: Jul–Sep
NOTES: Formerly *Haplopappus whitneyi*.

Helenium bigelovii

Bigelow's Sneezeweed

Asteraceae (Sunflower family)

Erect perennial herb 3–1.3 dm; erect, hairless or sparsely hairy, and branched toward top if branched at all.

LEAVES: Basal and alternate; cauline leaves decurrent.

FLOWERS: Disk globose, brownish or yellowish, with 250–500+ pistillate disk flowers. Ray flowers 14–20, rays 13–25 mm, 3-lobed, yellow.

FRUITS: 1.8–2.4 mm, hairy; pappus scales 6–8, 1.3–2.2 mm, awned.

SIERRA DISTRIBUTION: Both slopes, all.

ELEVATION: < 3400 m

HABITAT: Wet areas, meadows, streamsides; montane and subalpine zones.

FLOWERING TIME: Jul–Aug

NOTES: A related species called rosilla, *Helenium puberulum*, is found in wetlands < 1200 m from Madera County north, mostly in the foothill zone. The inflorescence looks like a ball because the ray flowers are either absent or much shorter (< 10 mm) than those of *H. bigelovii*.

Helianthella californica var. *nevadensis*

California Helianthella

Asteraceae (Sunflower family)

Hairless to hairy perennial herb 3–6 dm tall.

LEAVES: Basal and cauline 1.5–4 cm wide.

FLOWERS: Heads 1 or 2–3 per stem, erect. Pappus of 2 delicate awns. Involucres 15–25 mm wide; outer phyllaries few, usually longer than others, leaflike.

FRUITS: Winged, strongly compressed, with ciliate margins.

SIERRA DISTRIBUTION: All.

ELEVATION: < 2600 m

HABITAT: Grassy openings and slopes; foothills through montane zone.

FLOWERING TIME: Apr–Jul

NOTES: Compare with *Wyethia angustifolia*. *Helianthella* is most easily distinguished by the long, irregular outer phyllaries.

Helianthus annuus

Annual Sunflower

Asteraceae (Sunflower family)

Rough-hairy, highly branched annual herb that reaches heights of 3 m (approximately 10 ft!).

LEAVES: Alternate, petioled; blades 10–40 cm, ovate to lanceolate, with base wedge-shaped.

FLOWERS: Heads large, often with several per stem. Involucres 1.5–4 cm wide; phyllary 13–25 mm, 5–8 mm wide, with tips tapering to slender points.

FRUITS: 4–5 mm; pappus scales 2 longer and 0–4 shorter.

SIERRA DISTRIBUTION: West slope.

ELEVATION: < 2000 m

HABITAT: Disturbed areas, roadsides; foothill and montane zones.

FLOWERING TIME: Jun–Oct

NOTES: Native to most of North America, this is the ancestral species for the large domesticated sunflower. The seeds attract many species of birds, and the flowers provide nectar for a variety of insects, including the monarch butterfly.

Helianthus californicus

California Sunflower

Asteraceae (Sunflower family)

Erect perennial herb 15–35 dm; from short rhizomes, with stem hairless, glaucous.

LEAVES: Mostly alternate, sessile or with petioles < 3 cm; blades 10–20 cm, ± lanceolate, entire or sparsely toothed, with base wedge-shaped. Leaf underside bristly hairy and gland-dotted.

FLOWERS: Heads 3–10; involucres 10–25 mm wide; phyllaries 10–25 mm x 3–5 mm wide, outer phyllaries reflexed. Paleae (bracts) on receptacle 10–11 mm, 3-lobed. Ray flowers 12–21, sterile, rays 15–30 mm. Disk flowers numerous; corollas 6–8 mm, yellow.

FRUITS: 4–5.5 mm; pappus of 2 deciduous scales, 3–4 mm.

SIERRA DISTRIBUTION: West slope, sporadic, Butte to Kern Counties.

ELEVATION: < 1850 m

HABITAT: Streamsides, springs, marshy areas; foothill and montane zones.

FLOWERING TIME: Jul–Oct

NOTES: A striking, late-season bloomer, easily distinguished by its height and moist or wet habitat.

Hemizonella minima

Miniature Tarweed

Asteraceae (Sunflower family)

Erect, glandular-hairy annual herb 1–20 cm; larger plants are branched.

LEAVES: Basal and cauline, with lower leaves opposite and upper alternate, 5–25 mm, 1–2.5 mm wide, ± linear.

FLOWERS: Ray flowers 3–5, small, enveloped by phyllaries; rays < 1 mm. Disk flowers 1–2, anthers yellow.

FRUITS: Ray fruits 1.8–2.8 mm, flattened front-to-back, bristly or hairless; disk fruits cylindrical. No pappus.

SIERRA DISTRIBUTION: All.

ELEVATION: 300–2900 m

HABITAT: Gravelly or sandy flats, rocky outcrops; foothill through subalpine zones.

FLOWERING TIME: Jul–Aug

NOTES: Formerly *Madia minima*. Often in full bloom when < 1 in tall, thus easy to overlook.

Heterotheca grandiflora

Telegraph Weed

Asteraceae (Sunflower family)

Erect, fall-blooming annual or biennial herb 1–2.5 m; branched above, hairy. Aromatic.

LEAVES: Basal in rosettes, alternate; blades 5–8 cm, oblong to oblanceolate with basal lobes. Lower leaves have bases clasping stems; upper are sessile and much more glandular.

FLOWERS: Heads in ± flat-topped clusters; involucres 6–10 mm; phyllaries in 4–6 series. Ray flowers 25–40, rays 5–8 mm; disk flowers 30–75.

FRUITS: 2–5 mm; ray fruit hairless, without pappus; disk fruit is hairy with reddish pappus.

SIERRA DISTRIBUTION: West slope.

ELEVATION: < 1200 m

HABITAT: Disturbed areas, roadsides, dry stream courses; foothill and montane zones.

FLOWERING TIME: Jun–Oct (year-round in some places)

NOTES: Extremely common, and especially noticeable in the late summer and fall when it flowers abundantly. Telegraph weed is native to CA, Nevada, and Baja CA. It is considered an invasive pest in the Hawaiian Islands.

Hieracium albiflorum

White Hawkweed

Asteraceae (Sunflower family)

Erect perennial herb 2–12 dm; usually single-stemmed, with leaves and lower stems long-hairy. Milky sap.

LEAVES: Mostly basal, 8–15 cm, oblong to oblanceolate, long-hairy; stem leaves 1–5, near stem base, often smaller.

FLOWERS: Heads numerous, liguliflorous, in open panicles; hairs on peduncles branched; involucres 7–9 mm. Each head has 5–30 flowers, ligules white, anthers and styles yellow.

FRUITS: 2–4 mm; pappus 5–7 mm, bristles tan or dull white.

SIERRA DISTRIBUTION: All.

ELEVATION: < 3300 m

HABITAT: Floor of coniferous forest; montane and subalpine zones.

FLOWERING TIME: May–Sep

NOTES: This is one of the most common native herbs of SN coniferous forests.

Hieracium horridum

Shaggy Hawkweed

Asteraceae (Sunflower family)

Perennial herb 1–4.5 dm. Plant densely white- or brown-hairy. Sap milky.

LEAVES: Basal and alternate, 3–10 cm, oblong to oblanceolate, covered with long, soft hairs, margins smooth.

FLOWERS: Peduncles with stellate hairs; bearing numerous heads, each head with 6–12 yellow, ligulate flowers. Involucres 6–9 mm; with stellate hairs as well as long, unbranched hairs with black bases.

FRUITS: 3 mm; pappus bristles 4–6 mm, tan.

SIERRA DISTRIBUTION: All.

ELEVATION: 1350–3300 m

HABITAT: Rocks and crevices, cliffs; montane through alpine zones.

FLOWERING TIME: Jun–Oct

NOTES: Shaggy hawkweed is a distinctive plant of rock outcrops and cliffs. Plants are often more shaggy-hairy than those in the photo. The long phyllary hairs with black bases are diagnostic.

Holocarpha heermannii

Heermann's Tarweed

Asteraceae (Sunflower family)

Erect annual herb < 1.2 dm; glandular, aromatic, and much branched above. Stem is densely covered with stalked glands.

LEAVES: Basal in rosettes that wither by flowering time, stem leaves alternate; linear or wider.

FLOWERS: Involucres spheric or sometimes bell-shaped; each phyllary has 25–50 gland-tipped protuberances, along with much smaller, sessile or stalked glands and glandless hairs. Ray flowers 3–13, disk flowers 9–22; anthers yellow to brownish.

FRUITS: Ray achenes 2.5–4 mm, cylindrical; disk achenes club-shaped; no pappus.

SIERRA DISTRIBUTION: West slope, Kern to Madera Counties.

ELEVATION: < 1400 m

HABITAT: Dry, open areas, disturbed sites, roadsides; foothill and montane zones.

FLOWERING TIME: May-Nov

NOTES: The scent of these strongly aromatic annuals permeates the air in late summer and early fall in the foothills. This tarweed dominates overgrazed or otherwise disturbed foothill landscapes and can be problematic for livestock as it coats animals' legs and faces with a sticky, gluey resin and is unpalatable. However, *Holocarpha* spp. provide abundant nectar and pollen for native insects during the late summer and fall, when most other flowers are long faded. Narrow tarweed, *H. virgata* ssp. *virgata* (with red or purple anthers), grows on the west slope mostly from Mariposa County north, mostly < 800 m. It has stiff, wandlike inflorescences, and 3–8 ray flowers. *Holocarpha* seeds were historically important in the diet of Native Americans—the tiny seeds are harvested by beating fruiting heads into baskets, then ground into a flour.

Hulsea algida

Pacific Hulsea

Asteraceae (Sunflower family)

Perennial herb 10–40 cm; herbage has varying degrees of long, soft hairs as well as glandular hairs.

LEAVES: Basal < 10 cm, wavy-margined and toothed; cauline leaves few and smaller.

FLOWERS: Heads 1–2 per stem; involucres 10–25 mm wide, hemispheric to obconic; phyllaries 8–15 mm x 1–3 mm wide, narrowly oblong. Ray flowers 25–60, yellow, ray corollas 10–15 mm, rays 2–4 mm, wide and puberulent. Disk flowers numerous, corollas 5–9 mm, yellow, hairless, with style tips linear-oblong and anther tips ovate.

FRUITS: Achenes 6–10 mm, cylindric or club-shaped, and sparsely hairy; pappus scales in 2 pairs, < 1.5 mm, ± equal, deeply cut, and translucent.

SIERRA DISTRIBUTION: Both slopes, Tulare and Inyo Counties north to El Dorado County.

ELEVATION: 3000–4000 m

HABITAT: Rocky sites, talus; subalpine and alpine zones.

FLOWERING TIME: Jul–Aug

NOTES: Studies of alpine flora in relation to climate change (GLORIA plots) predict that *H. algida* is likely to be lost from mountain peaks currently occupied by alpine flora. (If vegetation zones move upward 500 m or more in elevation, there would be no more alpine flora where it now exists.)

Hulsea heterochroma

Redray Alpinegold

Asteraceae (Sunflower family)

Biennial or perennial herb 50–150 cm; herbage densely glandular-hairy.

LEAVES: Basal 10–20 cm, toothed, and broadly oblanceolate. Cauline alternate, sessile, becoming gradually smaller toward top, and upper leaves ovate to lanceolate.

FLOWERS: Heads radiate, in clusters of 3–15+; involucres 11–18 mm diameter; phyllaries 10–14 mm, long-acuminate. Ray flowers 30–75+, brick red, corollas 6–10 mm, ray < 2 mm wide, hairy. Disk corollas many, 5–9 mm, yellow often tinged with red.

FRUITS: Achenes 6–8 mm; pappus scales unequal; shorter scales 1–2 mm, and longer 2–3 mm.

SIERRA DISTRIBUTION: All, distribution irregular.

ELEVATION: 300–2700 m

HABITAT: Open areas, recent burns; foothill, montane, and sagebrush zones.

FLOWERING TIME: May–Aug

NOTES: This striking plant occurs sporadically in time and space on the west and east slopes; it's never common but can emerge in locally abundant patches after fire.

Hymenoxys hoopesii

Orange Sneezeweed, Owl's-Claws

Asteraceae (Sunflower family)

Perennial herb 30–100 cm; from black rhizomes, with stems 1–4, branched above, sometimes tinged reddish purple, and hairless or hairy; flowers more than once each year.

LEAVES: Basal 5–30 cm, oblong to oblanceolate, entire; cauline leaves alternate, smaller toward top, linear to lanceolate, and hairless or hairy.

FLOWERS: Heads 1–12 per plant, in ± flat-topped or paniclelike clusters. Peduncles 3–16 cm, white-woolly below involucre; involucres 12–17 mm x 19–26 mm, hemispheric to widely bell-shaped; phyllaries in 2 series, subequal, the outer being fused at base ¼–⅓ their lengths, and the inner free. Disk flowers 100–325, most bisexual and a few staminate; ray flowers 14–26, corollas 21–45 mm, yellow to yellowish-orange.

FRUITS: 3.5–4.5 mm, obconic, 5-angled; pappus scales 5–7, 2.9–4.1 mm, ± lanceolate, not awned.

SIERRA DISTRIBUTION: Both slopes, Tulare and Inyo Counties north, patchy.

ELEVATION: 1500–3700 m

HABITAT: Meadows, streamsides, conifer forest; sagebrush, montane, and subalpine zones.

FLOWERING TIME: May–Nov

NOTES: Formerly *Dugaldia hoopesii*, before that *Helenium hoopesii*. Compare with Bigelow's sneezeweed, *Helenium bigelovii*, which has more pronounced dome-shaped disks and decurrent leaves.

Hypochaeris glabra (non-native)

Smooth Cat's-Ear

Asteraceae (Sunflower family)

Erect annual herb 1–6 dm; often many-stemmed, ± hairless, but leaves may have widely spaced prickly hairs.

LEAVES: Basal in rosettes, 2–10 cm, smooth-margined or shallowly lobed, hairless or with widely spaced prickly hairs and teeth.

FLOWERS: Involucres 8–16 mm; ray flowers scarcely protrude above involucre, rays 5–8 mm; disk flowers absent.

FRUITS: Achenes 3–4 mm; outer achenes lack beaks, inner have beaks ≤ body; both with pappus tan, plumose.

SIERRA DISTRIBUTION: West slope, all.

ELEVATION: < 1630 m

HABITAT: Dry, open areas, roadsides; foothill and montane zones.

FLOWERING TIME: Mar–Jun

NOTES: An exceedingly common weed, the rosettes often coat the ground in disturbed areas. The related non-native but less common *H. radicata*, rough cat's-ear, is a hairy perennial 1–8 dm, with lobed leaves 6–25 cm and much longer (10–15 mm) ray flowers. Fruiting plants can be distinguished readily: All achenes are beaked in *H. radicata*.

Iva axillaris

Poverty Weed

Asteraceae (Sunflower family)

Grayish, erect perennial herb 1–6 dm; rhizomatous.
LEAVES: Lower leaves opposite, upper alternate; blades 15–25 mm, elliptic or obovate, strigose or scabrous, and dotted with glands.
FLOWERS: Inverted heads of unisexual flowers have no ray flowers; male and female are in same head.
FRUITS: 2.5–3 mm.
SIERRA DISTRIBUTION: West slope, Tulare County south and Placer County north; east slope, Inyo and Mono Counties.
ELEVATION: < 2500 m
HABITAT: Moist areas, seasonally west areas, roadsides; foothill, montane, and sagebrush zones.
FLOWERING TIME: Apr–Oct
NOTES: Wind-pollinated, poverty weed produces enormous quantities of pollen grains coated with tiny spines, and is renowned as a severe allergen.

Lasthenia debilis

Greene's Goldfields

Asteraceae (Sunflower family)

Erect or sprawling annual herb < 30 cm; herbage hairy and villous toward top.
LEAVES: Opposite, simple, 1–8 cm, linear.
FLOWERS: Involucres 6–8.5 mm; phyllaries 5. Ray flowers 5–10, rays 3–5 mm, pale yellow to white; disk flowers 5-lobed.
FRUITS: < 3 mm, linear, black, hairy; pappus scales 2–4, lanceolate white or brown, with long, slender tips (pappus sometimes absent).
SIERRA DISTRIBUTION: West slope, Mariposa County south.
ELEVATION: < 500 m
HABITAT: Partly shaded areas in grasslands, oak woodlands; foothill zone.
FLOWERING TIME: Mar–May
NOTES: The species name *debilis* is from the Latin for "weak," referring to the somewhat floppy stems, not that evident in this photo. Also note the pale, short, widely spaced ray flowers and pale green foliage.

Lasthenia gracilis
Common Goldfields
Asteraceae (Sunflower family)

Erect annual herb < 40 cm, forming masses of bright yellow in good rain years.
LEAVES: Opposite, simple, 0.8–7 cm, ± linear, ± hairy.
FLOWERS: Heads 1 per stem or clustered; involucres 5–10 mm, bell- or bowl-shaped; phyllaries 4–13, not fused, and hairy. Receptacle lacks paleae, conic; ray flowers 6–13, rays 5–10 mm; disk flowers numerous.
FRUITS: < 3 mm, linear or wider at top; pappus scales 4 (sometimes 0, 2, or 6), broad, white, opaque.
SIERRA DISTRIBUTION: West slope, all.
ELEVATION: < 1500 m
HABITAT: Open areas and roadsides; foothill and lower montane zones.
FLOWERING TIME: Feb–Jun
NOTES: Separated from *L. californica* ssp. *californica* (which has pappus of 1–7 brownish, linear scales that are not expanded at the base). *L. gracilis* is the most common of the goldfields, but look at the pappus; you're bound to run into *L. californica* ssp. *californica*, California goldfields, in the same range but mostly < 1000 m. Both will occasionally lack pappus, in which case they cannot be distinguished in the field despite genetic data confirming that they are separate species.

Layia fremontii
Fremont's Tidytips
Asteraceae (Sunflower family)

Hairy, nonglandular annual herb < 40 cm.
LEAVES: Basal < 7 cm. Lower stem leaves opposite, deeply pinnately lobed, lobes < 30; upper leaves alternate.
FLOWERS: Heads have 3–15 deep yellow ray flowers with rays 5–18 mm, each enfolded by a phyllary, with tip white or lighter yellow; disk flowers 4–100, anthers purple.
FRUITS: Fruits hairless. Disk fruits have pappus scales lanceolate; ray fruits have 9–12 white to brownish pappus scales with long, slender tips.
SIERRA DISTRIBUTION: West slope, Tulare County north.
ELEVATION: < 800 m
HABITAT: Open grassy areas or sites with shallow or heavy soil, rocky sites, serpentine; foothill zone.
FLOWERING TIME: Feb–May
NOTES: Often found just upslope of vernal pools in great numbers, spectacular!

Layia glandulosa

White Layia

Asteraceae (Sunflower family)

Glandular-hairy annual herb 3–60 cm, with stem often purplish (but not purple-streaked). May be spicy-scented or unscented.

LEAVES: Basal < 10 cm; lower stem leaves opposite, toothed or lobed; upper leaves alternate.

FLOWERS: Heads have ray flowers 3–14, white (sometimes yellow), rays 3–22 mm, each enfolded by a phyllary; disk flowers 17–105, anthers yellow or brownish.

FRUITS: Disk fruits short-hairy; pappus scales 10–15, long, plumose at tips. Ray fruits hairless.

SIERRA DISTRIBUTION: All, sporadically occurring on both slopes (see notes).

ELEVATION: < 2700 m

HABITAT: Open, sandy, gravelly areas; foothill, montane, and sagebrush zones.

FLOWERING TIME: Feb–Jul

NOTES: No records exist for counties between Madera and Plumas (except for Nevada County at Nevada City and Truckee) on the west slope. If you find localities within this gap, document and post on Calflora or iNaturalist!

Layia pentachaeta ssp. *pentachaeta*

Sierra Tidytips

Asteraceae (Sunflower family)

Erect, glandular-hairy annual herb 5–100 cm. Strongly scented.

LEAVES: Basal < 11 cm; lower stem leaves opposite, upper alternate, irregularly pinnately lobed.

FLOWERS: Heads have 4–14 deep yellow ray flowers with ligules 3–26 mm, each enfolded by a phyllary; disk flowers 7–125, anthers yellow to brownish.

FRUITS: Ray fruits have no pappus; disk fruits have 1–22 bristles, and are plumose on inner side (but sometimes pappus is absent).

SIERRA DISTRIBUTION: Placer County south.

ELEVATION: < 1200 m

HABITAT: Open areas and roadsides; foothill and lower montane zones.

FLOWERING TIME: Mar–Jun

NOTES: Sometimes confused with *Madia elegans*, but the toothed leaf margin is one clear difference. *Layia pentachaeta* ssp. *albida* is a white-rayed form of this species that grows in Tulare and Kern Counties < 900 m.

Leptosyne bigelovii

Bigelow's Tickseed

Asteraceae (Sunflower family)

Hairless annual herb 10–30 cm tall with leafless (or nearly leafless) stems 1 to many, branched from base.

LEAVES: Basal or alternate (if present on stem), and 1–2 times pinnately lobed, with lobes linear. Petioles 1–5 mm; blades 2–8 cm, 1–2 mm wide. Leaf lobes have grooves running the length of the upper surface.

FLOWERS: Heads solitary and radiate. Ray flowers 5–10, rays 5–25 mm, spreading horizontally, and with no pappus. Disk flowers 2–50, with paleae attached to base of fruits, falling with fruits; pappus scales 2.

FRUITS: Ray fruits have no pappus; disk fruits ciliate with 2 short pappus scales 1.7–2.8 mm.

SIERRA DISTRIBUTION: West slope, Mariposa County south.

ELEVATION: < 2000 m

HABITAT: Open areas and roadsides in foothill woodland and chaparral; foothill zone.

FLOWERING TIME: Mar–Jun

NOTES: Formerly *Coreopsis bigelovii*.

Leptosyne stillmanii

Stillman's Tickseed

Asteraceae (Sunflower family)

Hairless annual herb 10–30 cm tall with basal and alternate leaves. Stems 1 to few.

LEAVES: 1–2 times pinnately lobed; lobes 1–3 mm wide, broader than bases.

FLOWERS: Heads solitary, with 5–8 ray flowers with ligules 8–15 mm, ovate; disk flowers 2–50, with paleae not fused to base of fruits; black-striped hairs at base of phyllaries.

FRUITS: Hairless, having no pappus or with 1–2 short awns < 1 mm.

SIERRA DISTRIBUTION: West slope, Tulare to Placer Counties.

ELEVATION: < 900 m

HABITAT: Open areas in grasslands, foothill woodland, and chaparral; foothill zone.

FLOWERING TIME: Mar–May

NOTES: Formerly *Coreopsis stillmanii*. Similar to *L. bigelovii*, but black-striped hairs at the base of the phyllaries only occur in *L. stillmanii*, which also ranges further to the north than *L. bigelovii*.

Lessingia glandulifera var. *peirsonii*

Pierson's Lessingia

Asteraceae (Sunflower family)

Erect annual herb 5–40 cm, openly branched.

LEAVES: Simple, 0.2–2.5 cm; basal leaves petioled; stem leaves alternate, sessile, woolly, and much smaller toward top of plant, margins with beadlike glands.

FLOWERS: Involucres 4–7 mm, obconic to cylindric, and woolly, with phyllary tips erect and margins with beadlike glands. Flowers 10–30 per head, yellow, all disk flowers but outer ± bilateral, with outer corolla lobes greatly enlarged; flowers appear as ray flowers.

FRUITS: Achenes 1.5–3.5 mm; pappus > body.

SIERRA DISTRIBUTION: Kern County (see notes).

ELEVATION: 300–1850 m

HABITAT: Dry sites, often sandy soils; foothill and montane zones.

FLOWERING TIME: May–Sep

NOTES: Not pictured is the more common and very similar-looking *L. glandulifera* var. *glandulifera*, which has few woolly leaves and phyllaries, and is found in similar habitats from Calaveras County south < 2200 m in the foothill and montane zones.

Lessingia leptoclada

Sierra Lessingia

Asteraceae (Sunflower family)

Erect annual herb 5–9 dm, branched above.

LEAVES: Simple; basal leaves withered by flowering time; stem leaves alternate, sessile, 0.5–5 cm, and hairless to woolly (hairiness variable). Leaves much smaller toward top of plant, with sunken glands that may be obscured by hairs.

FLOWERS: Heads radiant, in open clusters at ends of stiff, slender branchlets. Involucres 5–10 mm, hemispheric or obconic; phyllaries graduated in 4–8 series and woolly. Flowers 6–25 per head, lavender, all technically disk flowers, but peripheral flowers with corolla lobes greatly enlarged and showy.

FRUITS: 3–4 mm; pappus tan.

SIERRA DISTRIBUTION: Nevada County south.

ELEVATION: < 2100 m

HABITAT: Dry, open areas, roadsides; foothill and montane zones.

FLOWERING TIME: Jul–Oct

NOTES: A welcome splash of pink in the late summer and fall, when most other wildflowers have faded. Compare with *L. nemaclada*. Has occasionally been confused with the noxious/invasive spotted knapweed, *Centaurea stoebe* ssp. *micranthos*.

Lessingia nemaclada

Slenderstem Lessingia

Asteraceae (Sunflower family)

Erect annual herb 5–6 dm, branching from base and above, with stem tan to brown and hairless above.

LEAVES: Simple; basal leaves withered by flowering time; stem leaves alternate, sessile, 0.2–3 cm, hairless to woolly (hairiness variable), and much smaller toward top of plant. Stem leaves have sunken glands.

FLOWERS: Involucres 5–6 mm, elongate, and glandular-hairy but not woolly. Flowers 3–10 per head, white or light lavender; all disk flowers, but outer ones ± bilateral, with outer corolla lobes greatly enlarged, thus flowers appear as ray flowers.

FRUITS: 3–4 mm; pappus tan.

SIERRA DISTRIBUTION: West slope, Fresno County north.

ELEVATION: 50–2100 m

HABITAT: Dry, open areas, roadsides; foothill and lower montane zones.

FLOWERING TIME: Jul–Oct

NOTES: Compare with *L. leptoclada*.

Leucanthemum vulgare (non-native)

Ox-Eye Daisy

Asteraceae (Sunflower family)

Perennial herb 10–100 cm; erect, rhizomatous, hairless.

LEAVES: Basal, simple, 3–7 times pinnately lobed or toothed; blades 12–35 mm x 8–20 mm wide (sometimes larger). Stem leaves irregularly toothed.

FLOWERS: Involucres 12–20 cm wide; phyllaries have brown margins. Heads have 13–34 white ray flowers and 120– 200 yellow disk flowers.

FRUITS: Ray fruits 1.5–2.5 mm, and tips have small projections; no pappus.

SIERRA DISTRIBUTION: Fresno and Mono Counties north (but range expanding).

ELEVATION: < 2600 m

HABITAT: Dry to moist, open areas, roadsides; foothill, montane, and sagebrush zones.

FLOWERING TIME: Jun–Aug

NOTES: Ox-eye daisy was introduced from Europe and sometimes escapes cultivation, becoming a wildland weed that outcompetes native plants, especially in wet or moist sites. Cal-IPC notes that this daisy is expanding its range and that it is still widely sold in wildflower seed mixes. The similar-looking Shasta daisy, *L. maximum*, also non-native, is shorter (< 80 cm) and has toothed but not pinnately lobed basal leaves. Shasta daisy foliage has a foul odor, often reported as similar to cat urine.

Madia elegans

Common Madia

Asteraceae (Sunflower family)

Erect, sticky annual herb 1–25 dm, with stems simple to much branched above. Covered densely or sparsely with black or yellow, sticky, glandular hairs.

LEAVES: Opposite toward base of plant (sometimes very densely arranged) and alternate above, linear or wider, entire or somewhat toothed, and glandular-hairy.

FLOWERS: Heads showy, in open cymes. Receptacle has a fused ring of paleae between the ray and disk flowers. Ray flowers 5–21, yellow or yellow with a maroon spot at base. Disk flowers 25–50, yellow or maroon, anthers yellow or black.

FRUITS: Ray achenes 2.5–5 mm, laterally compressed or 3-angled, and hairless; disk achenes slender. Neither has pappus.

SIERRA DISTRIBUTION: All.

ELEVATION: < 2300 m

HABITAT: Dry areas in many plant communities; foothill, montane zone, and sagebrush/pinyon-juniper zones.

FLOWERING TIME: May–Sep

NOTES: One of the most variable, common, and showy annuals in the Sierra, formerly separated into several sspp., which are no longer recognized. The flower heads close at midday. Seeds are eaten by Native Californians, and fire was used to stimulate higher seed production.

Madia exigua

Small Tarweed

Asteraceae (Sunflower family)

Glandular-hairy annual herb 1–30 cm, with glands yellow (or purple); lateral branches sometimes longer than main stem.

LEAVES: Opposite toward base of plant and alternate above, < 4 cm x 0.5–2 mm wide, linear.

FLOWERS: Heads in open clusters, not showy, with slender peduncles. Ray flowers 1–8, light yellow, rays 0.7–1 mm. Disk flowers 1 (sometimes 2); corollas 1–1.8 mm, hairless, and bisexual, with anthers yellow or brown. Receptacle has half-fused paleae between the ray and disk flower, which appear as bracts clasping each disk flower.

FRUITS: Ray achenes compressed side-to-side, hairless, with tiny, curved, offset beaks; disk achenes obovoid. Neither has pappus.

SIERRA DISTRIBUTION: West slope, all.

ELEVATION: < 2500 m

HABITAT: Open grassy, sandy, clay, or gravelly areas, and often found in disturbed areas; foothill and montane zones.

FLOWERING TIME: Apr–Jul

NOTES: Compare with the similar *Hemizonella minima*, which always has yellow anthers (not brown), and hairy ray fruits that are compressed front-to-back rather than side-to-side.

Madia glomerata
Mountain Tarweed
Asteraceae (Sunflower family)

Annual herb 0.5–12 dm; stem soft-hairy to stiff-haired at base, glandular and pubescent distally, with glands yellowish or black; lateral branches often exceed main stems. Strongly aromatic.
LEAVES: Opposite below and alternate above, 2–10 cm x 2–7 mm wide, linear to lance-linear.
FLOWERS: Heads not showy and tightly grouped (glomeruli). Involucres 5.5–9 mm, narrowly ovoid or ellipsoid, long-hairy, and glandular-pubescent, with glands yellow or black. Ray flowers either absent or 1–3, rays 1–3 mm, greenish yellow to purplish. Disk flowers 1–5, corollas hairy, yellow, anthers purple. All flowers bisexual and fertile.
FRUITS: Ray and disk fruits similar, dull black, compressed; no pappus.
SIERRA DISTRIBUTION: Both slopes, all.
ELEVATION: 1000–2700 m
HABITAT: Meadows, open areas, often in sandy/gravelly soils; montane and subalpine zones.
FLOWERING TIME: Jun–Sep
NOTES: Ray flowers in the photo are longer than the usual 1–3 mm. The narrowly ovoid, tightly grouped involucres are diagnostic for *M. glomerata*.

Madia gracilis
Slender Tarweed
Asteraceae (Sunflower family)

Erect annual herb 6–100 cm, with long, soft hairs at base and yellow, purple, or black glandular hairs at top of plant.
LEAVES: Opposite and nonglandular-hairy below, and alternate and glandular-hairy above; 1–10 cm, linear or oblong.
FLOWERS: Heads in clusters, with lateral flowering branches no longer than the main stem. Involucres 5–10 mm, urn-shaped, and glandular-hairy with glands yellow, purple, or black; sometimes nonglandular hairs also present. Ray flowers 3–10, lemon yellow to greenish-yellow, rays 1.5–8 mm. Disk flowers 2–16, corollas 2.5–5 mm, hairy, and bisexual, with anthers dark purple. Receptacle has a ring of paleae between the ray and disk flowers, fused at least ½ their length.
FRUITS: Ray and disk achenes compressed side-to-side, black, purple, or mottled, hairless, with no beaks.
SIERRA DISTRIBUTION: All.
ELEVATION: < 2500 m
HABITAT: Open or partially shaded sites, often found in disturbed areas; foothill through montane zones.
FLOWERING TIME: Apr–Aug
NOTES: The lemon-yellow color of the ray flowers and the purple anthers are helpful diagnostic features.

Malacothrix californica

California Desert Dandelion

Asteraceae (Sunflower family)

Annual herb < 45 cm, with many "stems," and with all leaves basal and shaggy-hairy at base.

LEAVES: 6–14 cm, hairy at base, pinnately lobed, lobes threadlike.

FLOWERS: Heads solitary; involucres hairy; phyllary tips reddish-purple. Heads have ray flowers only; corollas 17–22 mm.

FRUITS: 2–3 mm; pappus bristles white.

SIERRA DISTRIBUTION: Southern foothills and desert edge in Tulare, Kern, and Inyo Counties; records exist for low elevations of Madera County.

ELEVATION: < 1700 m

HABITAT: Open sites in sandy soil, roadsides; foothill zone and desert edge.

FLOWERING TIME: Mar–May

NOTES: The related *M. glabrata* looks similar but is much less hairy, often branched, often has 1+ head per stem, and has leaves on the lower stem.

Microseris nutans

Nodding Silverpuffs

Asteraceae (Sunflower family)

Perennial herb 10–70 cm; from taproots, branched, and leafy at base. Sap milky.

LEAVES: Basal and on lower stem, 5–30 cm, entire or pinnately lobed.

FLOWERS: Heads liguliflorous; involucres 8–22 mm, generally mealy and black-hairy (need a 10x lens); outer phyllaries much shorter than inner, linear to triangular. Flowers 13–75, yellow.

FRUITS: Achenes 3.5–8 mm, not beaked; pappus of 15–30 scales 1–3 mm, silvery, with feathery tips 4–7 mm.

SIERRA DISTRIBUTION: Both slopes, Tulare and Inyo Counties north.

ELEVATION: 1000–3400 m

HABITAT: Moist, rocky sites and forest openings; montane, subalpine, and sagebrush zones.

FLOWERING TIME: Apr–Jul

NOTES: Distinguished by flower heads that nod when in bud (arrow in photo) and unique pappus.

Microseris sylvatica

Woodland Silverpuffs

Asteraceae (Sunflower family)

Perennial herb 15–75 cm; from taproot, mostly unbranched. Sap milky.

LEAVES: Basal and on lower stem, 10–35 cm, distinctively wavy-margined with pointed lobes.

FLOWERS: Outer phyllary bracts much smaller than inner, with rolled tips; flowers 25–100.

FRUITS: Achenes not beaked; pappus scales 5–10, tan with feathery tips.

SIERRA DISTRIBUTION: Butte, Placer, Amador, Fresno, Tulare, Kern, and Inyo Counties.

ELEVATION: < 1700 m

HABITAT: Grasslands and open areas; foothill and montane zones.

FLOWERING TIME: Mar–Jun

NOTES: Distinguished by nodding buds, recurved outer phyllaries, and unusual pappus.

Nothocalais alpestris

Alpine Lake False-Dandelion

Asteraceae (Sunflower family)

Perennial herb 3–45 cm; from thick caudex, scapose, and hairless below single flower head. Sap milky.

LEAVES: Basal 3–20 cm, entire to pinnately toothed or lobed, and hairless or white-ciliate.

FLOWERS: Heads liguliflorous; involucres 10–20 mm, hairless; phyllaries 8–35, minutely purple-dotted, with outer phyllaries wider than inner. Receptacle without paleae, hairless. Flowers 13–75, ligules yellow and longer than involucres, often reddish beneath, and withering rapidly as seeds mature.

FRUITS: 5–10 mm, brown, not beaked but sometimes narrowed at top 1–3 mm, with 30–50 soft, white, minutely barbed bristles, 6–10 mm.

SIERRA DISTRIBUTION: Tulare and Mono Counties north.

ELEVATION: 1800–3400 m

HABITAT: Forest openings, rocky slopes, meadows; subalpine and alpine zones.

FLOWERING TIME: Jul–Sep

NOTES: Compare with mountain agoseris, *Agoseris monticola*, which has the same distribution at elevations 1500–3800 m.

Oreostemma alpigenum var. *andersonii*

Anderson's Mountaincrown

Asteraceae (Sunflower family)

Perennial herb to 40 cm (sometimes taller), sparsely to densely covered with nonglandular hairs; stems often have a purple cast.

LEAVES: Mostly linear in basal rosettes, 30–120 cm x 1–6 mm, grasslike; with nonglandular hairs.

FLOWERS: Heads radiate, 1 per stem, with 10–40 ray flowers and 30–55 disk flowers; pappus bristles 25–40.

FRUITS: Cylindric, ribs 5–10, brownish, and hairless or sparsely hairy; pappus ± white, with bristles minutely barbed.

SIERRA DISTRIBUTION: Both slopes, Tulare County north.

ELEVATION: 1300–3500 m

HABITAT: Meadows, moist areas; upper montane through alpine zones.

FLOWERING TIME: Jul–Aug

NOTES: A very common plant of high-elevation meadows, formerly included in the genus *Aster*. *Oreostemma peirsonii*, endemic to the southern Sierra, appears similar but has glandular hairs, shorter stature, and occurs in rocky areas and dry meadows > 3000 m.

Orochaenactis thysanocarpha

California Mountain-Pincushion

Asteraceae (Sunflower family)

Hairy to glandular annual herb 1–25 cm, with simple or branched slender stems.

LEAVES: Opposite below and alternate above, 1–4 cm, linear.

FLOWERS: Heads have disk flowers only, each head with 4–9 disk flowers. Pappus of scales with expanded, fringed tips, fused at the base and falling as a ring. What at first appears to be 1 head is several subtended by leafy bracts.

FRUITS: 2.5–4 mm, club-shaped, ribbed; pappus falling in a ring of scales united at base.

SIERRA DISTRIBUTION: Inyo, Kern, and Tulare Counties.

ELEVATION: 2100–3800 m

HABITAT: Dry meadows, open gravelly areas; montane through alpine zones, sagebrush/pinyon-juniper zones.

FLOWERING TIME: Jun–Aug

NOTES: *Orochaenactis* is one of the few monotypic genera (only 1 species in the genus) endemic to the SN.

Packera cana

Woolly Groundsel

Asteraceae (Sunflower family)

Woolly perennial herb 10–40 cm tall, growing in tufts.

LEAVES: Basal < 4 cm, linear-oblong with margins rolled under; petioles may be indistinct. Stems leaves bractlike. All leaves gray-woolly and feltlike.

FLOWERS: Heads radiate, in clusters of < 6 heads; ray flowers 0, 8, or 13, yellow, rays 5–10 mm. Bracts below phyllaries, if present, are indistinct.

FRUITS: Cylindric, ribbed, 2.5–3.5+ mm, glabrous; pappus of white bristles with tiny barbs.

SIERRA DISTRIBUTION: Both slopes, Tulare and Inyo Counties north.

ELEVATION: 1300–3600 m

HABITAT: Flats in sagebrush plains, dry rocky areas; montane through alpine zones, sagebrush/pinyon-juniper zone.

FLOWERING TIME: Jun–Aug

NOTES: Dwarfed alpine forms might be confused with *P. werneriifolia*.

Packera pauciflora

Alpine Groundsel

Asteraceae (Sunflower family)

Perennial herb 2–4+ dm; stems 1 or in clusters of 2–3, hairless or nearly so.

LEAVES: Basal, petioled; blades 2–4 cm x 1–3 cm wide; elliptic-ovate, ovate, or somewhat kidney-shaped, with margins dentate to crenate. Lower stem leaves petioled, and upper cauline sessile.

FLOWERS: Heads usually discoid, in clusters of 4–8; phyllaries 13 or 21, 7–10 mm, green or deep reddish, and hairless. Disk flowers 60–80+, yellow to deep reddish-orange; ray flowers 8–13 (if present), rays 5–7 mm, deep orange-yellow.

FRUITS: 1–1.5 mm, cylindric, ribbed, hairless; pappus of white bristles.

SIERRA DISTRIBUTION: Both slopes, Tulare and Inyo Counties north.

ELEVATION: 1800–3300 m

HABITAT: Moist areas; subalpine and alpine zones.

FLOWERING TIME: Jul–Aug

NOTES: The SN populations are disjunct by 1200 km (746 mi) from the primary range of alpine groundsel further to the north (Wyoming, Alaska, Canada).

Packera streptanthifolia var. *streptanthifolia*

Rocky Mountain Groundsel

Asteraceae (Sunflower family)

Perennial herb 2–5 dm, with stems 1 or in clusters of 2–5, hairless or woolly at base and in axils.

LEAVES: Basal and lower cauline petioled, 2–4 cm x 1–3 cm wide, spoon-shaped to ovate, and dentate to crenate, thick and fleshy. Upper cauline leaves smaller, sessile.

FLOWERS: Heads radiate, in clusters of 6–20+; phyllaries 13 or 21, 4–7 mm, green or red-tipped, and hairless. Disk flowers 35–60+, yellow; ray flowers 8 or 13, ray 5–10 mm, deep orange-yellow.

FRUITS: 1–2.5 mm, cylindric, ribbed, and hairless; pappus bristles white.

SIERRA DISTRIBUTION: Both slopes, uncommon, Tulare to Plumas County.

ELEVATION: 2600–3800 m

HABITAT: Moist to wet sites, and damp rocky areas; subalpine and alpine zones.

FLOWERING TIME: May–Aug

NOTES: Formerly *Senecio streptanthifolius*. Occurs in wet meadows, marshy areas, and seeps in the subalpine and alpine areas of Yosemite National Park, e.g., Tuolumne and Dana Meadows, and Lyell Canyon.

Packera werneriifolia

Hoary Groundsel

Asteraceae (Sunflower family)

Woolly perennial herb 7–15 cm tall; rhizomatous, growing in tufts.

LEAVES: Basal < 4 cm, linear-oblong with margins rolled under, and petioles indistinct; stem leaves bractlike.

FLOWERS: Heads radiate, in clusters of < 6. Ray flowers 0, 8, or 13, yellow, rays 5–10 mm; bracts below phyllaries and phyllary tips purplish.

FRUITS: 1.5–2 mm, hairless; pappus bristles white, minutely barbed.

SIERRA DISTRIBUTION: Both slopes and crest, Tulare and Inyo Counties to El Dorado County.

ELEVATION: 3000–4000 m

HABITAT: Fellfields, talus slopes, sandy openings above timberline; alpine zone.

FLOWERING TIME: Jun–Aug

NOTES: Formerly *Senecio werneriifolius*. Compare with *P. cana*.

Phalacroseris bolanderi

Bolander's Dandelion

Asteraceae (Sunflower family)

Perennial herb 10–45 cm; scapose. Sap milky.

LEAVES: Basal 6–20 cm, linear to narrowly oblanceolate, entire.

FLOWERS: Heads 1 per scape, liguliflorous; involucres 7–13 mm, green and hairless. Flowers 13+, yellow, ligule 10–18 mm, withering quickly after flowering.

FRUITS: Achenes 3–4 mm; no pappus.

SIERRA DISTRIBUTION: West slope, Tulare to Nevada Counties; also on east slope, Mono County.

ELEVATION: 1600–3300 m

HABITAT: Wet meadows, fens; montane and subalpine zones.

FLOWERING TIME: Jun–Aug

NOTES: The grasslike basal leaves and 1-headed stems (scapes) help to distinguish this dandelion, as well as its habitat of saturated areas and peatlands in meadows.

Pseudobahia heermannii

Foothill Sunburst

Asteraceae (Sunflower family)

Erect, woolly annual herb 1–4 dm.

LEAVES: Alternate, 1–2 times pinnately lobed, woolly.

FLOWERS: Peduncles 2–5 cm (sometimes red-tinged); phyllaries fused together about ½ their length, with a hard crest protruding at the junction of each phyllary bract. Ray and disk flowers lack a pappus and have rings of hairs below the corolla lobes.

FRUITS: 2–2.5 mm, oblanceolate; no pappus.

SIERRA DISTRIBUTION: West slope.

ELEVATION: < 1600 m

HABITAT: Dry openings; foothill and montane zones.

FLOWERING TIME: Mar–Jun

NOTES: Forms splashes of yellow on foothill slopes in early spring. Resembles goldfields, *Lasthenia gracilis*, but foothill sunburst has alternate, lobed leaves while goldfields have opposite, entire leaves.

Pseudognaphalium beneolens

Cudweed

Asteraceae (Sunflower family)

Annual to perennial herb 3–11 dm; herbage is tomentose, giving the plant a grayish cast; stalked-glandular beneath woolly stem hairs. Scented.

LEAVES: Basal and alternate, 3–6 cm x 1.5–3.5 mm wide, smaller toward top of stem, linear, gray on both sides, with base not clasping, decurrent 5–15 mm, and stalked-glandular beneath woolly hairs. Leaves coil and curve in age, giving plant a distinctive look.

FLOWERS: Heads in loose clusters, disciform; involucres 5–6 mm, bell-shaped to inversely conical; phyllaries in 5–6 series, white, opaque, and hairless. Pistillate flowers 44–69, disk flowers 5–8.

FRUITS: Ridged, smooth or slightly roughened; pappus of ± 12 bristles that fall early.

SIERRA DISTRIBUTION: West slope.

ELEVATION: < 1950 m

HABITAT: Dry to moist open areas; foothill and montane zones.

FLOWERING TIME: Jun–Oct

NOTES: The similar-looking small-headed cudweed, *P. thermale*, is common < 2500 m on the west slope and the east slope around Tahoe. It has stems 2–7 dm that are tomentose, but not glandular beneath the woolly stem hairs, and has leaves on average slightly longer and wider (3–8 cm x 3–6 mm wide) and oblanceolate, with sessile, rather than stalked, glands beneath the woolly hairs. Heads have 35–55 pistillate flowers, 2–7 disk flowers. Where both occur, *P. thermale* grows at higher elevations than *P. beneolens*.

Pseudognaphalium californicum

California Rabbit Tobacco

Asteraceae (Sunflower family)

Perennial (sometimes annual) herb 2–13 dm; the stem has stalked glands, ± tomentose. The plant has a maple syrup scent.

LEAVES: Basal and alternate, 4–10 cm x 5–10 mm wide, green on both sides; bases often clasping and often decurrent. Leaves are stalked-glandular and sticky, and often wavy margined.

FLOWERS: Heads in rounded or flat-topped terminal clusters, disciform. Involucres 5.5–7 mm, bell-shaped to round; phyllaries in 7–10 series, white (to pink), opaque, and hairless. Heads with 105–140 pistillate flowers and 7–12 disk flowers.

FRUITS: Ridged, smooth (not bumpy), pappus of ± 12 bristles that fall early.

SIERRA DISTRIBUTION: West slope.

ELEVATION: < 2400 m

HABITAT: Dry to moist, open areas; foothill and montane zones.

FLOWERING TIME: Apr–Jun

NOTES: This plant can be identified by smell alone once you get to know it! At all life stages, even post-flowering when dry, it has a strong, sweet aroma reminiscent of maple syrup. The green foliage separates this common tall herb from other *Pseudognaphalium* spp. in the SN, all of which have gray-woolly herbage.

Pseudognaphalium luteoalbum (non-native)

Weedy Cudweed

Asteraceae (Sunflower family)

Annual herb 1–6 dm; gray-tomentose, and erect or decumbent; herbage nonglandular and unscented.

LEAVES: Basal and alternate, 1–6 cm x 2–8 mm wide, more crowded at base of stem, and smaller toward top of stem. Leaves gray-tomentose on both sides or the upper surface is hairless, with base ± clasping, decurrent 1–2 mm, and margins curled under.

FLOWERS: Heads in compact clusters 1–2 cm diameter, disciform. Involucres 3–4 mm, bell-shaped; phyllaries in 3–4 series, silvery to yellowish, papery, and hairless. Heads have 135–160 pistillate flowers, surrounding 5–10 disk flowers; corollas pink or red-tipped, the color visible through the semitransparent phyllaries.

FRUITS: Not ridged, covered with short papillate hairs.

SIERRA DISTRIBUTION: West slope.

ELEVATION: < 1100 m (likely higher)

HABITAT: Moist (to dry) disturbed areas, abundant on mudflats next to water bodies; foothill and montane zones.

FLOWERING TIME: Apr–Oct

NOTES: The native cotton-batting plant, *P. stramineum*, occurs in similar habitats < 2500 m throughout the SN, and is much like *P. luteoalbum*, but flower heads have 18–28 yellow-tipped disk corollas and 160–200 pistillate flowers. *P. luteoalbum* is native to Eurasia and has spread to the US and much of the world.

Pyrrocoma apargioides

Alpine Flames

Asteraceae (Sunflower family)

Perennial herb from a woody taproot, stem 5–18 cm, decumbent to erect.

LEAVES: Basal 4–10 cm, with irregularly cut or toothed margins (sometimes entire); stem leaves smaller, alternate, entire.

FLOWERS: Heads usually 1 per stem; phyllaries in 3–4 series. Ray flowers 11–40, 7–11 mm, often reddish below; disk flowers 45–90, corollas 5–7 mm, with pappus.

FRUITS: 3–4 angled, hairy; pappus bristles 15–60, rigid, unequal, and tan.

SIERRA DISTRIBUTION: Both slopes, Tulare and Inyo Counties north.

ELEVATION: 2200–3800 m

HABITAT: Open rocky areas, flats, and meadows; subalpine and alpine zones.

FLOWERING TIME: Jul–Sep

NOTES: A common summer wildflower in the High Sierra. Distinguished by the rosette of sharply toothed basal leaves and solitary flower heads with phyllaries arranged in 3–4 rows.

Rafinesquia californica

California Chicory

Asteraceae (Sunflower family)

Erect annual herb 2–15 dm; branched mostly above the middle, hollow stem. Sap milky.

LEAVES: Basal and also cauline, alternate, sessile, clasping, and smaller toward top of plant.

FLOWERS: Heads in cymelike or paniclelike clusters, liguliflorous; involucres 12–20 mm, cylindric to obconic; outer phyllaries 8–14, ½ the length of the inner ones, spreading to reflexed. Flowers 15–30, with ligule 5–8 mm, longer than phyllaries, white or cream, and often rose-tinged below.

FRUITS: Achenes 4–5 mm, beak 5–7 mm, slender; pappus 6–10 mm, plumose throughout bristle.

SIERRA DISTRIBUTION: West slope, sporadic from Kern to Butte Counties.

ELEVATION: < 1500 m

HABITAT: Open sites in chaparral, woodlands, conifer forest; burned areas; foothill and montane zones.

FLOWERING TIME: Apr–Jul

NOTES: In Kern and Inyo Counties at similar elevations grows *R. neomexicana*, desert chicory. Desert chicory is found in sandy soils and has rays 15–20 mm > than phyllaries, with beak < achene body, and stout, pappus bristles that are plumose only on lower 65%–80% of pappus bristles.

Raillardella argentea

Silky Raillardella

Asteraceae (Sunflower family)

Low-growing perennial herb 1–15 cm; from branched rhizomes.

LEAVES: Basal 0.7–8 cm, silky-hairy, sometimes with glands.

FLOWERS: Heads usually 1 per stem, phyllaries absent, and paleae appearing as phyllaries. No ray flowers, disk flowers 7–26, corollas 6–11 mm.

FRUITS: 5–9.5 mm; pappus scales 16–30, flattened.

SIERRA DISTRIBUTION: Tulare and Inyo Counties north.

ELEVATION: 1800–3900 m

HABITAT: Open rocky or gravelly sites; subalpine and alpine zones.

FLOWERING TIME: Jul–Sep

NOTES: Can be seen in alpine areas of Yosemite National Park.

Raillardella scaposa

Green-Leaved Raillardella

Asteraceae (Sunflower family)

Erect, glandular perennial herb 6–53 cm; from branched rhizomes.

LEAVES: Basal 1–16 cm, linear to oblanceolate, entire, and glandular, sometimes also with nonglandular hairs.

FLOWERS: Heads usually 1 per stem, can be radiate or discoid; phyllaries 0–7; paleae 8–20, in a ring within phyllaries. Ray flowers 0–5, with corollas yellowish-orange (if present), rays 5–25 mm. Disk flowers 7–44, corollas 7.5–12 mm.

FRUITS: 4.5–11 mm; pappus scales 8–30, ray fruits may have no pappus.

SIERRA DISTRIBUTION: Mostly west slope, but occurs on both; Tulare and Inyo Counties north

ELEVATION: 2000–3500 m

HABITAT: Sandy sites (often moist), meadow edges, rocks, flats, slopes; montane, subalpine and alpine zones.

FLOWERING TIME: Jun–Sep

NOTES: The long, leafless flowering scapes are distinctive.

Rudbeckia californica

California Coneflower

Asteraceae (Sunflower family)

Erect perennial herb 6–18 dm; from stout rhizomes, usually unbranched, hairless.

LEAVES: Basal and alternate, petioled toward base and sessile toward top of stem, 10–60 cm x 5–15 cm wide. Blades lanceolate to ovate or elliptic, toothed to lobed, and hairy underneath.

FLOWERS: Heads usually 1 per stem; involucres disk-shaped; outer phyllaries 1–2 cm, larger than inner. Receptacle 1.5–6 cm, conic or columnar. Ray flowers 8–21, rays 2–6 cm, often reflexed; disk flowers numerous, corollas green-yellow to dark brownish-purple.

FRUITS: 4–5 mm, hairless; pappus scales 1–1.5 mm.

SIERRA DISTRIBUTION: West slope, Tulare County north (missing from Nevada and Placer Counties).

ELEVATION: 800–2600 m

HABITAT: Meadows, streamsides; montane zone.

FLOWERING TIME: Jul–Aug

NOTES: Spectacular plants, unmistakable. No other sunflower has these unique, exaggerated cones of disk flowers.

Senecio clarkianus

Clark's Ragwort

Asteraceae (Sunflower family)

Erect perennial herb 5–12 dm; stems single, herbage sparsely tomentose.

LEAVES: Alternate, evenly distributed; lower leaves 9–16 cm x 1–8 cm wide and pinnately lobed; upper leaves smaller, weakly clasping.

FLOWERS: Heads radiate, in ± flat-topped clusters of 8–20; phyllaries 5–8 mm and green-tipped. Ray flowers 8 or 13, rays 8–12 mm; disk flowers 35–45.

FRUITS: 2.5–3 mm, cylindric, hairless; pappus bristles white, minutely barbed.

SIERRA DISTRIBUTION: Mostly west slope, Kern County north; east slope, Mono County.

ELEVATION: 1200–2700 m

HABITAT: Meadows, streamsides; montane zone.

FLOWERING TIME: Jul–Sep

NOTES: One of the easiest *Senecio* spp. to identify, this tall woolly plant with pinnately lobed leaves stands out.

Senecio flaccidus var. *douglasii*

Douglas' Threadleaf Ragwort

Asteraceae (Sunflower family)

Shrub or subshrub 4–12 dm; herbage tomentose, gray.

LEAVES: Alternate, evenly distributed, short-petioled or sessile, 3–10 cm, and narrowly linear or pinnately linear-lobed.

FLOWERS: Heads radiate, in ± flat-topped clusters of 5–10; phyllaries 13 or 21, 7–10 mm, green-tipped. Ray flowers 8, 13, or 21, rays 10–15 mm; disk flowers 40–50.

FRUITS: Cylindric, densely hairy; pappus bristles minutely barbed.

SIERRA DISTRIBUTION: Mostly west slope, Placer County south.

ELEVATION: < 1500 m

HABITAT: Dry rocky or sandy sites, roadsides; foothill and montane zones.

FLOWERING TIME: Jul–Sep

NOTES: This attractive shrub is found along roadsides, cheerfully blooming well into the winter in favorable years, providing a source of nectar for butterflies and other insects when little else is in flower.

Senecio hydrophilus

Water Ragwort

Asteraceae (Sunflower family)

Erect biennial or perennial herb 4–10 dm; stems 1–4; herbage is blue-green and glaucous, from buttonlike caudex.

LEAVES: Alternate and fleshy. Lower leaves 5–20 cm x 2–10 cm wide, tapered to unwinged petiole; upper stem leaves smaller and sessile, elliptic to oblanceolate, and entire to denticulate.

FLOWERS: Heads discoid or radiate, in rounded or ± flat-topped clusters of 20–40+. Involucres cylindric; phyllaries 8 or 13, 5–8 mm, and black-tipped. Ray flowers 0 or 5, rays 3–5 mm; disk flowers 20–35.

FRUITS: 2.5–3 mm, hairless, ribbed; pappus of minutely barbed bristles, white or tan.

SIERRA DISTRIBUTION: Both slopes, Fresno and Inyo Counties north.

ELEVATION: 1200–3200 m

HABITAT: Meadows, marshy or swampy areas, standing water, alkaline sites; montane zone.

FLOWERING TIME: May–Sep

NOTES: The glaucous stems distinguish this plant from sweet marsh ragwort, *S. hydrophiloides*, which looks similar but is not glaucous and has red-tinged herbage, lower leaves have winged petioles, flower heads have 30–45 disk flowers, and if ray flowers are present, rays are 5–10 mm. Sweet marsh ragwort occurs sparingly in the northern and eastern SN in moist to wet, but not alkaline, sites 1200–2200 m.

Senecio integerrimus var. *exaltatus*

Columbia Ragwort

Asteraceae (Sunflower family)

Erect biennial or perennial herb 2–7 dm, cobwebby-hairy; from buttonlike caudexes, not rhizomes.

LEAVES: Basal and lower leaves nearly sessile (petioles indistinct), 6–25 cm, 1–6 cm wide, and lanceolate, oblanceolate, or elliptic; upper leaves much smaller.

FLOWERS: Heads 6–15, discoid or radiate; when radiate, heads have ± 5 ray flowers, rays 5–15 mm. Phyllaries 13 or 21, with black tips; outer phyllaries < ⅔ inner, fewer.

FRUITS: 2.5–3 mm, hairless with hairy ribs; pappus of barbed bristles.

SIERRA DISTRIBUTION: Tulare and Inyo Counties north.

ELEVATION: 500–3200 m

HABITAT: Open areas and forests; montane and sagebrush zones.

FLOWERING TIME: May–Aug

NOTES: Extremely variable, sometimes with ray flowers and sometimes without, as shown in photo. Lambstongue ragwort, *S. integerrimus* var. *major*, occurs in the same range, and has even more densely tomentose, somewhat cobwebby herbage; it usually has 5 yellow ray flowers with rays 5–8 mm, though sometimes discoid, and linear phyllaries without black tips.

Senecio scorzonella

Sierra Ragwort

Asteraceae (Sunflower family)

Perennial herb 1–5 dm; 1–4 stems from stout rhizome, woolly to tufted-woolly, sometimes becoming unevenly hairless with age.
LEAVES: Basal and lower cauline 4–12 cm x 1.5–3 cm wide, oblanceolate to lanceolate, base tapered to winged petiole, dentate. Leaves smaller toward top of stem and sessile.
FLOWERS: Heads discoid or radiate, in clusters of 10–24; involucres bell-shaped; phyllaries ± 13, 3–5 mm, green with black tips. Ray flowers 0 or 5+, ray 5–8 mm; disk flowers 10–20.
FRUITS: 2.5–3 mm, hairless; pappus of barbed bristles, tan or white.
SIERRA DISTRIBUTION: Both slopes, Tulare and Inyo Counties north to Sierra County.
ELEVATION: 1600–3500 m
HABITAT: Open areas in forests, subalpine meadows; montane, subalpine, and alpine zones.
FLOWERING TIME: Jul–Aug
NOTES: The rhizomatous growth resulting in connected patches of leaf clusters and stems, black-tipped phyllaries, tight clusters of flower heads, and the tendency for most flower heads to be discoid all help in distinguishing this from other *Senecio* spp.

Senecio triangularis

Arrowleaf Ragwort

Asteraceae (Sunflower family)

Hairless to slightly hairy, erect perennial herb to 1.5 m, with stems often clustered.
LEAVES: Elongate-triangular shape; blades 3–10 cm, sometimes slightly smaller on upper stem, with margins toothed.
FLOWERS: Heads radiate, in clusters of 10–30; phyllaries black-tipped. Ray flowers 5–12, disk flowers 35–45.
FRUITS: 2.5–3.5 mm, hairless; pappus of white or tan barbed bristles.
SIERRA DISTRIBUTION: All.
ELEVATION: < 3400 m
HABITAT: Moist to wet sites; montane, subalpine, and alpine zones.
FLOWERING TIME: Jun–Sep
NOTES: Easy to recognize by the bright green, triangular leaves and preference for wet to moist sites.

Sericocarpus oregonensis ssp. *californicus*

White-Topped Aster

Asteraceae (Sunflower family)

Erect perennial herb 4–10 dm; from fibrous caudexes; stem is hairless or hairy, and branched toward top.

LEAVES: Mostly on stem, 3–8 cm, oblanceolate or elliptic, tips acute, and hairy; leaves near base wither first.

FLOWERS: Heads radiate in flat-topped clusters on each branch; phyllaries graduated and keeled. Ray flowers ± 5, white, corollas 4–7 mm; disk flowers 3–5 mm, white to cream, anthers purplish.

FRUITS: 2.5–3.5 mm, hairy; pappus bristles 25–45, 5–7 mm.

SIERRA DISTRIBUTION: West slope, Tulare to Butte and Plumas Counties.

ELEVATION: 500–2200 m

HABITAT: Dry to moist, open sites, sometimes along streams; foothill and montane zones.

FLOWERING TIME: Jul–Sep

NOTES: Formerly *Aster oregonensis* ssp. *californicus*. The widely spaced, white ray flowers and purple anthers make this species easily recognizable. Ssp. *oregonensis* is known from Placer and Plumas Counties; it has less hairy herbage but otherwise looks similar.

Solidago elongata

Meadow Goldenrod

Asteraceae (Sunflower family)

Erect perennial herb 25–170 cm; rhizomatous, forming clones of numerous stems that form large patches in drier meadows. Stems sparsely strigose to long-soft-hairy proximally, distally more densely so.

LEAVES: Alternate, sessile, 5–15 cm, 3-nerved, with uppermost and lowermost ± oblanceolate, entire to toothed. The largest leaves occur at midstem and are lanceolate.

FLOWERS: Heads 12–500+, arranged in paniclelike arrays atop each stem; each array 10–20 cm x 4–8 cm wide (length 2–3 x width), club-shaped or narrowly pyramidal. Involucres 2.5–3.5 mm; phyllaries lanceolate, with the outer ¼–⅓ length of the inner ones. Ray flowers 8–15, ray 1.5–2 mm; disk flowers 5–12, corollas 2.5–3.5 mm.

FRUITS: 1–1.5 mm, obconic, compressed, and strigose; pappus bristles 25–45, barbed.

SIERRA DISTRIBUTION: All.

ELEVATION: 500–3200 m

HABITAT: Moist areas, dry parts of meadows, streamsides; all zones but foothill and alpine.

FLOWERING TIME: May–Oct

NOTES: Formerly *Solidago canadensis* ssp. *elongata*. The most common *Solidago*, often forming large continuous stands in meadows. *S. spectabilis*, showy or Nevada goldenrod, occurs in the southern Sierra and east slope, often in alkaline sites; it differs as follows: not rhizomatous, largest leaves are < 25 cm and at base of stem, herbage is ± hairless, and the plant is up to 2 m tall. *S. altissima* ssp. *altissima*, late goldenrod, is an escaped cultivar that could be mistaken for *S. elongata*, but flower heads are more numerous (100–1200+) in wide, pyramidal arrays 5–20 cm x 2–25 cm wide; branches of arrays are divergent to recurved, with flowers usually attached on one side only (secund). Found in disturbed areas < 2800 m.

Solidago multiradiata

Alpine Goldenrod

Asteraceae (Sunflower family)

Perennial herb 10–30 cm; from woody, branched caudexes, hairless or sparsely hairy near base of plant, and hairy toward top of plant.

LEAVES: Basal and alternate, ciliate, with largest (≤ 12 cm) leaves at base of stem, linear to spatulate with winged petioles. Upper leaves smaller, sessile to clasping, linear to ovate, and long-ciliate.

FLOWERS: Heads 4–75, in flat to rounded terminal inflorescences, or inflorescences can be more elongated and paniclelike. Flowering branches arise from leaf axils, and such branches do not exceed the main stem. Involucres 4–8 mm; phyllaries subequal or unequal, outer phyllaries linear-lanceolate to lanceolate and ⅔ the length of inner ones, with margins ciliate and acute to acuminate. Ray flowers 12–18, rays 3–4 mm x 0.5–1 mm; disk corollas 3–5 mm.

FRUITS: 2–3 mm, obconic, and strigose; pappus bristles 25–45, barbed.

SIERRA DISTRIBUTION: Both slopes, Tulare and Inyo to Plumas Counties.

ELEVATION: 1400–4150 m

HABITAT: Dry to moist areas, sometimes rocky; subalpine and alpine zones, occasionally in montane zone on west slope.

FLOWERING TIME: Jun–Sep

NOTES: Arrangement of flower heads is variable, usually in headlike or flat-topped clusters, but they can be more elongated and look quite different; the length and width of ray flowers are also variable. *Solidago* spp. are important for bees and serve as host plants for many species of butterflies and moths.

Solidago velutina ssp. *californica*

California Goldenrod

Asteraceae (Sunflower family)

Perennial herb with stems 2–150 cm; ascending to erect, rhizomatous; herbage densely short-hairy to strigillose.
LEAVES: Basal and alternate. Basal and lower leaves ≤ 14 cm, linear-oblanceolate to oblanceolate, serrate, with leaf bases tapering to winged petioles, and often still green at flowering time; upper stem leaves much smaller and entire.
FLOWERS: Heads 30–500 per inflorescence; inflorescences ± 1-sided, pyramidal, or wand-shaped. Involucres 3–5 mm; phyllaries in 3–5 series, clearly unequal, with the outer phyllaries ¼–⅓ length of inner ones, and sparsely strigillose. Ray flowers 6–11, rays 3–6 mm; disk flowers 5–17, corollas 3–5 mm.
FRUITS: 0.7–1.5 mm, strigose.
SIERRA DISTRIBUTION: Both slopes, all.
ELEVATION: < 2500 m
HABITAT: Dry to moist areas, grassy openings or sandy and gravelly areas, in woodlands, grasslands; foothill and montane zones.
FLOWERING TIME: Jul–Oct
NOTES: Great Basin goldenrod, *S. velutina* ssp. *sparsiflora*, occurs primarily in the Great Basin and east of the SN, but it has been recorded in the Sierra in Inyo, Alpine, Placer, and Plumas Counties. Distinguished from California goldenrod by its less hairy herbage and hairless phyllaries.

Stenotus acaulis

Stemless Goldenweed

Asteraceae (Sunflower family)

Mat-forming subshrub 3–21 cm; leaves and stems hairless or scabrous, and sometimes with stalked glands.
LEAVES: Basal and alternate, oblanceolate, stiff.
FLOWERS: Heads 1–4 per stem. Ray flowers 5–15, 6–12 mm; disk flowers 17–20, corollas 5–9.5 mm, with pappus of bristles.
FRUITS: 2–4.5 mm, hairless to silky; pappus tan or white.
SIERRA DISTRIBUTION: Fresno and Inyo Counties north.
ELEVATION: 1900–3660 m
HABITAT: Dry, open, often rocky sites; subalpine, alpine, and sagebrush/pinyon-juniper zones.
FLOWERING TIME: May–Aug
NOTES: The common name notwithstanding, this plant does have stems!

Stephanomeria exigua ssp. *coronaria*

Small Crown Wirelettuce

Asteraceae (Sunflower family)

Annual herb 1–20 dm; 1 main stem with spreading branches, hairless or sparsely pubescent. Sap milky.

LEAVES: Basal 3–10 cm, linear to oblanceolate, pinnately lobed with lobes pointing toward base, and withered at flowering time. Stem leaves much smaller, alternate, and bractlike.

FLOWERS: Heads single or in clusters along branches, liguliflorous; peduncles 2–10 mm. Involucres 5–7 mm; outer phyllaries appressed (not reflexed). Flowers 5–11 per head, ligules white to pink.

FRUITS: 2.3–3.1 mm; pappus bristles 7–20, white, basal half of each bristle slightly widened and fused at base in 2s or 4s; upper portion of bristle plumose from about midway to tip.

SIERRA DISTRIBUTION: West slope, Kern to Madera County; East slope, Inyo and Mono Counties.

ELEVATION: < 2800 m

HABITAT: Open areas, often disturbed; foothill, montane, and sagebrush/pinyon-juniper zones.

FLOWERING TIME: Jun–Nov

NOTES: Compare with *S. virgata* ssp. *pleurocarpa*—the differences in mature fruits are diagnostic: *S. virgata* ssp. *pleurocarpa* has slightly larger achenes, without grooves, pappus of 23–18 white bristles, which are entirely plumose from base to tip, each bristle free, falling individually rather than in 2s to 4s. If you have a hard time seeing the grooves, focus on the pappus.

Stephanomeria lactucina

Woodland Wirelettuce

Asteraceae (Sunflower family)

Perennial herb 5–60 cm; from rhizomes, stem 1 (up to 4), with branches erect or ascending, and located toward top of plant if present, hairless. Sap milky.

LEAVES: Basal and alternate, 3–8 cm, linear to narrowly lanceolate, with margin sparsely toothed or entire. Leaves green at flowering and not reduced above.

FLOWERS: Heads 1 per branch, peduncles 1–5 cm; involucres 12–14 mm, and outer bracts unequal, ½ phyllary length. Flowers 7–10 per head, ligules pink.

FRUITS: Seeds 5–6 mm, light tan, smooth-faced but grooved; pappi 25–30, plumose from base to tip.

SIERRA DISTRIBUTION: Mostly west slope, Lassen County to Kern County.

ELEVATION: 1100–2300 m

HABITAT: Dry, sandy openings in upper montane forests; montane zone.

FLOWERING TIME: Jun–Aug

NOTES: *S. lactucina* could be mistaken for another perennial, the more widespread narrow-leaved wirelettuce, *S. tenuifolia*. The latter has grayish, slenderer leaves that are reduced toward top of stem, 1–5 stems, and 4–5 flowers in each head.

Stephanomeria virgata ssp. *pleurocarpa*

Wand Wirelettuce

Asteraceae (Sunflower family)

Annual herb with stems 5–20 dm; branches wandlike, and hairless (occasionally tomentose). Sap milky.

LEAVES: Basal 3–10 cm, oblanceolate to spatulate, pinnately lobed, with lobes pointing toward base, and dried up by flowering time. Stem leaves much smaller, alternate, and bractlike.

FLOWERS: Heads single or in clusters on branches, liguliflorous, peduncles 3–10 mm, with bracts. Involucres 5–6 mm, outer phyllaries appressed, involucres 6–8 mm, flowers 5–6 per head.

FRUITS: Achenes 2.2–3.6 mm, light tan to dark brown, smooth to bumpy, and not grooved; pappus bristles 23–18, white, entirely plumose from base to tip, each bristle free, falling individually rather than in 2s to 4s.

SIERRA DISTRIBUTION: West slope, all.

ELEVATION: < 1800 m

HABITAT: Roadsides, open areas, often disturbed; foothill and montane zones.

FLOWERING TIME: Jun–Nov

NOTES: Compare with *Stephanomeria exigua* ssp. *coronaria*—the differences in mature fruits are diagnostic: *S. exigua* ssp. *coronaria* has smaller achenes, with a long groove on each face, 7–20 white pappus bristles that are only plumose on the distal 60%–85% (not from base to tip), bases widened and connate in groups of 2–4, which fall together. If you have a hard time seeing the grooves, focus on the pappus.

Symphyotrichum spathulatum var. *spathulatum*

Western Mountain Aster

Asteraceae (Sunflower family)

Perennial herb 2–6 dm; from long rhizomes, forming patches; hairless to ± hairy.

LEAVES: Basal and alternate, 5–15 cm, margins entire, lanceolate to narrowly elliptic, tips acute; stem leaves smaller toward top and sessile.

FLOWERS: Heads radiate, in cymelike inflorescences of 3–10 heads. Involucres bell-shaped, 4–7 mm, phyllaries in 3–5 series. Ray flowers 15–40, rays 9–15 mm x 1–2 mm wide, lavender to violet; disk flowers 30–80, corollas 4.5–6 mm, yellow.

FRUITS: 2.5–3.5 mm, cylindric to obovoid, brown, hairy; pappus bristles with tiny barbs, 5–7 mm, ± white.

SIERRA DISTRIBUTION: Both slopes, Tulare and Inyo Counties north.

ELEVATION: 1200–2800 m

HABITAT: Dry, open, often rocky sites; montane, subalpine, and sagebrush/pinyon-juniper zones.

FLOWERING TIME: Jun–Aug

NOTES: Formerly *Aster occidentalis* var. *occidentalis*. Other varieties of *S. spathulatum* found in the SN are: var. *intermedium*, with stems 4–8 dm, inflorescences of 10–50 heads, occurring in meadows on both slopes from Fresno County north between 1200–2000 m; and var. *yosemitanum*, with stems 3–8 dm, linear leaves, heads 3–25, in meadows 1200–2100 m, Tulare and Mono Counties north.

Taraxacum officinale (non-native)

Dandelion

Asteraceae (Sunflower family)

Perennial herb; taprooted. Sap milky.

LEAVES: In rosettes, < 30 cm, irregularly pinnately lobed or toothed.

FLOWERS: Heads solitary, 2–5 cm wide, yellow, on scapes up to 40 cm, liguliflorous; outer phyllaries reflexed; flowers 40–100.

FRUITS: Brown or olive-brown to grayish; surface with sharp bumps at top ⅓: beak >> fruit body, pappus of 50–150 minutely barbed, white bristles.

SIERRA DISTRIBUTION: All.

ELEVATION: < 3400 m

HABITAT: Moist to dry disturbed areas; all zones.

FLOWERING TIME: Year-round, depending on the site

NOTES: Introduced from Europe and present as far up as the subalpine zone, in this case mostly in meadows with a history of past disturbance.

Tetradymia axillaris var. *longispina*

Longspine Horsebrush

Asteraceae (Sunflower family)

Shrub < 15 dm; stem woolly, spines 2–5 cm.

LEAVES: Main leaves function as spines and are alternate, 2–5 cm; leaves clustered in axils, 2–12 mm, linear to club-shaped, green, ± hairless.

FLOWERS: Heads 1–3 appear in axils of spines; involucres 7–9 mm, obconic; phyllaries 5, woolly, keeled. No ray flowers, disk flowers 5–7.

FRUITS: 4–5 mm, densely white-hairy, with hairs 9–11 mm; pappus slender scales ± 25, 6–7.5 mm.

SIERRA DISTRIBUTION: East slope and southern foothills, mostly Kern and Inyo Counties.

ELEVATION: 1200–2300 m

HABITAT: Sandy or gravelly soils; sagebrush/pinyon-juniper zone.

FLOWERING TIME: Apr–Jun

NOTES: This is more of a desert species, but it does extend into the eastern and southern Sierra.

Tetradymia canescens

Spineless Horsebrush

Asteraceae (Sunflower family)

Shrub or subshrub 1–8 dm; stem woolly, with hairless stripes on older plants, and no spines.

LEAVES: Alternate, < 4 cm, lanceolate to oblanceolate, and woolly; leaves clustered in axils are similar but smaller.

FLOWERS: Heads in ± flat-topped clusters of 3–6; involucres 6–8 mm, obconic or cylindric; phyllaries 4, keeled. No ray flowers, disk flowers 4, with long, showy corolla lobes.

FRUITS: 2.5–5 mm; pappus bristles 6–11 mm.

SIERRA DISTRIBUTION: East slope, from Lassen County south; west slope, Tulare and Kern Counties (Kern Plateau).

ELEVATION: 1000–3400 m

HABITAT: Dry, open, often rocky areas; montane, subalpine, and sagebrush zones.

FLOWERING TIME: Jul–Oct

NOTES: Toxic to sheep.

Tonestus peirsonii

Peirson's Serpentweed

Asteraceae (Sunflower family)

Tufted perennial herb ≤ 20 cm; from branched, woody caudexes, with stems erect, and unbranched.

LEAVES: Basal and alternate, ≤ 8 cm, entire to coarsely serrate.

FLOWERS: Heads radiate, 1–4 per stem; involucres 14–28 mm; outer phyllaries leaflike. Ray flowers 16–20, disk flowers 44–63, corollas 7.5–10.5 mm.

FRUITS: 2.5–6 mm, hairy; pappus bristles many, 6–11 mm, white.

SIERRA DISTRIBUTION: Sierra crest in Fresno, Inyo, and Mono Counties.

ELEVATION: 2900–3700 m

HABITAT: Rocky sites; subalpine and alpine zones.

FLOWERING TIME: Jul–Aug

NOTES: Formerly *Haplopappus peirsonii*; currently treated as *Lorandersonia peirsonii* in the *Flora of North America North of Mexico.*

Tragopogon dubius (non-native)

Yellow Salsify

Asteraceae (Sunflower family)

Erect annual or biennial herb 3–10 dm; simple or sparsely branched. Sap milky.

LEAVES: Basal and cauline 2–5 dm; cauline leaves sessile, sheathing, and grasslike.

FLOWERS: Heads solitary on peduncles that widen to base of flower heads, liguliflorous, closed by midday. Involucres urn-shaped in bud, bell-shaped when flowers open; phyllaries 8–13, 1.5–4 cm, longer than flowers. Flowers pale yellow (ligule).

FRUITS: Achenes 25–35 mm, ribs 5–10, beak > body; pappus tan (white), plumose; forming perfect spheres prior to dispersal.

SIERRA DISTRIBUTION: All.

ELEVATION: < 2700 m

HABITAT: Moist to dry disturbed areas; foothill and montane zones.

FLOWERING TIME: May–Sep

NOTES: An increasingly common weed. Native to Europe and present as far up as the subalpine zone. The widened peduncle is diagnostic for this species.

Uropappus lindleyi

Lindley's Silverpuffs

Asteraceae (Sunflower family)

Erect annual herb 5–70 cm. Sap milky.

LEAVES: Basal only or basal and alternate, 5–30 cm, linear or pinnately lobed; lobes linear, soft-hairy.

FLOWERS: Heads erect in bud, 1 per scape/stem, liguliflorous, hairless; involucres 10–40 mm; phyllaries in 3–4 series; flowers 5 to many, ligules 2–19 mm, equal to or barely longer than phyllaries at full bloom.

FRUITS: Achenes 7–17 mm, ribs 10, tapered to beaklike tip; pappus scales 5, 5–15 mm, smooth, silvery, and tipped by bristle 4–6 mm.

SIERRA DISTRIBUTION: Both slopes, all.

ELEVATION: < 2300 m

HABITAT: Grasslands and open areas; foothill and montane zones.

FLOWERING TIME: Mar–May

NOTES: Formerly *Microseris lindleyi*.

Wyethia angustifolia

Narrow-Leaved Mule Ears

Asteraceae (Sunflower family)

Perennial herb 3–9 dm; rough-hairy.

LEAVES: Basal leaves ± length of stems, linear-lanceolate, and tapered at both ends; stem leaves smaller.

FLOWERS: Heads usually 1 per stem, 3–4 cm wide. Ray flowers 8–21.

FRUITS: 7–8 mm, hairy; pappus scales 1–4, tipped by awns.

SIERRA DISTRIBUTION: Fresno County north.

ELEVATION: < 2050 m

HABITAT: Open areas; foothill, montane, subalpine, sagebrush zones.

FLOWERING TIME: Apr–Aug

NOTES: *Wyethia* spp. are important to a variety of insects, including bees and butterflies.

Wyethia elata

Hall's Wyethia

Asteraceae (Sunflower family)

Erect, branched perennial herb 5–10 dm; soft-hairy, from extensive rhizome system.

LEAVES: Cauline only present, alternate, blades 8–20 cm, broadly lanceolate to somewhat heart-shaped, and densely glandular-hairy below.

FLOWERS: Heads many per plant, 2.5–4 cm wide; phyllaries 25–25 mm; rays 10–20, up to 5.5 cm.

FRUITS: 8–12 mm, hairless or hairy at tip; pappus a short crown of uneven scales.

SIERRA DISTRIBUTION: West slope, Tulare to Mariposa Counties.

ELEVATION: 500–1400 m

HABITAT: Open areas and roadsides; foothill and montane zones.

FLOWERING TIME: May–Aug

NOTES: Although the range of this plant only encompasses 4 counties, it is abundant within much of its range, especially after disturbance such as fire or clearing and along roadsides. The robust rhizome system allows for rapid recovery after ground disturbance.

Wyethia helenioides

Gray Mule Ears

Asteraceae (Sunflower family)

Woolly, grayish perennial herb 2–7 dm, often becoming greener and hairless in age.

LEAVES: Basal blades 25–45 cm, elliptic to oblong-ovate, densely woolly, gland-dotted, and becoming greener and hairless in age. Cauline leaves similar to basal but smaller, alternate.

FLOWERS: Heads 1 (or 2); involucres 3.5–6 cm wide; phyllaries 16–24, with outer phyllaries 4–8 cm, leaflike, and much longer than inner ones. Ray flowers 13–21, rays 20–35 mm.

FRUITS: 12–15 mm, strigose at tips.

SIERRA DISTRIBUTION: West slope, Mariposa County north.

ELEVATION: < 1700 m

HABITAT: Open areas, grassy slopes; foothill, montane, and sagebrush zones.

FLOWERING TIME: Mar–May

NOTES: *W. helenioides* resembles *W. mollis* but is distinguished primarily by the long, leaflike outer phyllaries that far exceed the inner phyllaries and often exceed the ray flowers.

Wyethia mollis

Mule Ears

Asteraceae (Sunflower family)

Woolly perennial herb 3–5 dm (up to 10 dm), sometimes becoming hairless in age.

LEAVES: Basal blades 20–40 cm, woolly, gland-dotted, and hairless in age. Cauline leaves smaller and alternate.

FLOWERS: Heads 1–3; involucres 2–3 cm wide; phyllaries 12–22, subequal to unequal, with outer phyllaries 20–40 mm, equal or slightly longer than inner ones. Ray flowers 6–15, rays 15–45 mm.

FRUITS: 9–11 mm, hairless or minutely hairy near tip, with a short crown of pappus scales with 2–4 slender awns.

SIERRA DISTRIBUTION: Both slopes, Fresno and Inyo Counties north (plus a few locations in Kern County).

ELEVATION: 300–2750 m

HABITAT: Open areas; sagebrush, foothill, montane, and subalpine zones.

FLOWERING TIME: May–Sep

NOTES: Aptly named! *W. mollis* and *W. helenioides* are the species with leaves most like the ears of a mule.

Berberis aquifolium var. *dictyota*

Dwarf Barberry

Berberidaceae (Barberry family)

Evergreen shrub < 1 m, usually < 0.5 m, with yellow inner bark and wood.

LEAVES: Pinnately compound, 9–15 cm, entire, dull bluish-green, especially below. Leaflets 5–9, 2–6 cm, 2–3.5 cm wide, margins wavy, with 6–10 spines on each side of leaflet (with the exception of plants in the central SN that have fewer, as shown in the photo), 2–5 mm.

FLOWERS: Inflorescences dense, of > 10 flowers, 3–6 cm. Sepals arranged in 3 whorls of 3 below the petals, yellow, petal-like. Petals total 6, in 2 whorls of 3.

FRUITS: 4–7 mm, round to ovoid, blue-purple, and glaucous.

SIERRA DISTRIBUTION: West slope, Butte County south.

ELEVATION: 90–2200 m

HABITAT: Open or partially shaded areas; foothill and lower montane zones.

FLOWERING TIME: Mar–May

NOTES: Flowers have a strong, sweet fragrance. Two other varieties occur in the SN; both differ by having 10+ spines per side on each leaflet. Var. *aquifolium* tends to be a taller shrub, reaching heights of 2 m, has shiny, darker green leaves, and is confined to Plumas County; var. *repens* is a ground-hugging subshrub < 0.8 m, leaflets tend to be shiny above and dull below, and its range overlaps with var. *dictyota*.

Alnus incana ssp. *tenuifolia*

Mountain Alder

Betulaceae (Birch family)

Monoecious shrub to 9 m; usually not that tall, often forming impenetrable thickets along streams, deciduous.

LEAVES: Alternate, with deciduous stipules, and margins double-toothed (the edge is toothed, and the larger teeth have smaller teeth). Upper leaf surface has indented veins.

FLOWERS: Male and female flowers held in separate catkins on the same plant; both form the season before they bloom and overwinter, exposed, until they open prior to leaves emerging in spring. Neither have petals or sepals. Staminate flowers in clusters of long, hanging catkins 2–5 cm; stamens 4 per flower. Female catkins much shorter (5–20 mm).

FRUITS: In woody, persistent conelike structures.

SIERRA DISTRIBUTION: Both slopes, Tulare and Mono Counties north.

ELEVATION: 1200–2800 m

HABITAT: Meadow edges, streamsides; montane zone.

FLOWERING TIME: Apr–Jun

NOTES: Leaves look like those of white alder, *A. rhombifolia*, a common riparian tree, but are darker green, and the veins of white alder leaves are not indented on the upper leaf surface. White alders are erect trees with whitish bark, growing along perennial streams.

Corylus cornuta ssp. *californica*

California Hazelnut

Betulaceae (Birch family)

Deciduous shrub to small tree 4+ m; monoecious.
LEAVES: Alternate, 4–10 cm, soft-velvety, with margins double-toothed.
FLOWERS: Male and female flowers in separate catkins on the same plant; staminate catkins 4–6 cm, usually in clusters of 2–3, flowers 3 per bract, stamens 4 per flower. Pistillate catkins in small clusters of flowers and bracts, with bright red styles protruding.
FRUITS: Each female catkin produces 1–2 typical hazelnuts, each 2–3 cm wide, surrounded by velvety bracts (involucres).
SIERRA DISTRIBUTION: West slope, Tulare County north.
ELEVATION: < 2100 m
HABITAT: Moist to wet areas in shade, often along streams; montane zone.
FLOWERING TIME: Jan–Mar
NOTES: Once you feel the soft, velvety leaves of California hazelnut, you'll know this species forever. Related to the common hazel, *C. avellana*, of Europe, it's especially good with chocolate.

Adelinia grandis

Adelinia, Western Hound's Tongue

Boraginaceae (Borage family)

Erect perennial herb 3–9 dm; stem hairless and at least somewhat glaucous.
LEAVES: Basal and alternate; petioles 8–15 cm. Blades 8–15 cm x 3–10 cm wide, ovate or elliptic, with base squared off or slightly lobed (not tapered to petiole); hairless above and hairy below.
FLOWERS: Inflorescences much taller than basal leaves, pedicels 10–25 mm; calyx deeply 5-lobed; corollas 8–12 mm long, 10–15 mm diameter when looking at it face on, 5-lobed, with lobes blue, and appendages at top of corolla tube white.
FRUITS: Nutlets 4, 5–8 mm diameter.
SIERRA DISTRIBUTION: West slope, Tulare County north.
ELEVATION: < 1700 m
HABITAT: Dry sites; foothill and montane zones.
FLOWERING TIME: Feb–May
NOTES: Formerly *Cynoglossum grande*. This attractive plant blooms early, one of the first heralds of spring in the western foothills.

Amsinckia eastwoodiae

Eastwood's Fiddleneck

Boraginaceae (Borage family)

Annual herb 2–12 dm.
LEAVES: Alternate, entire, no stipules.
FLOWERS: Coiled cyme of showy orange-yellow corollas 10–20 mm, 10–14 mm wide at the widest point, with 5 lobes, calyx 5-lobed. Corolla tubes 10-veined; stigmas longer than stamens; anthers held away from the stigma.
FRUITS: Nutlets 4, 2.5–4 mm, with sharp tubercles; dorsal surface may or may not be ridged.
SIERRA DISTRIBUTION: West slope; all.
ELEVATION: < 1500 m
HABITAT: Grassy slopes, valleys; many habitats; foothill zone.
FLOWERING TIME: Jan–May
NOTES: One of the most common early bloomers in the foothills, forming masses of orange along roadsides and in disturbed areas. Of the SN fiddlenecks, Eastwood's fiddleneck is distinguished by having the longest (up to 2 cm) corollas with the widest (up to 14 mm) limbs. This species is definitely putting on a show to attract pollinators. Compare with *A. intermedia*.

Amsinckia intermedia

Common Fiddleneck

Boraginaceae (Borage family)

Annual herb 2–12 dm; herbage is green to brownish, bristly, branched or not.
LEAVES: Basal and alternate, entire, green, and coarsely hairy (to bristly).
FLOWERS: Coiled cyme of corollas 7–11 mm, orange, with or without 5 darker spots at the base of lobes; calyx 5-lobed, lobes ± equal. Corolla limbs 4–10 mm, tube exserted from calyx; anthers held close to stigmas.
FRUITS: Nutlets 4, 2–3.5 mm, sharp tubercles, with dorsal surface sometimes ridged; similar to nutlets of *A. eastwoodiae*.
SIERRA DISTRIBUTION: West slope, all.
ELEVATION: < 1700 m
HABITAT: Grassy slopes, valleys, disturbed areas; foothill and lower montane zones.
FLOWERING TIME: Feb–May
NOTES: Along with *A. eastwoodiae*, one of the most common early bloomers in the foothills, *A. intermedia* is pollinated by insects (outcrossing rather than self-pollinating), which explains the somewhat larger, showier flowers and dark spots on the corollas, which serve to guide pollinators to the nectar at the base of the corolla tube, in contrast to the tiny flowers of *A. retrorsa*.

Amsinckia retrorsa

Rigid Fiddleneck

Boraginaceae (Borage family)

Erect, grayish annual herb 2–12 dm; the softer, nonbristly hairs on the lower plant are appressed (pointing upward) or reflexed, and mixed with bristly hairs throughout.

LEAVES: Basal and alternate, entire, grayish, and bristly-hairy.

FLOWERS: Inflorescences of coiled cymes of corollas 5–8 mm, small, yellow, with 5 unmarked lobes; calyx 5-lobed, lobes ± equal. Corolla limbs 1.5–3 mm, tube does not protrude beyond calyx; styles not exserted, anthers held close to stigma.

FRUITS: Nutlets 4, 2–3 mm, densely sharp-tubercled, with dorsal surface sometimes ridged; generally similar in appearance to nutlets of *A. eastwoodiae*.

SIERRA DISTRIBUTION: West slope.

ELEVATION: < 1700 m

HABITAT: Sunny, open, often disturbed areas; foothill and montane zones.

FLOWERING TIME: Feb–May

NOTES: Self-pollinated. This small-flowered fiddleneck could be confused with the less common *A. menziesii*, if indeed *A. menziesii* occurs in the SN (this is in doubt; see comments in the Jepson eFlora descriptions for both species). Because the presence of *A. menziesii* in the SN is suspect, it is not included here as a separate entry. *A. menziesii* is distinguished by its decumbent to ascending stems that are multibranched from base, smaller corollas (4–7 mm, limbs 1–3 mm), and dark-spotted lobes that can be yellow to orange. Many records for *A. menziesii* in Calflora are likely actually *A. retrorsa*.

Andersonglossum occidentale

Western Hound's Tongue

Boraginaceae (Borage family)

Erect perennial herb 1.5–5 dm, with stems clustered, hairy, not glaucous.

LEAVES: Basal 5–15 cm x 1–4 cm wide, oblanceolate, with base tapered to winged petiole; cauline leaves numerous, sessile.

FLOWERS: Inflorescences taller than basal leaves, pedicels 4–8 mm; calyx deeply 5-lobed. Corollas 6–9 mm, 4–9 mm diameter, bell-shaped, with lobes blue (often with pink or brown tones), and bulges at base (appendages) white.

FRUITS: Nutlets 4, 5–8 mm diameter.

SIERRA DISTRIBUTION: West slope, all.

ELEVATION: 900–2600 m

HABITAT: Dry slopes in conifer forest; montane zone.

FLOWERING TIME: May–Jul

NOTES: Formerly *Cynoglossum occidentale*.

Cryptantha affinis

Side-Grooved Cryptantha

Boraginaceae (Borage family)

Annual herb 5–30+ cm; unbranched or branched above middle, strigose, with some hairs upcurved or spreading.

LEAVES: Few-leaved, with the lowest leaves opposite, 1–4 cm, and the upper alternate and reduced toward top of stem; oblong to oblanceolate, with hairs bristly and bulbous-based.

FLOWERS: Cymes single or in 2s, few-flowered, 1 to few bracts at base; calyx 1.5–2.5 mm (in fruit), somewhat constricted at top, lobes ± lance-linear, margins with ascending coarse hairs, midvein with tan bristles. Corollas tiny, 1–2 mm wide, late deciduous, with appendages miniscule.

FRUITS: Nutlets 4, 1.8–2.5 mm, ovate, asymmetric, brown to mottled, shiny, with attachment scar obviously off-center.

SIERRA DISTRIBUTION: Both slopes, all.

ELEVATION: 630–2600 m

HABITAT: Open sandy or gravelly sites in conifer forest or sagebrush scrub; montane and sagebrush zones.

FLOWERING TIME: May–Aug

NOTES: Mature nutlets are a must for ID of all *Cryptanthas*; in this case the off-center attachment scar is diagnostic.

Cryptantha echinella

Hedgehog Cryptantha

Boraginaceae (Borage family)

Erect annual herb 5–35 cm; unbranched to branched throughout, but usually few-branched; herbage is pale green, hairy, and sometimes fine-bristly.

LEAVES: Alternate, 1–5 cm, oblong to linear, sparse; upper leaf surface is strigose and ascending, rough-hairy, and lower surface has bulbous-based hairs.

FLOWERS: Sepals 1.5–2.5 mm (4–6 mm in fruit), constricted at top, and deciduous; corollas tiny, limb 1–2 mm wide.

FRUITS: Nutlets 3–4, 1.5–2.2 mm, ovate, densely covered with small, sharp spines, and dull gray, sometimes mottled, with attachment scar forked at base, not raised.

SIERRA DISTRIBUTION: Both slopes, mostly Nevada County south; a few scattered locations to the north.

ELEVATION: 700–3200 m

HABITAT: Dry, open sandy or rocky areas, disturbed sites in conifer forest; montane, subalpine, alpine, and sagebrush/pinyon-juniper zones.

FLOWERING TIME: May–Aug

NOTES: Species name from *echinos*, meaning "spines or hedgehog," and *-ella*, "diminutive," referring to the spiny nutlets.

Cryptantha flaccida

Weakstem Cryptantha

Boraginaceae (Borage family)

Erect annual herb 10–50 cm; covered with stiff white hairs, stems branched.

LEAVES: Opposite below and alternate above, 0.5–5 cm, linear, oblong, or slightly oblanceolate, strigose.

FLOWERS: Inflorescences condensed in flower and more open in fruit. Sepals 1.5–3 mm (longer in fruit), densely covered with curved hairs, especially toward the tips, often forming a characteristic tuft at base of sepals. Corollas 1–4 mm wide.

FRUITS: Nutlet 1, smooth, shiny.

SIERRA DISTRIBUTION: West slope, all.

ELEVATION: < 2300 m

HABITAT: Dry, open sandy or rocky areas; foothill woodland, chaparral, lower coniferous forest (sometimes at meadow edges).

FLOWERING TIME: Apr–Jun

NOTES: The fruiting calyx appears slim because it contains only 1 nutlet. Look for the little "beard" of white hairs at base of sepal lobes between each lobe.

Cryptantha simulans

Pine Cryptantha

Boraginaceae (Borage family)

Slender annual herb 5–45 cm; pale green, usually single-stemmed with few branches; stem is strigose, and ascending coarse hairs may be present.

LEAVES: Alternate, 1–5 cm, linear or narrowly oblanceolate, with bristles on margins and undersurface ascending, bulbous-based.

FLOWERS: Inflorescences often in 2s or 3s, flowers sparse. Calyx 2–3 mm (4–7 mm in fruit), constricted, deciduous, with calyx lobes lance-linear, margins with fine-bristly ascending hairs, and midveins thickened with numerous short, recurved bristles. Corollas tiny, 1.5–2 mm wide, deciduous, and appendages tiny.

FRUITS: Nutlets 3–4, 1.9–2.2 mm, ovate, dull gray; surface grainy with widely spaced tubercles, tips pointed to ± beaked.

SIERRA DISTRIBUTION: Both slopes but mostly west slope in the southern Sierra, all.

ELEVATION: 250–2650 m

HABITAT: Dry, gravelly or sandy sites in open conifer forest; montane zone.

FLOWERING TIME: May–Aug

NOTES: Compare with *C. echinella*. Plants look similar, but mature nutlets distinguish the 2 spp.

Greeneocharis circumscissa var. *circumscissa*

Cushion Cryptantha

Boraginaceae (Borage family)

Cushionlike annual herb 2–8 cm x 4–15 cm diameter (wider than tall); with herbage bristly-hairy; the root and sometimes the stem base are red-purple.

LEAVES: Alternate, 0.3–1.5 cm, linear, and bristly-hairy, with margins rolled under.

FLOWERS: Inflorescences single or in 3s, of 1–5 flowers, bracted (bracts are leaflike); pedicels ± 0.5 mm, and sepals fused below middle, forming a tube; corolla limbs 1–2 mm diameter.

FRUITS: Calyx 3–4 mm in fruit; nutlets 3–4, 1.2–1.7 mm, triangular-ovate or slenderer, dull or shiny, and light brown or mottled brown and gray, with tips not beaked.

SIERRA DISTRIBUTION: Common on east slope < 3650 m; less common with a spotty distribution on the west slope.

ELEVATION: 800–3700 m

HABITAT: Sandy and open rocky areas; mostly montane, subalpine, and sagebrush zones.

FLOWERING TIME: Mar–Aug

NOTES: Formerly *Cryptantha circumscissa* var. *circumscissa*. The name *circumscissa* refers to the circumscissile calyx: The fused tube formed by the lower part of the calyx breaks off along a line around its circumference, leaving a cuplike structure containing the nutlets. The other variety, Rosette cushion cryptantha, var. *rosulata*, is a tiny (< 3.5 cm), prostrate rare plant of the alpine zone in Inyo and Tulare Counties 2950–3650 m, with flat (not rolled under) leaf margins; it blooms Jul–Aug.

Gruvelia pusilla

Little Combseed

Boraginaceae (Borage family)

Erect annual herb 3–20 cm, branched near base.

LEAVES: Lower stem leaves opposite and fused at base; upper leaves alternate.

FLOWERS: Calyx lobes in fruit ± equal, longer than nutlets. Corollas tiny, limbs 0.5–2.5 mm wide, white.

FRUITS: Nutlets 1.5–3 mm, flattened, diamond-shaped, and not winged, with hooked bristles at tips.

SIERRA DISTRIBUTION: West slope, all.

ELEVATION: < 1800 m

HABITAT: Dry areas, roadsides; foothill and montane zones.

FLOWERING TIME: Mar–Jun

NOTES: Formerly *Pectocarya pusilla*. Compare with *Pectocarya penicillata*, which has differently shaped nutlets that are clearly arranged in a lower and upper pair, while *G. pusilla* nutlets are radially symmetrical in their arrangement.

Hackelia mundula

Pink Stickseed

Boraginaceae (Borage family)

Densely hairy or strigose perennial herb 4–9 dm.

LEAVES: Lower leaves oblanceolate; upper leaves ovate or lanceolate, often clasping stem.

FLOWERS: 10–18 mm wide, pink with whitish centers, fading blue.

FRUITS: Nutlets 4.5–6.5 mm. Marginal and facial prickles are similar and numerous.

SIERRA DISTRIBUTION: Both slopes, Tulare and Mono Counties north (patchy north of Mariposa County).

ELEVATION: 1600–2800 m

HABITAT: Open or shady areas; montane zone.

FLOWERING TIME: Jun–Aug

NOTES: The only truly pink-flowered stickseed in the SN, though it does fade blue as fruits mature. If flower color is confusing you, look to see that the prickles on the body of the fruit are as long as the marginal prickles. North of Alpine County you could see *H. amethystina*, which has blue to pink flowers, but they're smaller (7–12 mm wide), with smaller (4–5 mm) nutlets with < 12 facial prickles.

Hackelia nervosa

Sierra Stickseed

Boraginaceae (Borage family)

Erect to spreading perennial herb 4–7 dm; hairless above, sparsely hairy below.

LEAVES: Basal 3–12 cm, oblong to oblanceolate; lower stem leaves smaller and sessile; stem leaves further up the stem ovate to lanceolate, and the uppermost ± clasping.

FLOWERS: Inflorescences open and branched, pedicels 6–12 mm (in fruit), calyx 1.5–3.5 mm. Corolla limbs 8–18 mm, corolla tube larger than calyx, and throat not closed by the corolla appendages. Appendages 5, about the same width and length, indented at top, white.

FRUITS: Nutlets 4–6 mm, with prickles covering entire nutlet.

SIERRA DISTRIBUTION: Both slopes, Tulare and Inyo Counties north.

ELEVATION: 1600–3100 m

HABITAT: Open or shady areas in coniferous forest; montane through subalpine zones.

FLOWERING TIME: May–Aug

NOTES: Often grows with *H. micrantha*, which has smaller corollas (limbs 5–8 mm) and nutlets with prickles, mostly located at margins.

Hackelia velutina

Velvety stickseed

Boraginaceae (Borage family)

Perennial herb 4–8 dm; herbage covered with soft, spreading hairs.
LEAVES: Basal, petioled, 5–17 cm x 0.2–2 cm wide, narrowly elliptic to oblanceolate, velvety hairy. Stem leaves smaller, sessile, and the uppermost clasping.
FLOWERS: Inflorescences of coiled cymes, dense, and branched, each branch few-flowered; pedicels 6–20 mm (in fruit), calyx 2–3 mm. Corollas generally larger than calyx; tube ± purple, throat ± closed, limb 12–20 mm wide, blue to purple; appendages white, arising near base of corolla lobes, much longer than wide, long-exserted, recurved, and 2-pronged at tip, hiding anthers.
FRUITS: Nutlets 4.5–6.5 mm; prickles cover entire nutlet.
SIERRA DISTRIBUTION: Nevada County south.
ELEVATION: 1350–2750 m
HABITAT: Open or shaded areas in coniferous forest; montane through subalpine zones.
FLOWERING TIME: Jun–Aug
NOTES: The striking, white corolla appendages distinguish *H. velutina* from all other SN stickseeds. The degree to which the corolla tube exceeds the calyx varies, but the appendages give it away. Compare with *H. nervosa*.

Mertensia ciliata var. *stomatechoides*

Streamside Bluebells

Boraginaceae (Borage family)

Hairless perennial herb with stems 4–15 dm; sometimes glaucous, from branched caudexes.
LEAVES: Basal and alternate; upper leaves sessile.
FLOWERS: Inflorescences paniclelike, with calyx 1.5–4 mm, deeply lobed; corollas 10–17 mm, cylindric, blue (pinkish in bud); styles exserted.
FRUITS: Nutlets, wrinkled.
SIERRA DISTRIBUTION: Tulare County north.
ELEVATION: 90–2100 m
HABITAT: Moist areas, meadows, streamsides; montane and subalpine zones.
FLOWERING TIME: Jun–Aug
NOTES: Sagebrush bluebells, *Mertensia oblongifolia* var. *nevadensis*, is a much shorter (< 2 dm) plant, with flowers that look similar but the style does not protrude from the corolla. Found in scattered locations from Mono and Amador Counties north, elevation 1760–2510 m, blooming from Apr–Jun in moist areas and meadow edges in the montane and sagebrush zones.

Oreocarya confertiflora

Yellow-Flowered Cryptantha

Boraginaceae (Borage family)

Erect, branched perennial herb 13–45 cm; from woody caudexes; herbage is gray-green, woolly below, strigose, and sparsely bristly toward top of stem.

LEAVES: Basal in rosettes, leaves 3–12 cm, oblanceolate; stem leaves smaller and alternate.

FLOWERS: Inflorescences dense and headlike above, with smaller clusters below; not bracted. Calyx 6–10 mm (9–14 mm in fruit), strigose and bristly; corollas yellow, limbs 8–12 mm diameter.

FRUITS: Nutlets 4, 3.5–4 mm, ovate-triangular, shiny and smooth, gray, featuring margins with sharp angles.

SIERRA DISTRIBUTION: East slope and southern Sierra, Mono, Inyo, Tulare, and Kern Counties.

ELEVATION: 1400–3400 m.

HABITAT: Dry, rocky areas; montane, subalpine, alpine, and sagebrush zones.

FLOWERING TIME: May–Jul

NOTES: Formerly *Cryptantha confertiflora*. The only CA perennial within the *Cryptantha*/*Oreocarya* group with bright yellow flowers.

Oreocarya nubigena

Sierra Cryptantha

Boraginaceae (Borage family)

Biennial or perennial herb 3–30 cm; from branched, woody caudexes; stems are several, branched; herbage is strigose to densely tomentose.

LEAVES: Basal in rosettes, alternate, 2–5 cm, spoon-shaped or narrowly oblanceolate.

FLOWERS: Inflorescences are dense spheres at apex and smaller axillary clusters below; pedicels 0.5–1 mm; calyx 2.5–4 mm, bristly-hairy; corollas white, with yellow appendages.

FRUITS: Nutlets 2–4, 2.4–4.5 mm, dorsal surface rough but shiny, ventral surface smooth, with rimmed margins; fruiting calyx 4.5–8 mm.

SIERRA DISTRIBUTION: Both slopes, Tulare and Inyo Counties north to Tuolumne and Mono Counties.

ELEVATION: 2400–3900+ m

HABITAT: Sandy and rocky areas, mostly granitic, but found on volcanics in northern populations; subalpine and alpine zones.

FLOWERING TIME: Jun–Aug

NOTES: Formerly *Cryptantha nubigena*.

Pectocarya penicillata

Sleeping Combseed

Boraginaceae (Borage family)

Annual herb with stems 2–25 cm, tending to grow horizontally (decumbent), sometimes with the tips erect (and sometimes plants are entirely erect).

LEAVES: Alternate, 0.6–4 cm, linear.

FLOWERS: Calyx with sepals spreading in pairs, the upper 2 sepals longer than the lower; corollas 0.8–3 mm, white, radial.

FRUITS: Nutlets 4, in 2 pairs, margins winged, tips bristly.

SIERRA DISTRIBUTION: West slope, El Dorado County south; distribution patchy. Kern County has the most records, likely under collected north of there.

ELEVATION: 90–2100 m

HABITAT: Disturbed areas, openings; foothill and montane zones.

FLOWERING TIME: Feb–May

NOTES: Compare with *Gruvelia pusilla*.

Plagiobothrys canescens var. *canescens*

Valley Popcorn Flower

Boraginaceae (Borage family)

Prostrate or erect annual herb with stems 1–6 dm; canescent (covered with fine, white hairs), and not staining red upon crushing of stems.

LEAVES: Basal in rosettes, alternate, 1.5–5 cm.

FLOWERS: Inflorescence has leafy bracts throughout; calyx 4–6 mm, usually not circumscissile in fruit; corolla limbs 2–3 mm wide; appendages at base of corolla lobes, pale yellow.

FRUITS: Nutlets 3–4, 2–2.5 mm, round-ovate, and arched, with flattened areas between ridges on back of nutlets much larger than ridges.

SIERRA DISTRIBUTION: West slope.

ELEVATION: < 1400 m

HABITAT: Open grassy sites, disturbed areas; foothill and montane zones.

FLOWERING TIME: Mar–Jun

NOTES: Intergrades with *P. nothofulvus*. Compare with *P. tenellus*.

Plagiobothrys nothofulvus

Popcorn Flower, Foothill Snowdrops

Boraginaceae (Borage family)

Erect annual herb with branched stems 20–70 cm; herbage is villous, with hairs often rusty-reddish when young. Stains red upon crushing.

LEAVES: Basal in rosettes, leaves 3–10 cm, oblanceolate; stem leaves smaller toward top of stem and alternate.

FLOWERS: Inflorescences 2–3 times forked, with few to no bracts, in coiled racemes that elongate in fruit; calyx 2–3 mm, circumscissile, with rust-colored hairs; flowers white with white centers, limbs 3–9 mm wide.

FRUITS: Calyx circumscissile (opening at a line of weakness around midcircumference). Nutlets 1–4, generally round-ovate, wrinkled, and narrowed abruptly toward top, with prominent keel.

SIERRA DISTRIBUTION: West slope, all.

ELEVATION: < 1500 m

HABITAT: Open grassy sites in many plant communities; foothill and lower montane zones.

FLOWERING TIME: Jan–May

NOTES: One of the first wildflowers to bloom each spring in the western foothills. In a good rain year, the blossoms can blanket the foothills in white, which doubtless invoked the colloquial name "snowdrops." This is the tallest and largest-flowered of the *Plagiobothrys*, also distinguished by its circumscissile calyx (in fruit) and the fact that all stem leaves are alternate.

Plagiobothrys stipitatus var. *stipitatus*

Showy Great Valley Popcornflower

Boraginaceae (Borage family)

Ascending to erect annual herb with stems 1–5 dm; fleshy, ± hollow, with strigose herbage.

LEAVES: Cauline, alternate, the lower 2–11 cm, and smaller toward top of stem.

FLOWERS: Inflorescences open, with bracts below middle; pedicels 0–3 mm, not hollow; calyx 3 mm, and lobes linear to lanceolate. Corollas 5-lobed, white; appendages at base of lobes, yellow; limbs 5–12 mm diameter.

FRUITS: Nutlets 4, 1–2 mm, lanceolate to lance-ovate, tubercled; dorsal ridge only present toward tip of nutlet; ventral ridge V-shaped in *x's*; attachment scar basal, on a short peg.

SIERRA DISTRIBUTION: West slope, Tulare County north; east slope, Plumas County.

ELEVATION: < 400 m

HABITAT: Vernal pools, swales, wet sites; foothill, montane, and sagebrush zones.

FLOWERING TIME: Mar–Jun

NOTES: *Plagiobothrys stipitatus* var. *micranthus* is similar looking but with much smaller (2–4 mm diameter) corolla limbs. Above 400 m, var. *micranthus* is the taxon of the foothills and mountains, found up to 2130 m in vernal pools and wet sites in a variety of vegetation types. Nutlets do not differ between the 2 varieties—the nutlets shown are from var. *micranthus*.

Plagiobothrys tenellus

Pacific Popcorn Flower

Boraginaceae (Borage family)

Erect annual herb 0.5–3 dm, with spreading hairs.

LEAVES: Basal in rosettes, alternate, 1–5 cm.

FLOWERS: Coiled racemes of sparse flowers, uncoiling in maturity; calyx 3–4 mm; flowers 1.5–3 mm wide, white with yellow centers.

FRUITS: Nutlets 1–4, cross-shaped, and marked by wide, lateral stripes; papillate, with keel above nutlet scar prominent, elevated.

SIERRA DISTRIBUTION: West slope, all.

ELEVATION: < 2000 m

HABITAT: Open grassy sites, disturbed areas; foothill and montane zones.

FLOWERING TIME: Mar–Jun

NOTES: It may help to remember that the cross-shaped nutlets resemble a *t* (for *tenellus*).

Plagiobothrys torreyi var. *diffusus*

High Sierra Popcorn Flower

Boraginaceae (Borage family)

Decumbent to ascending annual herb with stem 0.5–3 dm, staining red, and with stiff, spreading hairs.

LEAVES: Basal in rosettes, alternate, 1–8 cm, oblong or elliptic; stem leaves dense.

FLOWERS: Bracts densely arranged throughout inflorescence, similar to leaves; calyx 1.5–4 mm; flowers 1.5–3 mm wide, white (stained pink) with pale yellow centers.

FRUITS: Nutlets 1.5–2.2 mm, dull gray, round/ovate, with tip narrowed, horizontal ridges obscure, and bumpy.

SIERRA DISTRIBUTION: West slope (no records for Calaveras, Amador, Placer Counties in CCH).

ELEVATION: 1200–3400 m

HABITAT: Moist or dry meadows, forest edges, open areas; foothill and montane zones.

FLOWERING TIME: Jun–Aug

NOTES: Two other varieties exist in the SN: Var. *perplexans* is found in gravel soils in chaparral of the central Sierra, blooming Apr–Jun, often after fires, and has pointed nutlet bases with prominent rows of round translucent bumps on the ridges; var. *torreyi* has erect stems, sparse leaves and inflorescence bracts, and is restricted to a few central Sierran counties (rare enough to be CRPR 1B.2).

Athysanus pusillus

Common Sandweed

Brassicaceae (Mustard family)

Delicate annual herb with stems 20–30 cm; simple or branched at base; herbage has simple and/or 2- to 4-rayed hairs.

LEAVES: Basal and alternate, short-petioled or sessile, 4–24 mm, entire or dentate, with simple and/or short-stalked, 2- to 4-rayed hairs.

FLOWERS: Inflorescence is 1-sided, with sepals 0.5–1 mm; petals 1–3 mm, white, spatulate, and occasionally absent.

FRUITS: Silicles 2–2.5 mm, round, indehiscent (or late dehiscent) on reflexed pedicels; hairs hooked, smaller hairs branched. Seed 1, ≤ 1.2 mm.

SIERRA DISTRIBUTION: Mostly west slope.

ELEVATION: < 2700 m

HABITAT: Many microhabitats, dry to moist sites; foothill, subalpine, and sagebrush zones.

Flowering time: Feb–Jun

NOTES: *Athysanus* from Greek means "without fringe" (*a* meaning "without," *thysanos* meaning "fringe"). Compare with *Thysanocarpus*, with similar, round silicles bordered by a conspicuous fringe. This small, rather elegant plant is easily overlooked but fun to find and easy to identify. In good rain years and/or in burned areas, plants can reach heights of 50 cm.

Barbarea orthoceras

American Wintercress

Brassicaceae (Mustard family)

Erect, mostly hairless perennial herb with stems 2–6 dm; rhizomatous.

LEAVES: Basal 1.5–12 cm, pinnately lobed, lateral lobes usually in 2–4 pairs, with a much larger terminal lobe that is variously margined (entire to toothed). Stem leaves are clasping and vary from entire to deeply lobed; the basal lobes that clasp the stem are usually hairy.

FLOWERS: Terminal racemes; sepals 2.5–3.5 mm, tips hairy; petals 5–8 mm.

FRUITS: 2.5–4.5 cm, linear, erect or ascending; styles 0.2–1.2 mm. Seeds 24–36, 1.2–1.5 mm.

SIERRA DISTRIBUTION: All.

ELEVATION: < 3400 m

HABITAT: Damp, moist, and wet areas in most zones.

FLOWERING TIME: Mar–Jul

NOTES: The non-native yellow rocket, *Barbarea vulgaris* (native to Eurasia), is similar in appearance but has no hairs on the basal leaf lobes or tips of sepals, and has a longer (1.5–3.5 mm), slenderer style. It is documented from a few scattered SN locations but may have been overlooked and actually be more common.

Boechera arcuata

Arching Rockcress

Brassicaceae (Mustard family)

Perennial herb 3–8 dm; from woody base, with simple and 2-rayed hairs toward base of stem.

LEAVES: Basal 2–10 cm, 2–12 mm wide, 2- to 5-rayed, with hairs 0.1–0.3 mm (need a 10x lens), short-stalked. Cauline leaves clasping, basal lobes 2–6 mm.

FLOWERS: Inflorescences of 12–70 flowers; sepals hairy; petals 9–14 mm x 2–4 mm wide, purple.

FRUITS: 6–13 cm, 1.5–2.2 mm wide, spreading, arched, hairless or sparsely hairy; fruiting pedicels have branched hairs. Seeds in 1 row.

SIERRA DISTRIBUTION: West slope; found in a few locations on east slope.

ELEVATION: 300–2000 m

HABITAT: Rocky areas, cliffs, roadsides; foothill and montane zones.

FLOWERING TIME: Mar–Jun

NOTES: Formerly *Arabis sparsiflora* var. *arcuata*. One of the showiest and most striking early spring wildflowers in the foothills.

Boechera lemmonii

Lemmon's Rockcress

Brassicaceae (Mustard family)

Perennial herb 0.5–2.5 dm; from woody caudexes, hairless at base or with short-stalked stellate hairs 0.1–0.2 mm, with 2–6 rays.

LEAVES: Basal 1–2 cm, entire or minutely dentate, with hairs 0.1–0.2 mm (need a 10x lens), short-stalked, 3- to 9-rayed. Cauline leaves 2–12, upper lanceolate, hairless or with sparse hairs; bases unlobed or with tiny basal lobes 0.1–0.5 mm.

FLOWERS: Inflorescences of 3–17 flowers; sepals hairless or with few hairs; petals 3.5–6 mm, 1.0–1.5 mm wide, purple to lavender.

FRUITS: Inflorescence becomes 1-sided in fruit, with pedicels spreading, ascending, or horizontal (not appressed), 2–6 mm, hairless or sparsely hairy. Siliques 1.6–4.4 cm, 1–1.5 mm wide, hairless. Seeds 28–44 in 1 row, 1.3–2 mm, with wings 0.1–0.5 mm wide.

SIERRA DISTRIBUTION: Both slopes and crest; Tulare and Inyo Counties north.

ELEVATION: 2000–4350 m

HABITAT: Rocky areas, talus slopes, cliffs, sandy or gravelly soil; subalpine and alpine zones.

FLOWERING TIME: Jun–Aug

NOTES: Formerly *Arabis lemmonii*. The 1-sided fruiting arrangement is a distinctive feature of Lemmon's rockcress.

Boechera lyallii

Lyall's Rockcress

Brassicaceae (Mustard family)

Tufted perennial herb ≤ 25 cm; from basal rosette.

LEAVES: Basal 1–3 cm x 1–5 mm wide, with entire margins usually with simple or hairs 0.1–0.3 mm (need a 10x lens), 2- to 3-rayed; leaf surface hairless. Cauline leaves few, lanceolate, hairless, with basal lobes 0.5–1.5 mm.

FLOWERS: Inflorescences of 2–10 flowers; sepals hairless; petals 6–8.5 mm, 1.5–3 mm wide, lavender to purple.

FRUITS: 3–5.6 cm x 1.5–2.5 mm wide, erect, appressed, hairless. Seeds in 2 rows.

SIERRA DISTRIBUTION: Both slopes, Tulare and Inyo Counties north.

ELEVATION: 2100–3720 m

HABITAT: Talus slopes, rocky areas, sandy or gravelly soil; subalpine and alpine zones.

FLOWERING TIME: Jun–Aug

NOTES: Formerly *Arabis lyallii*. Compare with *B. platysperma*, which has wider fruits with seeds in 1 row.

Boechera platysperma

Pioneer Rockcress

Brassicaceae (Mustard family)

Tufted perennial herb ≤ 4 dm; from basal rosette.

LEAVES: Basal 2–6 cm, 3–10 mm wide, entire, surfaces have hairs 2- to 5-rayed. Cauline leaves 3–12; distal leaves sometimes hairless, sessile, without basal lobes.

FLOWERS: Inflorescences of 2–7 flowers; sepals have branched hairs; petals 4–6 mm x 1–2 mm wide, white to purple.

FRUITS: 2.5–8.5 cm x 3–5.5 mm wide, ascending but not appressed, hairless, and wavy when fully mature. Seeds in 1 row.

SIERRA DISTRIBUTION: All.

ELEVATION: 1600–3000 m

HABITAT: Rocky areas, sandy or gravelly soil; montane through alpine zones.

FLOWERING TIME: Jun–Aug

NOTES: Formerly *Arabis platysperma*. Compare with *B. lyallii*, which has hairless sepals and narrower fruits. Another similar-looking plant is *B. howellii*, with stems 0.6–2 dm, basal leaves 1–7 mm wide, entire, 2-rayed and with simple hairs present on leaf margins but surfaces hairless; cauline leaves 2–4; sepals hairless, flowers white to purple, petals 4–8 mm, fruits 2.5–6.5 cm x 3–7 mm wide, ascending to erect. *B. howellii* occurs from Tulare and Inyo Counties north on both slopes, in rocky sites 1800–3800 m in alpine and subalpine zones.

Boechera pratincola

Meadow Rockcress

Brassicaceae (Mustard family)

Short-lived perennial herb with stems 2–6 dm; from rosettes of nonwoody caudexes, with sessile, 2- to 3-rayed hairs 0.15–0.6 mm toward base of stem.

LEAVES: Basal, sessile, 1.5–7 mm wide, entire or denticulate, with hairs 0.1–0.45 mm (need a 10x lens), 2- to 6-rayed. Cauline leaves 13–42, the uppermost sometimes hairy at tip, and basal lobes 1–4 mm.

FLOWERS: Inflorescences of 14–45 flowers; sepals hairless; petals 6–10 mm x 1.5–2 mm wide, lavender.

FRUITS: Siliques 4–6.5 cm x 1.5–2 mm wide; pedicels erect; fruits erect, appressed. Ovules 60–140; seeds in 2 rows, 1.4–2 mm, and wings 0.1–0.35 mm wide.

SIERRA DISTRIBUTION: Both slopes, erratic distribution, Tulare and Inyo Counties north to El Dorado County.

ELEVATION: 1900–3200 m

HABITAT: Meadows, rocky/gravelly areas, forest edges; montane, subalpine, alpine, and sagebrush/pinyon-juniper zones.

FLOWERING TIME: Jun–Aug

NOTES: Formerly *Arabis pratincola*. Uncommon. Apomictic, meaning that seeds are formed without fertilization (different from self-pollination).

Boechera rectissima

Bristly Leaf Rockcress

Brassicaceae (Mustard family)

Erect biennial or perennial herb 2–10 dm; from basal rosette.

LEAVES: Basal 1–3 cm x 3–13 mm wide, surface with hairs that are simple, < 0.8 mm, and 2- to 3-branched (< 0.5 mm); stem leaves clustered near base, with basal lobes 1–1.5 mm.

FLOWERS: Inflorescences of 10–50 flowers; pedicels 4–10 mm, turned downward, hairless; sepals hairy; petals 3–4 mm, 0.7–1.2 mm wide, white, inconspicuous.

FRUITS: 5–9 cm, 1.8–2.5 mm wide, reflexed, usually appressed to stem, hairless. Seeds in 1 row.

SIERRA DISTRIBUTION: Mostly west slope, Tulare County north.

ELEVATION: 1100–2800 m

HABITAT: Rocky areas, sandy or gravelly soil; montane through alpine zones.

FLOWERING TIME: May–Jul

NOTES: Distinguished from other *Boechera* spp. by its glabrous stems combined with large, mostly simple hairs (not branched) on basal leaves, short white petals, and markedly reflexed and appressed fruits.

Boechera retrofracta

Reflexed Rockcress

Brassicaceae (Mustard family)

Erect biennial or perennial herb; from basal rosettes, caudexes branched and not woody; stems 1 per rosette 1.5–10 dm, short-stalked; 2- to 8-rayed hairs 0.1–0.2 mm are on lower part of stem.

LEAVES: Basal 2–7 mm wide, entire or toothed, hairs 0.5–0.2 mm, short-stalked, 5- to 10-rayed. Cauline leaves numerous (15–40), alternate, uppermost hairy, usually with basal lobes 0.5–2.5 mm.

FLOWERS: Inflorescences of 15–140 flowers, hairs appressed, branched; flowers 4-merous, with sepals hairy, and petals 4–8 mm x 0.8–2.2 mm wide, white to lavender.

FRUITS: Strongly reflexed, usually appressed; inflorescences sometimes ± 1-sided in fruit. Fruiting pedicels 7–18 mm, reflexed, abruptly recurved at base. Siliques 3.5–9 cm x 0.9–1.8 mm wide, hairless or sparsely hairy, styles 0.05–0.3 mm. Ovules/seeds 60–116 in 1 row, 1–1.4 mm, and wings ≤ 0.1 mm.

SIERRA DISTRIBUTION: Both slopes, Tulare and Inyo Counties north.

ELEVATION: 900–3700 m

HABITAT: Rocky areas, grassy slopes; all zones.

FLOWERING TIME: Apr–Aug

NOTES: Formerly *Arabis holboellii* var. *retrofracta*.

Brassica nigra (non-native)

Black Mustard

Brassicaceae (Mustard family)

Annual herb 2–20 dm, with stiff, simple hairs, branched toward top.

LEAVES: Basal in rosettes, petioled, and pinnately lobed, with margins toothed; cauline leaves like basal but sessile with tapered base, and smaller toward top.

FLOWERS: Racemes terminal, flowers yellow; sepals 4, 4–7 mm, erect to ascending; petals 4, 7–13 mm x 1.5–5.5 mm wide, clawed.

FRUITS: Siliques 1–2.7 cm x 1.5–3 mm wide, erect and appressed, ± 4-angled, with valves 1-veined, terminal segment 3–6 mm with no seeds, lower segment with 2–8 seeds in 1 row, and pedicels 2–6 mm. Seeds 1.5–2 mm, spheric, brown to black.

SIERRA DISTRIBUTION: West slope.

ELEVATION: < 2000 m

HABITAT: Disturbed sites; foothill through upper montane zones.

FLOWERING TIME: Apr–Sep

NOTES: Native to Europe. Invasive. Seeds are used to make spices and condiments, making this one of the more pungent mustards. Easily confused with *Hirschfeldia incana*, which has 1–2 seeds in the terminal fruit segment, so any visible swelling in the terminal segment of maturing fruits takes you to *Hirschfeldia*.

Capsella bursa-pastoris (non-native)

Shepherd's Purse

Brassicaceae (Mustard family)

Erect annual herb 1–5 dm; stem has stellate (3- to 5-rayed) hairs except at base, where hairs are generally simple.

LEAVES: Basal 3–6 cm, in a rosette, pinnately lobed, toothed, or entire. Cauline leaves sessile, 1.5–6 cm, lobed at base or clasping stem, margins pinnately lobed, toothed, or entire.

FLOWERS: Sepals 1.5–2 mm; petals 2–4 mm, obovate, and white (sometimes tinged pink).

FRUITS: Silicles 4–9 mm, shaped like an upside-down triangle, hairless; pedicels 5–15 mm, spreading to ascending.

SIERRA DISTRIBUTION: All.

ELEVATION: < 2900 m

HABITAT: Disturbed areas; all zones, except alpine zone.

FLOWERING TIME: Jan–Oct

NOTES: Native to Eurasia, now distributed throughout much of the world. Although ubiquitous, shepherd's purse doesn't seem to be especially threatening to the native flora.

Cardamine breweri

Brewer's Bittercress

Brassicaceae (Mustard family)

Perennial herb with stems 2–6 dm; rhizomatous, of wet places, trailing, and rooting at nodes.

LEAVES: Rhizome leaves lacking; stem leaves 3–8, either simple or with 3–5 leaflets, irregularly toothed or lobed.

FLOWERS: Sepals 2–3 mm; petals 3.5–7 mm x 1.5–2.5 mm wide, white, with pedicels ascending to spreading.

FRUITS: 1.5–3.5 cm, erect. Seeds 14–28.

SIERRA DISTRIBUTION: Both slopes, Tulare and Inyo Counties north.

ELEVATION: < 3200 m

HABITAT: Wet sites; montane and subalpine zones.

FLOWERING TIME: Jun–Jul

NOTES: Compare with *Nasturtium officinale*.

Cardamine californica

Milkmaids, Toothwort

Brassicaceae (Mustard family)

Perennial herb with stems 2–6 dm; erect, from deeply buried tuberlike rhizomes; "rhizome leaves" that arise from the same rhizome are usually visible near flowering individuals.

LEAVES: Rhizome leaves simple (top inset) or with 3 (up to 7) leaflets; stem leaves 2–5, alternate, each one simple (lower inset) or with 3 (up to 5) leaflets; leaf shape highly variable.

FLOWERS: Sepals 3.5–4.5 mm; petals 8–13 mm, white to pale pink.

FRUITS: 2.2–5 cm, linear; styles 3–6 mm, with pedicels ascending. Seeds 12–22.

SIERRA DISTRIBUTION: West slope.

ELEVATION: < 1400 m

HABITAT: Shaded, dry areas; foothill and montane zones.

FLOWERING TIME: Jan–May

NOTES: One of the first wildflowers to bloom in the montane zone, emerging just after snow melts. Rock toothwort, *C. pachystigma*, occurs in a patchy distribution from Tulare County north < 2900 m in rocky (often serpentine) sites; plants are 1–3 dm, rhizome and cauline leaves are simple, and cauline leaves are high on stem below short inflorescences; petals are 14–18 mm, pink to purple, occasionally white; fruits are erect, 3.2–5.4 cm, with 10–14 seeds.

Cardamine hirsuta (non-native)

Hairy Bittercress

Brassicaceae (Mustard family)

Erect annual herb 1–4.5 cm; stiff-hairy toward base of stem. Extremely prolific!

LEAVES: Basal in rosettes, not withered at flowering time, petioles ciliate, pinnately compound with 4–7 (or 2–11) pairs of leaflets, kidney-shaped to roundish, with margins entire, wavy, dentate, or with 3–5 lobes, the terminal leaflet the largest. Stem leaves 1–4, similar to basal but smaller.

FLOWERS: Inflorescences raceme; pedicels (occasionally absent) 1.5–2.5 cm x 1–1.4 mm wide, erect, appressed, ascending or spreading, hairless. Styles 0.4–1 mm; stamens usually 4 (sometimes 6).

FRUITS: Siliques erect, appressed, and hairless. Seeds 14–90, 0.9–1.3 mm, oblong and margined.

SIERRA DISTRIBUTION: West slope, expected throughout the SN eventually.

ELEVATION: < 1600 m

HABITAT: Disturbed areas including after fires, and increasingly observed in plant communities that are not particularly disturbed; foothill and montane zones.

FLOWERING TIME: Jan–Jul

NOTES: Native to Europe. Seeds burst en masse out of capsules with a popping sound. Since a large plant can produce up to 5,000 seeds that land 0.25–5 m from the parent plant, this annual plant has spread across SN landscapes quickly and has become ubiquitous in the central SN in the last 12 years. Few-seeded bittercress, *Cardamine oligosperma*, is a native annual (or biennial) that occurs from Tulare County north on the west slope < 3300 m, in damp, shady sites. *C. oligosperma* has ascending to erect, rather than appressed, siliques, with 16–32 oblong seeds 1–1.6 mm, without margins, and always has 6 stamens.

Caulanthus coulteri

Coulter's Jewelflower

Brassicaceae (Mustard family)

Erect annual herb 1–16 dm; stem has both 2-rayed (forked) and simple hairs at base.

LEAVES: Basal 1–17 cm, oblong or oblanceolate, pinnately toothed or lobed; stem leaves sessile, entire or toothed, with base lobed or clasping stem.

FLOWERS: Inflorescences topped by a cluster of maroon, sterile flower buds. True flowers have sepals 5–15 mm, yellow-green; petals 8–25 mm, with wavy margins, white to purplish, veins purple. Filaments 3 pairs of unequal length.

FRUITS: Siliques 3.5–15 cm, cylindric, spreading to reflexed.

SIERRA DISTRIBUTION: West slope, Madera County south; east slope, Kern and Inyo Counties.

ELEVATION: < 2100 m

HABITAT: Open grassy or rocky areas; foothill and sagebrush zones.

FLOWERING TIME: Mar–Jul

NOTES: The apical cluster of maroon sterile flowers adds panache to these handsome plants.

Draba densifolia

Dense-Leaved Draba

Brassicaceae (Mustard family)

Cushion-forming perennial herb with stems 1.5–10 cm; scapose; stems (scapes) are few to many, unbranched, and mostly hairless (occasionally simple and/or 2- to 4-rayed hairs present).

LEAVES: Basal 2.5–9 mm, linear to oblanceolate, entire, ciliate, with midvein prominent, and leaf surface hairless or with hairs 2- to 4-rayed.

FLOWERS: Inflorescences of 2–10 flowers (up to 22); inflorescence axis hairless or hairy like the stem, with pedicels 1.5–10 mm, usually hairless and sometimes hairy. Sepals 4, 2–3 mm; petals 4, 2–5 mm x 1–1.7 mm, yellow.

FRUITS: Silicle 3–6 mm x 2–3 mm wide, ovate, flat, and plane (not curved or twisted when mature), with hairs simple and branched (rays 2–5 on short stalks).

SIERRA DISTRIBUTION: Both slopes, Tulare and Inyo to Nevada Counties.

ELEVATION: 1900–3650 m

HABITAT: Rocky outcrops, talus slopes, crevices, alpine ridges; alpine zone.

FLOWERING TIME: Jun–Aug

NOTES: This species is apomictic, meaning that seeds are formed without fusion of a male gamete (sperm in pollen) and a female gamete (egg in ovary). Normal-looking fruits are produced with viable seeds that are identical genetically to the parent plant. The anthers contain sterile pollen and do not dehisce.

Draba longisquamosa

Granite Draba

Brassicaceae (Mustard family)

Perennial herb 1.5–9 cm; tufted and scapose; herbage has simple (not branched) hairs.

LEAVES: Basal, rosetted, 5–20 mm, oblong, and ciliate, with both surfaces simple-hairy.

FLOWERS: Inflorescences of 4–16 flowers, with axis hairy like the stem, and pedicels 3–7 mm, hairy. Sepals 1.5–2.2 mm; petals 3.5–5 mm x 1.5–2 mm wide, yellow. Styles 0.4–1 mm.

FRUITS: Silicle 3.5–7 mm x 2.5–5 mm wide, ovate to ± round, flat, and simple-hairy.

SIERRA DISTRIBUTION: Both slopes, Fresno, Inyo, and Tulare Counties.

ELEVATION: 3000–3900 m

HABITAT: Gravelly and rocky areas; alpine zone.

FLOWERING TIME: Jul

NOTES: Formerly *D. lemmonii* var. *lemmonii*. Southern Sierra endemic.

Draba sierrae

Sierra Draba

Brassicaceae (Mustard family)

Cushion-forming perennial herb with stems 1–4 cm; scapose, with stem hairs 2- to 4-rayed and mixed with simple hairs.

LEAVES: Basal 2–6 mm, oblong; both surfaces densely covered with hairs stalked, 7- to 12-rayed.

FLOWERS: Inflorescences of 2–6 flowers; inflorescence axis hairy like the stem; pedicels 2–5 mm, hairy. Sepals 4, 2.5–3.5 mm; petals 4–5 mm x 1.2–1.8 mm wide, yellow. Styles 0.3–0.7 mm.

FRUITS: Silicle 4–8 mm x 3–4.5 mm wide, lanceolate to widely ovate, flat, and twisted when mature. Hairs short-stalked with rays 2–4, sometimes mixed with simple hairs.

SIERRA DISTRIBUTION: Fresno, Inyo, and Mono Counties.

ELEVATION: 3500–4000 m

HABITAT: Rocky outcrops, crevices; alpine zone.

FLOWERING TIME: Jun–Aug

NOTES: The dramatic twisting of mature fruits is diagnostic.

Draba verna (non-native)

Spring Draba

Brassicaceae (Mustard family)

Early blooming annual herb 2–20 cm; stems leafless.
LEAVES: Basal 2–18 mm, narrowly obovate to linear, and toothed or entire; both surfaces have hairs simple or 2- to 4-branched.
FLOWERS: 4–30 per stem; flowers close up in the evening. Sepals 1–1.5 mm; petals 2–4.5 mm, white, and deeply 2-lobed. Ovary superior; stamens 2.
FRUITS: Inflorescence axis is wavy in fruit. Fruits 4–9 mm x 1.5–2.5 mm, roundish, flattened, shape various (usually obovate to oblanceolate), wide and hairless.
SIERRA DISTRIBUTION: West slope, all counties.
ELEVATION: < 2500 m
HABITAT: Open grassy areas, disturbed sites; below alpine zone.
FLOWERING TIME: Feb–May
NOTES: The first plant to bloom in late winter/early spring. *Verna* means "spring," and indeed *Draba verna* heralds spring. Although this non-native plant is originally from Eurasia and northern Africa and is ubiquitous, it doesn't appear to cause any serious problems for the native flora.

Erysimum perenne

Sierra Wallflower

Brassicaceae (Mustard family)

Short-lived perennial or biennial herb with stem 0.4–6.5 dm; rough-pubescent, from short caudexes.
LEAVES: Basal 2–10 cm x 3–10 mm wide, spatulate to oblanceolate, entire or dentate, with hairs 2- to 5-rayed.
FLOWERS: Racemes often headlike early on, elongating with maturity. Sepals 4, 8–12 mm; petals 4, 15–22 mm x 3.5–6 mm wide, claws 8–14 mm, yellow.
FRUITS: Silique 3.8–14 cm x 1.2–3 mm wide; constricted between seeds, fruiting pedicels 4–12 mm, spreading to ascending. Seeds 26–44, 2–3.4 mm, unwinged or winged only at tips.
SIERRA DISTRIBUTION: Both slopes, Tulare and Inyo Counties north.
ELEVATION: 2000–4000 m
HABITAT: Decomposing granite, marble, gravelly areas, talus, alpine fell fields; upper montane through alpine zones.
FLOWERING TIME: Jun–Sep
NOTES: Look for constrictions between seeds in mature fruits; this is one major distinction from *E. capitatum* var. *purshii*, which can have yellow flowers.

Erysimum capitatum var. *capitatum*

Western Wallflower

Brassicaceae (Mustard family)

Short-lived perennial or biennial herb; with stem 1.2–10 dm, usually single and unbranched. Flowers are remarkably fragrant!

LEAVES: Basal 3–10 cm x 0.3–1.5 mm wide, linear to oblanceolate, entire or dentate; upper surfaces of basal and lowest cauline leaves have hairs 3-rayed (sometime up to 7-rayed). Upper cauline leaves sessile and smaller, with hairs 2- to 4-rayed.

FLOWERS: Racemes often headlike early on, elongating with maturity. Sepals 4, 7–14 mm; petals 4, 12–25 mm, claws 8–16 mm, usually orange, sometimes yellowish.

FRUITS: Silique 3.5–11 cm x 1.3–3.3 mm wide, erect, ascending, or spreading, and 4-angled. Seeds 54–82, 2–4 mm, winged at tips.

SIERRA DISTRIBUTION: All.

ELEVATION: < 4000 m

HABITAT: Open and partially shaded areas; foothill through alpine zones.

FLOWERING TIME: Mar–Sep

NOTES: In the Sierra National Forest, 4 years after the 2020 Creek Fire, western wallflower was so abundant in the burned area at about 1600 m that it qualified as a superbloom! There were tens of thousands of orange wallflowers, as far as the eye could see (Whisky Ridge, Madera County). In Mono and Inyo Counties, on the east slope, you may encounter var. *purshii.* This variety has mostly 2-rayed hairs on the upper surfaces of the basal and lower stem leaves, yellow or purple flowers, and slightly shorter, wingless seeds 1.5–2 mm. Found on dry slopes and in meadows < 4000 m; blooming time is Apr–Sep.

Isatis tinctoria (non-native, highly invasive)

Dyer's Woad

Brassicaceae (Mustard family)

Erect biennial herb 4–10 dm (up to 15 dm); glaucous, and hairless (may be hairy at base), with stem branched toward top.

LEAVES: Basal in rosettes, with petioles 0.5–5.5 cm; blades 15–35 cm, oblong to oblanceolate. Cauline leaves are sessile with sagittate or lobed bases, entire.

FLOWERS: Panicle of racemes. Sepals 4, 1.5–2.8 mm, and not saclike at bases; petals 4, 2.5–4 mm, oblanceolate, and yellow. Stamens 6, with 2 somewhat shorter.

FRUITS: Inflorescence and pedicels elongated in fruit. Silique 1.1–2 cm, wider at tip, winged, black or brown (inset).

SIERRA DISTRIBUTION: Mostly Tuolumne County north, but a few plants were found at 2200 m (and eradicated) in Madera County after the 2020 Creek Fire. Expected elsewhere.

ELEVATION: < 2200 m

HABITAT: Disturbed areas; montane zone and below.

FLOWERING TIME: Apr–Jul

NOTES: Native to Eurasia and northern Africa, introduced to most US states as well as South America. The leaves have been used for millennia to produce a blue dye called woad, similar to indigo but not as bright. Unfortunately, as pretty as the dye is, this plant is a devastating noxious weed, taking over enormous swaths of land to the exclusion of native plants.

Lepidium virginicum ssp. *virginicum*

Wild Pepper Grass

Brassicaceae (Mustard family)

Annual herb 1.5–6.5 dm; puberulent, with early deciduous leaves, stem branched toward top.

LEAVES: Basal and alternate, 2.5–10 cm, obovate to oblanceolate, toothed or pinnately lobed. Midcauline leaves 1–6 cm, linear or wider, toothed or pinnately lobed, with bases tapered (not lobed or clasping).

FLOWERS: Racemes elongated, with pedicels 0.15–0.2 mm wide, cylindrical. Sepals 0.7–1 mm; petals 1–2.5 mm, white. Stamens 2.

FRUITS: 0.3–0.4 mm wide, ± round, flattened, hairless, and winged toward tips, with tips notched and styles shorter than notches.

SIERRA DISTRIBUTION: It is safe to say that *L. virginicum* (both sspp., see notes) occurs throughout the SN.

ELEVATION: 200–2900 m

HABITAT: Grassy areas, disturbed sites; below alpine zone.

FLOWERING TIME: Mar–Sep

NOTES: Highly variable. A recent taxonomic reworking of the species reduced the number of sspp. from 4 to 2. *Lepidium virginicum* ssp. *menziesii* also occurs throughout the SN (< 2900 m); it has wider (0.3–0.4 mm), flattened pedicels attached to hairless or hairy fruits, and blooms Mar–Jun.

Nasturtium officinale

Watercress

Brassicaceae (Mustard family)

Aquatic perennial herb with branched stems 1–6 dm; often roots at nodes, hairless or with simple (unbranched) hairs.

LEAVES: Pinnately compound, 2–20 cm, leaflets 3–11.

FLOWERS: Dense clusters at tip of stem. Petals 2.8–4.5 mm, white (sometimes pink).

FRUITS: 2.2–5 cm x 2–3 mm wide, linear, with pedicels 1–3 cm, ascending. Seeds 12–22.

SIERRA DISTRIBUTION: All.

ELEVATION: < 1400 m

HABITAT: Wet areas including streams, ponds, springs, lakes, and marshes in many plant communities; all zones but alpine.

FLOWERING TIME: Mar–Nov

NOTES: Formerly *Rorippa nasturtium-aquaticum*. Watercress occurs worldwide, and has been used for millennia for edible greens, including to this day by Native Americans.

Phoenicaulis cheiranthoides

Daggerpod

Brassicaceae (Mustard family)

Perennial herb 1.2–2.5 dm; from thick caudexes covered with remnants of previous year's leaves; stem is hairless.

LEAVES: Basal 3–15 cm, tapered to petioles, densely covered with multibranched, treelike hairs.

FLOWERS: Racemes, with sepals 3–6 mm, 2 with saclike bases, and petals 9–13 mm, pink to purplish.

FRUITS: Silique 2–6 cm x 3–5 mm wide, dehiscent, flat, parallel to septum, and hairless, with midrib prominent; style 0.5–2 mm, ovate to oblong.

SIERRA DISTRIBUTION: Mostly Sierra crest and east slope, Inyo County north.

ELEVATION: 1300–3500 m

HABITAT: Dry, open slopes, talus, fellfields; subalpine, alpine, and sagebrush zones.

FLOWERING TIME: Mar–Nov

NOTES: *Phoenicaulis* means "purple stem." It is a monotypic genus, meaning that *cheiranthoides* is the only species in the genus. Another, much smaller alpine perennial called "daggerpod" is *Anelsonia eurycarpa* (formerly *Phoenicaulis e.*), a scapose perennial 1–4 cm, with dense clusters of canescent basal leaves, white flowers in umbel-like clusters, and fatter and shorter (1.5–3 cm x 5–9 mm wide) siliques, found along SN crest from Tulare to Alpine Counties.

Physaria occidentalis ssp. *occidentalis*

Western Bladderpod

Brassicaceae (Mustard family)

Perennial herb; from caudex, densely covered with stellate hairs.
LEAVES: Basal in rosettes, spoon-shaped; cauline leaves smaller, alternate.
FLOWERS: Sepals 4.5–7 mm; petals 7–9 mm, yellow.
FRUITS: Elliptical to obovoid, flattened, acute-tipped. Seeds 4–16.
SIERRA DISTRIBUTION: Mono and Alpine Counties north, few occurrences recorded.
ELEVATION: 600–3350 m
HABITAT: Gravel, talus on ridges and rocky (metamorphic) slopes; montane and subalpine zones.
FLOWERING TIME: Apr–Aug
NOTES: Formerly *Lesquerella occidentalis* ssp. *occidentalis*. Uncommon. The photo shows relatively short flowering stems, but they can be as tall as 15 cm (sometimes even 30 cm).

Raphanus raphanistrum (non-native)

Jointed Charlock

Brassicaceae (Mustard family)

Annual herb 3–8 dm; sparsely to densely hairy.
LEAVES: Basal and cauline, alternate, with blades 3–15 cm x 1–5 cm wide, oblong to oblanceolate, and pinnately lobed. Upper stem leaves smaller, sessile.
FLOWERS: Sepals 7–11 mm; petals 15–25 mm x 4–7 mm wide, yellow. Stamens 6, 4 long and 2 short.
FRUITS: Cylindric to narrowly lanceolate, with proximal segment 1–1.5 mm and terminal segment 2–11 cm x 8 mm wide, constricted greatly between seeds; pedicels 0.7–2.5 cm, ascending or spreading.
SIERRA DISTRIBUTION: West slope, Tulare to Butte Counties.
ELEVATION: < 1100 m
HABITAT: Disturbed areas; foothill and montane zones.
FLOWERING TIME: Apr–Jul (or earlier)
NOTES: Wild radish, *Raphanus sativus*, blooms at the same time and has stems 4–13 dm, with petals pink, purple, or white, and fruits with no constriction between seeds or weakly constricted. These 2 spp. are found growing together and often hybridize to form stands of plants with varied fruit constriction and flower color. Both are native to the Mediterranean region of Europe.

Sisymbrium altissimum (non-native)

Tumble Mustard

Brassicaceae (Mustard family)

Erect annual herb 4–12 dm (up to 16 dm); many-branched distally, hairless distally, and sparsely-densely hairy proximally. The overall look is of a grayish-green plant with lighter yellow flowers than other annual mustards.

LEAVES: Basal leaves rosetted, with petioles 5–20 cm (up to 35 cm); leaf shape lanceolate to oblanceolate in overall outline, deeply pinnately divided into 4–6 pairs of lobes, appearing compound. Stem leaves similar, and distal leaves have linear, threadlike.

FLOWERS: Racemes of yellow flowers; sepals 2–2.5 mm, not saclike at base, with tips horned; petals 6–8 mm x 2.5–4 mm wide, spatulate, claws 3.5–6 mm.

FRUITS: Silique 6–9 cm (up to 12 cm) x 1–2 mm wide, linear; pedicels 6–10 mm (up to 13 mm), ± same width as fruit (not narrower), spreading.

SIERRA DISTRIBUTION: All.

ELEVATION: < 2700 m

HABITAT: Disturbed areas; foothill and montane zones.

FLOWERING TIME: May–Oct

NOTES: Native to Europe, introduced to most US states as well as South America.

Sisymbrium officinale (non-native)

Hedge Mustard

Brassicaceae (Mustard family)

Erect annual herb 2.5–7.5 dm; unbranched or branched distally, and sparsely to densely reflexed-hairy with simple hairs.

LEAVES: Basal and lower cauline 3–10 cm, oblanceolate to oblong-ovate in overall outline, pinnately divided into 3–4 pairs of lobes.

FLOWERS: Racemes of yellow flowers. Sepals 2–2.5 mm, not saclike at base; petals 2.5–4 mm x 1–2 mm wide, spatulate, claws 1–2 mm.

FRUITS: Silique 1–1.4 cm x 1–1.5 mm wide, narrowly awl-shaped, not segmented, appressed to stalk, with stout pedicel 1.5–3 mm.

SIERRA DISTRIBUTION: West slope.

ELEVATION: < 2200 m

HABITAT: Disturbed areas, including burns; foothill and montane zones.

FLOWERING TIME: Apr–Sep

NOTES: Native to Eurasia, introduced throughout North America, also Central and South America, and Australia. The short, appressed fruits distinguish *S. officinale* from other spp. in the genus.

bracts
lower leaves

Streptanthus diversifolius

Varied-Leaved Jewelflower

Brassicaceae (Mustard family)

Erect annual herb 2–10.5 dm; hairless, branched toward top.

LEAVES: Basal not in rosettes and pinnately divided into threadlike lobes; upper cauline leaves sessile, clasping, and entire.

FLOWERS: Inflorescences open, no terminal sterile flower cluster; bracts below and/or between bottom 1–2 flowers. Calyx urn-shaped, with 4 sepals, 5–7 mm, keeled, yellow or purplish; petals 4, 8–16 mm x 1.5–3.5 mm wide, recurved, crinkled, and white or pale yellow with purple veins. Filaments are in 3 pairs of unequal length, free.

FRUITS: Silique 3–9 cm x 1–1.5 mm wide, strongly reflexed, straight without constrictions between seeds. Seeds 38–80, 1–2.2 mm, oblong, winged or not at tips.

SIERRA DISTRIBUTION: West slope, Tulare to Butte Counties.

ELEVATION: 200–1900 m

HABITAT: Open areas, usually gravelly and often on or near rocky outcrops; foothill and montane zones.

FLOWERING TIME: Apr–Jul

NOTES: The species name *diversifolius* is apt: This plant has dramatically different lower leaves with threadlike pinnate lobes in contrast to upper stem leaves, which are lance-linear to cordate, entire, and clasp the stem.

Streptanthus farnsworthianus

Farnsworth's Jewelflower

Brassicaceae (Mustard family)

Erect, hairless annual herb with stem 2–10 dm; often branched distally, hairless; herbage is strikingly green-purple.

LEAVES: Basal in rosettes, bipinnately lobed, withering early; midcauline leaves 3–16 cm x 1–4 cm wide, with basal lobes obviously clasping, lanceolate, toothed or entire toward tips; distal leaves reduced, entire.

FLOWERS: Open racemes with large, maroon-purple, leaflike bracts below or between the lowest 1 or 2 flowers. Calyx urn-shaped; sepals 6–10 mm, light purple; petals 8–11 mm x 1.5–2 mm wide, white with purple veins, clawed, and slender. Stamens in 3 pairs of unequal lengths, the longest 2 pairs exserted with petals.

FRUITS: Silique 6–12 mm x 1.7–2.5 mm wide, ascending, straight or somewhat curved, and flattened. Seeds 60–100, 2–2.5 mm, with a wing 0.1–0.3 mm wide at tip.

SIERRA DISTRIBUTION: West slope, Madera County south to northern Kern County.

ELEVATION: 400–1400 m

HABITAT: Rocky, gravelly, and sandy soils; foothill and lower montane zones.

FLOWERING TIME: May–Jun

NOTES: This stunning plant would not be confused with any other *Streptanthus*, with its large purple leaves and bracts, and exquisite jewel-like flowers. Named by J. T. Howell for Evalyn L. Farnsworth, a Kern County cattle rancher who collected the type specimen. Mrs. Farnsworth collected regularly beginning in 1962 and contributed significantly to knowledge of the flora of Kern and Tulare Counties.

Streptanthus tortuosus var. *foliosus*

Wawona Jewelflower

Brassicaceae (Mustard family)

Hairless, short-lived perennial herb 4–8.4 dm; 1–3 stems from woody base; branched distally.

LEAVES: Basal in rosettes, petioled, withering early; midcauline sessile and clasping or lobed at base, 2.3–3.3 cm (shorter than basal and lower leaves); spoon-shaped to lance-ovate; entire to dentate.

FLOWERS: Inflorescences with numerous showy, yellowish bracts crowded densely at base; bracts 1.8–2.1 cm, circular, mucronate or rounded. Calyx urn-shaped, purple or yellow, tips recurved and acute-acuminate; sepals 7.7–11.3 mm; petals 12.7–13.4 mm, purplish or yellow-white with purple veins. Stamens 6, in 3 pairs of unequal lengths, the longest filament pair 7–11 mm.

FRUITS: Spreading, 8–13.5 cm, straight or arched. Seeds 30–75, 1.8–2 mm, winged at tips.

SIERRA DISTRIBUTION: Both slopes, all SN counties.

ELEVATION: 1220–2050 m

HABITAT: Rocky sites; gravelly and sandy soils; montane zone.

FLOWERING TIME: Jun–Jul

NOTES: *S. tortuosus* was recently separated into 6 varieties in CA, and 4 are in the SN: Sierran foothill jewelflower, var. *optatus*, has 1–3 stems 4.6–10 dm from a woody base, with ovate to oblong bracts 2–7.5 cm that are neither crowded nor round, as in var. *foliosus*, and have acute tips, blooming Apr–Jul. Mountain jewelflower, var. *tortuosus*, is similar, 2–10 dm, with round to oblong inflorescences and uncrowded bracts with obtuse tips, blooming Apr–Jun in the foothill and lower montane zones. High mountain jewelflower, var. *orbiculatus*, is a much shorter (1–3 dm), bushy-looking annual plant of the upper montane and subalpine zones, blooming Jun–Sep. All are host plants for many butterflies and moths.

Thelypodium crispum

Wavy-Leaved Thelypodium

Brassicaceae (Mustard family)

Biennial herb 1–7 dm; often glaucous, hairs lacking except near base, simple or branched.

LEAVES: Basal and alternate; basal blades 2–15 cm, dentate to pinnately lobed, occasionally entire; leaves wither once fruits have formed; petioles ciliate. Midcauline leaves sessile, with lobed or sagittate bases, entire or dentate.

FLOWERS: Inflorescence a dense raceme; flowers 4-merous, petals 6–11 mm x 0.5–0.7 mm wide, linear to narrowly oblanceolate, white or pale lavender, and crisped.

FRUITS: Silique 1–2.5 cm x 0.7–1 mm wide, cylindric, constricted between seeds. Stalk above receptacle is 0.5–1.5 mm, slender, with style 0.5–1.5 mm, fruiting pedicel erect to ascending, appressed to stalk at least initially, slender. Seeds 22–50, 1.5–2 mm, plump.

SIERRA DISTRIBUTION: East slope, all.

ELEVATION: 1200–3000 m

HABITAT: Lake margins, alkaline soils; subalpine and pinyon-juniper zones.

FLOWERING TIME: Jun–Aug

NOTES: In Mono County, the similar-looking short-podded thelypodium *Thelypodium brachycarpum*, may be encountered; the white petals are slightly longer and narrower (8–12 mm x 0.3–0.5 mm wide), fruits are slightly longer and wider (1.2–2.7 cm x 1–1.5 mm wide), and fruiting pedicels are much shorter (1–2 mm) and stout rather than slender. Flowering is from Apr–Aug, and elevation range in the SN is 2100–3000 m (for CA, 800 m is the lower limit).

Tropidocarpum gracile

Dobie Pod

Brassicaceae (Mustard family)

Annual herb 1–4.5 dm; stem prostrate to erect, branched, with hairs simple or stalked-forked.

LEAVES: Basal not in rosettes; lowermost stem leaves 2.5–10 cm, deeply pinnately lobed, with lobes entire or themselves lobed or dentate, and lobes in 3–8 pairs on leaf.

FLOWERS: Inflorescence has leaflike bracts. Sepals 4, 2.5–4 mm; petals 4, 3–6 mm x 1.5–4 mm wide, yellow (sometimes tinged purple). Style 0.5–2.5 mm; stamens 6.

FRUITS: Silique 3–6 cm x 1.5–2 mm wide, narrowly linear; pedicels 6–17 mm, erect to ascending.

SIERRA DISTRIBUTION: West slope, Butte County south.

ELEVATION: < 1550 m

HABITAT: Open grassy areas, oak woodlands, roadsides; foothill zone.

FLOWERING TIME: Mar–May

NOTES: The leaflike bracts throughout the inflorescence are diagnostic.

Thysanocarpus curvipes ssp. *elegans*

Lace Pod

Brassicaceae (Mustard family)

Delicate, erect annual herb with stem 1–6 dm; hairy at base of stem and hairless above.

LEAVES: Basal in rosettes, 1–6 cm, oblanceolate to obovate, usually wavy dentate but not pinnatifid; cauline leaves alternate, sessile, lanceolate, clasping, and entire.

FLOWERS: Inflorescence an open raceme; flowers 4-merous, with sepals ascending; petals at least as large as sepals, white or tinged purple and not clawed. Ovary superior, style 1, stigmas entire; anthers purple when fertile.

FRUITS: Capsule ≥ 6 mm x 5.5–7 mm wide, 2-valved, indehiscent, pendent, hairy, with a perforated wavy margined wing, style 0.4–0.9 mm.

SIERRA DISTRIBUTION: Both slopes, all (few records for Inyo County).

ELEVATION: < 2500 m

HABITAT: Roadsides, grasslands and savannahs, woodlands, forests, streambanks; foothill and montane zones.

FLOWERING TIME: Feb–Jun

NOTES: Due to the enormous variability within *T. curvipes*, the *Flora of North America North of Mexico* treatment of *Thysanocarpus* does not recognize infraspecific taxa (e.g., sspp. such as *elegans.)* The Jepson eFlora recognizes 6 sspp., including *elegans*; *curvipes* and *longistylus* are also found in the SN (refer to the eFlora for distinguishing characters). Also in the Sierra < 1800 m is *T. laciniatus*, with pinnatifid basal leaves and fruits 2.5–5 mm wide, deeply crenate or entire, perforated or not. *T. radians* occurs < 800 m adjacent to the San Joaquin Valley; the fruits are larger than other spp. at 7–10 mm wide, not perforated, and fruiting pedicels are straight with an abrupt bend at the tips.

Turritis glabra

Tower Mustard

Brassicaceae (Mustard family)

Erect biennial herb 3–15 dm; usually simple but occasionally branched toward top, glaucous toward top, with hairs on lower herbage simple to forked to stellate.

LEAVES: Basal in rosettes, petioled, pinnately lobed or toothed leaves 5–12 cm x 1–3 cm wide; stem leaves alternate, sessile, 2–9 cm x 1–2.5 cm wide, with bases lobed or clasping stem, lanceolate or ovate, entire or toothed.

FLOWERS: Inflorescence is elongated; sepals 2.5–5 mm; petals 5–8.5 mm, cream-colored (sometimes purplish).

FRUITS: Silique 4–10 cm x 0.7–1.5 mm wide, tightly appressed to stem, erect, hairless.

SIERRA DISTRIBUTION: Mostly west slope, all.

ELEVATION: < 2800 m

HABITAT: Open areas in forests, woodlands, meadows; below alpine zone.

FLOWERING TIME: Apr–Jul

NOTES: Formerly *Arabis glabra*. Unmistakable with its towering habit and clasping stem leaves. Often found in isolated patches of dozens to hundreds of plants.

Opuntia polyacantha var. *erinacea*

Grizzlybear Pricklypear

Cactaceae (Cactus family)

Shrub < 5 dm; flattened, succulent stem sections, each 5.5–20 cm, surfaces green and hairless; covered with areoles (defined small areas from which spines and glochids arise), each with 4–24 spines. Glochids are small, fine, barbed bristles in *Opuntia* spp.

LEAVES: Ephemeral—small, conic, and fleshy; deciduous soon after appearing on young stems and ovaries (yes, ovaries!).

FLOWERS: Arising from areoles (1 per areole), bisexual, radially symmetrical, sepals and petals alike (tepals). Tepals numerous; outer tepals green to yellow, with margins tinged the color of inner tepals; inner tepals pale yellow to orange. Ovary inferior; stamens numerous.

FRUITS: 2.5–4 cm; fleshy, berrylike, and green when young; tinted red and dry in age.

SIERRA DISTRIBUTION: Kern, Tulare, Inyo, and Mono Counties.

ELEVATION: 900–2200 m

HABITAT: Woodlands; pinyon-juniper zone.

FLOWERING TIME: May–Jun

NOTES: This cactus barely makes it into the southeastern Sierra; it is primarily a desert species. Flowers of the genus *Opuntia* are fragrant and attract a variety of bees, butterflies, birds, and other pollinators. Humans have used *Opuntia* for millennia for food, beverages, and treatments for a litany of ailments.

Downingia cuspidata

Toothed Calicoflower

Campanulaceae (Bellflower family)

Hairless annual herb 10–40 cm; stem decumbent to erect.

LEAVES: Alternate, 0.5–4 mm wide, sessile, entire, lanceolate to awl-like, often deciduous by flowering time.

FLOWERS: Inflorescence is a spike, with terminal flowers that tend to abort. Fertile flowers sessile, 7–15 mm, bilateral, and hairless; lower lip 3-lobed, pale to bright blue or lavender at tips, with a white area towards the mouth with 2 yellow spots near throat. Ovary inferior, hairless to minutely scabrous; anthers included.

FRUITS: Capsule 20–72 mm, dehiscent on sides by 3–5 slits. Seeds have spiral lines, giving them a twisted look.

SIERRA DISTRIBUTION: West slope, Tulare County north.

ELEVATION: < 1700 m

HABITAT: Vernal pools, meadows, lake margins; foothill zone.

FLOWERING TIME: Mar–Jun

NOTES: *D. cuspidata* tends to grow in a variety of moisture conditions at the periphery of vernal pools, compared to other *Downingia* spp. with stricter preference for specific water depths toward the center of the pool. *D. cuspidata* is visited by a diverse array of insects, many of which are surely pollinators: beetles, bees, flies, and butterflies.

Nemacladus interior

Sierra Threadplant

Campanulaceae (Bellflower family)

Erect annual herb 7–25 cm; erect, branching starting at 3–6 cm above ground.

LEAVES: Basal in rosettes, petioled or sessile, 10–20 mm, oblanceolate to elliptic, irregularly toothed, hairless.

FLOWERS: Borne on a strongly zigzagged axis, racemelike, with 1 bract per flower, 1–3 mm, leaflike; pedicels 7–13 mm, threadlike. Hypanthium 0.5 mm; sepals 5, ± 2 mm, narrowly triangular, and erect. The corolla is half-inferior, clearly 2-lipped, with 3 upper lobes 2.5 mm, erect, white with maroon arches over yellow spots; lower lobes 3 mm, white and unmarked. The floral appendages look like tiny glass rods and are quite beautiful.

FRUITS: Capsule 2–2.5 mm, acute at base and tip, opening by 2 valves at top. Seeds many, 0.5 mm, elliptic with pitted rows.

SIERRA DISTRIBUTION: West slope, all.

ELEVATION: < 2700 m

HABITAT: Gravelly/sandy openings in foothill chaparral and ponderosa pine forests; prolific after wildfire and other ground disturbance; foothill and montane zones.

FLOWERING TIME: Jun–Sep

NOTES: Plants tend to blend into gravelly background, with their tiny flowers and slender, zigzagging stems. After fire, plants often grow larger than normal and can be abundant years 1–2 post-burn.

Smithiastrum prenanthoides

California Harebell

Campanulaceae (Bellflower family)

Erect or decumbent, hairless to sparsely hairy perennial herb 20–150 cm.

LEAVES: Alternate, short-petioled or sessile, 1–6 cm, lanceolate to ovate with toothed margins.

FLOWERS: Racemes, with 2–4 flowers per node, 5-merous, with pedicels 2–6 mm; calyx lobes spreading to reflexed; corollas 7–14 mm, bell-shaped, lobes linear, cut ¾ to fused base, and light to dark blue. Ovary inferior; style 15–18 mm, stigmas 3-lobed.

FRUITS: Capsule hemispheric, ribbed, seeds are released through 2–3 lateral pores positioned below middle of fruit. Seeds numerous, 2 mm, oblong.

SIERRA DISTRIBUTION: West slope, Tulare County north.

ELEVATION: < 2130 m

HABITAT: Shaded edges and interior of coniferous forest; montane zone.

FLOWERING TIME: Jun–Sep

NOTES: Formerly *Asyneuma prenanthoides* (prior to that, in *Campanula*). In Butte, Plumas, and Sierra Counties < 1900 m along streams and in shaded woodlands, Scouler's harebell, *Campanula scouleri*, could be mistaken for *A. prenanthoides*; this plant is shorter (20–30 cm), leaves have petioles 1–2 cm, racemes have only 1 flower per node, corollas are 8–15 mm, and styles are longer (12–25 mm).

Lonicera conjugialis

Double Honeysuckle

Caprifoliaceae (Honeysuckle family)

Erect shrub 6–18 dm; puberulent.

LEAVES: Opposite; petioles 2–8 cm; blades 2–8 cm, elliptic or ovate, tips acute.

FLOWERS: Occur in pairs, 4–8 mm, 2-lipped, red to maroon, with ovaries fused together > halfway.

FRUITS: 5–8 mm wide, bright red, fused.

SIERRA DISTRIBUTION: Tulare and Inyo Counties north.

ELEVATION: 1200–3300 m

HABITAT: Moist to wet areas; upper montane to subalpine zones.

FLOWERING TIME: May–Jul

NOTES: Compare with *L. involucrata*, which also has paired flowers but the ripe fruits are shiny black and are completely separate (not fused). *L. cauriana* (not pictured) also has red fruits that appear completely fused, though they are actually held tightly together by an encircling bract. *L. cauriana* has white or light yellow flowers that are barely 2-lipped (they appear radially symmetrical). It is found in most west slope counties 2200–3200 m and blooms May–Jul.

Lonicera interrupta

Chaparral Honeysuckle

Caprifoliaceae (Honeysuckle family)

Evergreen shrub with vinelike, climbing branches up to 3 m.
LEAVES: Opposite, 2–2.5 cm, round-oblong, without stipules; upper 1–3 leaf pairs are fused into a disk surrounding stem.
FLOWERS: Inflorescences 3–12 cm; flowers 8–10 mm, 2-lipped, hairless, yellow or creamy.
FRUITS: ± 10 mm, spherical, red.
SIERRA DISTRIBUTION: West slope, all.
ELEVATION: 240–1400 m
HABITAT: Chaparral, woodlands, riparian areas; foothill and montane zones.
FLOWERING TIME: Apr–May
NOTES: Below 1100 m on the west slope, the pink-flowered vine pink honeysuckle, *L. hispidula*, may be encountered. It has fused upper leaves like chaparral honeysuckle, and also has bright red fruits (slightly larger at 10 mm). When not in flower, it can be distinguished by its larger (4–8 cm) leaves with stipules present on those below the fused pairs, and a densely glandular fruiting inflorescence.

Lonicera involucrata var. *involucrata*

Twinberry

Caprifoliaceae (Honeysuckle family)

Erect, glandular-hairy shrub 6–30 dm.
LEAVES: Opposite; petioles 2–10 cm; blades 3–12 cm, elliptic or ovate, tips acute.
FLOWERS: Occur in pairs, within bracts leaflike, turning dark reddish-purplish with age. Corollas 12–18 mm, narrowly bell-shaped or cylindric, yellow, with tiny lobes ± 2 mm. Ovary inferior.
FRUITS: A striking pair of fruits 6–10 mm, separate, shiny, black, held within the enlarged involucre bracts reddish-purple.
SIERRA DISTRIBUTION: All.
ELEVATION: 600–2900 m
HABITAT: Moist to wet areas; upper montane to subalpine zones.
FLOWERING TIME: Jun–Aug
NOTES: Compare with *L. conjugialis*, which also has paired flowers, but the ovaries are fused, ripening into red (not black) berries without the showy involucres.

Symphoricarpos mollis

Creeping Snowberry, Trip Vine

Caprifoliaceae (Honeysuckle family)

Vinelike shrub 1.5–6 dm; stems ± horizontal, with soft hairs (sometimes hairless); branches root at nodes to form large patches.

LEAVES: Opposite; blades 0.5–3 cm, elliptic or round, irregularly lobed, with short, soft hairs.

FLOWERS: Inflorescences of 2–8 flowers; corollas 4–5 mm, bell-shaped, bright pink; lobes erect, about ½ corolla length, hairy on inner surfaces. There are 5 nectaries found at the base of corolla tube. Ovary inferior.

FRUITS: Berry 8 mm, round, white.

SIERRA DISTRIBUTION: West slope, all.

ELEVATION: < 3000 m

HABITAT: Forest floor and moist meadow edges; montane zone.

FLOWERING TIME: May–Aug

NOTES: The name trip vine is apropos. Watch your step around it!

Symphoricarpos rotundifolius var. *rotundifolius*

Mountain Snowberry

Caprifoliaceae (Honeysuckle family)

Woody, deciduous shrub/vine 6–12 dm; ± erect, puberulent, with older bark shredding.

LEAVES: Opposite; blades 0.8–2 cm, oval, irregularly lobed, paler and with more prominent veins on underside.

FLOWERS: Inflorescences of 1–2 flowers in leaf axils; corollas 6–10 mm, funnel- to bell-shaped, white to pink; lobes ⅕–⅓ the length of corolla and hairless on inner surfaces. Corolla tube hairy in the inner third; nectaries 5, found at the base of the tube. Ovary inferior.

FRUITS: 8–12 mm, ovoid, ultimately turns white.

SIERRA DISTRIBUTION: Both slopes, Tulare and Inyo Counties north.

ELEVATION: 1200–3300 m

HABITAT: Dry, open slopes, rocky areas; montane and subalpine zones.

FLOWERING TIME: Jun–Aug

NOTES: At about the same elevations, var. *parishii* is mostly in the southern Sierra and along the crest. In contrast to var. *rotundifolius,* it is a vinelike, trailing plant, often with arched branches, rooting at tips; and rather than being hairy for only the middle third, the inside of the corolla tube is sparsely hairy throughout. *Symphoricarpos* spp. are important plants for birds, bees, butterflies, and moths, providing nectar and berries.

Eremogone congesta var. *suffrutescens*

Suffrutescent Sandwort

Caryophyllaceae (Pink family)

Tufted perennial herb to subshrub 8–40 cm; hairless or glandular-hairy.

LEAVES: Basal and opposite, 15–80 mm x 0.5–1.5 mm wide, needlelike, 1-veined; cauline leaves in 3–5 pairs and smaller.

FLOWERS: Umbellate heads with membranous bracts clustered at base, with pedicels 1–12 mm. Sepals 5, 3–4 mm, obtuse; petals 5, 5–8 mm, white. Styles 3; stamens 10.

FRUITS: Capsule 3.5–6 mm, 6-valved. Seeds 4–8, tubercled, reddish-brown.

SIERRA DISTRIBUTION: Both slopes, Tulare and Inyo Counties north.

ELEVATION: 1200–3300 m

HABITAT: Open gravelly and sandy areas, rocky outcrops; subalpine zone.

FLOWERING TIME: Jun–Aug

NOTES: Suffrutescent means "subshrub"—stems with woody bases but herbaceous otherwise. Additional varieties occur in the Sierra: Var. *congesta* has more condensed, ball-like heads (pedicels < 0.2 mm) and occurs from Butte and Sierra Counties north; var. *simulans* occurs in Lassen County, has more open heads, but the bracts are scattered throughout rather than at the base of the head, and leaves are 20–35 mm; and var. *subcongesta* is also more open, with smaller (10–30 mm) leaves and acute sepals, occurring mostly from Nevada and Sierra Counties north.

Eremogone kingii var. *glabrescens*

King's Sandwort

Caryophyllaceae (Pink family)

Tufted perennial herb 1–20 cm; glandular hairy (sometimes hairless).

LEAVES: Basal and opposite, 3–20 mm, linear, sharp-pointed, 1-veined.

FLOWERS: Few to many per open head, with pedicels 2–15 mm. Sepals 2.5–4 mm, glandular, white-margined; petals 5, 4–7 mm, white. Styles 3.

FRUITS: Capsule 4.5–7 mm, 6-valved. Seeds 2–5, reddish-brown or black, bumpy.

SIERRA DISTRIBUTION: Mostly east slope and crest of Sierra, Inyo County north.

ELEVATION: 2100–4050 m

HABITAT: Open gravelly and sandy areas, rocky outcrops; montane, subalpine, and alpine zones.

FLOWERING TIME: Jun–Sep

NOTES: Compare with *E. congesta*, which has hairless sepals and a much more condensed inflorescence.

Eremogone macradenia var. *macradenia*

Desert Sandwort

Caryophyllaceae (Pink family)

Tufted perennial herb 20–40 cm; glandular hairy or hairless.
LEAVES: Cauline and opposite, in 5–8 pairs, 20–60 mm x 0.8–1.2 mm wide, ascending rather than curved downward, linear, sharp-pointed (or dull-pointed), 1-veined.
FLOWERS: Few to many per open head, with pedicels 3–55 mm. Sepals 5, 4.5–7.2 mm, acute to acuminate, hairless; petals 5, 6–11 mm, white; nectaries 0.7–1.5 mm, 2-lobed or truncate, rectangular. Styles 3.
FRUITS: Capsule 6–7 mm, 6-valved. Seeds 4–9, reddish-brown or black, with bumpy conical tubercles.
SIERRA DISTRIBUTION: Both slopes in Southern Sierra, Kern, Tulare, Inyo, and Mono Counties.
ELEVATION: 1100–2200 m
HABITAT: Open gravelly and sandy areas, rocky outcrops; montane, subalpine, and alpine zones.
FLOWERING TIME: Jun–Sep
NOTES: Also found in the southern Sierra is var. *arcuifolia*, with downward-curving cauline leaves.

Petrorhagia dubia (non-native)

Hairy Pink

Caryophyllaceae (Pink family)

Erect annual herb 10–60 cm.
LEAVES: Opposite; sheaths 4–9 mm; blades 10–60 mm, linear.
FLOWERS: Terminal clusters subtended by papery involucres; sepals 5, petals 5, obcordate to 2-lobed, with up to 5 dark veins.
FRUITS: Capsule 4-valved. Seeds 1–1.4 mm, conic-papillate.
SIERRA DISTRIBUTION: West slope, Tulare County north.
ELEVATION: < 1800 m
HABITAT: Disturbed areas; foothill and montane zones.
FLOWERING TIME: Apr–Jul
NOTES: Introduced to the SN from the Mediterranean in the 1920s. *P. prolifera*, also non-native, is found from Mariposa County north on the west slope < 800 m. It differs by having a much shorter (1–2 mm) leaf sheath and 1-veined petals.

Sabulina californica

California Sandwort

Caryophyllaceae (Pink family)

Delicate, hairless annual herb < 12 cm tall.

LEAVES: Opposite, 2–5 mm, fused at base, linear to oblong with obtuse tips; leaves much shorter than internodes.

FLOWERS: Solitary or in open cymes of up to 9 flowers; sepals 2.5–2.8 mm, petals longer. Styles 3.

FRUITS: Capsule to 3.5 mm, 3-valved, longer than sepals. Seeds reddish-brown, papillose, without winglike margins.

SIERRA DISTRIBUTION: West slope, all.

ELEVATION: < 1600 m

HABITAT: Open gravelly and sandy areas, rocky outcrops; foothill and montane zones.

FLOWERING TIME: Mar–May

NOTES: Formerly *Minuartia californica*. Compare with *S. douglasii*.

Sabulina douglasii

Douglas' Stitchwort

Caryophyllaceae (Pink family)

Delicate, glandular-hairy annual herb 4–30 cm tall; simple or (usually) highly branched, stems may be green or purple.

LEAVES: Opposite, 5–40 mm, < 0.3 mm wide, very slender.

FLOWERS: 7–15 per open cymes. Sepals 2.5–3.7 mm; petals about twice as long, white, may be notched or not.

FRUITS: Capsule to 4 mm, 3-valved, longer than sepals. Seeds reddish brown, with winglike margins.

SIERRA DISTRIBUTION: All, except east side.

ELEVATION: < 2000 m

HABITAT: Open gravelly or sandy areas, rocky outcrops; foothill and montane zones.

FLOWERING TIME: Apr–Jun

NOTES: Formerly *Minuartia douglasii*. Compare with *Sabulina californica*. *S. douglasii* is generally taller, has longer, thinner leaves, and glandular hairs on the inflorescence.

Sabulina nuttallii var. *gracilis*

Nuttall's Sandwort

Caryophyllaceae (Pink family)

Mat-forming, densely glandular-hairy perennial herb 0.2–2 dm.
LEAVES: Opposite or fascicled, 5–12 mm x 0.3–1.1 mm wide, linear, needlelike, and straight or recurved.
FLOWERS: Sepals 3–7 mm, with tips pointed, margins membranous, 1- to 3-nerved; petals slightly shorter. Styles 3.
FRUITS: Capsule 3-valved. Seeds reddish-brown to dark brown, with margins thick.
SIERRA DISTRIBUTION: Both sides of crest, Tuolumne County south.
ELEVATION: 1500–3800 m
HABITAT: Open sandy and gravelly areas, talus, rocky outcrops and cliffs; upper montane through alpine zones.
FLOWERING TIME: Jul–Aug
NOTES: Formerly *Minuartia nuttallii* var. *gracilis*.

Silene bernardina

Palmer's Catchfly

Caryophyllaceae (Pink family)

Erect to decumbent perennial herb 1.5–5 dm; few-branched from woody base, puberulent to short-hairy, and often glandular.
LEAVES: Opposite, ± linear; lower leaves 2–8 cm, the upper smaller; stem leaves appear in 2–4 pairs.
FLOWERS: Inflorescences axillary and terminal. Calyx 12–25 mm, tubular, glandular-puberulent, lobes 2–3.5 mm, with 10 parallel purplish veins. Petals 5, 4–6 mm, 4-lobed, with petal claws and stamens ciliate at base, white to pink; appendages 2, deeply cut or fringed, shorter than petals. Style and stamens exserted.
FRUITS: Capsule elliptic, puberulent, opening by 6–8 ascending teeth. Seeds many, 1.5–2 mm, brown.
SIERRA DISTRIBUTION: Both slopes, all.
ELEVATION: 1220–3660 m
HABITAT: Dry openings and rocky areas; montane through alpine zones.
FLOWERING TIME: Jun–Aug
NOTES: Small forms of *S. bernardina* could be mistaken for *S. sargentii*, which has linear leaves and a more bell-shaped calyx, with cross-veins between the purple calyx veins. See entry for *S. sargentii*.

Silene bridgesii

Bridges' Catchfly

Caryophyllaceae (Pink family)

Erect to decumbent perennial herb 3.0–8.0 dm; puberulent; glandular-hairy above.

LEAVES: Basal 3–6 cm x 5–15 mm wide, with petioles; stem leaves smaller, sessile, and glandular, with 3–6 pairs below inflorescence.

FLOWERS: Nodding, with calyx 7–14 mm, ovoid, 10-veined, glandular-hairy, and calyx lobes 2–3 mm. Corolla limb is white; petals 14–28 mm, 2-lobed, appendages 2; claws hairless. Stamens and 3-lobed styles strongly exserted.

FRUITS: Capsule ovoid and puberulent, with stalk 2–3 mm, opening by 6 triangular teeth.

SIERRA DISTRIBUTION: West slope, Tulare County north (no records for Placer or Nevada Counties).

ELEVATION: 600–2400 m

HABITAT: Open areas; montane coniferous zone.

FLOWERING TIME: Jun–Aug

NOTES: Compare with the similar-looking Lemmon's catchfly, *Silene lemmonii*, which has 4 lobes per petal rather than 2. Bridge's catchfly is also distinguished from Lemmon's by its overall larger size, with larger leaves and flowers.

Silene douglasii var. *douglasii*

Douglas's Catchfly

Caryophyllaceae (Pink family)

Erect to decumbent perennial herb 1.0–7.0 dm; puberulent, not glandular.

LEAVES: Basal and lower leaves 2–6 cm x 2–8 mm wide; upper stem leaves smaller, 1–4 cm x 1–5 mm wide, and more linear.

FLOWERS: Erect or spreading; calyx 10–14 mm, hairless or puberulent (not glandular), 10-veined, with calyx lobes 1.5–2.5 mm. Corolla limb 4–11 mm, white (to pink or purplish), 2-lobed, appendages 2; claws ciliate at base. Stigmas 3, not exserted; stamens barely exserted.

FRUITS: Capsule oblong to ovoid, with truncate base, puberulent, with stalk 3–4 mm, opening by 3–5 ascending triangular teeth.

SIERRA DISTRIBUTION: Alpine County north.

ELEVATION: 1400–2900 m

HABITAT: Open areas; montane coniferous zone.

FLOWERING TIME: Jun–Aug

NOTES: Although the petals are 2-lobed in both *S. douglasii* and *S. bridgesii*, *S. douglasii* does not have nodding flowers, the stamens are not dramatically exserted, and the base of the calyx is truncate (more or less squared off).

Silene gallica (non-native)

Common Catchfly, Windmill Pink

Caryophyllaceae (Pink family)

Erect to decumbent, glandular-hairy annual herb 1–4 dm.
LEAVES: Opposite; lower leaves 1–3.5 cm x 3–5 mm wide, oblanceolate; upper leaves smaller, yielding to leaflike inflorescence bracts.
FLOWERS: Inflorescences racemose, with leafy bracts; calyx 6–10 mm with lobes 2–2.5 mm, glandular hairy on veins and with 2 mm, nonglandular hairs throughout. Petals 5, longer than calyx, white to pink, sometimes with dark spot, slightly twisted (hence the name "windmill pink"), and sometimes notched; petal claw has 2 appendages (visible in photo). Styles and stamens more or less included.
FRUITS: Capsule ovoid, puberulent, opening by 6 narrow, recurved teeth in a ring at top.
SIERRA DISTRIBUTION: West slope, all.
ELEVATION: < 1220 m (possibly higher)
HABITAT: Disturbed areas, roadsides; foothill and montane zones.
FLOWERING TIME: Jun–Aug
NOTES: Native to Europe, Great Britain, and north Africa. This common weed of disturbed sites in CA is of conservation concern in Britain, with an 80% population decline noted in 2006. Possible reasons are shifting agricultural practices and/or climate change.

Silene laciniata ssp. *californica*

California Pink

Caryophyllaceae (Pink family)

Erect to decumbent perennial herb 1.5–4 dm, branched from base.
LEAVES: Opposite, 3–8 cm, ovate or oblanceolate.
FLOWERS: Calyx 15–25 mm, cylindrical, glandular-hairy, lobes 2.5–5 mm, 10-veined; corolla lobes 2–3 cm, 4-lobed, bright red-orange to bright red. Styles and stamens exserted.
FRUITS: Capsule ovoid, hairless, opening by 6–10 teeth.
SIERRA DISTRIBUTION: West slope, all.
ELEVATION: < 1700 m
HABITAT: Openings and shady areas; foothill and lower montane zones.
FLOWERING TIME: Mar–May
NOTES: Formerly *Silene californica*. Unmistakable, as this is one of the most electric-red wildflowers in the SN. The color varies from somewhat faded to almost unreal fluorescent red.

Silene lemmonii

Lemmon's Catchfly

Caryophyllaceae (Pink family)

Erect to decumbent perennial herb 1.5–4.5 dm, glandular-hairy above.

LEAVES: Basal 2–3.5 cm x 8–10 mm wide; stem leaves smaller and sparse.

FLOWERS: Nodding; calyx 6–10 mm, ovoid, glandular-hairy, with lobes 1–2 mm, 10-veined. Corolla limb is creamy white to yellowish to pinkish; each petal has 4 long, slender, recurved lobes; petal claw has woolly hairs, appendages 2. Styles and stamens strongly exserted.

FRUITS: Capsule oblong, glandular-puberulent, opening by 6–10 teeth.

SIERRA DISTRIBUTION: West slope, all.

ELEVATION: 850–2800 m

HABITAT: Dry, open forests; montane zone.

FLOWERING TIME: Jun–Aug

NOTES: The only *Silene* with 4 linear, recurved, tangled-looking lobes on each petal.

Silene sargentii

Sargent's Catchfly, Alpine Campion

Caryophyllaceae (Pink family)

Decumbent to erect perennial herb 10–15 cm; branched from base, glandular above.

LEAVES: Basal 1.5–3 cm, tufted, linear-oblanceolate, fleshy; stem leaves smaller, linear.

FLOWERS: Calyx 9–15 mm, glandular-puberulent, lobes 2–3 mm, with 10 purplish veins; corolla lobes 5, 4–6 mm, white to red-purple, with each petal limb 2- to 4-lobed. Petal claw has 2 appendages at the level of the anthers, and the claw is hairy at base. Styles 3 and stamens included.

FRUITS: Capsule ovoid, glandular-puberulent, opening by 6–10 teeth.

SIERRA DISTRIBUTION: Tulare and Inyo Counties north.

ELEVATION: 2400–3800 m

HABITAT: Dry, grassy, rocky, or gravelly openings; subalpine and alpine zones.

FLOWERING TIME: Jul–Aug

NOTES: Compare with *S. bernardina*. Sargent's catchfly has a less tubular calyx, and the purple calyx veins are less strictly parallel; notice the cross-walls of purple between the main calyx veins.

Stellaria longipes ssp. *longipes*

Long-Stalked Starwort

Caryophyllaceae (Pink family)

Erect to ascending perennial herb 5–35 cm; rhizomatous, and mostly hairless.

LEAVES: Opposite, 10–40 mm, linear or slightly wider, flat, and often with marginal hairs near base.

FLOWERS: Inflorescences of 1–7 flowers, with each flower on a long, slender, erect pedicel. Sepals 5, 3–5.5 mm, hairless with wide, white margins; petals 5, longer than sepals and notched. Styles 3; stamens 10.

FRUITS: Capsule 6-valved, surrounded by 3-nerved sepals. Seeds 0.7–1 mm, reddish-brown, minutely roughened.

SIERRA DISTRIBUTION: Tulare and Inyo Counties north.

ELEVATION: 1250–3500 m

HABITAT: Moist to wet sites; montane, subalpine, and alpine zones.

FLOWERING TIME: May–Aug

NOTES: *Stellaria borealis* ssp. *sitchana* may appear similar initially, but its petals are shorter than sepals (or absent), leaves are lanceolate rather than linear with papillate margins, and pedicels turn downward in fruit. It occurs sporadically in the montane zone 1000–2100 m on the west slope from Tulare County north.

Stellaria media (non-native)

Common Chickweed

Caryophyllaceae (Pink family)

Sprawling to erect annual herb 7–50 cm, with stem hairs in a longitudinal line (best seen with a 10x lens)

LEAVES: Opposite, 8–45 mm, ovate, and ciliate at base.

FLOWERS: Inflorescences few-flowered; sepals 5, with pre-fruiting length 3–4.5 mm (< 6 mm in fruit); petals 5, 70%–90% length of sepals and deeply notched.

FRUITS: Capsule 6-valved on reflexed pedicels. Seeds 0.9–1.3 mm, brown, tubercled.

SIERRA DISTRIBUTION: West slope, all.

ELEVATION: < 1500 m

HABITAT: Moist, shady sites; foothill and montane zones.

FLOWERING TIME: Feb–Sep

NOTES: An extremely common weed of gardens and disturbed sites in wildlands; native to Europe.

Torreyostellaria jamesiana

Tuber Starwort

Caryophyllaceae (Pink family)

Erect or sprawling, glandular-hairy perennial herb 12–45 cm; stem 4-angled, arising from rhizomes with tuberlike thickenings 3–12 mm wide.

LEAVES: Opposite, sessile, 15–150 mm, narrowly to moderately lanceolate, and smaller above.

FLOWERS: Inflorescences are leafy cymes of 1 to many white flowers; sepals 5, 3–7 mm, densely glandular-hairy; petals are much longer, notched < ⅕ to base. Styles 3, 2–4.5 mm; stamens 10.

FRUITS: Capsule spheric, 3-valved, fruiting pedicels becoming reflexed or recurved. Seeds 1–2, 2–3.4 mm, red-brown.

SIERRA DISTRIBUTION: West slope, all.

ELEVATION: 1400–2700 m

HABITAT: Understory and edges of conifer forest, meadow edges, sagebrush/grasslands; montane and sagebrush zones.

FLOWERING TIME: Jul–Aug

NOTES: Formerly *Pseudostellaria jamesiana*, then *Schizotechium jamesianum*. Compare with *Stellaria longipes* ssp. *longipes*: In contrast, *T. jamesiana* is glandular-sticky, has rhizomes with tubers, and has lanceolate rather than linear leaves. Sierra starwort, *Hartmaniella sierrae* (formerly *Pseudostellaria sierrae*), is a SN endemic found from Mariposa County north in the montane zone of the west slope 1400–2000 m. *H. sierrae* is hairless (not sticky), has only 1 flower per stem tip and 5 (not 10) stamens, and its petals are narrower with shallower notches.

Chenopodium album (non-native)

Lamb's Quarters

Chenopodiaceae (Goosefoot family)

Erect, branched annual herb 18–100 cm; stems are grooved, often appear striped, and are covered with white, powdery-appearing hairs.

LEAVES: Alternate; blades 15–70 mm, roughly triangular or ovate, and irregularly toothed; leaves paler below due to white, powdery covering.

FLOWERS: Inflorescences are branched axillary or terminal spikes. Sepals 5, white-margined, and white-powdery; petals absent. Styles 2; stamens 5.

FRUITS: 1–1.5 mm, enclosed by calyx. Seeds are horizontally arranged.

SIERRA DISTRIBUTION: All.

ELEVATION: < 1800 m

HABITAT: Disturbed, open areas, all zones except alpine (scarce in subalpine).

FLOWERING TIME: Jun–Oct

NOTES: An exceedingly variable species believed to be native to Europe, it now occurs in temperate regions worldwide.

Grayia spinosa

Hop-Sage

Chenopodiaceae (Goosefoot family)

Rounded shrub < 1 m; many-branched, with older twigs spinelike. The bark is striped, and plants are either male or female.

LEAVES: Alternate, 5–25 mm, oblanceolate or wedge-shaped, white-tipped, and mealy to hairless.

FLOWERS: Male plants produce spikelike inflorescences 7–18 mm with 2–5 flowers and subtending, leaflike bracts; sepals 4, no petals, stamens 4–5. Female plants produce spikelike inflorescences 6–18 cm, bracts 2 (round), no petals, stigmas 2.

FRUITS: Each fruit is enclosed by bracts bright pink, round, appearing saclike.

SIERRA DISTRIBUTION: East slope and southern Sierra (Kern Plateau, Walker Pass area), Lassen County south.

ELEVATION: 300–2900 m

HABITAT: Sandy or gravelly soils; sagebrush zone.

FLOWERING TIME: Mar–Jun

NOTES: Hop-sage is mostly a desert species, but it extends into the Sierra on the upper east slope and on the Kern Plateau (Tulare County), around Walker Pass, and in many areas of Kern County.

Crocanthemum scoparium

Peak Rush-Rose

Cistaceae (Rock-Rose family)

Erect, evergreen perennial herb or subshrub 12–65 cm; hairs stellate.

LEAVES: Alternate, 5–40 mm, 0.5–6 mm wide.

FLOWERS: Raceme has inflorescence hairs that are mostly nonglandular; sepals 5, with outer 2 linear; petals 3–11 mm. Style < 2 mm, stigma roundish; stamens 5–45.

FRUITS: Capsule 3-valved.

SIERRA DISTRIBUTION: Mariposa to El Dorado Counties.

ELEVATION: < 1500 m

HABITAT: Dry sandy or rocky soil on ridges and slopes; foothill and montane zones.

FLOWERING TIME: Feb–Sep

NOTES: Much more common near the coast. Especially abundant after fire, but also in showy stands in the central SN, primarily in nongranitic soils.

Cleomella arborea var. *globosa*

Bladderpod

Cleomaceae (Spiderflower family)

Many-branched, evergreen shrub 5–20 dm; hairless or nearly so.
LEAVES: Alternate; petioles 1–3 cm. Leaflets 3, 15–45 mm, oblong to elliptic.
FLOWERS: Bracted racemes, with bracts resembling a leaflet, 2–15 mm; pedicels 7–15 mm; sepals 4, 4–7 mm, fused ± halfway. Corollas bilateral; petals 4, 8–14 mm, yellow. Ovary superior, born on a gynophore (stalk), pistil 1; stamens 6, yellow; anthers coil with release of pollen.
FRUITS: Capsule inflated, spheric, on elongated, recurved gynophore, dehiscent; emit a somewhat unpleasant smell when opened. Seeds 5–25, dark brown and smooth.
SIERRA DISTRIBUTION: West slope, Kern County.
ELEVATION: < 1830 m
HABITAT: Open areas, roadsides, washes; foothill zone.
FLOWERING TIME: Oct–Jul
NOTES: Formerly placed in Capparaceae, the Caper family. This unique shrub occurs naturally only in Kern County; plants growing in Fresno County foothills near the San Joaquin River were planted long ago.

Comandra umbellata ssp. *californica*

Bastard Toadflax

Comandraceae (Bastard Toadflax family)

Perennial herb or subshrub 10–40 cm; rhizomatous, glabrous and often glaucous, partial root parasite.
LEAVES: Alternate, sessile, 15–55 mm, elliptic, ovate, lanceolate, or linear with acute or sharp-pointed tips; sometimes slightly paler on undersides.
FLOWERS: Inflorescences in simple or compound cymes. Ovary inferior—the ovary wall is fused to the perianth tube, with 5 white, petal-like perianth lobes spreading above the ovary.
FRUITS: 5–7.5 mm, ovoid to spheric, brown; persistent perianth lobes remaining on the mature fruit.
SIERRA DISTRIBUTION: West slope, most counties.
ELEVATION: 300–3000 m
HABITAT: Dry, often rocky sites; foothill, montane, and subalpine zones.
FLOWERING TIME: Apr–Aug
NOTES: *Comandra umbellata* is a hemiparasite—it is green and photosynthetic, but plants are more vigorous when roots connect to roots of a variety of host species, perennial and woody, and extract nutrients via structures called haustoria.

Calystegia longipes

Paiute False Bindweed

Convolvulaceae (Morning Glory family)

Subshrub with woody base, and hairless stems that are erect, stiff rather than trailing, and often intertwined.

LEAVES: Alternate, < 6 cm, linear or sometimes narrowly sagittate.

FLOWERS: Peduncles < 20 cm; inflorescence bracts alternate, 3–17 mm x 0.2–3 mm wide, linear, situated well below the calyx; sepals 5, 8–11 mm; corollas 28–36 mm, white, cream, sometimes pale pink.

FRUITS: Capsule roundish.

SIERRA DISTRIBUTION: Tulare, Inyo, and Kern Counties.

ELEVATION: 600–2600 m

HABITAT: Dry rocky sites; foothill, montane, and sagebrush zones; also desert scrub where it extends into the southern and eastern Sierra.

FLOWERING TIME: May–Jul

NOTES: The photo doesn't show the inflorescence bracts well, but they are alternate and far enough below the calyx that you wouldn't confuse them with a lower sepal.

Calystegia malacophylla ssp. *malacophylla*

Sierra Morning Glory

Convolvulaceae (Morning Glory family)

Vinelike perennial herb with stems 10–100 cm; rhizomatous, herbage densely hairy to tomentose.

LEAVES: Alternate, 3–6 cm, 4–9 cm wide, widely triangular or irregularly kidney-shaped, with the tips acute or notched, and lobes with 2 tips or 1 blunt tip.

FLOWERS: Peduncles shorter than subtending leaves; bracts 2, leafy, directly below and nearly hiding the calyx of 5 sepals (thus could appear to be 7 sepals); calyx 9–15 mm; corollas 20–45 mm. Stigmas 2-lobed.

FRUITS: Capsule round.

SIERRA DISTRIBUTION: West slope, all.

ELEVATION: 600–2600 m

HABITAT: Dry areas; foothill and montane zones.

FLOWERING TIME: Jun–Aug

NOTES: To tell this plant apart from *C. occidentalis*, note the soft-hairy leaves and the more rounded leaf shape with 2 (rather than 1) points on the lobes. The root *malaco* means "soft," and *phyllum* means "leaf."

Calystegia occidentalis ssp. *fulcrata*

Western Morning Glory

Convolvulaceae (Morning Glory family)

Trailing or vinelike perennial herb with stems < 1 m; spreading along the ground; herbage finely hairy, rising from woody caudexes.

LEAVES: Alternate, 1.5–4 cm, arrowhead-shaped but overall outline is triangular, lobes usually 1-tipped.

FLOWERS: Peduncles 1-flowered; bracts 2, 5–30 mm x 2–7 mm wide, set 2–15 mm below the calyx, and leafy, resembling miniature leaves. Sepals 9–15 mm, unequal in size (outer ones smaller); corollas 20–48 mm. Stigmas 2-lobed.

FRUITS: Capsule round.

SIERRA DISTRIBUTION: West slope, Kern to Butte Counties.

ELEVATION: 300–2700 m

HABITAT: Dry grassy openings, chaparral, conifer/oak forests; foothill and montane zones.

FLOWERING TIME: Jun–Aug

NOTES: *C. occidentalis* ssp. *occidentalis* occurs up to 1800 m and is distinguished by its more vinelike habit (climbing up to 2 m on other woody vegetation) and inflorescence bracts not resembling the leaves (the bracts are located 1–7 mm below the calyx, 4–18 mm x 1–5 mm wide, and linear to narrowly oblong), as well as by having 2–4 (rather than 1) flowers per axil.

Convolvulus arvensis (non-native, invasive)

Field Bindweed, Orchard Morning Glory

Convolvulaceae (Morning Glory family)

Prostrate, twining perennial herb; from a deep root system.

LEAVES: Alternate, petioled, arrowhead-shaped, blades 2–3 cm.

FLOWERS: In axils, with 2 linear bracts set well below the calyx (never hiding it). Calyx < 10 mm; corollas < 4 cm, funnel-shaped, white (not cream, as in *Calystegia*) and often tinged pink, especially along pleats on outside. Stigma lobes 2, flat, with tips rounded.

FRUITS: Capsule to 8 mm, round. Seeds up to 4, roughened brown, borne on reflexed peduncles.

SIERRA DISTRIBUTION: All.

ELEVATION: < 2700 m

HABITAT: Disturbed areas, many plant communities; all zones but subalpine and alpine.

FLOWERING TIME: Mar–Oct

NOTES: The plant was introduced to Virginia in 1739 and had spread to all 50 states by 1900. Field bindweed is a serious agricultural pest that aggressively uses soil water and binds up harvesting machinery with its wiry stems. The complex system of roots and rhizomes can extend to depths of 3 m (10 ft), and new plants can form from segments as small as 5 cm. Seeds can live up to 60 years.

Cuscuta subinclusa

Canyon Dodder

Convolvulaceae (Morning Glory family)

Orange-stemmed annual or perennial vine; parasitic, not rooted in soil but attached to host plants via numerous specialized roots called haustoria.

LEAVES: Absent or reduced to tiny, alternate scales.

FLOWERS: Clusters or dense heads of 5–20 (up to 40) white to cream-colored flowers with 5 corolla lobes, sessile or with pedicels < 1 mm. Corollas 5–7 mm (up to 9 mm), corolla lobes ¼–⅓ the length of the tube, with lobes straight or recurved but not pointing inward. Calyx lobes 5, < ½ length of corolla tube, with bases overlapping.

FRUITS: Capsule 1.5–3 mm, ovoid, capped by the dried corolla. Seed 1.

SIERRA DISTRIBUTION: West slope, all.

ELEVATION: < 2440 m

HABITAT: Found on a variety of shrubs and herbs (not host-specific); foothill and montane zones.

FLOWERING TIME: Jun–Nov

NOTES: *C. subinclusa* is the most common native dodder in the SN, and in a good year it can drape native shrubs like something in a horror movie! There are other native dodders in the Sierra; some are rare, and some are very host-specific. These spp. are important components of Sierra ecosystems, and in addition to providing nectar and pollen to native insects, they likely play a role in modulating the abundance of host species. Although historically all dodder spp. have been listed as federal and state noxious weeds, this is unnecessary as native species rarely cause economic harm. The non-natives, especially Japanese dodder, *Cuscuta japonica* var. *formosa*, are another matter. This plant has thick stems like spaghetti and can cause serious problems for agriculture.

Cornus nuttallii
Mountain Dogwood
Cornaceae (Dogwood family)

Deciduous shrub to small tree < 25 m; with young twigs green and hairy, and older bark dark red.

LEAVES: Opposite; petioles 5–17 mm; blades 6–15 cm, elliptic, ovate, or obovate. Leaf veins in 4–6 pairs, most originating from basal half of blade.

FLOWERS: Inflorescences 1.5–3 cm diameter, headlike, convex, of 40–75 flowers; subtended by 4–7 bracts, 4–6 cm x 3–6 cm wide, petaloid, white. Flowers 4-merous; sepals 1–2.5 mm; petals 3–4 mm, greenish to white. Ovary inferior, style 2 mm; stamens 4.

FRUITS: Drupe 1–1.7 cm, red. Stone is smooth.

SIERRA DISTRIBUTION: West slope, all.

ELEVATION: < 2000 m

HABITAT: Moister microhabitats in mixed conifer forest; montane zone.

FLOWERING TIME: Apr–Jul

NOTES: Flowers generally occur before leaves, and a grove of mountain dogwoods in full bloom with their showy white bracts on display is a beautiful sight. Fall brings another show, with foliage turning bright red, orange, or yellow prior to leaf drop.

Cornus sericea ssp. *occidentalis*
Western Red-Osier Dogwood, American Dogwood
Cornaceae (Dogwood family)

Deciduous shrub 1.5–4 m; with red-purple branches graying in age, ± hairless.

LEAVES: Opposite; blades 5–10 cm, hairy and paler below, with veins in 4–7 pairs.

FLOWERS: Cymes of numerous flowers, each with 4 petals 3–4.5 mm, cream-colored to white. Ovary inferior, style 2.5–3 mm; stamens 4.

FRUITS: Berry 7–9 mm, white to cream-colored. Stone generally has 3 ridges.

SIERRA DISTRIBUTION: All. Appears to be sparsely distributed across the SN, but more abundant than current records indicate.

ELEVATION: < 3200 m

HABITAT: Moist sites, streamsides; montane, subalpine, and sagebrush/pinyon-juniper zones.

FLOWERING TIME: May–Jul

NOTES: *Cornus sericea* ssp. *sericea* is as common as ssp. *occidentalis* and intergrades with it. Technically speaking, ssp. *sericea* is distinguished by having hairless to sparsely strigose leaf undersurfaces, along with smaller (2–3 mm) petals, shorter (1–2 mm) styles, and smooth stones (seeds) lacking the 3 ridges. In nature you may encounter specimens with features of both sspp. Ssp. *sericea* does occur higher (< 2800 m) and in every SN county.

Crassula connata

Pygmyweed

Crassulaceae (Stonecrop family)

Erect succulent annual herb 2–6 cm, reddish in age.
LEAVES: Opposite; leaf bases 1.3 mm, connate.
FLOWERS: 2 per node, parts in 4s; sepals 0.5–2 mm, lanceolate, acute-acuminate; petals 0.6–1.2 mm, narrowly triangular.
FRUITS: Follicles 4, ovoid, ascending.
SIERRA DISTRIBUTION: West slope, all.
ELEVATION: < 1500 m
HABITAT: Open, moist areas, especially sandy or gravelly flats on rocky outcrops; foothill and lower montane zones.
FLOWERING TIME: Feb–May
NOTES: Forms large, reddish patches where conditions are ideal. The photo shows fruiting plants, but flowers are not showy, so they look about the same in fruit.

Dudleya cymosa ssp. *cymosa*

Dudleya, Liveforever

Crassulaceae (Stonecrop family)

Fleshy succulent perennial herb 5–45 cm, stems 1–3.25 cm diameter.
LEAVES: In 1 to several rosettes, sessile, succulent, 3–17 cm x 10–60 mm wide, of variable shape but often deltate, sometimes glaucous, tips acuminate to mucronate; cauline alternate, smaller.
FLOWERS: Cymes with at least 3 primary branches, peduncles 5–30 cm; terminal branches 1–17 cm, 4–20 flowers per branch; sepals 5, 1.5–5 mm, triangular or ovate; petals 5, 7–14 mm, fused at base 1–3 mm, yellow, orange, or red. Stamens 10, anthers 0.2–0.4 mm.
FRUITS: Follicles 5. Seeds many, < 1 mm, brown and striate.
SIERRA DISTRIBUTION: West slope.
ELEVATION: < 2700 m
HABITAT: Rocky outcrops and cliffs, talus slopes; foothill and montane zones.
FLOWERING TIME: May–Jul
NOTES: By far the most common *Dudleya* taxon in the SN. The gray-green stems, inflorescence stalks, and foliage often turn reddish in age, and since corollas can be yellow, orange, or red, one might think they're different species. There are a few sspp. of *D. abramsii* in the southern SN; petals of these taxa are cream, white, or pale yellow, often with jagged margins. Refer to the Jepson eFlora for further details.

Rhodiola integrifolia ssp. *integrifolia*

Western Roseroot

Crassulaceae (Stonecrop family)

Perennial herb 3–15 cm; from short, thick caudexes; hairless, fleshy.

LEAVES: Cauline and alternate, sessile, 5–30 mm x 2–14 mm wide; leaf shape oblanceolate to elliptic; sometimes glaucous.

FLOWERS: Dense, hemispheric heads of 7–50 mostly unisexual flowers. Sepals and petals usually 4, sometimes 5; sepals 1.5–3 mm; petals 2–4 mm, dark red to reddish-purple, spoon-shaped and spreading in staminate flowers, linear and erect in pistillate flowers. Pistils 4–5; stamens twice the number of petals, anthers brown to reddish-purple.

FRUITS: Follicles 4–5, 3–6 mm, erect. Seeds 1.4–2 mm, brown and striate.

SIERRA DISTRIBUTION: Both slopes, Tulare and Inyo Counties north to Nevada County.

ELEVATION: 1800–4000 m

HABITAT: Rocky outcrops and cliffs, talus slopes, edges of lakes and meadows; subalpine and alpine zones.

FLOWERING TIME: May–Aug

NOTES: Formerly *Sedum roseum* ssp. *integrifolium*. A 2015 study concluded that climate change is predicted to decrease alpine habitat for western roseroot by up to 31% in the Rocky Mountains if forests continue to extend the elevation of tree line upward; similar outcomes are likely for the SN.

Sedella pumila

Sierra Mock Stonecrop

Crassulaceae (Stonecrop family)

Succulent annual herb 2–17 cm; often with reddish stems and leaves, 0 to several branches from base.

LEAVES: Opposite at base, alternate above, 2–7 mm x 1–3 mm wide.

FLOWERS: Sepals 5, 0.5–0.8 mm; petals 2.5–5 mm, yellow. Stamens 10, anthers 0.2–0.4 mm.

FRUITS: 5, indehiscent, erect or outcurved. Seeds 0.8–1.5 mm.

SIERRA DISTRIBUTION: West slope, all.

ELEVATION: < 1500 m

HABITAT: Moist to wet sites on rocky outcrops, clay soils; foothill and montane zones.

FLOWERING TIME: Mar–May

NOTES: *Sedella pumila* combines 2 spp. that were recognized as distinct in the 1993 edition of *The Jepson Manual* as *Parvisedum pumilum* and *P. congdonii*.

Sedum lanceolatum

Spearleaf Stonecrop

Crassulaceae (Stonecrop family)

Succulent perennial herb 3–20 cm, with herbage often glaucous.
LEAVES: Leaves crowded at base, alternate on stem; blades 5–30 mm, 1.5–2 mm thick, nearly cylindrical but slightly flattened, linear to ovate (wider below middle), with tips acute.
FLOWERS: Inflorescences 1–4 cm, of 3–24 flowers, with inflorescence bracts 3–10 mm, mostly shed by flowering. Flowers 5-merous, with sepals and petals free to base; sepals 4 mm; petals 5–8 mm, narrowly lanceolate, acute to acuminate, yellow. Stamens 10, anthers yellow or reddish-brown.
FRUITS: Follicles, 4–9 mm, tips outcurved. Seeds ± 1 mm.
SIERRA DISTRIBUTION: Both slopes, Tulare to Plumas Counties.
ELEVATION: 1800–2800 m
HABITAT: Rocky sites, outcrops; montane through alpine zones.
FLOWERING TIME: May–Aug
NOTES: A host plant for the SN endemic snow apollo butterfly (*Parnassius behrii*). Another narrow-leaved and narrow-petaled stonecrop, *Sedum stenopetalum*, wormleaf stonecrop, occurs sparingly in the south and central Sierra; plants are 5–27 cm, and often have plantlets in place of upper leaves (which makes them look very different from *S. lanceolatum*). Leaves are 6–16 mm, ± lanceolate, with tips long-tapered, often 3-lobed at base, and red and papery with age. Inflorescences are racemelike, of 1–44 flowers; petals are 4–10 mm, long-tapered, lanceolate, and yellow, often with reddish midrib, with anthers 10, yellow or brown.

Sedum obtusatum ssp. *obtusatum*

Sierra Stonecrop

Crassulaceae (Stonecrop family)

Succulent perennial herb 3–22 cm; new rosettes are produced at the tips of offsets that arise from leaf axils.
LEAVES: Basal in dense rosettes 1–6 cm wide; blades 5–30 cm, 1–4 mm thick, and wider above the middle, obovate to spoon-shaped; stem leaves alternate, obtuse, 30–70 mm (most are shriveled in photo).
FLOWERS: Inflorescences 2–12 cm, of 8–26+ flowers, inflorescence bracts 4–9 mm. Flowers 5-merous; sepals and petals united at base ± ¼ length; sepals 1.8–5 mm; petals 4–10 mm, rounded or obtuse, often mucronate, yellow, red-veined or tinged red with age. Stamens 10.
FRUITS: Follicles 5–10 mm, erect.
SIERRA DISTRIBUTION: Both slopes, Tulare to Plumas Counties.
ELEVATION: 1200–3700 m
HABITAT: Rocky sites, gravelly flats; montane through alpine zones.
FLOWERING TIME: Jun–Aug
NOTES: *Sedum obtusatum* ssp. *boreale* occurs sparingly from Mono County north at 1400–2700 m; it has longer (8–15 mm) inflorescence bracts, and lighter yellow or white corollas tinged or veined with pink.

Sedum spathulifolium

Broadleaf Stonecrop

Crassulaceae (Stonecrop family)

Succulent perennial herb 5–22 cm; herbage often glaucous; produces new rosettes via offsets.

LEAVES: Basal in dense rosettes 1–6 cm wide, with outer rosette leaves much larger than inner; blades 11–22 mm, wider above middle, and tip rounded to obtuse. Stem leaves smaller, alternate, and elliptic to oblong.

FLOWERS: Inflorescences 3–8 cm, of 5–48 flowers, with inflorescence bracts 6–11 mm. Flowers 5-merous, with sepals and petals free to base; sepals 2–4 mm; petals 4–10 mm, narrowly lanceolate, acute, and yellow. Stamens 10, anthers yellow or reddish-brown.

FRUITS: Follicles 4–8 mm, spreading. Seeds ± 1 mm.

SIERRA DISTRIBUTION: Mostly west slope, Tulare to Plumas Counties.

ELEVATION: < 2500 m

HABITAT: Rocky sites, outcrops, streamsides; foothill and montane zones.

FLOWERING TIME: Apr–Aug

NOTES: Forms attractive mats of rosettes via offsets. Flowers have a sweet fragrance.

Cucurbita foetidissima

Calabazilla, Buffalo Gourd

Cucurbitaceae (Gourd family)

Trailing perennial vine with tendrils, from a large, tuberlike root; can extend to a patch 8 m wide.

LEAVES: Alternate, 15–30 cm, triangular (sometimes ovate), grayish, entire or somewhat lobed; foul-smelling.

FLOWERS: Unisexual, 1 per node; male and female flowers located at different nodes on same plant. Corollas 9–12 cm, bell-shaped, yellow. Ovary inferior.

FRUITS: Gourd 7–8 cm wide (or larger), round, mottled, with white stripes.

SIERRA DISTRIBUTION: West slope, Madera County south.

ELEVATION: < 1300 m

HABITAT: Sandy, gravelly, disturbed sites; foothill zone.

FLOWERING TIME: Jun–Aug

NOTES: Related to squash, zucchini, cucumbers, and watermelon. Buffalo gourd's native range is from the Midwestern US westward, and south to Mexico. Uses by a variety of Native American tribes include eating young gourds like squash (older ones are too tough), using hollowed-out gourds for musical instruments (rattles), and making medicinal remedies.

Marah horrida

Sierra Man-Root

Cucurbitaceae (Gourd family)

Perennial vine with branched tendrils and large, underground tuber; herbage is scabrous, not glaucous; monoecious or sometimes temporarily dioecious.

LEAVES: Alternate, palmately 5- to 7-lobed, tendrils 1 per node.

FLOWERS: Unisexual, sepals absent. Corollas cup-shaped, with tubes green and limbs white, > 8 mm wide at opening. Staminate corollas in racemes, and pistillate corollas borne singly; ovary inferior.

FRUITS: 9–20 cm, oblong, with long, straight prickles. Seeds 6–16, oval, and flattened at one end.

SIERRA DISTRIBUTION: West slope, El Dorado County south.

ELEVATION: < 1600 m

HABITAT: Chaparral, dry areas; foothill and lower montane zones.

FLOWERING TIME: Mar–Apr

NOTES: Although called Sierra man-root, the range extends southward into Los Angeles County. California man-root, *M. fabacea*, occurs on the west slope < 1600 m; it is told easily from Sierra man-root by the rotate, or wheel-shaped, rather than cup-shaped, corollas and the smaller (4–5 cm), spheric fruits. Co-occurring in the northern Sierra < 1200 m, from Tuolumne County north, is taw man-root, *Marah watsonii*, which has staminate flowers < 8 mm wide, glaucous leaves, and small, round fruits 2–3.5 cm, with or without hooked prickles and often striped like a watermelon. Another common name for *Marah* is wild cucumber, but none of the spp. are edible; rather, all parts are toxic. *Marah* tubers with wet weights of 91 kg (201 lbs) have been recorded, and a 14-year-old *M. fabacea* plant attained a wet weight of 58 kg (128 lbs), as big as a human!

Datisca glomerata

Durango Root

Datiscaceae (Datisca family)

Erect perennial herb 1–2+ m, with clustered stems.

LEAVES: Mostly alternate, 10–20 cm, deeply and irregularly pinnately lobed.

FLOWERS: 2 kinds of flowers, neither with petals, are borne in clusters in leaf axils: Male flowers have a calyx 2 mm and 8–12 sessile yellow anthers; bisexual flowers have a calyx 5–8 mm, with forked styles; stamens 0–4.

FRUITS: Capsule ± 8 mm. Seeds 1 mm, brown.

SIERRA DISTRIBUTION: West slope.

ELEVATION: < 2300 m

HABITAT: Along streams and in moist areas; foothill and montane zones.

FLOWERING TIME: May–Aug

NOTES: All parts of the plant are toxic. Cases of cattle poisoning and deaths have been documented after ingestion of large amounts of Durango root. Superficially, it resembles marijuana, but *Cannabis sativa* has palmately lobed leaves.

Drosera rotundifolia

Round-Leafed Sundew

Droseraceae (Sundew family)

Perennial herb; carnivorous, with a rosette of insectivorous leaves.

LEAVES: Leaf blades 3–12 mm, 4–20 mm wide, ± round, with red, gland-tipped hairs.

FLOWERS: Flowering stalks 5–35 cm, emerging in late summer, appearing 1 or more 1-sided racemes. Sepals 5, 4–6 mm; petals 5, 4–6 mm, white or pink. Styles 2, stamens 5.

FRUITS: Capsule 3-valved.

SIERRA DISTRIBUTION: West slope, Tulare County north.

ELEVATION: < 2700 m

HABITAT: Fens in wet meadows and lakeshores; often associated with *Sphagnum* moss; montane zone.

FLOWERING TIME: Jun–Aug

NOTES: The sticky, gland-tipped hairs trap insects and release enzymes to digest them. With a 10x hand lens you can often see insects in varying stages of disintegration on the leaves. The plants get up to 30% of their nitrogen from digested insects.

Allotropa virgata

Sugarstick

Ericaceae (Heather family)

Striking, red-and-white, candy cane–like perennial herb < 5 dm; erect, with no green tissue.

LEAVES: None.

FLOWERS: The "stem" is technically the inflorescence; there is no true stem. Inflorescences are racemes of numerous cup-shaped, white corollas. Sepals 0–4; petals 5, concave. Ovary superior, styles < 2 mm, stigmas disk-shaped; stamens 10, anthers maroon.

FRUITS: Locules, splitting from tips first.

SIERRA DISTRIBUTION: West slope, patchy distribution.

ELEVATION: < 3000 m

HABITAT: Floor of coniferous forest, usually in thick humus; montane zone.

FLOWERING TIME: Jun–Aug

NOTES: The only plant that could possibly be mistaken for sugarstick is a young pinedrops, which can have faint red striping, but this would only be if you had never seen a sugarstick. They are flabbergastingly gorgeous and different from anything else. Most look like a candy cane.

Arctostaphylos manzanita ssp. *manzanita*

Common Manzanita

Ericaceae (Heather family)

Shrub or small tree 2–8 m, with no burl. Twigs, petioles, and inflorescence axes are finely tomentose.

LEAVES: Erect, hairless, and shiny; blades 2–5 cm, 1–3.5 cm wide, ovate, oblong-ovate, or obovate, with rounded or wedge-shaped bases. Upper and lower leaf surfaces are alike, bright green.

FLOWERS: Inflorescence is an open panicle 15–25 mm, with 5–7 slender branches. Corollas 7–8 mm, pale pink or white. Ovary hairless or hairy.

FRUITS: 8–12 mm wide, hairless to hairy.

SIERRA DISTRIBUTION: West slope, Mariposa County north.

ELEVATION: < 2000 m

HABITAT: Dry slopes, rocky areas in chaparral, coniferous forest; foothill and montane zones.

FLOWERING TIME: Nov–Mar

NOTES: Identifying this species is best done by a process of elimination: If it's not *A. viscida*, *A. mewukka*, or *A. patula*, consider *A. manzanita*. Also, be aware that hybrids between *A. patula* and *A. viscida* have frequently been mislabeled as *A. manzanita*. The flowers attract hummingbirds and insects; other kinds of birds eat the fruit.

Arctostaphylos mewukka ssp. *mewukka*

Indian Manzanita

Ericaceae (Heather family)

Spreading shrub 1–2 m; smooth, deep red to purple bark, burl present (enabling sprouting after fire).

LEAVES: Erect, hairless, and dull, with blades 3–7 cm, 2–4 cm wide. Upper and lower leaf surfaces alike, gray, glaucous, and smooth to the touch.

FLOWERS: Inflorescence is an open panicle with 3–5 branches, hairy, hairless, or glandular. Corollas 6–7 mm, pink or white. Ovary hairless.

FRUITS: 12–15 mm wide, reddish-brown, hairless.

SIERRA DISTRIBUTION: West slope, Fresno County north.

ELEVATION: < 2000 m

HABITAT: Dry slopes in chaparral, coniferous forest; foothill and montane zones.

FLOWERING TIME: Mar–Apr

NOTES: Can be overlooked when growing with *A. viscida*, which also has grayish-green leaves, but distinguished by the basal burl and hairless twigs, leaves, and fruits.

Arctostaphylos nevadensis

Pinemat Manzanita

Ericaceae (Heather family)

Low-growing shrub < 6 dm; mat-forming, no burl.

LEAVES: Bright green, shiny, and entire, with blades 1–3 cm, 1–1.5 cm wide, obovate to oblanceolate, with sharp points at the tips.

FLOWERS: Inflorescence is a raceme of white flowers. Ovary hairless.

FRUITS: 6–8 mm wide, round, dark red, hairless. Stones easily separable.

SIERRA DISTRIBUTION: All.

ELEVATION: 600–3200 m

HABITAT: Open rocky or sandy areas; upper montane and subalpine zones.

FLOWERING TIME: May–Jul

NOTES: Pinemat manzanita is exceedingly common in the upper elevations of the montane zone into the subalpine, often forming vast fields.

Arctostaphylos patula

Greenleaf Manzanita

Ericaceae (Heather family)

Large shrub 1–2 m, with burl present or absent (present in central and southern Sierra plants).

LEAVES: Oval to round, erect, hairless, with blades 2.5–6 cm, 1.5–4 cm wide; upper and lower surfaces similarly shiny dark green.

FLOWERS: Inflorescence is a panicle with 4–8 branches. Corollas 5–8 mm, pink. Ovary hairless.

FRUITS: 7–10 mm wide, round, brown, hairless.

SIERRA DISTRIBUTION: All.

ELEVATION: 750–3350 m

HABITAT: Open areas; montane and subalpine zones.

FLOWERING TIME: Apr–Jun

NOTES: One of the most common shrubs in the montane zone of the SN, overlapping with *A. viscida* ssp. *mariposa* in the lower-elevation ponderosa pine forest and becoming the main large manzanita species in the mixed conifer, Jeffrey pine, and red fir forest. Flowering begins early in the spring while snow is still on the ground in most years.

Arctostaphylos uva-ursi

Bearberry, Kinnickinnick

Ericaceae (Heather family)

Low-growing shrub < 5 dm; mat-forming, with trailing stems that root at nodes, may or may not have a burl.

LEAVES: Blades 1–2.5 cm, 0.5–1.5 cm wide, obovate to oblanceolate, with tips rounded; upper surface shiny, dark green, and lower surface lighter, duller.

FLOWERS: Inflorescence a short raceme of pinkish or white flowers, corollas 4–5 mm. Ovary hairless.

FRUITS: 6–12 mm wide, round, pink to bright red, shiny, hairless.

SIERRA DISTRIBUTION: Mono County.

ELEVATION: 2400–3300 m (for SN, lower elsewhere in CA)

HABITAT: Rocky outcrops, sandy or gravelly soils in coniferous forest; montane and subalpine zones.

FLOWERING TIME: May–Jul

NOTES: This circumboreal species is more common along the Northern CA coast; in the SN it is limited to the east slope in the Convict Creek drainage.

Arctostaphylos viscida ssp. *mariposa* ✿

Mariposa Manzanita, Sticky Whiteleaf Manzanita

Ericaceae (Heather family)

Large shrub 1–5 m, burl absent. Recovery after fire is entirely from seedlings.

LEAVES: Erect, glandular-bristly, and rough to the touch; blades 2–5 cm, 1.5–4 cm wide, oval to round, with upper and lower surfaces a similar shiny dark green.

FLOWERS: Inflorescence is an open panicle with 4–7 branches; twigs and branches densely glandular hairy. Corollas 6–7 mm, pink to whitish. Ovary glandular-bristly.

FRUITS: 6–8 mm wide, round, reddish-brown, sticky due to glandular hairs.

SIERRA DISTRIBUTION: West slope, all.

ELEVATION: 600–1850 m

HABITAT: Dry, open areas, forests; foothills and lower montane zones.

FLOWERING TIME: Mar–Apr

NOTES: A major component of the chaparral in much of the SN, forming dense, impenetrable fields over many acres. Grows with the superficially similar *A. mewukka* ssp. *mewukka*, but that species has hairless leaves and fruits, and a basal burl. Overlaps with *A. patula* in the lower-elevation ponderosa pine forest. *A. viscida* ssp. *viscida* is similar but does not have glandular-hairy twigs or fruits, though the inflorescence axis is similarly sticky.

Cassiope mertensiana

White Heather

Ericaceae (Heather family)

Mat-forming, evergreen shrub ascending 30 cm, with branches that creep along the ground.

LEAVES: Opposite, 3–6 mm, ovate to lanceolate and with a keel on the back, arranged densely in 4 ranks so as to mostly or completely obscure the stem.

FLOWERS: Nodding on long pedicels toward stem tip; calyx 4- to 5-lobed; corollas 5–6 mm, 4- to 5-lobed, bell-shaped, white or light pink.

FRUITS: Capsule 5-chambered.

SIERRA DISTRIBUTION: Fresno and Inyo Counties north.

ELEVATION: 1800–3500 m

HABITAT: Moist areas, often around rocks; subalpine and alpine zones.

FLOWERING TIME: Jul–Aug

NOTES: One of the most common subalpine and alpine groundcovers.

Chimaphila menziesii

Little Prince's Pine

Ericaceae (Heather family)

Woody, evergreen perennial herb to 15 cm; rhizomatous.

LEAVES: 1–5 cm, lanceolate to elliptic, toothed or entire, sometimes with white mottling on upper surface.

FLOWERS: Inflorescences of 1–3 flowers, hairless or with minute papillae, with bracts widely ovate or obovate that persist during flowering. Flowers nodding, 5-merous; sepals ± 5 mm; petals ± 6 mm, white, maturing pink. Anther filaments hairy all around at base.

FRUITS: Capsule erect, 5-chambered, with valves opening from tips.

SIERRA DISTRIBUTION: West slope, spotty distribution from Tulare County north.

ELEVATION: 500–2600 m

HABITAT: Shaded areas in conifer forest; montane zone.

FLOWERING TIME: Jun–Aug

NOTES: Compare with *C. umbellata*. There are no *C. menziesii* records for Mariposa or Madera Counties.

Chimaphila umbellata

Pipsissewa, Prince's Pine

Ericaceae (Heather family)

Woody, evergreen perennial herb 15–30 cm; rhizomatous.
LEAVES: In whorls of 3–8, 3–7 cm, oblanceolate, and toothed.
FLOWERS: Inflorescences of 3–10 flowers, glandular-hairy, with bracts narrow, deciduous. Flowers nodding, 5-merous; sepals < 2 mm; petals 5–6 mm, pink; base of filament hairy only on margins.
FRUITS: Capsule erect, 5-chambered, valves open from tips.
SIERRA DISTRIBUTION: West slope, Tulare County north.
ELEVATION: 500–3000 m
HABITAT: Dry areas in coniferous forest; montane zone.
FLOWERING TIME: Jun–Aug
NOTES: More common in northern Sierra; south of El Dorado County only vouchered from Mariposa and Tulare Counties.

Gaultheria humifusa

Alpine Wintergreen

Ericaceae (Heather family)

Mat-forming, evergreen shrub < 2 dm, with a wintergreen aroma and hairless stems (sometimes puberulent).
LEAVES: Alternate, 0.5–3 cm, round to oval, leathery, with margins entire or minutely toothed.
FLOWERS: 1 per leaf axil; sepals 5, fused, red with age, hairless; corollas bell-shaped, white, lobes 5.
FRUITS: Capsule 5–7 mm diameter, round, enclosed in calyx fleshy and red.
SIERRA DISTRIBUTION: West slope to crest, Tulare to El Dorado Counties.
ELEVATION: 2450–3300 m
HABITAT: Moist sites; subalpine zone.
FLOWERING TIME: Jun–Aug
NOTES: Another species, slender wintergreen, *G. ovatifolia*, occurs 400–1900 m on the west slope from El Dorado County north; stem is ≤ 3.5 dm, with the calyx hairy.

Kalmia polifolia ssp. *microphylla*

Bog Kalmia

Ericaceae (Heather family)

Low-growing, evergreen shrub 1–3 dm; rhizomatous.
LEAVES: Opposite, 4–20 mm, 3–6 mm wide, ovate to oblong; margins often rolled under. Upper surface is dark green and hairless; lower surface is densely hairy.
FLOWERS: Erect in upper axils, 8–12 mm wide, dark pink to purple, and shallowly bowl-shaped with 5 rounded lobes.
FRUITS: Capsule 4–5 mm wide, 5-chambered.
SIERRA DISTRIBUTION: Tulare and Inyo Counties north.
ELEVATION: 1800–3500 m
HABITAT: Moist to wet areas; montane, subalpine, and alpine zones.
FLOWERING TIME: Jul–Aug
NOTES: Common in meadows and along streams of the subalpine and alpine zones. Contains andromedotoxin.

Leucothoe davisiae

Sierra Laurel

Ericaceae (Heather family)

Erect, evergreen shrub to 1.5 m, typically forming extensive patches.
LEAVES: Alternate, 2–6 cm, 1–2 cm wide, entire, and leathery.
FLOWERS: Racemes < 15 cm. Corollas white, urn-shaped, longer than calyx, and nodding.
FRUITS: Capsule.
SIERRA DISTRIBUTION: West slope, Tulare County north.
ELEVATION: 900–2600 m
HABITAT: Moist areas at meadow edges, along streams; montane through subalpine zones.
FLOWERING TIME: Jun–Aug
NOTES: A beautiful shrub, rarely encountered. This plant is toxic to all livestock; 1 oz of leaves can kill a sheep.

Orthilia secunda

One-Sided Wintergreen

Ericaceae (Heather family)

Evergreen perennial herb < 20 cm; from long, creeping rhizomes.

LEAVES: Crowded toward base of stem; 1.5–6 cm, ovate, entire to finely toothed, thin and shiny green.

FLOWERS: Inflorescence is a 1-sided, arching raceme; flowers are erect in bud, then nodding when open. Sepals 5, much shorter than corollas; petals yellow-green to creamy white, each with 2 basal tubercles (bumps) on the inner surface.

FRUITS: Capsule, opening from base.

SIERRA DISTRIBUTION: Both slopes, Tulare and Inyo Counties north.

ELEVATION: 1000–3200 m

HABITAT: Dry, shaded areas in conifer forest; montane through alpine zones.

FLOWERING TIME: Jun–Aug

NOTES: Distinctive with its 1-sided inflorescence and thin, bright green leaves. Circumboreal.

Phyllodoce breweri

Red Mountain Heather

Ericaceae (Heather family)

Low-growing, evergreen shrub up to 3 dm; rhizomatous.

LEAVES: Alternate, 6–15 mm, < 5 mm wide, needlelike, with margins strongly rolled under.

FLOWERS: Clustered near stem tips; sepals 5, corollas cup-shaped, with 5 lobes about the same length as the tubes. Stamens extremely long relative to petals, with filaments much longer than anthers.

FRUITS: Capsule 4–5 mm wide, 5-chambered.

SIERRA DISTRIBUTION: Both slopes, Tulare and Inyo Counties north.

ELEVATION: 1800–3600 m

HABITAT: Moist to dry, rocky areas and openings; montane through alpine zones.

FLOWERING TIME: Jul–Aug

NOTES: A characteristic species of the SN subalpine and alpine zones. Compare with *Kalmia polifolia*.

Pleuricospora fimbriolata
Fringed Pinesap
Ericaceae (Heather family)

Erect, stout, white or tan perennial herb 10–20 cm; hairless.
LEAVES: 8–12 mm, scalelike.
FLOWERS: Inflorescence is a spikelike raceme. Flowers are white at first and turn brownish later; sepals 4, 5–8 mm; petals 4 or 5, 8–10 mm, free, hairless, with jagged margins. Ovary superior; stamens 8, filaments hairless.
FRUITS: Berry < 1 cm wide, cream-colored to white.
SIERRA DISTRIBUTION: West slope, Tulare County north.
ELEVATION: < 2800 m
HABITAT: Floor of coniferous forest, usually in thick humus; montane zone.
FLOWERING TIME: Jun–Aug
NOTES: A nonphotosynthetic plant that relies on mycrorrhizal fungi to obtain carbon. A similar but much rarer plant is *Pityopus californicus*, California pinefoot. It has hairy filaments and petals with hairy inner surfaces.

Pterospora andromedea
Pinedrops
Ericaceae (Heather family)

Erect, nongreen perennial herb 30–100 cm; sticky, due to dense, glandular hairs. The plant persists intact after seed dispersal well into the next year. Entire plant is light cream to reddish or purplish, sometimes with faint longitudinal striping at base.
LEAVES: Crowded at base, scalelike.
FLOWERS: Racemes with nodding cream to yellowish flowers resembling upside-down urns. Sepals 5; corollas of united petals with 5 small recurved corolla lobes, 6–9 mm. Ovary superior; stamens 10, held within corolla tube.
FRUITS: Capsule < 2.5 cm wide, indehiscent.
SIERRA DISTRIBUTION: All.
ELEVATION: < 2900 m
HABITAT: Floor of coniferous forest, usually in thick humus; montane zone.
FLOWERING TIME: Jun–Aug
NOTES: The tallest and the most common of the nonphotosynthetic Ericaceae. The plant can look quite different when first emerging, prior to elongation of the inflorescence, but the united corolla lobes distinguish it from *Pleuricospora* or *Pityopus*, both of which have 5 free (not united) petals. Also compare with *Allotropa*, which has distinct red-and-white striping, while pinedrops may occasionally have indistinct yellowish-and-reddish striping.

Pyrola aphylla

Leafless Wintergreen

Ericaceae (Heather family)

Perennial herb < 6 dm, with stalks often clustered; mycoheterotrophic (nonphotosynthetic).

LEAVES: Absent or vestigial, < 1 cm.

FLOWERS: Racemes of 7–25 flowers, with bract below each flower 0.5–1.5 x length of the pedicels. Flowers bilateral, with 5 sepals 2–3 mm, 5 petals 5–7.5 mm, cream to pink. Styles exserted and curved downward; anthers 10, 2.8–4.5 mm.

FRUITS: Capsule, opening from base; margins of valves are fibrous.

SIERRA DISTRIBUTION: West slope.

ELEVATION: 600–2300 m

HABITAT: Dry sites with thick duff in coniferous forest, hardwood forest; montane zone.

FLOWERING TIME: Jun–Aug

NOTES: Formerly considered a nonphotosynthetic form of *Pyrola picta* ("forma *aphylla*").

Pyrola asarifolia ssp. *asarifolia*

Bog Wintergreen

Ericaceae (Heather family)

Evergreen perennial herb 10–35+ cm; mat-forming, rhizomatous.

LEAVES: Round to ovate, petioles + blades 3–13 cm, often purplish or brownish underneath.

FLOWERS: Bracts below flowers are 8–15 mm, 1–1.5 x pedicel length, sharp-tipped. Flowers bilateral or nearly radial, pink or purple; sepals 2–3.5 mm; petals 4.8–8.6 mm. Style 6–8 mm, exserted, bent downward at base; anthers 2–3 mm.

FRUITS: Capsule, opening from base; margins of valves are fibrous.

SIERRA DISTRIBUTION: Both slopes, Tulare and Inyo Counties north.

ELEVATION: 1000–3300 m

HABITAT: Moist sites in conifer forest, streambanks, meadows, swampy areas; montane zone.

FLOWERING TIME: Jul–Sep

NOTES: Could be confused with *P. picta*, which occasionally has pink flowers, though they are usually white. *P. picta* differs by having white-veined leaves and flower bracts shorter than pedicels.

Pyrola minor

Common Wintergreen

Ericaceae (Heather family)

Evergreen perennial herb < 20 cm; rhizomatous.

LEAVES: Basal < 5 cm, round, thin, nonglossy, dark green.

FLOWERS: Inflorescence bracts 3–7 mm, longer than pedicels; flowers white to light pink, with corollas radial, appearing closed; sepals 1.5–2 mm; petals 3–5 mm. Styles included, straight; anthers < 1.5 mm.

FRUITS: Capsule, opening from base; margins of valves are fibrous.

SIERRA DISTRIBUTION: West slope, Tulare to El Dorado Counties; east slope, Inyo and Mono Counties.

ELEVATION: 1000–3300 m

HABITAT: Moist, shady areas in conifer forest; montane through alpine zones.

FLOWERING TIME: Jul–Sep

NOTES: Circumboreal, sparsely distributed in the SN. Could be confused with *Orthilia secunda*, but inflorescence is not 1-sided, and the axis is straight, not arched.

Pyrola picta

White-Veined Wintergreen, White-Veined Shinleaf

Ericaceae (Heather family)

Evergreen, perennial herb with inflorescence stalks 1–2 dm (10–20 cm); rhizomatous.

LEAVES: Ovate to elliptic, leathery, with upper surface dark green, usually with white veins; lower surface < 10 cm, purplish (and sometimes leaves are oblanceolate with no white veins).

FLOWERS: Inflorescence bracts shorter than pedicels, slender, and sharp-pointed. Flowers more bilateral than radial; sepals > 2 mm; petals 7–8 mm, creamy-white, greenish, or pinkish. Styles > 10 mm, exserted, downwardly curved; anthers 2.5–5 mm.

FRUITS: Capsule, opening from base; margins of valves are fibrous.

SIERRA DISTRIBUTION: All.

ELEVATION: 400–3000 m

HABITAT: Dry sites in coniferous forest; montane zone.

FLOWERING TIME: Jun–Aug

NOTES: This common dweller of the forest floor is the most abundant *Pyrola* in the SN.

Rhododendron columbianum

Western Labrador Tea

Ericaceae (Heather family)

Evergreen shrub 50–150 cm; stems erect or decumbent, sometimes rooting and forming large patches. Twigs are yellowish, glandular-pubescent.

LEAVES: Alternate, 1.5–3 cm, entire, leathery.

FLOWERS: Inflorescence is dense, occurs at branch tips. Calyx lobes 5, ciliate; corollas composed of 5 distinct petals (not fused). Petals 5–8 mm, creamy-white; anther filaments very long.

FRUITS: Capsule.

SIERRA DISTRIBUTION: All.

ELEVATION: 1800–3600 m

HABITAT: Meadows, wet areas, streamsides; montane, subalpine, and alpine zones.

FLOWERING TIME: Jun–Aug

NOTES: Formerly *Ledum glandulosum*. Despite the common name Labrador tea, it would be a bad idea to use it for tea as this plant contains andromedotoxin, a toxic compound that lowers blood pressure and causes muscle weakness, nausea, and vomiting.

Rhododendron occidentale

Western Azalea

Ericaceae (Heather family)

Deciduous, highly branched shrub 1–3 m.

LEAVES: Simple, alternate, 3–10 cm, 1–3 cm wide, ovate to oblanceolate, with margins minutely toothed.

FLOWERS: Inflorescences of 3–15 funnel-shaped flowers. Calyx composed of 5 fused sepals; corollas 3–5 cm wide, of 5 fused petals, < 5 cm; the upper petal usually has a central yellow blotch. Stamens 5, longer than petals.

FRUITS: Capsule 1–2 cm, 5-chambered.

SIERRA DISTRIBUTION: West slope, all.

ELEVATION: 400–3000 m

HABITAT: Moist to wet areas around meadows and along streams; montane zone.

FLOWERING TIME: May–Aug

NOTES: Flower color can vary from that shown in the photo to pink or salmon. Foliage often gives off a skunky odor, but flowers have a pleasant, sweet fragrance.

Sarcodes sanguinea

Snow Plant

Ericaceae (Heather family)

Entire plant is stout, red, fleshy, and glandular-hairy, 5–30 cm. Although single-stemmed, plants often emerge in dense clusters.

LEAVES: Clustered at base of plant, scalelike.

FLOWERS: Racemes dense, spikelike; flowers 12–18 mm, red, subtended by long, ciliate bracts; sepals 5, petals 5. Ovary superior, style < 8 mm; stigma 2–3 mm wide, capitate; stamens 10.

FRUITS: Capsule < 2.5 cm wide, indehiscent.

SIERRA DISTRIBUTION: Mostly west slope, all.

ELEVATION: 1200–2800 m

HABITAT: Floor of coniferous forest, often in thick duff or humus; montane zone.

FLOWERING TIME: May–Jul

NOTES: Formerly thought to be a saprophyte (an organism that obtains its nutrients from dead organic matter), but now better understood as a mycoheterotroph, a nonphotosynthetic plant that obtains its fixed carbon from living fungi that are also connected to nearby conifers. The snow plant is frequently associated with *Rhizopogon ellenae*, a nongilled underground mushroom that is a common mycorrhizal associate of red fir (the fungus helps red fir roots to obtain nutrients).

Vaccinium cespitosum

Dwarf Bilberry

Ericaceae (Heather family)

Ground-hugging, dwarfed, deciduous shrub < 5 dm, forming extensive mats from rhizomes.

LEAVES: Alternate, 1–3 cm, obovate with wedge-shaped bases; upper surface hairless, lower surface glandular; glaucous and margin-toothed.

FLOWERS: Solitary in axils, calyx lobes absent; corollas < 6 mm, urn-shaped, pink or white.

FRUITS: Berry < 9 mm wide, blue, glaucous.

SIERRA DISTRIBUTION: Tulare and Inyo Counties north.

ELEVATION: 1000–3700 m

HABITAT: Moist to wet areas; subalpine and alpine zones.

FLOWERING TIME: May–Aug

NOTES: *V. cespitosum* covers the ground over vast areas in the subalpine and alpine regions of the Sierra. The leaves turn brilliant red, orange, or purple in the fall. The berries are edible and tasty. The SN endemic Sierra green sulfur butterfly (*Colias behrii*) uses bilberry as its host plant, along with willows.

Vaccinium uliginosum ssp. *occidentale*

Western Blueberry

Ericaceae (Heather family)

Low-growing (but not usually prostrate), deciduous shrub 3–7 dm; twigs are round in x.s. (not angled) and not green; hairless.

LEAVES: Alternate, 1–2 cm, ovate or elliptic, paler below, with margins entire.

FLOWERS: Inflorescence is a raceme of 1–4 flowers; calyx lobes 4–5, longer than calyx tubes; corollas oblong, white or pink, with very short lobes.

FRUITS: Berry < 6 mm wide, blue-black, glaucous.

SIERRA DISTRIBUTION: Tulare and Inyo Counties north.

ELEVATION: 1500–3400 m

HABITAT: Wet, acidic meadows; montane through alpine zones.

FLOWERING TIME: May–Aug

NOTES: The presence of western blueberry in a wet meadow that feels bouncy to walk on likely indicates that you're in a fen, a special type of wetland where accumulation of organic matter (peat) exceeds decomposition and acidic conditions prevail. Several rare moss species grow in fens, and mapping and characterization of SN fens to ensure their conservation have been ongoing since 2000.

Croton setiger

Turkey Mullein

Euphorbiaceae (Spurge family)

Low-mounded annual herb < 2 dm, < 8 dm wide; densely hairy, with many hairs stellate. Sap clear.

LEAVES: Alternate, with petioles 1–5 cm; blades 1–6 cm, ovate.

FLOWERS: Monoecious. Male inflorescences terminal, sepals 5–6, no petals, stamens 6–10. Female inflorescences axillary, without sepals or petals, style 1.

FRUITS: ± 4 mm wide, round. Seed 1.

SIERRA DISTRIBUTION: West slope, all.

ELEVATION: < 1800 m

HABITAT: Dry, open areas (roadsides, disturbed areas) < 1000 m; foothill and montane zones.

FLOWERING TIME: May–Oct

NOTES: Formerly *Eremocarpus setigerus*. Toxic. Used by California Indians to catch fish by placing herbage in small streams to slow the fish down and allow them to be caught by hand. Seeds are eaten by mourning doves, hence another common name: "dove weed."

Euphorbia albomarginata

Rattlesnake Sandmat

Euphorbiaceae (Spurge family)

Prostrate perennial herb, with stems repeatedly forking. Sap milky.

LEAVES: Opposite, subsessile, with stipules connate, forming ciliate scales; blades 3–8 mm, round to oblong, and base asymmetric.

FLOWERS: Cyathia 1 per node; involucre < 2.5 mm, nectary glands 4, elongated parallel to rim, and maroon. Appendages white, petal-like, wider than glands, and scalloped or entire. Staminate flowers 15–30, pistillate flowers with the 3 styles each forked ½–¾ of their length.

FRUITS: Pedicels elongate with maturity. Capsule 2–2.5 mm, ovoid, angled, and hairless.

SIERRA DISTRIBUTION: Southern Sierra, primarily Kern and Tulare Counties (Inyo County localities are east of the SN).

ELEVATION: < 2300 m

HABITAT: Dry, often sandy, flats and slopes; foothill and pinyon-juniper zones.

FLOWERING TIME: Apr–Nov

NOTES: Previously known as *Chamaesyce albomarginata*, now in *Euphorbia*, subgenus *Chamaesyce*.

Euphorbia crenulata

Chinese Caps

Euphorbiaceae (Spurge family)

Erect, hairless annual (sometimes biennial) herb 1.2–6 dm tall. Sap milky.

LEAVES: Alternate, sessile or petioled, 1.5–3.5 cm, obovate, with entire margins.

FLOWERS: Cyathia in clusters subtended by leafy bracts; involucre approximately 2 mm, glands 4, 1–2 mm, crescent-shaped and with 2 horns. Male flowers 11–18, female flowers with styles divided < ½ its length.

FRUITS: 3–3.5 mm, oblong.

SIERRA DISTRIBUTION: West slope, Tulare County north.

ELEVATION: < 2200 m

HABITAT: Dry, open or shady areas; foothill and montane zones.

FLOWERING TIME: Mar–Aug

NOTES: Could be confused with young plants of the noxious perennial leafy spurge, *E. virgata*, but the latter reaches heights of 9 dm and forms extensive mats through rhizomes, and the horns on the gland are narrower and longer than wide in *E. crenulata*. Any sightings of leafy spurge should be passed on to the county agricultural commissioner as this is one of the worst noxious weeds in the West, and it has not yet been allowed to establish widely in the SN.

Euphorbia oblongata (non-native)

Oblong Spurge, Eggleaf Spurge

Euphorbiaceae (Spurge family)

Erect perennial herb 5–8 dm tall; from woody taproots, stems hairy. Sap milky.

LEAVES: Alternate, sessile, 4–6.5 cm, oblong with finely-toothed margins, hairless.

FLOWERS: Involucres 1.5–2.5 mm, with glands 1 mm, elliptic; cyathium with 20–40 male flowers and 1 female flower with the style divided < ½ its length. Bell-shaped, in clusters subtended by larger, bright yellow, leafy bracts.

FRUITS: 3–4.5 mm, round, bumpy, protruding from involucre when mature. Seeds 2.5 mm, ovoid, smooth, brown.

SIERRA DISTRIBUTION: Known from Madera, Tuolumne, Calaveras, Amador, El Dorado, and Nevada Counties, but will likely be found elsewhere.

ELEVATION: < 1000 m (but could be found higher)

HABITAT: Disturbed areas; so far in foothill and montane zones.

FLOWERING TIME: Jun–Aug

NOTES: Originally from southwestern Europe. A noxious weed that can form monocultures that prevent native ecosystems from functioning. Likely toxic to livestock and wildlife like its close relative leafy spurge, *Euphorbia virgata*. Any new locations should be reported to the local county agricultural department.

Acmispon americanus var. *americanus*

American Bird's-Foot Trefoil, Spanish Clover

Fabaceae (Pea family)

Hairy, prostrate or erect annual herb with stem 0.5–6 dm.

LEAVES: Pinnate (sometimes simple); leaflets 3, 10–20 mm.

FLOWERS: Solitary, with calyx lobes long and slender, much longer than the tube, and shaggy-hairy; corollas 5–9 mm, white to pink (or yellowish).

FRUITS: 1.5–3 cm, straight, flat with curved beak 0.9–1.7 mm, dehiscent.

SIERRA DISTRIBUTION: All.

ELEVATION: < 2400 m

HABITAT: Many and varied habitats, roadsides, disturbed areas; foothill and montane zones.

FLOWERING TIME: May–Oct

NOTES: Formerly *Lotus purshianus* var. *purshianus*, and historically called Spanish clover, but native to North America. Extremely common; an important native colonizer of freshly disturbed areas, and nitrogen-fixing, as are all legumes.

Acmispon argophyllus var. *fremontii*

Silver-Leaved Lotus

Fabaceae (Pea family)

Silvery-hairy, prostrate perennial herb with stems 1–6 dm (up to 10 dm).

LEAVES: Pinnately compound with 3–7 leaflets, 6–12 mm, with dense hairs that usually completely hide the leaf surface.

FLOWERS: Inflorescences ≥ 1.5 cm wide, of 10–15 flowers. Calyx shaggy-hairy, lobes no longer than tubes; flowers 8–12 mm, yellow, fading to copper orange.

FRUITS: Pods 1–1.5 cm x 1.8–2 mm wide. Bodies bent, with long-tapered beaks, exserted from the calyx, indehiscent.

SIERRA DISTRIBUTION: West slope.

ELEVATION: 600–1800 m

HABITAT: Gravelly or rocky openings, slopes, roadsides; foothill and montane zones.

FLOWERING TIME: Apr–Jul

NOTES: Compare with *Acmispon decumbens* var. *decumbens*, with which this species hybridizes. Another var., *A. argophyllus* var. *argophyllus*, has inflorescences < 1 cm wide with only 4–8 flowers and occurs < 1600 m from Mariposa County south in similar habitat. The varieties intergrade.

Acmispon brachycarpus

Foothill Deervetch, Short Podded Lotus

Fabaceae (Pea family)

Mat-forming annual herb with stems 0.5–4 dm; covered with soft, spreading hairs; sometimes fleshy, with stems prostrate to ascending.

LEAVES: Compound, usually with 4 alternate leaflets on a flattened axis. Leaflets alternate, 4–12 mm, obovate or elliptic.

FLOWERS: Sessile, borne singly in leaf axils; calyx 3–6 mm, with long lobes up to twice the length of the tube; corollas 5–9 mm, yellow, aging to red.

FRUITS: 6–12 mm, 3–4 mm wide, dehiscent, straight, flat, and with upturned beaks 0.5–1.5 mm. Seeds 2–5.

SIERRA DISTRIBUTION: West slope.

ELEVATION: < 1700 m

HABITAT: Dry slopes, roadsides; foothill and lower montane zones.

FLOWERING TIME: Mar–Jun

NOTES: Formerly *Lotus humistratus*. *A. wrangelianus* looks similar and occurs in the same range and elevations but can be strigose or long soft-hairy, has shorter (0.8–1.2 x calyx tube) calyx lobes and longer (10–18 mm) and narrower (2.3–3 mm wide) fruits with 3–7 seeds.

Acmispon decumbens var. *decumbens*

Sierra Nevada Lotus

Fabaceae (Pea family)

Hairy, prostrate perennial herb, forming mats up to 60 dm across (larger in good water years).

LEAVES: Pinnately compound with 3–5 leaflets, 4–12 mm, green with dense hairs that do not hide the leaf surface.

FLOWERS: Inflorescences of 5–12 flowers, with calyx hairless or strigose; flowers 5–10 mm, yellow to yellow-orange, wings longer than keel.

FRUITS: Body 6–10 mm x 1.8–2.2 mm wide, bent with long, tapered, recurved beaks, exserted from the calyx, indehiscent.

SIERRA DISTRIBUTION: West slope, all.

ELEVATION: < 2700 m

HABITAT: Dry openings, slopes, roadsides; montane zone.

FLOWERING TIME: May–Aug

NOTES: Formerly *Lotus nevadensis* var. *nevadensis*, then *Acmispon n.* var. *n.* Compare with *Acmispon argophyllus*, with which this species hybridizes. This mat-forming perennial quickly colonizes disturbed areas and burns, covering bare soil to slow erosion and providing nectar to native pollinators.

Acmispon glaber var. *glaber*

Deerweed, California Broom

Fabaceae (Pea family)

Bushy perennial herb to subshrub 5–20 dm; woody base (wood green), hairless or strigose.

LEAVES: Pinnately compound, 6–15 mm; upper leaflets 3, lower 3–6, and leaflets often deciduous (giving plant a bare look); stipules (if present) black and glandlike.

FLOWERS: Inflorescences of 2–7 flowers, sessile; calyx 2.5–5 mm, lobes 1–2 mm, < tube; corollas 7–12 mm.

FRUITS: 1–1.5 cm, indehiscent, exserted, curved; either spreading or pendent; beaks 2–3 mm, also curved.

SIERRA DISTRIBUTION: West slope, all.

ELEVATION: < 1500 m

HABITAT: Dry slopes, disturbed areas, roadsides; foothill and lower montane zones.

FLOWERING TIME: Mar–Aug

NOTES: Formerly *Lotus scoparius* var. *scoparius*. Sometimes confused with the noxious weed *Cytisus scoparius*, Scotch broom, which is much larger with larger flowers and fruit. A beneficial native colonizer of bare soil, and helps to minimize erosion and runoff.

Acmispon parviflorus

Small-Flowered Lotus

Fabaceae (Pea family)

Ascending to erect annual herb 0.5–4 dm, usually hairless (sometimes sparsely strigose).

LEAVES: Pinnately or palmately compound, with 3–5 leaflets, 4–12 mm. Leaf axis sometimes flattened.

FLOWERS: Borne singly, calyx lobes equal in length to tube; corollas 4–6 mm, creamy to dark pink.

FRUITS: 1.5–2.5 cm, dehiscent, straight with wavy margins, beaks curved.

SIERRA DISTRIBUTION: West slope, all.

ELEVATION: < 1300 m

HABITAT: Open areas, roadsides; foothill and montane zones.

FLOWERING TIME: Mar–May

NOTES: Formerly *Lotus micranthus.* Abundant and often overlooked, as it's a small, demure plant with tiny flowers.

Acmispon strigosus

Strigose Bird's-Foot Trefoil

Fabaceae (Pea family)

Prostrate annual herb; shiny, green fleshy leaves, strigose or hairless.

LEAVES: Pinnately compound, with 4–9 (usually 5) narrow leaflets up to 10 mm. Leaf axis is flattened.

FLOWERS: Borne singly, with calyx lobes shorter than tube; corollas 5–10 mm, yellow, turning orange or red with age.

FRUITS: 1–3.5 cm, straight with upturned tips, dehiscent.

SIERRA DISTRIBUTION: All.

ELEVATION: < 2300 m

HABITAT: Dry slopes, roadsides, often in gravelly/sandy soil; foothill and montane zones.

FLOWERING TIME: Mar–Jun

NOTES: Formerly *Lotus strigosus.* Abundant after wildfire, this plant is often the first native plant to germinate after fire or in burn piles.

Astragalus bolanderi
Bolander's Locoweed
Fabaceae (Pea family)

Erect or decumbent perennial herb 1.5–4 dm, lacking basal leaves, hairless to sparsely hairy.

LEAVES: Alternate, pinnately compound, 3–16 cm, leaflets 13–27.

FLOWERS: 5–30 per inflorescence, white to cream, sometimes tinged with purple; banner 13–17.6 mm, longer than wings and keel, recurved ± 40°.

FRUITS: Pods 13–30 mm x 5–12 mm, inflated, incurved; on a stalklike base (beyond the pedicels).

SIERRA DISTRIBUTION: West slope, Tulare County north.

ELEVATION: 1400–3300 m

HABITAT: Dry, sandy/gravelly or rocky openings in coniferous forest, at meadow edges; montane and subalpine zones.

FLOWERING TIME: Jun–Sep

NOTES: *A. bolanderi* has the most leaflets of any SN *Astragalus*.

Astragalus gambelianus
Gambel Milkvetch, Little Blue Loco
Fabaceae (Pea family)

Slender annual herb 2–30 cm; stems erect to decumbent, herbage with tiny, incurved hairs.

LEAVES: 1–4 cm, with 7–15 leaflets, 1–9 mm, tips blunt or notched.

FLOWERS: Inflorescences black-hairy, of 4–15 flowers; corollas white, tinged purple, banner 2.5–3.3 mm, recurved 25°, keel 2–2.5 mm. Flowers.dense and ascending at first, spaced farther apart as flowering proceeds, and ultimately reflexed.

FRUITS: Pods 2.8–4.2 mm x 2.4–3.6 mm wide, reflexed, minutely hairy, and wrinkled at edges.

SIERRA DISTRIBUTION: West slope.

ELEVATION: <1450 m

HABITAT: Grassy openings; foothill zone.

FLOWERING TIME: Mar–Jul

NOTES: A very cute, little annual *Astragalus*! Easy to overlook as it tends to blend in with annual grasses and other plants in its foothill habitat.

Astragalus gibbsii

Gibbs' Milkvetch

Fabaceae (Pea family)

Leafy, gray-green perennial herb 1.5–3.5 dm; ± prostrate, with long, spreading hairs 0.5–0.8 mm, and many stems.

LEAVES: 1.5–9.5 cm, with small stipules; leaflets 7–19, 4–20 mm.

FLOWERS: Inflorescences of 10–30. Flowers are reflexed, each flower with a pouch at base of calyx on upper side; banner 14–18 mm, keel 12–15 mm.

FRUITS: Pods 22–30 mm x 4–8 mm wide, oblong, swollen but laterally compressed, curved to ¼ or ½ circle at maturity, and densely wavy-hairy; pendent from stalklike base (not the pedicel) 7–22 mm.

SIERRA DISTRIBUTION: Mostly east slope, Mono County north.

ELEVATION: 1250–2400 m

HABITAT: Dry, sandy and rocky areas; montane and sagebrush zones.

FLOWERING TIME: May–Jul

NOTES: Scattered sparingly in sagebrush/ponderosa pine forests. Look for it in the Genessee Valley area of Plumas County.

Astragalus purshii var. *lectulus*

Woolly-Pod Locoweed

Fabaceae (Pea family)

Matted perennial herb < 10 cm; herbage silvery-hairy with long hairs 1–2.3 mm.

LEAVES: Pinnately compound, 1–5 cm, leaflets 3–11, 2–10 mm.

FLOWERS: Inflorescences are located at the level of the leaves, with 1–5 flowers; calyx 5.6–8.8 mm; corollas pink to purple; banner 10.3–15 mm, recurved ± 40º; keel 9.4–11.7 mm.

FRUITS: Pods 7.5–15 mm x 4–8 mm wide, densely covered with white and woolly hairs.

SIERRA DISTRIBUTION: East slope and crest, Tulare County (Monache Meadows), Inyo to Alpine Counties.

ELEVATION: 1500–3700 m

HABITAT: Dry flats; alpine and sagebrush/pinyon-juniper zones.

FLOWERING TIME: May–Aug

NOTES: Also present in the SN is var. *purshii*, which usually has white or cream-colored flowers.

Astragalus whitneyi var. *whitneyi*

Whitney's Locoweed

Fabaceae (Pea family)

Low-growing, openly branched perennial herb with stems 4–40 cm; greenish to grayish.

LEAVES: 3–11 cm; stipules have papery sheaths; leaflets 5–21, 2–21 mm, well-spaced. Hairs incurved-ascending to appressed.

FLOWERS: Inflorescences short and dense, of 3–15 pink to purple flowers; banner 8.3–16.5 mm.

FRUITS: Pods 1.5–3 cm, inflated, mottled red, hairless, on stalklike base (beyond the pedicel) 2.4–5 mm.

SIERRA DISTRIBUTION: East slope and mostly Sierra crest.

ELEVATION: 1550–3500 m

HABITAT: Dry, rocky flats and slopes; subalpine, alpine, and sagebrush zones.

FLOWERING TIME: May–Sep

NOTES: Look for this locoweed in Mono County around Sonora Pass, along the trail to Dorothy Lake in Convict Basin.

Cercis occidentalis

Western Redbud

Fabaceae (Pea family)

Deciduous shrub (or small tree) to 7 m; hairless, flowering before leaves emerge.

LEAVES: Alternate, < 10 cm, roundish, opening toward the end of flowering period.

FLOWERS: 2–5 per inflorescence, with many inflorescences per branch; calyx 5-toothed, reddish; corollas 8–12 mm, bilateral, light pink to magenta; petals 5, 3 upper and 2 lower, all free. Stamens 10, distinct.

FRUITS: Pods 5–8 cm. Seeds 2–8.

SIERRA DISTRIBUTION: West slope, all.

ELEVATION: < 1500 m

HABITAT: Dry slopes, streamsides; foothill and lower montane zones.

FLOWERING TIME: Mar–May

NOTES: The common name "redbud" is for the red to maroon calyx that encloses the pink petals in bud. Redbud is in a different subfamily from the other Pea family plants in this book, distinguished by simple (not compound) leaves, the 5 free (not fused at base) petals, and 10 free stamens. Redbud branches still are used extensively by many Native tribes for basketry. Strong, straight, supple shoots with no branches are kept abundant by burning or cutting, and collected in the fall. The reddish bark is used for red designs on baskets.

Cytisus scoparius (non-native, invasive)

Scotch Broom

Fabaceae (Pea family)

Shrub < 2.5 m; hairless when mature, with stems distinctly 5-angled or -ridged, ridges green.
LEAVES: Alternate, compound; leaflets 3, 5–20 mm, oblong-obovate, hairless.
FLOWERS: Axillary clusters of 1–2 flowers; calyx 2-lipped, hairless; corollas yellow-gold, petals 5, banner 15–20 mm.
FRUITS: Pods 25–40 mm, flattened, hairless except for a fringe of hairs at margins (ciliate). Seeds 5–12.
SIERRA DISTRIBUTION: West slope, Tulare County north.
ELEVATION: < 1800 m
HABITAT: Dry to moist disturbed areas, roadsides; foothill and montane zones.
FLOWERING TIME: Apr–Jul
NOTES: A CDFA noxious weed, introduced as an ornamental plant from southern Europe and Britain. This aggressive shrub expands at the expense of native vegetation, degrades wildlife habitat, and creates a highly flammable monoculture making it a terrible landscape plant in fire-prone areas. Seeds can be viable in soil for up to 30 years; flowers and possibly all plant parts are toxic. Early detection and removal of new plants/infestations is important. The plants are stimulated by fire to produce vigorous populations that quickly become impossible to control.

Genista monspessulana (non-native, invasive)

French Broom

Fabaceae (Pea family)

Evergreen shrub to 3 m; soft silvery-hairy, with stems green, angled.
LEAVES: Alternate, compound; leaflets 3, 10–15 mm (twice as long as wide), oblanceolate, soft-hairy below.
FLOWERS: In axils, 4–10 flowers per inflorescence; calyx 2-lipped, silky hairy; corollas yellow, petals 5, banner 10–15 mm.
FRUITS: Pods 15–25 mm, densely silky-hairy all over. Seeds 3–6.
SIERRA DISTRIBUTION: West slope, Tulare County north.
ELEVATION: < 1100 m
HABITAT: Disturbed areas, roadsides; foothill and montane zones.
FLOWERING TIME: Mar–Jun
NOTES: A CDFA noxious weed, introduced from the Mediterranean, Azores, and the Canary Islands. Flowers and possibly all plant parts are toxic. Plants form monocultures that are extremely difficult to control.

Hoita macrostachya

Leather-Root

Fabaceae (Pea family)

Erect perennial herb < 2 m; from large rhizomes, herbage gland-dotted, stems striated, usually hairy.

LEAVES: Alternate; leaflets 3, 2–10 cm, lanceolate; glandular, but may or may not be hairy on either or both surfaces.

FLOWERS: Spikelike racemes 5–11 cm, with a single bract, 3–11 mm, that falls off early. Flowers 9–10 mm, calyx 4.5–9 mm, with banner conspicuous.

FRUITS: Pods 6–8 mm, brown, with distinct veins. Seed 1, 5–7 mm.

SIERRA DISTRIBUTION: West slope.

ELEVATION: < 1200 m (in the SN; higher elsewhere)

HABITAT: Moist to wet areas; foothill and montane zones.

FLOWERING TIME: May–Aug

NOTES: Formerly *Psoralea macrostachya*. Native American uses include making sewing thread from fibers of the inner bark, making rope from fibers of roots (rhizomes), and boiling roots with basketry plants to dye them yellow. Several medicinal uses are also documented. Compare with *H. orbicularis*, which is much shorter and more of a groundcover.

Hoita orbicularis

Round-Leaved Leather-Root

Fabaceae (Pea family)

Prostrate (to decumbent) perennial herb < 60 cm, gland-dotted, and usually hairy; forms extensive mats held together by extraordinarily strong rhizomes.

LEAVES: Leaflets 3, 3–11 cm, roundish, both surfaces brown-glandular. Stipules 4.5–10 mm; petioles 10–50 cm.

FLOWERS: Spikelike raceme 6–35 cm, with a single bract, 7–27 mm, that falls off early; peduncles 20–70 mm. Flowers 12–23 mm, calyx 15–19 mm, banner 10–16 mm.

FRUITS: Pods 6–9 mm, brown, with distinct veins. Seed 1 per pod, 4–5 mm.

SIERRA DISTRIBUTION: West slope.

ELEVATION: < 1200 m (in the SN; higher elsewhere)

HABITAT: Moist to wet areas; foothill through upper montane zones.

FLOWERING TIME: Apr–Aug

NOTES: Formerly *Psoralea orbicularis*. The rhizomes of this plant are truly like leather! Several large beds became visible below the high water line at Bass Lake, Madera County, during the 2014–17 drought. The Bass Lake Dam inundated a major living area for the Western Mono beginning in 1901; it seems likely these beds may have provided a source of fiber for cordage.

Hosackia crassifolia var. *crassifolia*

Broad-Leaved Lotus

Fabaceae (Pea family)

Stout, hairless (or strigose) perennial herb to 1.5 m; usually erect, stems are hollow at base.

LEAVES: Large, with 9–15 leaflets, 2–3 cm. Stipules are inconspicuous scarious bracts.

FLOWERS: Inflorescences of 12–20 flowers; calyx 5–8 mm, lobes shorter than calyx tube; corollas 12–17 mm, yellowish-green at first, and blotched maroon with age.

FRUITS: 3.5–7 cm, hairless, in a downward-pointing cluster.

SIERRA DISTRIBUTION: Mostly west slope, a few locations on east slope.

ELEVATION: 300–2100 m

HABITAT: Dry slopes, openings, roadsides, burned areas; montane zone.

FLOWERING TIME: May–Aug

NOTES: Formerly *Lotus crassifolius* var. *crassifolius*.

Hosackia oblongifolia var. *oblongifolia*

Meadow Lotus

Fabaceae (Pea family)

Sprawling perennial herb with stems 1–6 dm, usually hairy.

LEAVES: Stipules scarious; leaflets 7–11, 1–2.5 cm, elliptic to oblong, hairy or hairless.

FLOWERS: Inflorescence is an umbel of 2–6 flowers; peduncle bracts absent or directly below umbels; calyx 4–6 mm, lobes shorter than tube; corollas 9–13 mm, banner yellow; other petals white.

FRUITS: 2.5–5 cm, 1.5–2 mm wide, hairless.

SIERRA DISTRIBUTION: All.

ELEVATION: < 2700 m

HABITAT: Moist to wet areas; mostly montane through subalpine zones.

FLOWERING TIME: May–Sep

NOTES: Formerly *Lotus oblongifolius* var. *oblongifolius*. Compare with *H. pinnata*, which looks similar but is hairless and not found > 1700 m.

Hosackia pinnata

Pinnate-Leaved Lotus

Fabaceae (Pea family)

Hairless, sprawling (not erect) perennial herb 1.5–5 dm.

LEAVES: Alternate, pinnately compound; leaflets 5–7, ± opposite, 1–2.5 cm, elliptic or oblong.

FLOWERS: Inflorescence is an umbel of 4–10 flowers; peduncle bracts absent or directly below umbels; calyx 5–7 mm; corollas 10–13 mm, banner yellow and strongly bent or recurved; other petals white.

FRUITS: 3–5 cm, 1.5–2.5 mm wide, linear, hairless.

SIERRA DISTRIBUTION: West slope, Tulare County north.

ELEVATION: 600–1700 m

HABITAT: Wet areas, including in streams; foothill through montane zones.

FLOWERING TIME: May–Jul

NOTES: Formerly *Lotus pinnatus*. Compare with *H. oblongifolia*, which looks similar but is hairy and ranges higher in elevation.

Lathyrus latifolius (non-native, invasive)

Perennial Sweet Pea

Fabaceae (Pea family)

Hairless, robust, vinelike perennial herb, with conspicuously winged stems.

LEAVES: Each leaf has only 2 leaflets, each 5–14 cm; terminal tendril branched, coiled.

FLOWERS: 4–15 per inflorescence; flowers 20–25 mm, large and showy, pink, purple, or red.

FRUITS: Pods hairless.

SIERRA DISTRIBUTION: All.

ELEVATION: < 1900 m

HABITAT: Dry or moist areas mostly near disturbed or urban areas; foothill and montane zones.

FLOWERING TIME: Apr–Jun

NOTES: Mostly found in disturbed areas, the plant is aggressively invasive and can blanket large areas at the expense of native vegetation. Compare with *L. tingitanus*, which also has leaves with only 2 leaflets and broadly winged petioles, but its leaflets are smaller, and inflorescences have 2–3 flowers each. It is distinguished from weedy *Vicia* spp. by only having 2 leaflets, which unroll rather than unfold.

Lathyrus nevadensis var. *nevadensis*

Sierra Nevada Pea

Fabaceae (Pea family)

Erect perennial herb 1.5–4 dm, with angled stems, pubescent.

LEAVES: Stipules < ½ length of leaflets; leaflets 4–8, ± opposite on leaf axis, 1.5–4 cm, elliptic to ovate; tendrils branched or bristlelike.

FLOWERS: Inflorescences of 2–4 flowers; calyx tube > lobes; corollas 13–25 mm, white to blue-purple.

FRUITS: Pods 3–7 cm, 3.6 mm wide, hairless.

SIERRA DISTRIBUTION: West slope, Fresno County north.

ELEVATION: 450–2300 m

HABITAT: Forest floor, slopes, openings; montane zone.

FLOWERING TIME: Apr–Jun

NOTES: Distinguished from the similar-looking and often co-occurring *Vicia americana* most easily by juvenile leaflets, which unroll rather than unfold. A similar-looking species, Brown's brush pea, *L. brownii* occurs from Sierra County north 1000–1700 m on streambanks and in open woodlands. This pea is hairless, has 6–10 leaflets that are not opposite each other, coiled to branched tendrils, and smaller, blue-purple to pink flowers, with corollas 10–14 mm.

Lathyrus sulphureus

Snub Pea, Sulphur Pea

Fabaceae (Pea family)

Climbing, vinelike perennial herb with angled stems 0.5–3 m, hairless or pubescent.

LEAVES: Alternate, pinnately compound, with wide stipules 1–2.5 cm. Leaflets 6–12, 2–5 cm, elliptic to ovate, often alternately arranged on leaf axis rather than paired; tendrils are branched, coiled.

FLOWERS: 10–15 per inflorescence, often densely grouped and tending to be clustered on one side of inflorescence axis. Flowers 11–14 mm, whitish-yellow to bronze orange, sometimes with a tinge of pink or purple; banner often with pink or purple veins.

FRUITS: Pods 4–7 cm, hairless.

SIERRA DISTRIBUTION: West slope, all.

ELEVATION: < 2500 m

HABITAT: Dry slopes, foothill and montane zones.

FLOWERING TIME: Apr–Jul

NOTES: This attractive orange pea is easy to identify; no other species in the SN has bronze-orange flowers. With its large leaflets, this is a good species for clearly seeing features that distinguish *Lathyrus* from *Vicia*: The leaves of *Lathyrus* clearly unroll rather than unfold as they open, and the stigma appears as a tuft of hairs, which are found on all sides of the stigma, even if they are unevenly distributed.

Lathyrus tingitanus (non-native)

Tangier Pea

Fabaceae (Pea family)

Hairless, showy annual herb.
LEAVES: Alternate, with 2 leaflets, 2–6 cm; tendrils branched, coiled. Stipules either small or wide and conspicuous.
FLOWERS: Few-flowered: 2–3 large maroon or crimson flowers per inflorescence. Corollas 25–30 mm.
FRUITS: Pods < 10 cm, hairless.
SIERRA DISTRIBUTION: West slope, patchy distribution.
ELEVATION: < 1300 m
HABITAT: Disturbed areas; foothill and montane zones (so far).
FLOWERING TIME: Apr–Jul
NOTES: Mostly found in disturbed areas, but not as aggressive as perennial sweet pea.

Lupinus albicaulis

Sicklekeel Lupine

Fabaceae (Pea family)

Erect, hairy perennial herb 3–12 dm.
LEAVES: Stems leafy, basal leaves absent; petioles 2–7 cm; leaflets 5–10, 20–70 mm, hairy on both surfaces; stipules slender, tan, and not leaflike.
FLOWERS: Inflorescences 10–44 cm, peduncles 2–12 cm; flowers are whorled, with pedicels 2–7 mm; corollas 12–16 mm, purple, yellowish-white, or light orange; banner back hairless, keel hairless, protruding from wings. Calyx deeply divided; upper lip 6–12 mm, 2-toothed; lower lip 7–13 mm, entire to 3-toothed.
FRUITS: Pods 2–5 cm, silky-hairy. Seeds 3–7, mottled.
SIERRA DISTRIBUTION: Mostly west slope, all.
ELEVATION: 500–3000 m
HABITAT: Dry slopes and flats; foothill through subalpine zones.
FLOWERING TIME: May–Jul
NOTES: This is the only tall perennial lupine with the keel exserted from the wings. Rely on the keel, and the variation in flower colors won't throw you off. From Mono County north, on both slopes 1000–3000 m, grows narrow-flowered lupine, *L. angustiflorus*, a tall, white- to yellow-flowered plant with many traits in common, except the hairy banner back, keel covered by wings, bloom time in Jun–Sep, and preference for volcanic soils in montane, subalpine, and sagebrush/pinyon-juniper zones.

Lupinus albifrons var. *albifrons*

Bush Lupine

Fabaceae (Pea family)

Evergreen, many-branched shrub 0.5–1.5 m (sometimes taller).
LEAVES: Leaflets 6–10, 10–45 mm, silvery-hairy; petioles 2–4 cm.
FLOWERS: Inflorescences 8–30 cm, with flowers blue or purple; banner spot white, aging to violet, keel ciliate on upper edges.
FRUITS: 3–5 cm, covered with long hairs.
SIERRA DISTRIBUTION: West slope, all.
ELEVATION: 300–2000 m
HABITAT: Open flats and slopes, often in rocky sites; foothill through lower montane zones.
FLOWERING TIME: Mar–Jun
NOTES: The only shrubby lupine native to the west slope of the SN. There is another variety in the foothills, var. *collinus*, a subshrub < 0.5 m with a horizontally branching habit that can look like a vigorous perennial with a woody base.

Lupinus andersonii

Anderson's Lupine

Fabaceae (Pea family)

Erect, green perennial herb 2–9 dm, hairy.
LEAVES: Stems leafy, basal leaves absent; petioles 2–6 cm; leaflets 6–9, 20–60 mm; stipules 3–15 mm, not leaflike.
FLOWERS: Inflorescences 2–23 cm, pedicels 1.5–5 mm. Flowers at least somewhat whorled, 9–12 mm, lavender to purple, sometimes yellow, with banner back hairless, and spot turning from white to purple, keel hairless.
FRUITS: 2–4.5 cm, silky
SIERRA DISTRIBUTION: All.
ELEVATION: 900–3200 m
HABITAT: Dry to moist, open areas in forests; montane through subalpine zones.
FLOWERING TIME: Jun–Sep
NOTES: Closely related to *L. albicaulis* and *L. fulcratus*, but keel is usually entirely within wings (unlike *L. albicaulis*) and stipules are not leaflike (as in *L. fulcratus*).

Lupinus arbustus

Spurred Lupine

Fabaceae (Pea family)

Erect, hairy perennial herb 3–7 dm, several-stemmed from base.
LEAVES: Leaflets 7–13, 20–70 mm, oblanceolate, greenish with hairs appressed.
FLOWERS: Inflorescences 7–12 cm, with flowers arranged in open whorls; flowers 8–14 mm, blue to purple, sometimes white or pink. Wings have a patch of hair toward the outer tips, keel ciliate on upper edges, and calyx has a distinct spur 1–3 mm.
FRUITS: Pods 2–2.5 cm, hairy.
SIERRA DISTRIBUTION: West slope, all.
ELEVATION: 500–3200 m
HABITAT: Dry soil in forests or partial shade; montane through subalpine zones.
FLOWERING TIME: May–Jul
NOTES: Distinguished from all other SN lupines by the bulging spur on the upper calyx, along with the patch of hair at the tip of the wing petals. If the wings have no patch of hairs at their tips, you're looking at *L. argenteus* var. *heteranthus*.

Lupinus argenteus var. *meionanthus*

Small-Flowered Sierran Lupine

Fabaceae (Pea family)

Erect or ascending perennial herb 2–9 dm.
LEAVES: Cauline, leaflets 5–9, 12–25 mm, oblanceolate, silvery-silky.
FLOWERS: Inflorescences 5–10 cm, dense; flowers 5–7 mm; banner back and wings are hairless, keel ciliate on upper margins; calyx bulge 0–3 mm (can be variable on the same plant).
FRUITS: Pods 1–3 cm, hairy.
SIERRA DISTRIBUTION: Both slopes, Inyo and Madera Counties north.
ELEVATION: 1500–3500 m
HABITAT: Dry, open areas; montane, subalpine, alpine, and sagebrush zones.
FLOWERING TIME: Jul–Aug
NOTES: Compare with *L. arbustus*. Silver-spurred lupine, *L. argenteus* var. *heteranthus*, grows primarily near the crest and on the east slope at 1000–3000 m. It has upper calyx bulges 1–2 mm, longer (8–14 mm) corollas that are faded blue, violet, or white.

Lupinus benthamii

Spider Lupine

Fabaceae (Pea family)

Erect annual herb 3–7 dm, 1-stemmed or branched from base.

LEAVES: Petioles 6–12 cm; leaflets 7–10, 20–50 mm, linear, hairless or with upper surface hairy.

FLOWERS: Inflorescences 6–30 cm, with flowers whorled or not. Flowers 10–18 mm, pedicels 3–9 mm, petals blue or purple; banner central spot yellow, aging dark pink; keel upper edges hairless.

FRUITS: Pods 5 cm, 5 mm wide, hairy. Seeds 5–8.

SIERRA DISTRIBUTION: West slope, all.

ELEVATION: < 1000 m

HABITAT: Open, often rocky areas in chaparral, oak woodlands; foothill zone.

FLOWERING TIME: Mar–May

NOTES: The name "spider lupine" derives from the way the slender leaflets radiate, spiderweblike, from the tip of the petiole. Spider lupine is a common foothill lupine covering hillsides with wall-to-wall purple in the early spring of good rain years.

Lupinus bicolor

Miniature Lupine

Fabaceae (Pea family)

Erect, hairy annual herb 1–4 dm, often many-branched from base, hairy.

LEAVES: Petioles 1–7 cm; leaflets 5–7, 10–40 mm, 1–5 mm wide (sometimes linear).

FLOWERS: Inflorescences 1–8 cm, with flowers usually in 2–5 loose whorls (sometimes not whorled). Pedicels usually < 3 mm. Flowers blue, with a white banner spot that turns pink with age; banner longer than wide (remove and flatten to determine this), keel ciliate toward the tip; tip of keel is moderately upturned, but not forming a right angle; calyx has upper lip 2–4 mm, lower lip 4–6 mm.

FRUITS: Pods 1–3 cm, 3–6 mm wide, hairy. Seeds 5–8.

SIERRA DISTRIBUTION: West slope, all.

ELEVATION: < 1600 m

HABITAT: Foothills up through the lower mixed conifer forest; foothill and montane zones.

FLOWERING TIME: Mar–May

NOTES: Abundant in springtime on slopes and roadsides, and often found growing intermixed with the showier *L. nanus*. Distinguished from *L. nanus* by the smaller flowers, shorter inflorescences, narrower banner, and the lack of fragrance. Seeds of miniature lupine are a favored food of mourning doves.

Lupinus breweri var. *breweri*

Brewer's Lupine

Fabaceae (Pea family)

Mat-forming, prostrate perennial herb with woody base to subshrub < 2.0 dm, silvery-silky.

LEAVES: Clustered near base of stem; leaflets 7–10, 6–20 mm, stipules 2–5 mm, oblanceolate.

FLOWERS: Inflorescences 1–2 cm, peduncles 1–3 cm, dense; flowers 6–9 mm, petals violet, with banner folded backward, and spot white or yellow; keel and banner back hairless.

FRUITS: 12 mm, silky. Seeds 3–4.

SIERRA DISTRIBUTION: Highest elevations of crest, both sides.

ELEVATION: 1700–3300 m

HABITAT: Dry openings and meadow edges; montane, subalpine, and alpine zones.

FLOWERING TIME: Jun–Aug

NOTES: Common, forms extensive mats on sandy flats in open forest and above tree line. Two other varieties occur in the high SN: Var. *bryoides* (2500–4000 m), which forms dense tufts with caudex showing above ground, flowers 4–7 mm, keel and banner hairless, and leaflets 3–5 mm; and var. *grandiflorus* (2000–3500 m), which is mat-forming and has the caudex showing, and has flowers 6–11 mm, keel ciliate, banner back silky, and leaflets 6–11 mm.

Lupinus citrinus var. *citrinus*

Orange Lupine

Fabaceae (Pea family)

Hairy annual herb 1–4 dm; hairs on stem and leaves up to 2 mm, foliage grayish.

LEAVES: Appear initially in rosettelike clusters, then alternate; petioles 2–7 cm; leaflets 6–9, 15–35 mm x 3–10 mm wide.

FLOWERS: Inflorescences 4–15 cm, peduncles 1–9 cm; flowers spiraled or whorled; pedicels 2.5–5 mm, reflexing with age; bracts 2.5–5 mm, deciduous; calyx 3–5 mm, lips equal; corollas 8.5–12 mm, petals golden-yellow, aging to orange.

FRUITS: 1–2 cm, hairless when mature, on reflexed pedicels. Seeds 3–8, speckled like grains of granitic gravel, perfectly camouflaged.

SIERRA DISTRIBUTION: West slope, Madera and Fresno Counties.

ELEVATION: 600–1700 m

HABITAT: Rocky outcrops and associated gravel/sand; foothill and lower montane zones.

FLOWERING TIME: Apr–Jul

NOTES: Unmistakable with its golden-yellow flowers and grayish, hairy foliage. Endemic to the central SN but abundant in good rain years within its limited range. Compare with harlequin lupine, *L. stiversii*, which always has pink wing petals. A rarer variety, the Mariposa lupine, *L. citrinus* var. *deflexus*, is endemic to the SN and grows in the foothills of Mariposa County 400–600 m and looks identical, except that flowers are white to light pink, aging darker pink. Both varieties have markedly reflexed pedicels after fertilization.

Lupinus concinnus
Bajada Lupine
Fabaceae (Pea family)

Decumbent or erect annual herb 1–3 dm, hairy.
LEAVES: Petioles 2–7 cm; leaflets 5–9, 10–30 mm x 1.5–8 mm wide.
FLOWERS: Inflorescences 1.5–9 cm, peduncles < 8 cm, dense, with flowers spiraled; calyx 3–5 mm, upper lip deeply lobed; corollas 5–12 mm, petals pink to purple or white, keel upper margins hairless.
FRUITS: 1–1.5 cm, hairy. Seeds 3–5.
SIERRA DISTRIBUTION: Tulare and Inyo Counties south, also found in El Dorado County.
ELEVATION: < 1700 m
HABITAT: Open disturbed areas, burned areas; foothill, montane, and sagebrush zones.
FLOWERING TIME: Mar–May
NOTES: Inflorescences are often longer than those shown in the photo.

Lupinus covillei
Coville's Lupine
Fabaceae (Pea family)

Erect perennial herb 2–9 dm, with shaggy, silvery foliage, and several stems arising from base.
LEAVES: Stems have light green leaves, petioles 1–6 cm; leaflets 4–9, 30–110 mm, 5–11 mm wide, with hairs > 1 mm.
FLOWERS: Inflorescences 2–6 cm, flowers generally not whorled. Flowers 10–14 mm, with banner back hairless, and keel ciliate on upper margin.
FRUITS: Pods 2.5–4 cm, woolly to densely silky. Seeds 4–6.
SIERRA DISTRIBUTION: Tulare to Tuolumne and Inyo Counties.
ELEVATION: 2500–3500 m
HABITAT: Moist areas in swales, meadows, rocky areas; montane and subalpine zones.
FLOWERING TIME: Jul–Sep
NOTES: Endemic to the central and southern SN, locally common in the high country of Yosemite National Park and surrounding areas, often forming masses of purple in dry meadows.

Lupinus dalesiae

Quincy Lupine

Fabaceae (Pea family)

Perennial herb 2–5 dm; stem erect, with long, dense, white, spreading hairs.

LEAVES: Stems leafy, no basal leaves; stipules 6–16 mm; petioles 1–3 cm; leaflets 6–9, 20–45 mm; tomentose.

FLOWERS: Inflorescences 5–16 cm, peduncles 2–5 cm. Flowers ± whorled, pedicels 2–5.5 mm, bracts 5–9 mm; calyx upper lip 4–7 mm and 2-toothed, lower lip 3–7 mm and 3-toothed. Corollas 9–12 mm, yellow or cream-colored, banner back hairy and keel hairless.

FRUITS: Pods 2–3 cm, strigose. Seeds 3–5, tan.

SIERRA DISTRIBUTION: Plumas and Sierra Counties.

ELEVATION: 1000–2500 m

HABITAT: Dry pine forest, roadsides, burned areas; montane zone.

FLOWERING TIME: May–Aug

NOTES: Abundant in Plumas County, but CRPR 4.2 due to its limited distribution and potential threats posed by logging, road maintenance, and lack of fire/overshading.

Lupinus formosus var. *robustus*

Summer Lupine

Fabaceae (Pea family)

Erect, bushy perennial herb 2–8 dm; densely hairy, with stout stems 5–7 mm diameter.

LEAVES: Stems leafy, no basal leaves; petioles 2–7 cm; leaflets 7–9, 25–70 mm. Leaves hairy on both surfaces; stipules slender, tan, not leaflike.

FLOWERS: Inflorescences 10–30 cm, peduncles 3–7 cm. Flowers ± whorled, pedicels 3–7 mm. Calyx has upper lip 7–11 mm, 2-toothed; lower lip 8–12 mm; corollas 16–18 mm, purple; banner back hairless, keel hairless and does not protrude from wings.

FRUITS: Pods 3–4.5 cm, hairy. Seeds 5–7, mottled brown.

SIERRA DISTRIBUTION: West slope, Kern to Tuolumne Counties.

ELEVATION: < 1500 m

HABITAT: Open grassy areas, roadsides; foothill and montane zones.

FLOWERING TIME: Apr–Sep

NOTES: Compare with *L. albicaulis*, which has the keel clearly exserted from the wings. Another variety, *L. formosus* var. *formosus*, generally grows in valleys in dry clay soils at lower (< 1000 m) elevations, has slenderer (3–4 mm diameter) stems, and shorter (10–14 mm) corollas.

Lupinus fulcratus

Green-Stipuled Lupine

Fabaceae (Pea family)

Erect or ascending, hairy perennial herb 3–8 dm, branched above base.

LEAVES: Stems leafy, no basal leaves; petioles 3–6 cm; leaflets 6–9, 20–60 mm; hairs stiff; stipules 6–30 mm, green, and leaflike.

FLOWERS: Inflorescences 3–20 cm; pedicels 2–7 mm. Flowers whorled, 10–14 mm, with banner back hairless, and keel hairless or upper margins have a few hairs near middle.

FRUITS: Pods 3–4 cm, with stiff hairs.

SIERRA DISTRIBUTION: Tulare to Placer Counties.

ELEVATION: 1000–3000 m

HABITAT: Dry forests; montane and subalpine zones.

FLOWERING TIME: May–Sep

NOTES: Closely related to *L. andersonii*. Large, green, leaflike stipules distinguish this species.

Lupinus grayi

Gray's Lupine

Fabaceae (Pea family)

Mat-forming, woolly-hairy perennial herb 2–3.5 dm.

LEAVES: Stems leafy, lower petioles 5–14 cm, and upper petioles shorter, 2–4 cm; leaflets 5–8, 35–80 mm, 2–5 mm wide.

FLOWERS: Inflorescences 10–16 cm, flowers ± whorled. Flowers 10–16 mm, petals purplish to light blue; banner back with or without hairs, and with a yellow spot that turns red; keel densely ciliate on upper margin, and lower keel also ciliate near base.

FRUITS: Pods 2–3.5 cm, hairy. Seeds 4–6, mottled.

SIERRA DISTRIBUTION: West slope, all.

ELEVATION: 500–2500 m

HABITAT: Montane zone.

FLOWERING TIME: May–Jul

NOTES: Common lupine of forest understory and roadsides in lower and upper coniferous forest, sometimes carpeting large areas of forest floor. Note the relatively long petioles.

Lupinus latifolius var. *latifolius*

Broad-Leaved Lupine

Fabaceae (Pea family)

Tall, mostly hairless, bright green perennial herb 0.3–2.4 dm.
LEAVES: Basal and cauline, with largest leaves at midstem, broadly oblanceolate; petioles 4–19 cm. Leaflets 6–9, 40–100 mm, at least somewhat hairy below if not on both surfaces.
FLOWERS: Inflorescences 16–60 cm, open. Flowers 10–14 mm; petals purple, blue, or pink; banner spot white, and banner back hairless; keel ciliate on upper margin toward tip.
FRUITS: 3 cm, hairy.
SIERRA DISTRIBUTION: Mostly west slope, Kern County north.
ELEVATION: 200–3350 m
HABITAT: Moist areas along streams, lakeshores, seeps, meadows; montane through alpine zones.
FLOWERING TIME: May–Sep
NOTES: The tallest herbaceous lupine in the SN. Compare with the other common tall, large-leaved lupine of moist sites, *L. polyphyllus*, which has a hairless keel.

Lupinus lepidus var. *confertus*

Meadow Lupine

Fabaceae (Pea family)

Erect to leaning, hairy perennial herb 2–6 dm.
LEAVES: Mostly cauline, but some leaves crowded at base of stems; leaflets 5–7, 15–40 mm.
FLOWERS: Inflorescences 5–30 cm, dense, elongate; extending well above leaves; bracts 4–15 mm, persistent. Flowers in > 7 whorls, corollas 7–9 mm, petals purple; banner spot pale yellow, fading brownish or reddish, and keel ciliate toward tip.
FRUITS: 1–1.8 cm, white-hairy.
SIERRA DISTRIBUTION: All (no records south of Lake Isabella in Kern County).
ELEVATION: 700–3200 m
HABITAT: Moist areas, meadow edges; montane through subalpine zones.
FLOWERING TIME: Jun–Aug
NOTES: Much taller and more erect than var. *lobbii*.

Lupinus lepidus var. *lobbii*
Lyall's Lupine
Fabaceae (Pea family)

Mat-forming, prostrate perennial herb < 1.0 dm, shaggy to appressed-hairy.
LEAVES: Mostly basal, but some leaves crowded at base of stems. Leaflets 5–7, 10–13 mm.
FLOWERS: Inflorescences 2–8 cm, ± oval, dense. Flowers 6–10 mm, petals purple; banner spot white, keel ciliate on upper margin toward tip.
FRUITS: 1–2 cm, silky-hairy, silver.
SIERRA DISTRIBUTION: Highest elevations of crest, and both sides.
ELEVATION: 2000–3800 m
HABITAT: Dry to moist meadows, fell fields; subalpine and alpine zones.
FLOWERING TIME: Jul–Aug
NOTES: A common high-country plant, it is easy to see in the vicinity of Tuolumne Meadows and Dana Meadows in Yosemite National Park, and the Road to Saddlebag Lake just east of the park.

Lupinus lepidus var. *sellulus*
Donner Lake Lupine
Fabaceae (Pea family)

Erect to prostrate, hairy perennial herb 1.2–3.5 dm.
LEAVES: Mostly basal, but some leaves crowded at base of stems. Leaflets 5–8, 10–30 mm.
FLOWERS: Inflorescences 4.5–11 cm, > leaves. Flowers 8–9 mm, petals purple; banner spot yellow or white, turning red; keel ciliate on upper margin toward tip.
FRUITS: 0.9–1.1 cm, silky.
SIERRA DISTRIBUTION: Both slopes, Tulare and Mono Counties north.
ELEVATION: 300–3100 m
HABITAT: Dry to moist, sandy or rocky areas; montane, subalpine, and sagebrush/pinyon-juniper zones.
FLOWERING TIME: Jun–Aug
NOTES: Flowering stalks are distinctively cylindrical, densely flowered, and erect, extending above leaves.

Lupinus microcarpus var. *densiflorus*

Whitewhorl Lupine

Fabaceae (Pea family)

Hollow-stemmed annual herb to 50 cm; stems are erect initially but bend horizontally as fruits ripen.

LEAVES: Petioles 3–15 cm. Leaflets 5–11, 10–50 mm, 1–12 mm wide.

FLOWERS: Inflorescences 2–30 cm, with bracts reflexed; flowers in clear whorls, pedicels 0.5–5 mm. Flowers white to yellowish-white, sometimes pinkish; keel upper margin hairy (ciliate) around middle.

FRUITS: Pods 1–1.5 cm, 1 cm wide, hairy; when ripe, pods migrate to one side of inflorescence stalk—a very distinctive characteristic. Seeds 2, mottled.

SIERRA DISTRIBUTION: West slope, all.

ELEVATION: < 1600 m

HABITAT: Dry slopes, open areas; foothill and montane zones.

FLOWERING TIME: Apr–Jun

NOTES: This common roadside plant is easily recognized by its distinctly whorled racemes of (usually) white flowers, then by the unique 1-sided fruiting stalks that bend horizontally as the pods mature. If you see a lupine with all the pods migrated to one side of the stalk, it is this lupine.

Lupinus nanus

Sky Lupine

Fabaceae (Pea family)

Erect, hairy annual herb 1–6 dm, 1 to several stems, flowers aromatic.

LEAVES: Petioles 2–8.5 cm. Leaflets 5–9, 10–40 mm, 1–12 mm wide (sometimes linear).

FLOWERS: Inflorescences 2–20 cm, usually with clearly visible whorls of flowers; pedicels usually > 3 mm. Flowers blue with a white banner spot that turns pink with age; banner wider than long, or width = length (remove and flatten to determine this); keel is ciliate toward the tip, and the tip of keel strongly upturned (forming a right angle). Calyx upper and lower lips 4–6 mm, approximately equal in length.

FRUITS: Pods 2–4 cm, 4–7 mm wide, hairy. Seeds 4–12.

SIERRA DISTRIBUTION: West slope, all.

ELEVATION: < 1600 m

HABITAT: Open sites; foothill and lower montane zones.

FLOWERING TIME: Mar–Jun

NOTES: Often forms masses of brilliant color in springtime on slopes and roadsides, wafting a distinctive fragrance. Sometimes intermixed with the less showy *L. bicolor*. *L. nanus* is distinguished from *L. bicolor* by the larger flowers, longer inflorescences with more definite, widely separated whorls, wider banner, and a strong, questionably pleasant smell reminiscent of sweet mothballs.

Lupinus obtusilobus

Blunt-Lobed Lupine

Fabaceae (Pea family)

Hairy perennial herb 1.5–3 dm, with stems erect or decumbent.
LEAVES: Cauline, silvery appressed-hairy; leaflets 6–7, 20–50 mm, < 10 mm wide, with stipules 7–14 mm, hairy.
FLOWERS: Inflorescences 3–7 cm, flowers ± whorled. Calyx upper lip 2-toothed, and lower lip 3-toothed; corollas 11–13 mm, petals blue to purple; back of banner hairy, keel upper margin hairy from middle to tip.
FRUITS: 2.5–4 cm, silky.
SIERRA DISTRIBUTION: West slope, Nevada County north.
ELEVATION: 1200–2300 m
HABITAT: Gravelly ridges, talus; montane zone.
FLOWERING TIME: Jun–Sep
NOTES: A plant of the northern SN, look for it in Plumas County along the Bucks Lake Road and around Bucks Lake.

Lupinus polyphyllus var. *burkei*

Broad-Leaved Lupine

Fabaceae (Pea family)

Tall, stout perennial herb 0.5–1 dm.
LEAVES: Basal and cauline, petioles 15–45 cm. Leaflets 5–11, 40–150 mm, 5–15 mm wide, hairless on both surfaces, or slightly hairy on undersurface.
FLOWERS: Inflorescences 6–40 cm, dense. Flowers 9–15 mm, petals purple; banner back and keel hairless.
FRUITS: 2.5–4 cm, hairy. Seeds 5–9.
SIERRA DISTRIBUTION: All.
ELEVATION: 1000–3300 m
HABITAT: Moist areas along streams, lakeshores, seeps, meadows; montane through alpine zones.
FLOWERING TIME: Jun–Aug
NOTES: Compare with the other tall, large-leaved lupine of moist sites, *L. latifolius*, which has a hairy upper keel margin and hairy upper leaflet surfaces.

Lupinus stiversii

Harlequin Lupine

Fabaceae (Pea family)

Erect to ascending annual herb 10–45 cm; often highly branched, resulting in a sprawling bushy plant.

LEAVES: Leaflets ± 7, 20–50 mm, up to 15 mm wide, wedge-shaped, and much wider at tips; sparsely hairy on both surfaces.

FLOWERS: Inflorescences 3–8 cm, peduncles longer than inflorescence. Flowers are spiraled (not whorled) and have a yellow banner; pink wings cover the white keel.

FRUITS: Pods 2 cm, 5–7 mm wide, bright green and hairless. Seeds ± 5.

SIERRA DISTRIBUTION: West slope, Kern to Plumas Counties.

ELEVATION: 300–2100 m

HABITAT: Dry, sandy, gravelly, and rocky areas, slopes and flats; foothill through montane zones.

FLOWERING TIME: Apr–Jul

NOTES: This is the only lupine with pink-and-yellow flowers. Once you are familiar with it, you can recognize it prior to blooming by its unique leaf shape (very wide wedge-shaped or rounded tips with a tiny point) and its preference for gravelly or sandy soils, often on or near rocky outcrops.

Melilotus albus (non-native)

White Sweetclover

Fabaceae (Pea family)

Annual to biennial herb with stem 0.5–2 m; plant is hairless or strigose.

LEAVES: Alternate, odd-pinnate. Leaflets 3, 1–2.5 cm, elliptic-oblong to obovate; margins usually toothed.

FLOWERS: Racemes terminal and axillary, axis 3–8 cm when flowering; calyx 2 mm, lobes ± equal; corollas 3.5–5 mm, papilionaceous, and white. Stamens 10, 9 fused and 1 free.

FRUITS: 3–5 mm, indehiscent, the surface bears a network of lines. Seed 1.

SIERRA DISTRIBUTION: All.

ELEVATION: < 1900 m (likely higher)

HABITAT: Disturbed areas, streamsides; all zones, except alpine.

FLOWERING TIME: May–Sep

NOTES: Introduced to the US in the late 1700s from the Mediterranean region of Europe, and now found in every US state; it is cultivated for honey, soil improvement in eroded agricultural sites, wildlife and livestock forage, and roadsides for soil stabilization, prior to our current understanding of its invasiveness and persistence. Yellow sweetclover, *Melilotus officinalis*, is closely related but usually shorter, has similar-looking, slightly larger (4.5–7 mm) yellow flowers, fruits with cross-ridges rather than just lines, and occurs > 2300 m. In some states, these 2 spp. are treated together as 1, *M. officinalis*, but not in CA.

Oxytropis borealis var. *viscida*

Sticky Oxytrope

Fabaceae (Pea family)

Tufted perennial herb, with herbage visibly sticky-glandular.
LEAVES: Basal and alternate, odd-pinnate, with 23–39 leaflets, each 3–10 mm. Cauline leaves alternate, reduced toward top of stem. Stipules fused to petioles.
FLOWERS: Racemes are spikelike, exserted above leaves, with 5–20 flowers. Calyx lobes 5, shorter than tube. Corollas 12–18 mm, papilionaceous, lilac, red-purple, or white, with purple keel tip. Style and stigma hairless; filaments 10, 9 fused and 1 free.
FRUITS: Pods 8–14 mm x 5–6 mm wide, lanceolate to ovate, partially 2-chambered, inflated, papery.
SIERRA DISTRIBUTION: East slope, Inyo County west of Big Pine.
ELEVATION: 3300–3900 m
HABITAT: Open rocky or gravelly sites, scree slopes; subalpine and alpine zones.
FLOWERING TIME: Jul–Aug
NOTES: Called locoweed, along with *Astragalus* spp., due to a variety of toxic alkaloids that cause neurological symptoms in ruminants such as changes in behavior, lack of coordination, and weight loss.

Pickeringia montana var. *montana*

Chaparral Pea

Fabaceae (Pea family)

Evergreen shrub 1–3 m; rhizomatous, with terminal or axillary spines; plant is hairless to inconspicuously strigose.
LEAVES: Simple or palmately compound with up to 3 leaflets; leaflets sessile, 1–2 cm, elliptic to obovate.
FLOWERS: Racemes terminal or axillary, and bracts leaflike. Calyx 5–7 mm; corollas 1.5–1.8 cm, papilionaceous, purple or pink, with keel petals not fused. Stamens 10, not fused.
FRUITS: Pods 3–6 cm x 4–5 mm wide, dehiscent, constricted between seeds. Seeds 1–8.
SIERRA DISTRIBUTION: West slope, sporadic from Mariposa County north.
ELEVATION: < 900 m
HABITAT: Chaparral, woodland; foothill and montane zones.
FLOWERING TIME: May–Aug
NOTES: *Pickeringia montana* is a monotypic genus endemic to CA and Baja CA. It is a paleoendemic that has likely been present in CA for at least 30 million years. Viable seed and seedlings are rarely observed. Distribution of var. *montana* in the SN is sparse compared to coastal CA, from Mendocino and Lake Counties south to the Mexico border.

Spartium junceum (non-native, invasive)

Spanish Broom

Fabaceae (Pea family)

Shrub to 3 m; with stems green, round, hairless; plant is usually in a leafless state.

LEAVES: Alternate, simple, < 25 mm, linear to lanceolate, and hairless above, ephemeral.

FLOWERS: Terminal racemes of 2–3 cm, large, fragrant flowers; calyx 2-lipped, split on top nearly to base, and silky hairy; corollas yellow, petals 5.

FRUITS: Pods 5–10 cm x 5 mm wide, hairy all over. Seeds 4–18.

SIERRA DISTRIBUTION: West slope, most (if not all) counties.

ELEVATION: < 1800 m

HABITAT: Dry, disturbed areas, roadsides; foothill and montane zones.

FLOWERING TIME: Apr–Jun

NOTES: This noxious weed was introduced from Mediterranean as an ornamental and planted widely in erosion control seed mixes decades ago. It forms monocultures that are extremely difficult to control and that decrease native biodiversity.

Trifolium albopurpureum

Indian Clover

Fabaceae (Pea family)

Decumbent to erect, soft-hairy annual herb 10–40 cm.

LEAVES: Cauline, leaflets 1–3 cm, obovate to wedge-shaped, toothed.

FLOWERS: Inflorescences 8–15 mm, longer than wide, with no involucre. Calyx 4–8 mm, lobes > tube, and longer than purple-and-white corollas.

FRUITS: Seeds 1–2.

SIERRA DISTRIBUTION: West slope, all.

ELEVATION: < 2100 m

HABITAT: Open grassy areas; foothill and montane zones.

FLOWERING TIME: Mar–Jun

NOTES: The elongate (sometimes almost cylindrical), "fuzzy," purplish inflorescence makes this an easy clover to recognize.

Trifolium appendiculatum

Large Variegated Clover

Fabaceae (Pea family)

Annual herb 1–5 dm, with stem usually ascending, usually hairless.

LEAVES: Cauline and alternate, palmately compound; leaflets 3, obovate or wedge-shaped, sometimes narrower; stipules deeply slashed.

FLOWERS: Inflorescences headlike, and involucres wheel-shaped and deeply cut. Heads 1.5–3 cm wide, of 10+ flowers; calyx 6–10 mm, lobes slender and usually longer calyx tube; corollas 9–17 mm, lavender to purple with tips white.

FRUITS: Legumes indehiscent, stalklike bases short or absent. Seeds 1–2.

SIERRA DISTRIBUTION: All.

ELEVATION: 50–2200 m

HABITAT: Perennially wet sites, meadows, streamsides, ditches; foothill, montane, and sagebrush/pinyon-juniper zones.

FLOWERING TIME: Apr–Jul

NOTES: Formerly *Trifolium variegatum* var. *major*. Intergrades with *T. variegatum*, which is also common throughout the SN < 2500 m. *T. variegatum* is exceedingly variable, but overall differs by having proximal stipules that are entire while the distal are deeply cut; smaller (0.5–2.5 cm wide) heads; fewer (1–10) flowers per head, and corollas 3–16 mm.

Trifolium bifidum

Notchleaf Clover

Fabaceae (Pea family)

Erect or sprawling, hairless or sparsely hairy annual herb 5–55 cm.

LEAVES: Cauline; leaflets 3, 1–2 cm, oblanceolate to obovate, length is 1.5–2.5x the width, usually notched at tip; stipules have short awns.

FLOWERS: Inflorescences 7–15 mm wide, ± spheric, no involucres; pedicels 1–2 mm, flowers reflex with age. Calyx 3.5–5 mm, tube 0.8–1.4 mm, hairless; lobes much longer than tube, awnlike and tipped with bristles; corollas 6–9 mm, pink-purple or dull yellow.

FRUITS: Seeds 1–2.

SIERRA DISTRIBUTION: West slope, Mariposa County north.

ELEVATION: < 1000 m

HABITAT: Open grassy areas; foothill zone.

FLOWERING TIME: Apr–Jun

NOTES: Until recently, treated as 2 varieties: var. *decipiens* and var. *bifidum*. The varieties are no longer recognized as distinct; they often grow together, seem to intergrade, and were not clearly differentiated.

Trifolium breweri

Brewer's Clover

Fabaceae (Pea family)

Perennial herb; stem decumbent to ascending, sprawling along the ground, puberulent.

LEAVES: Cauline; leaflets 5–20 mm, obovate, usually serrate and sometimes smooth-margined; stipules sheathing or lanceolate.

FLOWERS: Inflorescences 1–2 cm wide, umbel-like, of 5–15 flowers, no involucre, pedicels 2–5 mm. Calyx 3.5–5 mm, hairy, with lobes longer than tube; corollas 5–8 mm, yellowish-white or pinkish.

FRUITS: Seed 1.

SIERRA DISTRIBUTION: Tulare County north.

ELEVATION: 200–1800 m

HABITAT: Dry forest floor, open areas; montane zone.

FLOWERING TIME: May–Aug

NOTES: One of the few clovers found growing on the forest floor in mixed conifer forest, distinctive because of its ground cover habit. Compare with *T. monanthum*.

Trifolium ciliolatum

Foothill Clover

Fabaceae (Pea family)

Erect, hairless annual herb 2–5 dm.

LEAVES: Mostly cauline; leaflets 1–3 cm, oblong to obovate, and toothed; stipules 15–30 mm, bristle-tipped.

FLOWERS: Inflorescences 7–20 mm wide, ± spheric, no involucre; pedicels 3–6 mm, reflexing as flowers age, exposing the inflorescence axis. Calyx 5–6 mm, hairless except for margins, which are fringed with bristles (ciliate); corollas 6–7 mm, pinkish-purple.

FRUITS: Seed 1.

SIERRA DISTRIBUTION: West slope.

ELEVATION: 150–1700 m

HABITAT: Open grassy areas; foothill and montane zones.

FLOWERING TIME: Apr–Jun

NOTES: One of the most common native foothill clovers, distinguished easily with a 10x hand lens by looking at the fringed margins of the calyx lobes (they appear eyelash-like).

Trifolium depauperatum var. *truncatum*

Cowbag Clover

Fabaceae (Pea family)

Small, decumbent to erect, hairless annual herb.
LEAVES: Cauline; leaflets 0.5–2 cm, narrowly oblong to obovate, toothed, tips truncate, and sometimes lobed; lower stipules oblong, and upper bristle-tipped.
FLOWERS: Inflorescences 0.5–1 cm wide, ± headlike, of 3+ flowers. Involucre bracts ± free, with margins ± membranous; calyx 2.5–3 mm, hairless; corollas 4.5–7.5 mm, pink-purple, with white-tipped lobes, banner inflated in fruit.
FRUITS: 2–3 mm, with stalklike base 0.5–1 mm. Seeds 1–2.
SIERRA DISTRIBUTION: Lower west slope, patchy distribution.
ELEVATION: < 800 m
HABITAT: Open grassy areas, vernal pool edges; foothill zone.
FLOWERING TIME: Apr–Jun
NOTES: Most noticeable when banners have inflated into tiny balloons around maturing fruits.

Trifolium hirtum (non-native)

Rose Clover

Fabaceae (Pea family)

Erect or ascending, hairy annual herb.
LEAVES: Mostly cauline; leaflets 1–2.5 cm, obovate, toothed; stipules have bristles at tips.
FLOWERS: Inflorescence lacks an involucre, but a pair of leaves with stipules directly below the flower head appears as an involucre. Calyx 7–9 mm, with lobes much longer than tube, stiff and linear; corollas 11–15 mm, pink.
FRUITS: Inflorescence bristly in fruit due to calyx lobes. Seed 1.
SIERRA DISTRIBUTION: West slope, Fresno County north.
ELEVATION: < 2100 m
HABITAT: Disturbed areas, grassy sites; foothill and montane zones.
FLOWERING TIME: Apr–May
NOTES: A weedy clover native to Eurasia and used for decades in roadside seeding mixes for erosion control, it is now ubiquitous in CA and quite invasive.

Trifolium longipes ssp. *hansenii*

Long-Stalked Clover

Fabaceae (Pea family)

Finely hairy perennial herb 0.5–4 dm; spreading by rhizomes, stems decumbent or erect.

LEAVES: Basal and cauline; leaflets 2–5 cm, oblanceolate, toothed; stipules toothed.

FLOWERS: Inflorescences 20–30 mm wide, headlike, extending above the level of the leaves; no involucre. Flowers spreading to ascending, not reflexing in age; calyx 7–11 mm, mostly glabrous, 1–3x longer than tube; corollas 10–18 mm, creamy white, sometimes pink.

FRUITS: Seed 1.

SIERRA DISTRIBUTION: Both slopes, Tulare and Inyo Counties north.

ELEVATION: < 3000 m

HABITAT: Meadows, streamsides; montane and subalpine zones.

FLOWERING TIME: Jun–Sep

NOTES: Formerly *T. longipes* var. *nevadense*. Ssp. *atrorubens* is patchy in the SN, flowers often purple, calyx lobes with long, dense hairs.

Trifolium microcephalum

Small-Head Clover

Fabaceae (Pea family)

Finely hairy, erect to decumbent annual herb 2–4 dm, with herbage hairs soft and wavy.

LEAVES: Mostly cauline; leaflets 0.8–2 cm, obovate, notched, toothed; stipules bristle-tipped.

FLOWERS: Inflorescences 5–8 mm wide, ± headlike; involucres bowl-shaped and very hairy. Calyx 4–6 mm, hairy, with long lobes (> ½ length of tube); corollas 4–7 mm, pink, white, or lavender.

FRUITS: Pods rupturing the corollas. Seeds 1–2.

SIERRA DISTRIBUTION: Mostly west slope, all.

ELEVATION: 200–2700 m

HABITAT: Moist sites, roadsides, open grassy areas, meadows; foothill and montane zones.

FLOWERING TIME: Apr–Aug

NOTES: The soft, fuzzy, bowl-shaped involucre snugly encasing the small pink head of flowers distinguishes this clover.

Trifolium monanthum ssp. *monanthum*

Carpet Clover

Fabaceae (Pea family)

Mat-forming perennial herb with stem < 1 cm; from slender rhizomes, plant hairless to puberulent.
LEAVES: Mostly basal; leaflets 3, 2–12 mm, 2–4x longer than wide, obovate with rounded or truncate tips.
FLOWERS: Inflorescences of 1–4 flowers; involucres vestigial or of 2–5 separate bracts; calyx 4–5 mm, lobes about the same length as tube; corollas 7–10 mm, white or with lavender stripes.
FRUITS: Pods short, indehiscent. Seeds 1–2.
SIERRA DISTRIBUTION: All.
ELEVATION: 1700–3900 m
HABITAT: Moist to wet sites in meadows, streamsides, rocky areas; montane through alpine zones.
FLOWERING TIME: Jun–Aug
NOTES: Compare with *T. breweri*, which can look similar and also forms extensive mats; *T. breweri* is found in drier habitats and has 5–15 flowers per inflorescence.

Trifolium obtusiflorum

Clammy Clover

Fabaceae (Pea family)

Tall, erect, stout annual herb 3–8 dm; glandular-hairy.
LEAVES: Mostly cauline; leaflets 2–4 cm, elliptic, and sharply toothed; stipules large and deeply toothed.
FLOWERS: Inflorescence 15–30 mm wide, ± headlike; involucres flat, wheel-shaped. Calyx 9–11 mm, glandular-hairy, lobes no larger than tube; corollas 14–18 mm, white to pale lavender with lighter tips.
FRUITS: Seeds 1–2.
SIERRA DISTRIBUTION: West slope, Tulare County north.
ELEVATION: 200–1800 m
HABITAT: Moist areas; foothill and montane zones.
FLOWERING TIME: Apr–Jul
NOTES: Called "clammy clover" because the herbage has numerous glands that cause this plant to feel slimy or wet to the touch.

Trifolium pratense (non-native)

Red Clover

Fabaceae (Pea family)

Hairy perennial herb, with stem ascending.
LEAVES: Cauline and alternate; leaflets 1.5–3.5 cm, elliptic to obovate; stipules bristle-tipped.
FLOWERS: Inflorescences 2–3 cm wide, terminal and headlike, located above a pair of reduced leaves; peduncles and involucres absent. Calyx 4.5–7.5 mm, with tube 10-veined, and lobes bristlelike, straight, and sparsely plumose, with lower lobe twice the size of the others; corollas 11–15 mm, red-purple.
FRUITS: Circumscissile. Seeds 1–2.
SIERRA DISTRIBUTION: Both slopes, all.
ELEVATION: < 2900 m
HABITAT: Disturbed areas; foothill and montane zones.
FLOWERING TIME: Apr–Oct
NOTES: Native to Europe.

Trifolium productum

Productive Clover

Fabaceae (Pea family)

Hairless perennial herb 1–4 dm, with slender, ascending stems.
LEAVES: Basal and cauline; leaflets 1–5.4 cm, lanceolate to elliptic, and coarsely toothed; lower stipules entire, papery.
FLOWERS: Inflorescences 15–30 mm wide, ± headlike or slightly elongate, often turned to side, axis extending above inflorescence, often forked; no involucre. Calyx 3–4 mm, hairless, with lobes the same length as the tube; corollas 10–14 mm, pink or reddish-purple.
FRUITS: 2–4 mm, stalklike.
SIERRA DISTRIBUTION: Both slopes, Tulare and Mono Counties north.
ELEVATION: 1100–2400 m (possibly higher)
HABITAT: Moist openings, meadows, coniferous forest; montane and subalpine zones.
FLOWERING TIME: Jun–Aug
NOTES: Formerly *T. kingii* var. *productum*.

Trifolium repens (non-native)

White Clover

Fabaceae (Pea family)

Hairless perennial herb 1–3 dm; spreading by rooting at nodes.
LEAVES: Cauline and alternate, with petioles much longer than leaf blades; leaflets 0.5–2.5 cm, obovate, and finely toothed (appearing entire), with stipules 4–10 mm, white.
FLOWERS: Inflorescences 10–25 mm wide, headlike; no involucre. Calyx 3–6 mm, hairless, with irregular lobes; corollas 7–11 mm, white or pale pink.
FRUITS: Seeds 3–4.
SIERRA DISTRIBUTION: Tulare and Mono Counties north, probably all.
ELEVATION: < 1500 m (possibly higher)
HABITAT: Disturbed areas; foothill and montane zones.
FLOWERING TIME: Apr–Dec
NOTES: Native to Eurasia, a common garden weed as well as being pernicious in disturbed wildlands.

Trifolium willdenovii

Tomcat Clover

Fabaceae (Pea family)

Hairless annual herb 2–5 dm; stem horizontal to erect.
LEAVES: Cauline and alternate; leaflets 1–5 cm, linear to oblong-lanceolate (generally slenderer than other clovers), toothed, and stipules toothed.
FLOWERS: Inflorescences 13–30 mm wide, headlike, with involucres flat, wheel-like and irregularly toothed. Calyx 6–10 mm, hairless, with lobes smaller than the tube; corollas 8–15 mm, reddish, purple, or lavender, usually with white tips.
FRUITS: Seeds 1–2.
SIERRA DISTRIBUTION: All.
ELEVATION: < 1700 m (possibly higher)
HABITAT: Moist to wet areas; foothill and montane zones.
FLOWERING TIME: Mar–Jun
NOTES: A very common and variable clover of moist sites, but the narrow, pointed leaflets are diagnostic. Compare with *T. appendiculatum*.

Trifolium wormskioldii

Cow Clover

Fabaceae (Pea family)

Hairless perennial herb 1–3 dm; spreading by rhizomes, stems decumbent or ascending.

LEAVES: Basal and cauline; leaflets 1–3 cm, oblanceolate, toothed, and stipules toothed.

FLOWERS: Inflorescences 20–30 mm wide, headlike, at or extending above the level of the leaves; involucres wheel-like and irregularly divided. Calyx 7–11 mm, with lobes tapered to bristles; corollas 12–16 mm, pink.

FRUITS: Seeds 2–4.

SIERRA DISTRIBUTION: All.

ELEVATION: < 2600 m

HABITAT: Meadows, streamsides; montane and subalpine zones.

FLOWERING TIME: May–Oct

NOTES: A common pink clover of SN montane meadows.

Vicia americana ssp. *americana*

American Vetch

Fabaceae (Pea family)

Extremely variable, hairless or hairy, low-growing perennial herb; erect or sprawling and somewhat vinelike.

LEAVES: Leaflets 8–16, 1–3.5 cm in length, with shape varying from wedge-shaped to oblong to markedly truncate; margin varies from completely entire to toothed toward leaf tip.

FLOWERS: Inflorescence has peduncles about the same length as subtending leaves; flowers in loose clusters of 3–9. Corollas 1.5–2.5 cm, more than twice as long as wide, varying in length. Flower color is variable, blue-purple to lavender. Styles with hairs tufted toward the tips and arranged around the entire circumference of the style columns (like a bottlebrush, not a toothbrush), though 1 side may have longer hairs.

FRUITS: 2.5–7 cm, 5–7 mm wide, hairless or hairy.

SIERRA DISTRIBUTION: All.

ELEVATION: < 2400 m

HABITAT: Forest floor, dry or moist soil, sometimes in disturbed areas; montane and sagebrush/pinyon-juniper zones.

FLOWERING TIME: Apr–Jun

NOTES: The leaves of *V. americana* are exceedingly variable in shape. Distinguished from *Lathyrus* by juvenile leaflets, which open by unfolding rather than unrolling, and the style, which has tufted hairs surrounding the tip. Compare with *Lathyrus nevadensis* var. *nevadensis.*

Vicia sativa ssp. *sativa* (non-native)

Spring Vetch

Fabaceae (Pea family)

Erect or decumbent annual herb 1–6 dm, hairless or hairy.
LEAVES: Leaflets 8–14, 1.5–3.5 cm x 4–10 mm wide, and wedge-shaped, often with 1 tooth at tip; stipules toothed.
FLOWERS: Inflorescence is a cluster of 1–3 flowers. Calyx 5-lobed, tube 6–7 mm, lobes 5–11 mm, linear; corollas 18–30 mm, pink-purple.
FRUITS: Pods 2.5–6 cm, 2.5–8 mm wide, hairy, becoming hairless.
SIERRA DISTRIBUTION: Madera County north, based on CCH2 and Calflora, but this no doubt underrepresents the range of this weedy vetch.
ELEVATION: < 1300 m
HABITAT: Disturbed areas, grassy areas, roadsides, pastures; foothill, montane, and sagebrush zones.
FLOWERING TIME: Mar–Jun
NOTES: Native to Europe. *Vicia sativa* ssp. *nigra* also occurs in the SN; the corolla is smaller (10–18 mm), and the calyx lobes are only 3–4.5 mm; it has been found as high as 1600 m.

Vicia villosa ssp. *varia* (non-native)

Hairy Vetch

Fabaceae (Pea family)

Sprawling or climbing annual herb with stems < 1 m; hairs on stems and foliage are sparse to dense, ± 1 mm.
LEAVES: Leaflets 12–18, 1–2.5 cm, narrowly oblong with pointed tips.
FLOWERS: Inflorescence is 1-sided, of 10–20 flowers; calyx ± 1–2.5 mm, linear-lanceolate, 5-lobed, with the lower lobe the longest; corollas 10–14 mm, bluish-purple, lavender, or sometimes white.
FRUITS: Pods 1.5–4 cm, 6–10 mm wide, with short hairs; basal stalk 2–3 mm.
SIERRA DISTRIBUTION: Fresno and Inyo Counties north (collectors have tended to overlook this weedy, common vetch, hence herbarium records do not accurately reflect its range; most likely it is in every county).
ELEVATION: < 2100 m
HABITAT: Disturbed areas, grassy areas, roadsides, pastures; foothill, montane, and sagebrush zones.
FLOWERING TIME: Apr–Jun
NOTES: Native to Europe, various species of non-native vetch have been planted as cover crops to control erosion, fix nitrogen, and enhance habitat of game species. Non-native vetch can be an aggressive plant, crowding out native plants. To date, it seems not to encroach into undisturbed wildlands excessively. The practice of seeding it on public lands is rarely done nowadays.

Chrysolepis sempervirens
Bush Chinquapin
Fagaceae (Oak family)

Evergreen shrub < 3 m; ± rounded in overall shape, with bark smooth, gray or brown, and twigs with scalelike hairs.
LEAVES: Alternate, petioled, blades 2–8 cm; gold below due to minute golden scales, and dull green above.
FLOWERS: Separate male and female inflorescences on the same plant (monoecious). Male inflorescence is a many-flowered spike; female inflorescence is positioned above the male and has 1 to few flowers, without petals, with calyx of 6 tiny lobes. Ovary inferior.
FRUITS: Spiny bur.
SIERRA DISTRIBUTION: All.
ELEVATION: 700–3300 m
HABITAT: Conifer forest and montane chaparral; upper montane and subalpine zones.
FLOWERING TIME: Jul–Aug
NOTES: Another species, *C. chrysophylla* var. *chrysophylla*, is a small tree called giant chinquapin, which primarily grows along the coast but is found in the Sierra only in El Dorado County. The bark is thick and furrowed, leaves are 5–15 cm and long-tapered, and plants have a conical shape.

Quercus chrysolepis
Canyon Live Oak
Fagaceae (Oak family)

Evergreen shrub to small tree 6–20 m; light, gray bark; mature plants have deeply furrowed, scaly bark, and young twigs are golden-hairy.
LEAVES: Alternate, petioles 3–10 mm; blades 1.5–6 cm, dark green above, and dull white with gold hairs (some stellate) below; margin entire to spiny-toothed. Leaves on young branches and seedlings are extremely spiny-toothed, hollylike.
FLOWERS: Monoecious. Male catkins 1 to several, pendulous, on inner twig; flowers have no petals, 4–10 stamens. Female inflorescence 1-flowered, appearing in upper leaf axils, without petals, with calyx of 6 tiny lobes. Ovary inferior, style 3-branched.
FRUITS: Acorns, which mature in year 2; caps bright gold.
SIERRA DISTRIBUTION: All.
ELEVATION: 2750 m
HABITAT: Chaparral, shaded slopes, rocky areas; mostly montane zone.
FLOWERING TIME: Apr–May
NOTES: Although this species overlaps with *Q. wislizeni*, it generally occurs at higher elevations. The underside of the leaf is hairy and dull, rather than bright green and hairless like interior live oak. If the acorns are present, the bright gold caps confirm *Q. chrysolepis* (*chryso* meaning "golden," and *lepis* meaning "scale," referring to the golden scales on the acorn cap).

Quercus vacciniifolia

Huckleberry Oak

Fagaceae (Oak family)

Evergreen shrub < 1.5 m; spreading to prostrate, forming large continuous patches. Twigs are hairless, flexible.

LEAVES: Alternate, petioles 3–6 mm; blades 1.5–4 cm, dull green, and both surfaces hairless; margin usually entire.

FLOWERS: Monoecious. Male catkins 1 to several, pendulous, on inner twig, and flowers have no petals, 4–10 stamens. Female inflorescence 1-flowered, in upper leaf axils, without petals, with calyx of 6 tiny lobes. Ovary inferior, style 3-branched.

FRUITS: Acorns, which mature in year 2. Caps 10–14 mm wide, cup-shaped; nuts 10–15 mm.

SIERRA DISTRIBUTION: All.

ELEVATION: 90–2000 m

HABITAT: Montane chaparral, rocky areas, conifer forest; upper montane and subalpine zones.

FLOWERING TIME: May–Jul

NOTES: The species name means "like leaves of Vaccinium"; compare the leaves with *Vaccinium uliginosum* ssp. *occidentale.*

Quercus wislizeni var. *frutescens*

Interior Live Oak

Fagaceae (Oak family)

Evergreen shrub 2–6 m; gray bark, shallowly furrowed to checkered in mature specimens.

LEAVES: Alternate, petioles 3–15 mm; blades 1.8–4 cm, shiny and dark green above, and lighter below, both surfaces hairless; margins entire to spiny-toothed. Leaves on young branches and seedlings extremely spiny-toothed, hollylike.

FLOWERS: Monoecious. Male catkins 1 to several, pendulous, on inner twig; flowers have no petals, 4–10 stamens. Female inflorescence 1-flowered, in upper leaf axils, without petals, with calyx of 6 tiny lobes. Ovary inferior, style 3-branched.

FRUITS: Acorns, which mature in year 2. Caps 12–18 mm wide, deeply cup-shaped; nuts 20–40 mm.

SIERRA DISTRIBUTION: All.

ELEVATION: 90–2000 m

HABITAT: Chaparral, woodlands, steep slopes, rocky areas; foothill and lower montane zones.

FLOWERING TIME: Mar–May

NOTES: Occurs as a shrub or as a small tree. *Q. wislizeni* var. *wislizeni* is a tree 10–22 m tall, with leaves 2–5 cm, occurring < 1600 m. Extremely common; the distinction between varieties is not very clear. *Q. wislizeni* is distinguished from the similar *Q. chrysolepis* by the bright green, hairless leaf undersurface.

Garrya flavescens
Ashy Silk Tassel Bush
Garryaceae (Silk Tassel family)

Evergreen shrub < 3 m, with separate male and female plants (dioecious).

LEAVES: Opposite, 19–75 mm, hairy on both surfaces, with hairs on undersurface ± straight.

FLOWERS: Male inflorescences up to 3–4 cm, of many small staminate flowers held in silky bracts fused at base (1–4 flowers per bract, 4 stamens per flower), the entire structure forming the "silk tassel." Female inflorescences 3–5 cm, with 1–3 flowers per bract, no sepals and petals. Ovary inferior, densely hairy; styles 2–3.

FRUITS: Densely hairy.

SIERRA DISTRIBUTION: West slope, sporadic, mostly Mariposa county south, but recorded from Nevada and possibly Butte Counties; east slope, Inyo County.

ELEVATION: 650–2350 m

HABITAT: Dry, open slopes, rocky areas; mostly montane zone.

FLOWERING TIME: Feb–Apr

NOTES: Two other spp. may be encountered: *G. fremontii* is distinguished by its hairless upper leaf surfaces and fruits; *G. congdonii* has leaf undersurfaces that are densely covered with wavy hair and usually grows on serpentine < 750 m.

Gentiana calycosa
Explorer's Gentian
Gentianaceae (Gentian family)

Leafy perennial herb 5–45 cm; from stout caudexes.

LEAVES: Opposite, 8–50 mm x 6–30 mm wide, ovate to roundish.

FLOWERS: Inflorescences are usually terminal and sometimes also in upper leaf axils, of flowers 1–3 per inflorescence. Calyx lobes 5, corolla lobes 5, with deep blue or violet with green spots on inner surfaces. Corollas 25–55 mm, with 2–3 slender, threadlike projections between each corolla lobe.

FRUITS: Capsule 2-valved. Seeds unwinged.

SIERRA DISTRIBUTION: Both slopes, Tulare and Inyo Counties north.

ELEVATION: 1300–3900 m

HABITAT: Meadows, streamsides; montane through alpine zones.

FLOWERING TIME: Jul–Sep

NOTES: A unique and beautiful plant, as nothing else looks like it (but when it hybridizes with *G. newberryi* var. *tiogana* in Tulare County, all bets are off!).

Frasera speciosa

Monument Plant

Gentianaceae (Gentian family)

Perennial herb 7–20 dm; lives for years as a rosette before flowering, then dies (monocarpic).

LEAVES: Basal in rosettes, 7–50 cm x 1–15 cm wide, oblanceolate to obovate; cauline leaves in whorls of 3–7, 15–25 cm, smaller toward top of stem.

FLOWERS: Inflorescences dense above, more open below; flowers 4-parted; calyx 12–25 mm; corollas 12–20 mm, greenish- or yellowish-white, flecked with purple; each corolla lobe has 2 nectary pits.

FRUITS: Capsule 16–18 mm.

SIERRA DISTRIBUTION: Both slopes, Tulare and Inyo Counties north to El Dorado County.

ELEVATION: 1500–3000 m

HABITAT: Meadows, meadow edges, riparian areas; sometimes open slopes; upper montane and subalpine zones.

FLOWERING TIME: Jul–Aug

NOTES: At first one might mistake this plant for a monocot, as the leaf veins appear parallel. Other *Frasera* taxa in the SN are: whitestem frasera, *F. albicaulis* var. *nitida*, occurring < 1900 m on the west slope from El Dorado County north, with stems 1–6.5 dm, leaves opposite (not whorled), margins white, and corollas 6–12 mm; and Kern frasera, *Frasera tubulosa*, found at 1800–2700 m in Kern, Tulare, and Inyo Counties, with stems 2–11 dm, rosettes absent or few, leaves white-margined, cauline whorled, inflorescence elongate, corollas 8–13 mm, white to pale blue with darker blue veins, and nectary pit bordered by basally tubular, distally fringed extensions. Neither of these are monocarpic; but Inyo frasera, *F. puberulenta*, is monocarpic like *F. speciosa*. *F. puberulenta* is found only on the east slope at 1700–3400 m, with stems 1–3 dm, leaves rosetted and opposite, white-margined, and corollas 7–13 mm.

Gentiana newberryi var. *tiogana*

Alpine Gentian

Gentianaceae (Gentian family)

Perennial herb with stems 2–10 cm; usually nestled close to the ground and spreading vegetatively via rhizomes.

LEAVES: Basal leaves opposite, upper leaves sessile.

FLOWERS: Usually solitary; corollas 25–55 mm, white (or pale blue) with dark, longitudinal stripes on outside of corolla, and green spots on inner corolla lobes; lobes 7–17 mm, with 2 long, slender appendages between each corolla lobe.

FRUITS: Capsule 2-valved. Seeds winged.

SIERRA DISTRIBUTION: West slope, Fresno County north; east slope, Inyo and Mono Counties.

ELEVATION: 1200–4000 m

HABITAT: Dry to wet meadows; upper montane, subalpine, and alpine zones.

FLOWERING TIME: Aug–Oct

NOTES: In the subalpine and alpine zones, this is one of the last wildflowers to bloom, commonly found in full flower when other plants have begun to shut down for the winter. Pollinated by bumblebees.

Gentianopsis holopetala

Sierra Gentian

Gentianaceae (Gentian family)

Annual herb with stems 3–45 cm; erect or decumbent stems may be simple or branched.

LEAVES: Basal and opposite; basal and lower stem leaves < 70 mm, < 15 mm wide, spoon-shaped; upper stem leaves < 60 mm and < 10 mm wide, elliptic to linear.

FLOWERS: 1 per stem; calyx 10–36 cm; corollas 25–55 mm, lobes 7–17 mm, dark blue. Corolla lobes entire (with smooth margins) or with a tiny, jagged area near the tip of the corolla lobe (compare with *G. simplex*).

FRUITS: Capsule 2-valved. Seeds papillate, not winged.

SIERRA DISTRIBUTION: Both slopes, Tulare and Inyo Counties north.

ELEVATION: 1800–4000 m

HABITAT: Wet meadows; upper montane, subalpine, and alpine zones.

FLOWERING TIME: Jul–Sep

NOTES: Distinguished from *G. simplex* by the entire or mostly entire corolla lobes, and the purple-striped midrib on each calyx lobe.

Gentianopsis simplex

Hikers' Fringed Gentian

Gentianaceae (Gentian family)

Perennial herb with stems 4–40 cm; erect, unbranched stems single, separate from each other. Forms clonal patches from root buds.

LEAVES: Basal and opposite; basal and lower leaves withering early; upper leaves < 25 mm x 6–8 mm wide.

FLOWERS: 1 per stem; calyx 9–27 mm; corollas 17–45 mm, blue. Corolla lobes with toothed or fringed margins on outer half (compare with *G. holopetala*).

FRUITS: Capsule 2-valved. Seeds longitudinally ridged, winged.

SIERRA DISTRIBUTION: Both slopes, all (only 1 record for northern Kern County).

ELEVATION: 1200–3400 m

HABITAT: Wet meadows; upper montane through alpine zones.

FLOWERING TIME: Jul–Sep

NOTES: Distinguished from *G. holopetala* by the jagged/fringed corolla lobes, and the completely green calyx with no purple veins or stripes.

Zeltnera venusta

California Centaury

Gentianaceae (Gentian family)

Erect, hairless annual herb 3–50 cm; branched above, blooms after most associated herbs have dried up.

LEAVES: Opposite, sessile, 5–25 mm.

FLOWERS: Open cyme of salverform corollas; pedicels 4–25 mm; corollas 16–30 mm, throats white, lobes to 20 mm, pink with ± pointed tips. Stigmas with 2 yellow, wedge-shaped lobes; anthers spirally twisted after pollen release.

FRUITS: Capsule 2-valved.

SIERRA DISTRIBUTION: Most common on west slope, all.

ELEVATION: < 2500 m

HABITAT: Damp or drying areas near seeps; foothill and montane zones.

FLOWERING TIME: May–Aug

NOTES: Formerly *Centaurium venustum*. A similar, less abundant species, Muehlenberg's centaury, *Z. muehlenbergii*, occurs < 1600 m from Madera County north on the west slope. It has shorter (< 10 mm) pedicels, smaller (12–19 mm) corollas, and corolla lobes 2–7 mm with rounded tips.

Erodium botrys (non-native)

Long-Beaked Filaree

Geraniaceae (Geranium family)

Prostrate to ascending annual herb 1–9 dm; hairy (nonglandular).

LEAVES: Basal and lower stem leaves 3–15 cm, shallowly or deeply pinnately lobed.

FLOWERS: Sepals 7–8 mm, bristle-tipped; corollas cup-shaped, with petals 10–15 mm (broader than *E. cicutarium*), and hairless at base.

FRUITS: Body 8–11 mm, with roundish pit at top of fruit, subtended by 2 deep, hairless furrows; beak 5–12 cm.

SIERRA DISTRIBUTION: West slope, all.

ELEVATION: < 1200 m

HABITAT: Open disturbed areas; foothill and lower montane zone.

FLOWERING TIME: Mar–Jul

NOTES: This filaree has by far the longest fruit of any of the 4 spp. found in the SN. A similar species is *E. brachycarpum*, with shorter (< 40 cm) stems, shorter fruits (beak 5–8 cm), and pits at the top of the fruit body subtended by only 1 hairy fold.

Erodium cicutarium (non-native)

Redstem Filaree

Geraniaceae (Geranium family)

Glandular-hairy annual herb; decumbent to erect, with reddish stems.

LEAVES: Basal and lower stem leaves pinnately compound, 3–10 cm; leaflets 9–13, deeply dissected and ending in outer segments only 1–2 mm wide.

FLOWERS: Sepals 3–5 mm, bristle-tipped; petals 5–7 mm, ovate, tips rounded, with conspicuous white hairs at base (ciliate). Petals more delicate than wedge-shaped petals of *E. botrys*.

FRUITS: Body 4–7 mm, pits at top of body are subtended by 2 hairless ridges; beak 2–5 cm.

SIERRA DISTRIBUTION: All.

ELEVATION: < 2000 m

HABITAT: Open disturbed areas; foothill and montane zones.

FLOWERING TIME: Feb–Sep

NOTES: Native to Europe. Another non-native filaree from Europe with pinnately compound leaves is greenstem filaree, *E. moschatum*, which has 11- to 15-lobed to shallowly divided leaflets in contrast to these deeply divided, ferny-looking leaflets; found < 1500 m, in foothills of the lower west slope.

Geranium californicum

California Geranium

Geraniaceae (Geranium family)

Erect, branched, perennial herb 10–60 cm; hairy.
LEAVES: Palmately 5- to 7-parted with wide, wedge-shaped lobes; blades 3–8 cm wide.
FLOWERS: Pedicels long (3–8 cm) and have white or yellow glandular hairs; sepals 8–11 mm; petals 10–15 mm, white to pink with purple veins, and hairy on lower inner surface. Stigma branches 6–8 mm.
FRUITS: Carpels 4.5–4 mm, beak 24–29 mm, stigmas 6–8 mm.
SIERRA DISTRIBUTION: Both slopes, El Dorado County south.
ELEVATION: 1000–3000 m
HABITAT: Moist areas, streamsides, meadows; montane zone.
FLOWERING TIME: Apr–Aug
NOTES: Distinguished from *G. richardsonii* by the white-to-yellow-tipped glandular hairs on the pedicels and the much longer pedicels and style branches.

Geranium dissectum (non-native)

Cut-Leaf Geranium

Geraniaceae (Geranium family)

Erect to ascending annual herb up to 70 cm; hairy, glandular above.
LEAVES: Opposite, 2.5–5.4 cm wide. Deeply palmately divided into 5–7 slender, branched segments (divisions exceed ¾ length of the leaf), with each segment itself further deeply dissected.
FLOWERS: 2 per inflorescence, loosely clustered; pedicels 6–13 mm; flowers 5-merous, sepals 4.2–7.1 mm and awn-tipped; petals 2.9–5.8 mm, notched.
FRUITS: Mericarps 2.3–2.9 mm, without wrinkles, covered with short, stiff hairs; beak 11.1–13.5 mm, with narrow tip 1.6–2.7 mm; stigmas < 1 mm, red-purple tipped.
SIERRA DISTRIBUTION: West slope, all.
ELEVATION: < 1450 m
HABITAT: Open disturbed areas; foothill through lower montane zones.
FLOWERING TIME: Mar–Jul
NOTES: Native to Europe, North Africa, and western Asia. Distinguished from the similar-looking native Carolina geranium, *G. carolinianum*, by the narrow (almost linear) leaf segments with pointed rather than rounded tips. *G. carolinianum* is found < 1700 m on the west slope from Tulare County north; mericarps are larger (3–4.5 mm), beaks are longer (15–19 mm) with narrow tips 1–2 mm, and stigmas 0.7–1.4 mm.

Geranium molle (non-native)

Dovefoot Geranium

Geraniaceae (Geranium family)

Erect to decumbent annual herb up to 45 cm, with long, soft, nonglandular hairs.

LEAVES: Basal and lower stem leaves shallowly palmately lobed (lobes ½–¾ the length of the leaf), up to 5.2 cm wide.

FLOWERS: Sepals 1–6 mm; petals 3–10.5 mm, notched. Stigmas 1–10 mm; stamens 10.

FRUITS: Carpels 1.8–2.1 mm, wrinkled; beak 7–14 mm; stigmas 1–2 mm. Fruiting pedicels bend downward.

SIERRA DISTRIBUTION: West slope foothills.

ELEVATION: < 1000 m

HABITAT: Open or shady, disturbed areas; foothill through lower montane zones.

FLOWERING TIME: Feb–Aug

NOTES: A small, delicate herb distinguished by the rounded, shallowly to moderately lobed leaves and small, pink flowers. Native to Europe, North Africa, and Asia.

Geranium richardsonii

Richardson's Geranium

Geraniaceae (Geranium family)

Erect perennial herb 30–90 cm tall.

LEAVES: Palmate in 5–7 parts, 3–15 cm wide, with narrow, pointed lobes.

FLOWERS: Pedicels short (1.5–3 cm) with red or purple glandular hairs; sepals 7–8 mm; petals 14–20 mm, white to pink with magenta or purple veins, hairy on inner surfaces. Stigma branches 3–4 mm.

FRUITS: Carpels 3.5–4 mm, beak 14–19 mm, stigmas 3–4 mm.

SIERRA DISTRIBUTION: El Dorado County south.

ELEVATION: 1000–3000 m

HABITAT: Moist areas, streamsides, meadows; montane zone.

FLOWERING TIME: Jun–Jul

NOTES: Of the tall perennial geraniums found in wet areas, this one is distinguished primarily by the white-to-yellow-tipped glandular hairs on the pedicels. Compare with *G. californicum*, which has shorter pedicels with white or yellow glands and longer style branches.

Ribes cereum var. *cereum*

Wax Currant

Grossulariaceae (Gooseberry family)

Deciduous shrub < 1.5 m; without nodal spines. Spicy aroma.

LEAVES: Alternate; blades 1–4 cm, roundish, shallowly palmately lobed, and finely glandular-hairy, glossy above.

FLOWERS: Inflorescences of 3–7 flowers, in axils; bract tip toothed. Sepals 1–2 mm, white or pink, hairy, and reflexed; petals 1 mm, whitish. Ovary inferior, hypanthium 6–8 mm (including ovary) and tubular, styles hairy; stamens inserted inside tube below petals, anthers with cuplike depression at tips.

FRUITS: Berry 6–7 mm, red, with glandular hairs.

SIERRA DISTRIBUTION: All.

ELEVATION: 875–2365 m

HABITAT: Dry to moist, often rocky areas; montane, subalpine, and alpine zones.

FLOWERING TIME: Jun–Jul

NOTES: Restricted primarily to the alpine zone 2100–3850 m on the east slope is the dwarfed *R. cereum* var. *inebrians*, which has hairless styles, entire bracts (no teeth), and hairless leaves.

Ribes inerme var. *inerme*

White-Stemmed Gooseberry

Grossulariaceae (Gooseberry family)

Sprawling shrub < 2 m; spines at nodes, stems white to gray.

LEAVES: Alternate, 2–3 cm, 3- to 5-lobed, with spines 0–3 per node; soft-hairy or hairless, not glandular.

FLOWERS: Inflorescence of 1–5 flowers; sepals 3–5 mm, petal-like, strongly reflexed, green to white, and sometimes purple at base; petals 1–2 mm, white. Ovary inferior; anthers nearly 2 mm longer than petals.

FRUITS: Berry 7–11 mm, purple, hairless.

SIERRA DISTRIBUTION: Tulare and Inyo Counties north.

ELEVATION: 1200–3300 m

HABITAT: Moist, shady areas, streamsides, meadow edges; montane, subalpine, and sagebrush zones.

FLOWERING TIME: May–Jun

NOTES: More common on and near the east slope.

Ribes montigenum

Western Prickly Gooseberry

Grossulariaceae (Gooseberry family)

Sprawling shrub < 1 m; spines at nodes.

LEAVES: Alternate, 1.5–2.5 cm, with spines 1–5 per node, glandular-hairy.

FLOWERS: Rotate (flat rather than tubular). Sepals petal-like, yellow; petals shorter, red.

FRUITS: Berry 4–5 mm, bright red, glandular.

SIERRA DISTRIBUTION: Tulare and Inyo Counties north.

ELEVATION: 1200–4000 m

HABITAT: Openings, rocky slopes, and forests; subalpine and alpine zones.

FLOWERING TIME: Jun–Jul

NOTES: Arguably the best-tasting berries in the High Sierra.

Ribes nevadense

Mountain Pink Currant

Grossulariaceae (Gooseberry family)

Deciduous shrub 1–2 m; without spines.

LEAVES: Alternate, 3–8 cm, glandular on upper surface and usually hairless below, sometimes hairy.

FLOWERS: Inflorescences of 8–20 flowers; sepals 4–5 mm, petal-like, erect, pink to red; petals 2–3 mm, white. Anthers lack cuplike depression at tips.

FRUITS: Berry 6–8 mm, blue-black, glaucous (appearing whitish-blue), glandular.

SIERRA DISTRIBUTION: All.

ELEVATION: 600–3050 m

HABITAT: Moist sites in shaded forests and along streams and meadow edges; montane and subalpine zones.

FLOWERING TIME: May–Jun

NOTES: Compare with *Ribes viscosissimum*. The leaves look similar, but *R. viscosissimum* leaves are stickier and have a unique smell; also, the berries are larger (10–12 mm).

Ribes roezlii var. *roezlii*

Sierra Gooseberry

Grossulariaceae (Gooseberry family)

Deciduous shrub < 1.5 m; 1–3 spines at nodes.

LEAVES: Alternate, palmately lobed, and hairless to hairy on upper surface (nonglandular); blades 1.2–2.5 cm.

FLOWERS: Inflorescence of 1–3 flowers, in axils; hypanthium hairy, and longer than wide. Sepals 7–9 mm, purple, petal-like, hairy, and reflexed; petals 3–4 mm, whitish. Ovary inferior, styles > petals, anthers > petals.

FRUITS: Gooseberry 14–16 mm, red when mature, fleshy, with nonglandular prickles.

SIERRA DISTRIBUTION: West slope, all.

ELEVATION: 200–2850 m

HABITAT: Understory and openings in conifer forest, abundant in disturbed areas; montane zone.

FLOWERING TIME: May–Jun

NOTES: A similar-looking, much less common species, *R. amarum* occurs sporadically in the foothills of the west slope and is distinguished by having glandular-hairy leaves and glandular prickles on the fruits. All gooseberries are intermediate hosts for an introduced fungus that kills sugar pine, *Pinus lambertiana*. Sierra gooseberry was targeted for eradication during the 1960s, an effort that has since been abandoned.

Ribes viscosissimum

Sticky Currant

Grossulariaceae (Gooseberry family)

Deciduous shrub < 1.5 m; without spines.

LEAVES: Alternate, 3–8 cm, sticky-glandular on both surfaces, dull or grayish-green, thick, and aromatic.

FLOWERS: Inflorescence of 4–15 flowers; sepals 6–7 mm, greenish to pinkish, petal-like, spreading to reflexed; petals 3 mm, white. Anther tips have minute, cuplike depressions (need hand lens to see).

FRUITS: Berry 10–12 mm, black to glaucous, smooth, bristly, or glandular.

SIERRA DISTRIBUTION: Tulare and Mono Counties north.

ELEVATION: 960–3030 m

HABITAT: Moist sites in shaded forests and along streams and meadow edges; montane and subalpine zones.

FLOWERING TIME: May–Jul

NOTES: Compare with *Ribes nevadense*.

Heliotropium curassavicum var. *oculatum*

Alkali Heliotrope

Heliotropiaceae (Heliotrope family)

Fleshy, hairless perennial herb 1–6 dm; prostrate or ascending, from short, rhizome-like roots.

LEAVES: Alternate, 1–6 cm, oblanceolate, hairless, fleshy.

FLOWERS: Coiled cymes of bell-shaped white flowers, corollas 3–5 mm, throat sometimes blue or purple, tube yellow.

FRUITS: Nutlets 4, smooth.

SIERRA DISTRIBUTION: West slope, Madera County south.

ELEVATION: < 1000 m (in SN)

HABITAT: Moist to wet areas, saline or alkaline sites; foothill zone.

FLOWERING TIME: Feb–Oct

NOTES: Transferred from Boraginaceae to Heliotropiaceae in 2022. Reported to be toxic. A related, non-native species, European heliotrope, *H. europaeum*, is a hairy annual plant < 4 dm, with much wider leaves; so far, this weed has not made it past a few locations in the lowest elevations of the west slope.

Jamesia americana var. *rosea*

Rosy-Petalled Cliffbush

Hydrangeaceae (Hydrangea family)

Densely hairy, deciduous shrub < 2 m; with gray bark.

LEAVES: Opposite, no stipules; blades 2–6 cm x 1.5–4 cm wide, ovate, pinnately veined, with margins toothed. Upper surface hairy to hairless, and undersurface woolly-hairy.

FLOWERS: Cymes of pink flowers; corollas ≤ 1.5 cm wide; sepals 5, 5–8 mm, petals 5, 5–8 mm. Ovary ½ inferior; stamens 10.

FRUITS: Capsule 1–1.3 cm, conic, woody.

SIERRA DISTRIBUTION: Fresno and Mono Counties south.

ELEVATION: 2000–3700 m

HABITAT: Rocky areas, cliffs; upper montane through alpine zones.

FLOWERING TIME: Jul–Aug

NOTES: You'll have to go to the high country of the southern Sierra to see this shrub, so if you need an excuse to see some beautiful country, here it is!

Carpenteria californica

Carpenteria, Tree Anemone

Hydrangeaceae (Hydrangea family)

Multistemmed, evergreen shrub < 3 m, with tan to gray bark, which peels off as papery sheets from older trunks.

LEAVES: Opposite, petioles 0.5–1 cm; blades 4–10 cm x 1–2.5 cm wide, narrowly elliptic, dark green and hairless above and pale and hairy below, except in late summer when leaves wilt, turn yellow, and hanging vertically with the margins rolled under.

FLOWERS: Racemes or cymes of 5–9 flowers; sepals 5–7, 8–10 mm; petals 5–8, 15–25 mm. Style 1, with 5–7 stigmatic regions; stamens numerous.

FRUITS: Capsule 1.2 mm wide. Seeds up to 1,000, tiny, brown to reddish.

SIERRA DISTRIBUTION: Natural range is limited to foothills of eastern Fresno and Madera Counties (other counties are displayed in Calflora map, but they represent planted shrubs).

ELEVATION: 340–1500 m

HABITAT: Streamsides, draws, and slopes; primarily in foothill chaparral; foothill and montane zones.

FLOWERING TIME: Apr–Jul

NOTES: A relict, which was likely more common millions of years ago when summer rains occurred. The dramatic wilting/ yellowing of leaves each summer is followed by regreening and resumption of photosynthesis within days of the first soaking fall rains. Although plants have been in cultivation since the 1800s, seedlings were not observed in nature until March 1990, after a fire burned through large stands in 1989, resulting in millions of tiny seedlings. Carpenter bees and bumblebees gather pollen and are effective pollinators, but as no nectar is produced, bees can be seen "tanking up" on nectar from nearby tincture plants or yerba santa (among others) to maintain their energy. Carpenteria sprouts vigorously after fire.

Philadelphus lewisii

Wild Mock Orange, Syringa

Hydrangeaceae (Hydrangea family)

Shrub to 3 m; loosely branched and often with peeling bark.
LEAVES: Opposite, with no stipules; blades 2.5–7.5 cm, 2–4 cm wide, ovate, with 3 distinct veins originating at base, margins finely toothed or entire. Undersurface sparsely covered with stiff hairs.
FLOWERS: Racemes of 1 to few flowers. Flowers 2–3 cm wide, white; sepals 4–5, 4–7 mm; petals 4–5, 8–12 mm. Ovary half inferior; stamens > 20.
FRUITS: Capsule woody.
SIERRA DISTRIBUTION: West slope, all.
ELEVATION: < 2500 m
HABITAT: Slopes and canyons; foothill and montane zones.
FLOWERING TIME: May–Jul
NOTES: Also known by the common name "syringa," although this is confusing because the genus name for lilac is *Syringa*, in Oleaceae (Olive family).

Draperia systyla

Violet Draperia

Hydrophyllaceae (Waterleaf family)

Perennial herb 1–4 dm; from woody caudexes, with stem decumbent or erect, sometimes rooting at nodes, and forming mats or mounds, groundcoverlike; herbage is long, soft-hairy.
LEAVES: Cauline and opposite; upper leaves sessile; blades 1–7 cm, ovate, entire.
FLOWERS: Inflorescences are coiled cymes; pedicels 1–2 mm; calyx lobes 4–6 mm, linear, and hairy; corollas 7–14 mm, funnel-shaped, white to lavender (pink), hairy. Stamens included, unequal in length.
FRUITS: Capsule 2–3 mm wide, spheric, long-hairy. Seeds dark brown, honeycombed, 1- to 4-angled.
SIERRA DISTRIBUTION: West slope, Tulare County north.
ELEVATION: < 2700 m
HABITAT: Forest openings, talus, rocky areas; montane zone.
FLOWERING TIME: May–Oct
NOTES: The only species in its genus, *D. systyla* is a common but not ubiquitous perennial of montane forests. Often seen in giant sequoia groves.

Emmenanthe penduliflora
Whispering Bells
Hydrophyllaceae (Waterleaf family)

Erect, sticky, glandular annual herb 5–85 cm. Strong odor.
LEAVES: Basal and alternate, pinnately toothed, 1–12 cm.
FLOWERS: Calyx 5-lobed, with lobes 4–11 mm, glandular; corollas bell-shaped, and lobes 6–15 mm, pale yellow or cream.
FRUITS: Capsule 7–10 mm. Seeds 6–15, net-surfaced. Papery petals remain on fruiting plants.
SIERRA DISTRIBUTION: West slope, Tuolumne County south < 1400 m; east slope < 2200 m.
ELEVATION: < 2200 m
HABITAT: Sandy soils mostly in chaparral, most likely to be seen year 1 post-fire; foothill and sagebrush zones.
FLOWERING TIME: Apr–Jul
NOTES: Research has demonstrated that nitrogen dioxide in smoke cues seeds to germinate, explaining the abundance of whispering bells after fire. The sound of the wind through the papery fruiting corollas is the reason for the charming common name.

Hesperochiron nanus
California Hesperochiron
Hydrophyllaceae (Waterleaf family)

Perennial herb 2–4 cm; from short, thick roots (caudexes), not rhizomatous.
LEAVES: Rosettes of > 6 leaves, each < 8 cm x < 3 cm wide, with both surfaces hairy.
FLOWERS: Inflorescences of < 5 flowers. Corollas 10–30 mm, bell- to funnel-shaped, with lobes 3–10 mm (lobes 0.8–1.5x the tube), and limbs 10–20 mm wide.
FRUITS: Capsule 5–11 mm, hairy.
SIERRA DISTRIBUTION: Both slopes, all; distribution patchy.
ELEVATION: 1000–2800 m
HABITAT: Moist to wet areas, meadows, often in alkaline sites; montane zone.
FLOWERING TIME: Apr–Jul
NOTES: Formerly *Hesperochiron californicus*. Note the obvious corolla tubes and leaves with hairs on upper and lower surfaces to rule out *H. pumilus*.

Hesperochiron pumilus

Dwarf Hesperochiron

Hydrophyllaceae (Waterleaf family)

Perennial herb; from slender rhizomes.

LEAVES: Rosettes of 2–10 leaves, each 1–7 cm, < 2 cm wide, with undersurface hairless.

FLOWERS: 1–8 per plant; corollas 5–15 mm, white with purple veins, and rotate (tube is miniscule); limbs 7–11 mm wide; corolla lobes 3–11 mm (lobes 2–4x the tube length), rounded.

FRUITS: Capsule 5–11 mm, hairy.

SIERRA DISTRIBUTION: Mostly west slope, Tulare County north.

ELEVATION: 150–3000 m

HABITAT: Moist to wet sites, slopes, flats, meadows; montane and subalpine zones.

FLOWERING TIME: Apr–Jul

NOTES: Compare with *H. nanus*. *H. pumilus* flowers are rotate (like a dinner plate), rather than with the distinct corolla tube of *H. nanus*.

Hydrophyllum alpestre

Alpine Waterleaf

Hydrophyllaceae (Waterleaf Family)

Short, hairy, ± stemless perennial herb.

LEAVES: Deeply lobed or compound with 5–7 leaflets, bright green without mottling. Petioles 3–15 cm; blades 4–12 cm.

FLOWERS: Inflorescences 4–10 mm, at ground level (well below level of leaves), purple to white, with purple veins. Stamens 1–2 mm, well-exserted.

FRUITS: Capsule 4 mm wide.

SIERRA DISTRIBUTION: Alpine County north.

ELEVATION: 400–2900 m

HABITAT: Moist shaded sites in forests; montane through subalpine zones.

FLOWERING TIME: May–Jul

NOTES: Formerly *Hydrophyllum capitatum* var. *alpinum*.

Hydrophyllum occidentale
Western Waterleaf
Hydrophyllaceae (Waterleaf family)

Erect, hairy perennial herb 6–60 cm.

LEAVES: Deep lobes or compound with 7–15 leaflets, usually mottled with watermarklike spots above, pale below. Petioles 2–15 cm; blades 5–40 cm.

FLOWERS: Inflorescences headlike, rising well above leaves; calyx divided to base; corollas 7–10 mm, bell-shaped, white with purple veins or entirely purple. Stamens well-exserted.

FRUITS: Capsule 4 mm wide.

SIERRA DISTRIBUTION: All, except Inyo County.

ELEVATION: 400–2900 m

HABITAT: Dry to moist shaded sites in forests; montane through subalpine zones.

FLOWERING TIME: May–Jul

NOTES: A distinctive plant, with its large, attractive leaves mottled as if splashed with water.

Nemophila heterophylla
Varied-Leafed Nemophila
Hydrophyllaceae (Waterleaf family)

Sprawling annual herb with curved stems 10–30 cm; hairless or sparsely hairy.

LEAVES: Lower leaves opposite, 1–4 cm, 5- to 7-lobed, with lobes round and appearing stalked; each lobe has up to 3 teeth. Upper leaves alternate, 0- to 5-lobed.

FLOWERS: In leaf axils; pedicels the length of leaves when plant is in flower, and up to 60 mm long when in fruit. Calyx lobes 2–4 mm, with appendages < 1 mm, reflexed; corollas 3–10 mm, 4–12 mm wide, bowl-shaped, white or light blue. Style 2–4 mm.

FRUITS: Capsule round, hairy, exceeding calyx in size. Seeds 2–5, yellow-brown.

SIERRA DISTRIBUTION: West slope, all.

ELEVATION: < 1700 m

HABITAT: Partial shade; foothill and montane zones.

FLOWERING TIME: Feb–Jun

NOTES: Compare with the similar *N. pedunculata*. *N. heterophylla* has alternate leaves on the upper stem, sometimes not until plants have been flowering for a while.

Nemophila maculata

Fivespot

Hydrophyllaceae (Waterleaf family)

Showy annual herb with weak, branched stems to 30 cm, with soft or bristly hairs.

LEAVES: Opposite, the lower ones petioled, with blades 8–30 mm and 5–9 lobes; upper leaves sessile, entire or toothed.

FLOWERS: Pedicels 10–20 mm when in flower, up to 70 mm when in fruit; calyx lobes 4–9 cm, appendages 1–4 mm. Corollas 8–20 mm wide, bowl-shaped or rotate, white with purple spots toward tips of lobes. Style 36 mm.

FRUITS: Capsule 4–6 mm wide. Seeds 2–12.

SIERRA DISTRIBUTION: West slope, all.

ELEVATION: < 3100 m

HABITAT: Moist areas in partial shade to full sun; foothill and montane zones.

FLOWERING TIME: Apr–Aug

NOTES: Forms solid carpets, especially in swales or where moisture is a bit higher than surrounding areas.

Nemophila menziesii var. *menziesii*

Baby Blue-Eyes

Hydrophyllaceae (Waterleaf family)

Annual herb with branched stems to 30 cm.

LEAVES: Opposite; lower leaves 1–5 cm, with 6–13 lobes; upper leaves sessile, entire or with a few shallow lobes (lower and upper leaves dissimilar).

FLOWERS: Pedicels 20–60 mm in flower, up to 70 mm in fruit; calyx lobes 4–8 cm, appendages 1–4 mm. Corollas 5–20 mm, 10–40 mm wide, bowl-shaped or rotate, sky-blue with white center, sometimes all white. Style 2–7 mm.

FRUITS: Capsule 5–12 mm wide. Seeds 10–20.

SIERRA DISTRIBUTION: West slope.

ELEVATION: < 2000 m

HABITAT: Partial shade to full sun; foothill and montane zones.

FLOWERING TIME: Apr–Jun

NOTES: The other variety in the SN is var. *integrifolia*, found in Inyo, Kern, and Tulare counties < 1900 m. Lower leaves are 5- to 7-lobed, upper entire to slightly toothed; flowers are smaller; 5–10 mm x 6–15 mm wide.

Nemophila parviflora var. *quercifolia*

Oak-Leaved Nemophila

Hydrophyllaceae (Waterleaf family)

Annual herb with weak stems; hairless to soft-hairy or bristly.
LEAVES: Alternate at base, opposite toward inflorescence; 5- to 7-lobed, long-hairy, with base of blade truncate or cordate.
FLOWERS: In leaf axils; pedicels 2–15 mm, calyx lobes 1–3 mm, appendages between calyx lobes 1 mm; corollas 3–5 mm wide, light blue.
FRUITS: Capsule 3–6 mm wide, hairy. Seeds 2–8.
SIERRA DISTRIBUTION: West slope, El Dorado County south.
ELEVATION: 700–2200 m
HABITAT: Partial shade; foothill and montane zones.
FLOWERING TIME: Apr–Jul
NOTES: Tends to show up after wildfires. The distinctive, oaklike leaves and tiny flowers are characteristic.

Nemophila pedunculata

Littlefoot Nemophila

Hydrophyllaceae (Waterleaf family)

Annual herb with stems 10–30 cm; hairless or sparsely hairy. Stems are weak, angled or winged.
LEAVES: All opposite, with wedge-shaped bases and 5–9 primary lobes or divisions (these sometimes have 1–2 teeth); leaf blades 0.5–3.5 cm.
FLOWERS: In leaf axils; pedicels ≤ leaf; calyx lobes 1–4 cm, appendages 0.5–1.5 mm, reflexed. Corollas 3–8 mm wide, bell-shaped, plain white or light blue, commonly with dark spots or veins, and sometimes with a purple spot on each corolla lobe. Style < 2 mm.
FRUITS: Capsule 3–6 mm wide, hairy. Seeds 2–8.
SIERRA DISTRIBUTION: West slope.
ELEVATION: < 2400 m
HABITAT: Partial to full shade; foothill and montane zones.
FLOWERING TIME: Feb–Jul
NOTES: Compare with similar-looking *N. heterophylla*, which has alternate leaves on the upper stem in mature plants. Plants with distinct purple spots on each corolla lobe look like mini fivespot (*N. maculata*)!

Nemophila pulchella var. *pulchella*

Pretty Nemophila

Hydrophyllaceae (Waterleaf family)

Annual herb, forming tangled masses.

LEAVES: Lower leaves opposite, 2–5 cm, with 5 well-separated, 1- to 5-toothed lobes; upper leaves alternate, sessile, entire or with shallow lobes.

FLOWERS: In leaf axils; pedicels 10–30 mm when in flower, and up to 50 mm when in fruit. Calyx lobes 2–4 cm, appendages < 0.5 mm. Corollas 5–12 mm wide, rotate (like a dinner plate, not a bell or a bowl), blue with white center; petals much longer than calyx lobes. Style < 3 mm.

FRUITS: Capsule. Seeds 2–4, yellow-brown.

SIERRA DISTRIBUTION: West slopes, mostly recorded from the southern part of Madera County south to Kern County.

ELEVATION: < 2200 m

HABITAT: Foothill woodland, chaparral, lower coniferous forest; foothill and montane zones.

FLOWERING TIME: Apr–Jun

NOTES: If you find a similar plant with a rotate (platelike) corolla that is all white, with corolla lobes barely longer than the calyx lobes, leaves that are all alternate, and fruits with only 1 seed, you're seeing var. *gracilis*, which occurs from Fresno County south.

Nemophila spatulata

Sierra Baby Blue Eyes

Hydrophyllaceae (Waterleaf family)

Annual herb 10–30 cm; with stems ascending, often weak.

LEAVES: Opposite, 5–30 mm, spoon-shaped or wedge-shaped; leaves toward top of plant 3- to 5-toothed, with winged petioles > blades.

FLOWERS: Pedicels 3–8 mm when in flower, up to 30 mm when in fruit; calyx lobes 2–5 mm, appendages 1–2 mm. Corollas 2–8 mm, bowl-shaped, white with purple spots or veins, or blue.

FRUITS: Capsule 4–6 mm wide. Seeds 2–12.

SIERRA DISTRIBUTION: Both slopes, all.

ELEVATION: 1100–3200 m

HABITAT: Meadows, other moist areas; montane and subalpine zones.

FLOWERING TIME: Apr–Jul

NOTES: A diminutive plant, recognizable by the charming spatulate or wedge-shaped upper leaves and the small, white or purple-and-white flowers.

Phacelia cicutaria

Caterpillar Phacelia

Hydrophyllaceae (Waterleaf family)

Ascending to erect annual herb 18–60 cm; stiff-hairy, with glands present.

LEAVES: Alternate, deeply lobed to compound, 20–150 mm, segments toothed or lobed.

FLOWERS: Coiled cymes, with 5 calyx lobes, 6–8 mm, linear, greenish-yellow, with long, stiff hairs and glandular hairs. Corollas 8–12 mm, bell-shaped, white-yellow (brownish), deciduous.

FRUITS: Capsule 3–4 mm, round, with stiff, bulbous-based hairs > 1 mm; calyx 9–12 mm in fruit.

SIERRA DISTRIBUTION: West slope, all counties.

ELEVATION: < 1400 m

HABITAT: Rocky and grassy sites; foothill and lower montane zones.

FLOWERING TIME: Feb–Jun

NOTES: One of the first plants to bloom each spring, forming masses in rocky sites along roads.

Phacelia egena

Rock Phacelia

Hydrophyllaceae (Waterleaf family)

Perennial herb 15–60 cm; several ascending stems, densely stiff-hairy, not glandular.

LEAVES: Basal leaves numerous, lanceolate-oblanceolate, dissected into 7–11 segments, with prominent veins; blades 10–25 cm, > petioles. Cauline leaves alternate, entire.

FLOWERS: Coiled cymes, flowers 5-merous; calyx lobes 4–6 mm, linear to oblanceolate. Corollas 5–9 mm, bell-shaped, white, with limbs 4–8 mm diameter. Style 9–15 mm, cleft ½ its length, hairless, and exserted; stamens 5, 9–14 mm, hairy, and exserted.

FRUITS: Capsule ± 3 mm, ovoid, stiff-hairy; fruiting calyx 9–12 mm. Seeds 1–2, 2.5–3 mm, surface pitted in ± longitudinal rows.

SIERRA DISTRIBUTION: West slope, all.

ELEVATION: < 2500 m

HABITAT: Sandy to rocky sites; foothill and montane zones.

FLOWERING TIME: Apr–Aug

NOTES: Intergrades with *P. imbricata*, *P. hastata*, *P. corymbosa* (in northern SN)—keep this in mind if faced with uncertainty about precise identification of perennial *Phacelia*.

Phacelia eisenii

Eisen's Phacelia

Hydrophyllaceae (Waterleaf family)

Annual herb 2–10 cm; puberulent to stiff-hairy, glandular.

LEAVES: Lower leaves opposite, upper alternate; 10–25 mm, oblanceolate or elliptic, smooth-margined, with blades equal to or slightly longer than petioles.

FLOWERS: Cymes, not strongly coiled; calyx lobes 5, 1–3 mm, linear, and unequal. Corollas 2–4 mm, bell-shaped, lavender or white, deciduous. Style 1–4 mm, divided deeply, nearly to base; stamens 3–5 mm, hairless.

FRUITS: Capsule 2–3 mm, ovoid or round, short-hairy; calyx 4–5 mm.

SIERRA DISTRIBUTION: Sierra crest and northern and eastern Sierra (records exist from Kern, Inyo, Mono, Alpine, Nevada, Sierra, and Plumas Counties to date).

ELEVATION: 1300–3400 m

HABITAT: Sandy or gravelly soils; montane, subalpine, and sagebrush zones.

FLOWERING TIME: May–Aug

NOTES: Compare with *Phacelia humilis*.

Phacelia hastata var. *compacta*

Compact Phacelia

Hydrophyllaceae (Waterleaf family)

Mat-forming perennial herb 5–25 cm; stem decumbent to ascending, herbage with stiff, spreading hairs.

LEAVES: Mostly basal, stem leaves alternate; blades 15–40 mm, no longer than petioles, tapered at base, lanceoloate to elliptic, and usually entire, but sometimes 2- to 4-lobed or compound with 3–5 leaflets; basal leaves have prominent veins.

FLOWERS: Coiled cymes, flowers 5-merous; calyx lobes 3–7 mm, linear to lanceolate, glandular. Corollas 4–7 mm, tubular or bell-shaped, creamy-white (lavender). Style 7–10 mm, cleft ½ its length, hairless, and exserted; stamens 5, 6–10 mm, hairless or hairy, exserted.

FRUITS: Capsule 2–4 mm, ovoid, stiff-hairy, fruiting calyx 5–9 mm. Seeds 1–3, 1.5–2.5 mm, with surface pitted.

SIERRA DISTRIBUTION: Both slopes, Tulare and Mono Counties north.

ELEVATION: 1500–4000 m

HABITAT: Sandy to rocky sites in open forest, scree slopes, fellfields; montane, subalpine, and alpine zones.

FLOWERING TIME: Jun–Sep

NOTES: *Phacelia hastata* var. *hastata* is common within the same geographical range in the SN < 3100 m; it has erect to ascending stems 20–50 cm, appressed hairs, and nonglandular calyx lobes.

Phacelia heterophylla var. *virgata*

Varied Leaf Phacelia

Hydrophyllaceae (Waterleaf family)

Biennial herb 20–120 cm; usually with 1 erect stem, the apex taller than the tips of lateral branches; herbage densely stiff-hairy with sparse glandular hairs toward the top.

LEAVES: Basal and alternate, lanceolate to ovate, pinnately dissected into 5–9 lobes, with prominent veins; blades 5–15 cm, > than petioles. Stem leaves smaller, entire to dissected.

FLOWERS: Coiled cymes elongating in maturity, flowers 5-merous; calyx lobes 3–6 mm; corollas 4–7 mm, bell-shaped, white to lavender. Style 8–10 mm, cleft ½ its length, hairless, and exserted; stamens 5, 8–10 mm, hairy, exserted.

FRUITS: Capsule 2.5–4 mm, ovoid, stiff-hairy, fruiting calyx 5–9 mm, fruiting calyx lobes 6–10 mm. Seeds 1–3, 1.5–2.5 mm, surface with longitudinal rows of pits.

SIERRA DISTRIBUTION: Both slopes, Tulare and Mono Counties north.

ELEVATION: < 2900 m

HABITAT: Steep, exposed slopes, roadcuts, rocky sites, often abundant after fires; foothill and montane zones.

FLOWERING TIME: May–Sep

NOTES: The variety name *virgata* means "wandlike," referring to the tendency to have a single, erect, flowering stalk taller than lateral branches. Sometimes intergrades with *P. mutabilis*.

Phacelia humilis var. *humilis*

Low Phacelia

Hydrophyllaceae (Waterleaf family)

Erect annual herb 5–20 cm; stiff hairs, often glandular.

LEAVES: Lower leaves opposite, upper alternate; 10–40 mm, ovate or elliptic, smooth-margined.

FLOWERS: Coiled cymes ± 1-sided; calyx lobes 5, 2–3 mm, linear, with dense, white hairs; corollas 3–5 mm, bell-shaped, violet, deciduous. Style 4–7 mm; stamens 4–6 mm, short-hairy.

FRUITS: Capsule 2–5 mm, ovoid, short-hairy, calyx 5–8 mm.

SIERRA DISTRIBUTION: Sierra crest and northern and eastern Sierra (records exist from Kern, Inyo, Mono, Alpine, Nevada, Sierra, and Plumas Counties to date).

ELEVATION: 1500–2800 m

HABITAT: Open gravelly flats, meadows; montane, subalpine, and sagebrush zones.

FLOWERING TIME: May–Jul

NOTES: Compare with *P. eisenii.*

Phacelia hydrophylloides

Waterleaf Phacelia

Hydrophyllaceae (Waterleaf family)

Ground-hugging to ascending perennial herb 10–30 cm; hairy but not glandular.

LEAVES: Alternate, ovate or oblong, toothed or lobed; blades 15–60 mm.

FLOWERS: In heads; calyx lobes 5, 3–5 mm, narrowly oblong; corollas 5–8 mm, bell-shaped, light purple or white, deciduous.

FRUITS: Capsule 5–7 mm, ovoid, strigose, calyx 9–12 mm. Seeds 3–8, 2–3 mm, surface pitted.

SIERRA DISTRIBUTION: Tulare and Mono Counties north.

ELEVATION: 1500–3100 m

HABITAT: Understory and openings in conifer forest; montane and subalpine zones.

FLOWERING TIME: Jun–Sep

NOTES: Petals often curl downward and outward at margins, a trait that helps distinguish this species.

Phacelia imbricata var. *imbricata*

Imbricate Phacelia

Hydrophyllaceae (Waterleaf family)

Erect perennial herb 20–120 cm; arising from 1+ central stems, stiff-hairy (with irritating hairs), frequently glandular.

LEAVES: Mostly basal; blades 5–15 cm, same length as petiole, deeply dissected into 7–15 segments, lanceolate to ovate; stem leaves smaller, with fewer or no lobes.

FLOWERS: Coiled cymes; calyx lobes 5, 3–6 mm; corollas 4–7 mm, cylindric or bell-shaped, white (rarely lavender). Style 9–14 mm, hairy at base; stamens 9–13 mm, hairy.

FRUITS: Capsule 3–4 mm, ovoid, stiff-hairy, calyx lobes 5–10 mm, overlapping (imbricate). Seeds 1–3, 2–2.5 mm, surface pitted in rows.

SIERRA DISTRIBUTION: West slope.

ELEVATION: 400–1400 m

HABITAT: Open areas, roadsides; foothill and montane zones.

FLOWERING TIME: Apr–Jul

NOTES: Foliage tends to be a bright yellowish-green in the SN. The cylindrical flowers with markedly longer stamens and styles are characteristic.

Phacelia mutabilis

Changeable Phacelia

Hydrophyllaceae (Waterleaf family)

Biennial or perennial herb 10–60 cm; stem erect to decumbent, with sparse, long, stiff hairs, not glandular.

LEAVES: Basal and cauline; basal leaves wither by flowering time. Blades 20–180 mm, same length as petiole, lanceolate to ovate, entire to compound; if compound, then leaflets 3.

FLOWERS: Coiled cymes; calyx lobes 5, 3–5 mm; corollas 4–6 mm, tubular to bell-shaped, white, cream, or lavender. Style 6–8 mm, cleft ½ its length; stamens 6–8 mm, hairy.

FRUITS: Capsule 2–3 mm, ovoid, stiff-hairy, calyx lobes 6–10 mm. Seeds 1–4, 1.5–2.5 mm, surface pitted.

SIERRA DISTRIBUTION: West slope, all.

ELEVATION: 900–3500 m

HABITAT: Open areas, often rocky; foothill through subalpine zones.

FLOWERING TIME: Apr–Jul

NOTES: Veins of basal leaves are not prominent in *P. mutabilis* as they are in *P. imbricata* and *P. heterophylla* var. *virgata*.

Phacelia platyloba

Broadlobe Phacelia

Hydrophyllaceae (Waterleaf family)

Annual herb 9–45 cm; simple or branched at base, puberulent with sparse, short, stiff hairs.

LEAVES: Alternate; blades 1–2 times pinnately compound, 15–90 mm, > petioles, elliptic to ovate in overall outline.

FLOWERS: Coiled cymes ± 1-sided; calyx lobes 5, 2–4 mm, lanceolate to roundish, noticeably unequal, entire to crenate, and ciliate. Corollas 4–5 mm, bell-shaped, lavender. Style 4–6 mm, cleft ⅔–¾, hairless; stamens 3–4 mm, hairless.

FRUITS: Capsule 2–3 mm, oblong, densely puberulent; calyx lobes 4–5 mm. Seed 1, 2–3 mm, pitted.

SIERRA DISTRIBUTION: West slope, common in Mariposa, Madera, and Fresno Counties.

ELEVATION: 300–1200 m

HABITAT: Rocky to grassy areas, often in at least partial shade; foothill and montane zones.

FLOWERING TIME: Apr–Jun

NOTES: Forms tangled masses of lacy-leaved plants with abundant light lavender flowers.

Phacelia quickii

Quick's Phacelia

Hydrophyllaceae (Waterleaf family)

Erect or decumbent annual herb 4–18 cm; puberulent, often glandular. Skunky odor.

LEAVES: Alternate, 8–50 mm, linear to lanceolate, smooth-margined.

FLOWERS: Coiled cymes ± 1-sided; calyx lobes 5, 2–4 mm, linear or oblanceolate, noticeably unequal; corollas 3–5 mm, broadly bell-shaped, purple to blue (sometimes white); the shriveled corolla remains on the flower during fruiting. Style 3–6 mm, hairy; stamens 3–6 mm, hairless; both longer than the corollas.

FRUITS: Capsule 2–2.5 mm, round, short-hairy, calyx 4–7 mm.

SIERRA DISTRIBUTION: West slope, Tulare to Plumas Counties.

ELEVATION: 1000–2400 m

HABITAT: Open granitic areas on rocks and in gravel; foothill, montane, and subalpine zones.

FLOWERING TIME: May–Jul

NOTES: Another small annual *Phacelia* may be found in similar habitats from Fresno County north: *P. racemosa* has opposite leaves below, smaller (2–4 mm), narrower, whitish or pale blue corollas, with stamens and styles shorter (not sticking out of the corolla).

Phacelia ramosissima

Branching Phacelia

Hydrophyllaceae (Waterleaf family)

Erect to prostrate, many-branched perennial herb 30–150 cm; hairless to densely hairy, glandular or not.

LEAVES: Alternate; blades pinnately compound, 40–200 mm, > petioles, oblong to ovate in overall outline; leaflets sessile, elliptic, toothed or lobed, and the lobes themselves often toothed.

FLOWERS: Calyx lobes 5, 3–6 mm (not elongating in fruit), and noticeably unequal; corollas 5–8 mm, funnel- to bell-shaped, white to lavender to blue. Style 7–10 mm, cleft ⅔ of its length, hairless; stamens 7–10 mm, unequal, hairless, with purple or white filaments.

FRUITS: Capsule 3–4 mm, ovoid, stiff-hairy. Seeds 2–4, 2–3 mm, surface pitted.

SIERRA DISTRIBUTION: All.

ELEVATION: < 3800 m

HABITAT: Extremely variable in habitat preference; all zones above foothill zone.

FLOWERING TIME: Apr–Oct

NOTES: Pollinators include native bumblebees, butterflies, wasps, and sawflies.

Phacelia tanacetifolia

Tansy-Leaved Phacelia, Lacy Phacelia

Hydrophyllaceae (Waterleaf family)

Erect perennial herb 15–100 cm, with short stiff hairs, glandular.
LEAVES: Alternate, 20–200 mm; bladess compound; leaflets toothed to deeply lobed (causing "lacy" appearance).
FLOWERS: Inflorescence is 2- to 4-branched; calyx lobes 5, 4–6 mm, linear and noticeably long-hairy; corollas 6–9 mm, bell-shaped, blue. Style 11–15 mm, hairless; stamens 9–15 mm, hairless.
FRUITS: Capsule 3–4 mm, ovoid, hairless at top, calyx lobe 6–8 mm. Seeds 1–2, 2–3 mm.
SIERRA DISTRIBUTION: Both slopes, Mariposa and Mono Counties south. Most abundant in Kern County.
ELEVATION: < 2500 m
HABITAT: Open sites in sandy soils, gravelly slopes; foothill and lower montane zones.
FLOWERING TIME: Mar–May
NOTES: Distribution is very sparse north of Tulare County. Perhaps some energetic citizen scientists will find more localities and fill in the map.

Phacelia vallicola

Mariposa Phacelia

Hydrophyllaceae (Waterleaf family)

Erect annual herb 6–27 cm; puberulent, sparsely stiff-hairy.
LEAVES: Opposite below, alternate above; blades 10–35 mm, > petiole, lanceolate to ovate, entire.
FLOWERS: Calyx lobes 5, 3–5 mm, linear to oblanceolate, noticeably unequal, covered with tiny glands, and long-ciliate. Corollas 4–6 mm, bell-shaped, tube white to lavender, and lobes usually purple (sometimes white). Style and stamens included: style 2–4 mm, cleft to base; stamens 2–3 mm, papillate.
FRUITS: Capsule 4–6 mm, ovoid, short-hairy, calyx 7–10 mm. Seeds 8–14, 1.5–2 mm, pitted.
SIERRA DISTRIBUTION: West slope, Fresno to Plumas Counties.
ELEVATION: 600–2400 m
HABITAT: Open areas on cliffs, on rocks and gravel; foothill and montane zones.
FLOWERING TIME: Mar–Jun
NOTES: Veins are sunken on leaves.

Pholistoma auritum var. *auritum*

Blue Fiesta Flower

Hydrophyllaceae (Waterleaf family)

Hairy, sprawling annual herb with angled stems 2–12 dm; forming masses that appear vinelike, often in shade, stems with downward-pointing prickles.

LEAVES: Deeply lobed with 7–13 pinnate lobes. Lower leaves opposite, 4–16 cm; upper alternate, smaller; petioles winged and clasping the stem.

FLOWERS: Terminal cymes of up to 6 flowers or single flowers in axils; corollas 7–15 mm x 10–30 mm wide, appendages at base of lobes dark purple with yellow nectaries; calyx lobes 3–9 mm, with auricles between each lobe 1–4 mm.

FRUITS: Capsule 5–10 mm wide, round, bristly, calyx remaining attached.

SIERRA DISTRIBUTION: West slope, Tuolumne County south.

ELEVATION: < 1700 m

HABITAT: Shaded or partially grassy or rocky sites; foothill and lower montane zones.

FLOWERING TIME: Mar–Jul

NOTES: You cannot miss this plant. It can form extensive tangles of viny stems, especially after fire, but in any good rain year. I have been told that the name "fiesta flower" comes from the possible use of these flowers as decoration—the downward-pointing prickles stick to clothing, and one could dress up most any outfit by adorning it with these purple flowers. Best not to go too crazy, as the flowers exist to perpetuate the species after all!

Pholistoma membranaceum

White Fiesta Flower

Hydrophyllaceae (Waterleaf family)

Fleshy, sprawling annual herb 5–90 cm; appears hairless but is dotted with bristly hairs, glaucous.

LEAVES: Lower leaves opposite, 2–13 cm; upper alternate, smaller; petioles winged but not clasping stem.

FLOWERS: Inflorescence of 2–10 flowers; calyx lobes 1–3 mm with no appendages between each lobe; corollas 3–6 mm x ± 10 mm wide, white, lobe usually purple-spotted.

FRUITS: Capsule 2–4 mm wide, round, bristly, calyx not enclosing fruit.

SIERRA DISTRIBUTION: Mariposa County south.

ELEVATION: < 1400 m

HABITAT: Grassy and wooded slopes; foothill and lower montane zones.

FLOWERING TIME: Feb–May

NOTES: Compare with the small, white-flowered *Nemophila* spp. (*heterophylla* and *pedunculata*), which have appendages (small projections) between each calyx lobe.

Hypericum anagalloides

Tinker's Penny

Hypericaceae (St. Johnswort family)

Prostrate to decumbent perennial (sometimes annual) herb with stems 3–30 cm; prostrate to decumbent, from stolons, forming mats up to 25 cm wide.

LEAVES: Opposite, 4–15 mm, round or elliptic, with gland dots on leaf surface clear or greenish.

FLOWERS: Inflorescence of 1–15 flowers per stem; sepals 5, 2–4 mm, unequal; petals 2–5 mm, about same length as sepal. Ovary superior.

FRUITS: 3 mm. Seeds 1 mm, yellowish-brown.

SIERRA DISTRIBUTION: All.

ELEVATION: < 3220 m

HABITAT: Seeps, wet meadows, wet ditches along roads, streamsides; montane through alpine zones.

FLOWERING TIME: May–Sep

NOTES: Tinker's penny is one of the most common species of mountain meadows. The tiny, orange-yellow flowers on low-growing, matlike foliage make it easy to recognize.

Hypericum concinnum

Goldwire

Hypericaceae (St. Johnswort family)

Erect, many-stemmed perennial herb 15–20 cm; from woody base.

LEAVES: Opposite, 1.5–4 cm, lanceolate or linear, at least somewhat folded, with some black dots.

FLOWERS: Inflorescence of 3–9 flowers per stem; sepals 5, lanceolate, black-dotted; petals 10–15 mm, golden-yellow, with black dots on margins. Ovary superior, style 10 mm.

FRUITS: Capsule 6–7 mm, 3-lobed. Seed 1.5 mm, greenish-brown.

SIERRA DISTRIBUTION: West slope, Mariposa County north (1 report from Kern County near Lake Isabella at Breckenridge Campground).

ELEVATION: < 1600 m

HABITAT: Dry, open ridges, rocky sites, often on metamorphic substrates; foothill and montane zones.

FLOWERING TIME: May–Aug

NOTES: Goldwire is the showiest of the native St. Johnsworts and is easy to identify due to its large flowers and folded leaves. Proliferates after fire.

Hypericum perforatum ssp. *perforatum* (non-native, invasive)

Klamathweed, St. Johnswort

Hypericaceae (St. Johnswort family)

Erect, many-stemmed perennial herb 3–12 dm; main root extends to depth of 5 ft, and lateral roots extend to 3 ft and give rise to new plants from underground buds; sterile axillary branches 2–10 cm.

LEAVES: Opposite, 1.5–2.5 cm, oblong, with margins recurved, black-dotted, and some black dots on leaf surface, base ± clasping stem. Leaf surface sparsely conspicuously dotted with transparent glands (best seen with hand lens or backlit by sunlight).

FLOWERS: Inflorescence of 25–100 flowers per stem; sepals 5, lanceolate, black-dotted; petals 8–12 mm, oblong, margins conspicuously black-dotted, and twisting as they dry post-flowering. Ovary superior; anthers black-dotted.

FRUITS: Capsule 7–8 mm, unlobed. Seeds numerous, 1 mm, brown.

SIERRA DISTRIBUTION: West slope, possibly every county due to recent rapid spread.

ELEVATION: < 1980 m

HABITAT: Open disturbed areas, moist roadside ditches; foothill and montane zones.

FLOWERING TIME: May–Aug

NOTES: Compare with *H. scouleri*. Klamathweed is an invasive/noxious weed that has been dramatically reduced in the wetter parts of northwestern CA by beetles introduced to control it; but further south, despite the presence of the beetles, spread is uncontrolled and becoming a challenge for public lands managers and property owners. Klamathweed is highly competitive once established, displacing native plants, and poisonous to livestock if eaten in large quantities.

Hypericum scouleri

Scouler's St. Johnswort

Hypericaceae (St. Johnswort family)

Erect perennial herb 2–7 dm; rhizomatous or taprooted. Stems are few, arising from base; sterile axillary branches are short, < 2 cm (contrast with *H. perforatum*).

LEAVES: Opposite, 1–3 cm, ovate or elliptic, base ± clasping stem, edged with black dots, leaf surface sparsely dotted with transparent glands (best seen with hand lens or backlit by sunlight).

FLOWERS: Inflorescence of 3–25 flowers per stem; sepals 5, oblong, black-dotted; petals 7–12 mm, obovate, black-dotted. Ovary superior; anthers black-dotted.

FRUITS: Capsule 6–7 mm, 3-lobed. Seed 1 mm, brown.

SIERRA DISTRIBUTION: All.

ELEVATION: 1000–2940 m

HABITAT: Seeps, wet meadows, wet ditches along roads, streamsides; montane zone.

FLOWERING TIME: Jun–Sep

NOTES: Formerly *Hypericum formosum* var. *scouleri*. Compare with the similar-looking non-native plant *H. perforatum* ssp. *perforatum*, Klamathweed. One key difference is that the petals of *H. scouleri* do not twist with age as do those of Klamathweed. Distinguishing these 2 spp. is crucial as well-intentioned weed warriors have been known to remove this lovely native, thinking it was Klamathweed.

Agastache urticifolia

Nettle-Leafed Horsemint

Lamiaceae (Mint family)

Erect perennial herb to 1 m, with 4-angled stem, often growing in dense patches. Aromatic.

LEAVES: Opposite, petioled, broadly lanceolate, with no stipules; margins scalloped or toothed.

FLOWERS: Dense spikes with leafy green bracts at base of spike. Calyx lobes 5, pink-purple, surrounding corollas 2-lipped, pink to purple. Style 1; stamens 4; both exserted.

FRUITS: Nutlets 4, brown, smooth, with hairs at tip.

SIERRA DISTRIBUTION: All.

ELEVATION: 400–3000 m

HABITAT: Moist to dry sites; foothill and montane zones.

FLOWERING TIME: Jun–Aug

NOTES: This is the only species of *Agastache* in the SN.

Lamium amplexicaule (non-native)

Henbit

Lamiaceae (Mint family)

Erect or decumbent annual herb 1–4 dm; branched toward base, hairless or hairy.

LEAVES: Opposite, sessile, 1–2.5 cm, rounded, clasping stem.

FLOWERS: In axils and terminal, with calyx 4–7 mm, 5-lobed. Corollas 10–18 mm, 2-lipped; upper lip forms a hood and lower lip 3-lobed. Style 2-lobed, pink to reddish-purple; stamens 4, anthers hairy.

FRUITS: Nutlets 4, triangular.

SIERRA DISTRIBUTION: West slope, all counties.

ELEVATION: < 1400 m

HABITAT: Disturbed areas, especially moist sites, meadows; foothill and montane zones.

FLOWERING TIME: Apr–Sep

NOTES: Henbit is a common garden weed originally from Eurasia that is now ubiquitous in disturbed areas; it does not appear to be especially ecologically harmful. The larger, "normal" flowers are exquisite; the plant also makes smaller (< 8 mm) flowers that never open but form seeds by self-pollination (cleistogamous).

Lepechinia calycina

Pitcher Sage

Lamiaceae (Mint family)

Shrub < 1.2 m, much-branched at base and above. Aromatic.

LEAVES: Opposite, 4–12 cm, cordate, entire or toothed.

FLOWERS: Raceme with 1 bract per flower, with pedicel 0–1 cm; calyx 5-lobed, 2-lipped, hairy; corollas 5-lobed, 2-lipped, lavender or white, with lower lip longer and upper lip 4-lobed.

FRUITS: Calyx inflated and spherical in fruit; lobes smaller than the tube. Fruit 2–4 mm, dark brown or black.

SIERRA DISTRIBUTION: West slope, Mariposa County north.

ELEVATION: 150–1300 m

HABITAT: Rocky sites and open slopes, chaparral; foothill zone.

FLOWERING TIME: Apr–Jun

NOTES: This is the only species of *Lepechinia* in the SN. Note the long lower lip.

Lycopus uniflorus
Northern Bugleweed
Lamiaceae (Mint family)

Erect or ascending perennial herb 1–5 dm; puberulent or with sparse, stiff hairs; from slender rhizomes with tuberlike tips.
LEAVES: Opposite, ± sessile, 2–6 cm, widely lanceolate, sometimes purplish, toothed.
FLOWERS: Set densely in axils; calyx 5-lobed, with lobes ovate; corollas 2.5–4 mm, > calyx, white, 4-lobed, with lower lobe the longest. Stamens only 2, produce pollen, exserted with style.
FRUITS: Nutlets 4, 1–2 mm, = calyx, tip finely toothed.
SIERRA DISTRIBUTION: Mariposa County north.
ELEVATION: 500–2000 m
HABITAT: Moist areas, springs, fens, bogs; foothill and montane zones.
FLOWERING TIME: Jul–Sep
NOTES: Formerly *Lycopus uniflorus* var. *uniflorus*. Another species, *Lycopus americanus*, American bugleweed, also occurs in the SN < 1000 m, in Kern County and from El Dorado County north: It differs by having awl-like calyx lobes > nutlets; corollas 2–3 mm, ± = calyx; leaves with more deeply toothed to lobed blades; and is generally taller (2–8 dm).

Marrubium vulgare (non-native)
Horehound
Lamiaceae (Mint family)

Bushy, woolly, multistemmed perennial herb 1–6 dm.
LEAVES: Opposite, round to ovate, with blades 1.5–5.5 cm.
FLOWERS: Clusters of flowers in leaf axils; calyx has 10 short-hairy lobes with curved, sharp tips; corollas 2-lipped, white.
FRUITS: Nutlets.
SIERRA DISTRIBUTION: All.
ELEVATION: < 2000 m
HABITAT: Disturbed sites, especially abundant on overgrazed land; all zones except subalpine and alpine.
FLOWERING TIME: Mar–Nov
NOTES: The sharp, curved tips of the calyx lobes are ideal for dispersal, causing clusters of fruits to catch onto animals or clothing, Velcro-like. Horehound has been used for millennia to make tea and lozenges to treat colds and sinus infections, and to make hard candy.

Mentha arvensis

Field Mint

Lamiaceae (Mint family)

Perennial herb 1–5 dm; rhizomatous, with 4-angled stems, leafy to tip of flowering stalk. Strongly aromatic (minty, a "spicy" scent).

LEAVES: Opposite, with petioles 3–8 mm; blades 2–4 cm, ovate, rounded to cordate at base (not tapered).

FLOWERS: Headlike clusters in leaf axils; calyx 1.5–2.5 mm, 5-lobed, hairless inside and out or with ciliate lobes; corollas 2.5–3.5 mm, with upper lip 2-lobed and lower lip 3-lobed, purple, pink, or white. Stamens 4, exserted.

FRUITS: Nutlets 4, ellipsoid.

SIERRA DISTRIBUTION: All.

ELEVATION: < 3960 m

HABITAT: Moist or wet disturbed sites; foothill through subalpine zones.

FLOWERING TIME: Jul–Oct

NOTES: An extremely variable plant, but distinguished from other *Mentha* spp. by the leaves being about the same size as axillary flower clusters. Some races are probably native while others have been introduced from Europe; the species itself is circumboreal. The plant is one source of medicinal menthol.

Mentha canadensis

American Cornmint

Lamiaceae (Mint family)

Hairy perennial herb 1–5 dm; rhizomatous, with 4-angled stems, leafy to tip of flowering stalk. Strongly mint-scented.

LEAVES: Opposite, and smaller toward top of stem; petioles 3–8 mm; blades 1.5–5 cm, lanceolate to linear, toothed, with bases tapered and tips acute, lower surface hairy, especially on veins.

FLOWERS: Headlike clusters in leaf axils; calyx 1.5–3 mm, 5-lobed, outer surface short-hairy; corollas 4–7 mm, with upper lip 2-lobed and lower lip 3-lobed, purple, pink, or white. Stamens 4, exserted.

FRUITS: Nutlets 4, ellipsoid.

SIERRA DISTRIBUTION: All.

ELEVATION: < 2400 m

HABITAT: Moist or wet sites; foothill through subalpine zones.

FLOWERING TIME: Jul–Oct

NOTES: Formerly *Mentha arvensis* var. *canadensis*. Native to North America, Asia, and Russia. Source of commercial mint oil used for pharmaceuticals.

Mentha pulegium (non-native)

Pennyroyal

Lamiaceae (Mint family)

Short-hairy perennial herb 1–3 dm; rhizomatous, with 4-angled stems. Strongly aromatic.

LEAVES: Opposite, sessile, 0.5–2.5 cm, smaller toward top, short-hairy beneath, and elliptic or narrowly ovate, with tips rounded.

FLOWERS: Headlike clusters, in leaf axils, leaves below clusters reflexed; calyx 2.5–4 mm, 5-lobed, short-hairy; corollas 5–8 mm, with upper lip 2-lobed and lower lip 3-lobed, lavender. Stamens 4, exserted.

FRUITS: Nutlets 4, ellipsoid.

SIERRA DISTRIBUTION: All.

ELEVATION: < 1420 m

HABITAT: Moist sites; foothill through montane zones.

FLOWERING TIME: Jul–Oct

NOTES: Introduced from Europe. Fatally toxic to humans.

Monardella breweri ssp. *lanceolata*

Mustang Mint

Lamiaceae (Mint family)

Erect, branched annual herb 9–65 cm; with 4-angled purple stem, short-hairy or glandular toward top.

LEAVES: Opposite, 15–60 mm (including petiole), 4–12 mm wide, narrowly lanceolate, margins scalloped or toothed.

FLOWERS: Headlike cluster 5–25 mm wide; bracts below head are 5–15 mm, 8–10 mm wide, acute, with visible cross-veins. Calyx 6.5–7.5 mm, 13- to 14-veined; corollas 12–15 mm, purple, with upper lip 2-lobed, each lobe with an enlarged gland, and lower lip recurved and 3-lobed. Stamens protrude.

FRUITS: Nutlets 4.

SIERRA DISTRIBUTION: West slope.

ELEVATION: < 3000 m

HABITAT: Open, often rocky or sandy sites; foothill and montane zones.

FLOWERING TIME: May–Oct

NOTES: Formerly *Monardella lanceolata*.

Monardella odoratissima ssp. *glauca*

Mountain Pennyroyal

Lamiaceae (Mint family)

Subshrub with stems 10–30 cm (up to 45 cm); ascending to erect; green to gray, and gland-dotted. Aromatic.

LEAVES: Opposite, 15–25 mm (including petiole) x 16–28 mm wide, elliptic to ovate, ± hairless to hairy, and often purple-tinged, with glands on upper surface in obvious pits.

FLOWERS: Headlike clusters 1 to several per stem, 16–28 mm wide; leaves 2, 8–10 mm, often positioned just below head, reflexed (sometimes absent, sometimes not reflexed); membranous bracts 9–15 mm x 5–9 mm wide form a bowl, usually rose- or purple-tinged. Calyx 8–10 mm, 5-lobed, and tube stiff-hairy; corollas 10–20 mm, lavender, purple, or red-purple; upper lip 2-lobed, and lower lip 3-lobed and recurved. Stamens 4, stigma unequally 2-lobed

FRUITS: Nutlets 4.

SIERRA DISTRIBUTION: Both slopes, Tulare and Inyo Counties north.

ELEVATION: 1000–3500 m

HABITAT: Rocky or sandy sites; montane, subalpine, and sagebrush zones.

FLOWERING TIME: Jun–Sep

NOTES: This plant is a pollinator smorgasbord! Over 50 insect taxa have been observed on one patch in a single day in the subalpine zone, including bees, flies, butterflies, and moths. The species is extremely variable in the density of hairs on leaves and bracts, margin toothing, leaf length-to-width ratio, and gray vs. green leaf appearance. Intergrades with the co-occurring *M. odoratissima* ssp. *pallida*, as well as 3 or 4 other taxa. Ssp. *pallida* has flower clusters subtended by longer reflexed leaves 10–20 mm; corollas are usually white (but can be pink- or lavender-tinged).

Pogogyne douglasii
Douglas' Beardstyle
Lamiaceae (Mint family)

Erect annual herb < 45 cm, gland-dotted. Aromatic.
LEAVES: Opposite, short-petioled, varying from linear to round, entire to toothed, and bristly-hairy or not.
FLOWERS: Inflorescences 1–3 cm wide, terminal and axillary; ≥ 2 inflorescence bracts are below each flower, often longer than flowers, linear to narrowly oblanceolate. Calyx lobes 5, with lower lobes 3–8 mm and upper 2–5 mm, generally bristly-ciliate; corollas 9–20 mm, 2-lipped, lavender, and exserted from bracts, calyces. Style 2–6 mm, hairy, below the 2 stigma lobes ("beardstyle"); fertile stamens 4, with 1 pair longer.
FRUITS: Nutlets 4, 1–2 mm.
SIERRA DISTRIBUTION: West slope, most counties from Kern to Butte Counties.
ELEVATION: < 1300 m
HABITAT: Vernal pools, vernally moist swales; foothill and montane zones.
FLOWERING TIME: Mar–Jul
NOTES: Other *Pogogyne* spp. in the western foothills of the SN are shorter plants with smaller corollas: *P. zizyphoroides* is found in vernal pools < 400 m, from Madera County north (stems < 16 cm, corollas 4–8 mm); *P. serpylloides* occurs in grassy areas on the west slope < 1200 m between Tulare and Placer Counties, with stems < 20 cm, prostrate to decumbent, and corollas 2.5–5 mm.

Prunella vulgaris var. *lanceolata*
Mountain Self-Heal
Lamiaceae (Mint family)

Erect (sometimes decumbent) perennial herb 1–5 dm; usually soft-hairy. Not aromatic.
LEAVES: Opposite, ± lanceolate; blades 2–7 cm, 3x width.
FLOWERS: Inflorescences terminal; flowers are densely clustered, about 3 per purplish bract. Corollas 10–20 mm, 2-lipped, with upper lip hoodlike, enclosing stamens.
FRUITS: Nutlets obovoid.
SIERRA DISTRIBUTION: Mostly west slope, Tulare County north.
ELEVATION: < 2500 m
HABITAT: Moist areas along streams and in meadows; primarily montane zone.
FLOWERING TIME: May–Sep
NOTES: Mountain self-heal is native to North America and has been used by Native peoples for a variety of medicinal purposes, including fever reduction, treatment of boils, and heart health. Ecologically, the nectar and pollen are used by native pollinators, especially butterflies and bees. Self-heal, var. *vulgaris*, is found in scattered locations throughout the SN. It is native to Europe and distinguished by its prostrate (sometimes decumbent or erect) stems and leaves only twice as long as wide.

Pycnanthemum californicum

Sierra Mint

Lamiaceae (Mint family)

Erect perennial herb 1–13 dm; stem hairless to tomentose, and branched above middle (if branched at all). Aromatic.

LEAVES: Opposite, tomentose; petioles < 3 mm or sessile; blades 2.5–7.5 cm, > internodes (overlapping), ovate, with bases rounded and tips acute, margins smooth or toothed.

FLOWERS: Inflorescences of 2–5 headlike clusters of flowers subtended by leaflike bracts; calyx 4–5.5 mm, with short-hairy lobes. Corollas 5–6.5 mm, 2-lipped; upper lip 2.5–3 mm, 2-lobed, white, lower lip 2–3 mm, 3-lobed, and purple spotted. Stamens 4.

FRUITS: Nutlets obovoid.

SIERRA DISTRIBUTION: West slope, Fresno County north.

ELEVATION: 500–1900 m

HABITAT: Moist areas along streams and in meadows; foothill and montane zones.

FLOWERING TIME: Jun–Sep

NOTES: An important plant for Native Americans; leaves and stems are steeped for tea, and patches were burned periodically to rejuvenate them. Sierra mint is sparser in the SN nowadays, but hopefully this can change as tribes and public land managers continue to incorporate Traditional Ecological Knowledge to restore important cultural plants

Salvia columbariae

Chia

Lamiaceae (Mint family)

Erect annual herb < 5 dm; sparsely short-hairy. Aromatic.

LEAVES: Mostly basal; stem leaves opposite and smaller above. Leaves deeply pinnately lobed, and lobes irregularly lobed or toothed.

FLOWERS: In dense, headlike whorls, 1–2 per stem; corollas 12–16 mm, 2-lipped, purple; calyx 2-lipped.

FRUITS: Seeds 1.5–2 mm, mottled tan to grayish.

SIERRA DISTRIBUTION: Both slopes, Tuolumne County south.

ELEVATION: < 2500 m

HABITAT: Open, dry areas; foothill, montane, and sagebrush zones.

FLOWERING TIME: Mar–Jun

NOTES: Chia seeds have been an important food for Native Americans in CA for millennia and are still prized for their high nutritional value. The abundance of chia on the landscape was orders of magnitude greater than today, based on my conversations with Western Mono elders as well as published accounts. SN tribes from around Amador County southward managed for chia with fire. The alteration of land for farming or development, the sweep of non-native annual grasses across the foothills, and the inability to use fire as before contributed to the paucity of chia and other native annuals.

Salvia dorrii var. *pilosa*

Purple Sage

Lamiaceae (Mint family)

Multibranched shrub < 7 dm; sometimes mat-forming, with herbage white-scaly. Strongly aromatic.

LEAVES: Opposite, 4–32 mm, spatulate to linear, entire.

FLOWERS: In ≥ 1 clusters; calyx reddish, and bracts and calyx abaxially shaggy-hairy, with margin hairs long; corollas blue or purple.

FRUITS: 1.8–3.5 mm, reddish-brown or gray.

SIERRA DISTRIBUTION: East slope, Tehachapi area only.

ELEVATION: 300–1900 m

HABITAT: Open areas; foothill, montane, and sagebrush zones.

FLOWERING TIME: May–Jul

NOTES: A good place to view this shrub is along Highway 178, about 5 mi west of Walker Pass.

Salvia sonomensis

Sonoma Sage

Lamiaceae (Mint family)

Mat-forming perennial herb < 4 dm. Aromatic, especially when flowering.

LEAVES: 3–6 cm, elliptic to obovate, with surfaces roughened, grayish-green.

FLOWERS: Clusters < 1.5 cm, roundish. Corolla tube 5–15 mm; upper corolla lip 1–3 mm, and 2-lobed; lower lip much longer, 3–7 mm.

FRUITS: 2.5 mm, brown.

SIERRA DISTRIBUTION: West slope, Fresno County north.

ELEVATION: < 2000 m

HABITAT: Open or shaded dry areas; foothill and montane zones.

FLOWERING TIME: Mar–Jul

NOTES: Not a plant you are guaranteed to run into, but when you do it's a treat. Bees love this plant! A favorite for native plant gardeners, Sonoma sage makes a wonderful groundcover and attracts copious native pollinators.

Scutellaria bolanderi ssp. *bolanderi*

Sierra Skullcap

Lamiaceae (Mint family)

Erect or sprawling perennial herb 2–10 dm; rhizomatous. Stem has long (1–2 mm), gland-tipped hairs.

LEAVES: Opposite, ovate or cordate; upper leaves 1–2.5 cm, edges usually scalloped (sometimes entire, as in photo).

FLOWERS: In axils; calyx 2-lipped; upper lip with a hoodlike ridge ("skull cap"); corollas mostly white, hairy; lower lip 15–19 mm, with purple blotches.

FRUITS: Nutlets 4, brown-black.

SIERRA DISTRIBUTION: West slope, Kern to Plumas Counties (no records for El Dorado, Placer, Nevada, or Butte Counties).

ELEVATION: 100–1500 m

HABITAT: Moist sites, streambanks; foothill and montane zones.

FLOWERING TIME: Jun–Jul

NOTES: Compare with *S. californica*, which occurs from Tuolumne County north and has no purple spots on the lower corolla lip.

Scutellaria californica

California Skullcap

Lamiaceae (Mint family)

Sprawling perennial herb 1.5–4 dm; rhizomatous. Stem has short (< 0.5 mm) hairs, some glandular.

LEAVES: Opposite, 1.5–2.5 cm, narrowly ovate to oblong, bases narrowed, entire.

FLOWERS: In axils; calyx 2-lipped, the upper one with a hoodlike ridge ("skull cap"); corollas 15–19 mm, white to creamy, hairy.

FRUITS: Nutlets 4, brown-black.

SIERRA DISTRIBUTION: Mostly Amador County north (also known from Tuolumne County).

ELEVATION: < 1980 m

HABITAT: Shaded areas, rocky slopes, roadcuts; foothill and montane zones.

FLOWERING TIME: Jun–Jul

NOTES: The only other white skullcap is *S. bolanderi*, which has wider leaves and purple spots on the lower corolla lip.

Scutellaria siphocampyloides

Gray-Leaved Skullcap

Lamiaceae (Mint family)

Perennial herb 2–5.5 dm; rhizomatous, densely short-glandular hairy.

LEAVES: Opposite, 1–2.5 cm, linear-oblong, densely glandular-hairy, entire.

FLOWERS: In axils; calyx 2-lipped, upper one > lower, with a hoodlike ridge ("skull cap"); corollas 25–35 mm, blue or purple, glandular-hairy.

FRUITS: Nutlets 4, granular black.

SIERRA DISTRIBUTION: West slope (no records for El Dorado, Placer, or Nevada Counties).

ELEVATION: < 2200 m

HABITAT: Open areas, seeps, riparian areas; foothill and montane zones.

FLOWERING TIME: May–Jul

NOTES: A similar-appearing species, *S. antirrhinoides*, might be encountered from Nevada County north; it has shorter (13–21 mm) flowers and the upper lip = lower lip.

Scutellaria tuberosa

Danny's Skullcap

Lamiaceae (Mint family)

Perennial herb < 2.5 dm; rhizomes with tubers, stem with long (1–3 mm) hairs.

LEAVES: Basal and opposite, 1–2 cm, ovate, usually coarsely toothed.

FLOWERS: 1 per axil; calyx 2-lipped, the upper one with a hoodlike ridge ("skull cap"), soft-hairy; corollas 13–20 mm, blue or purple, soft-hairy.

FRUITS: Nutlets 4, obconic, black.

SIERRA DISTRIBUTION: West slope (no records for Nevada, Sierra, or Lassen Counties).

ELEVATION: < 1650 m

HABITAT: Open, dry areas, especially burned sites; foothill and montane zones.

FLOWERING TIME: Mar–Jul

NOTES: After wildfire in foothill chaparral, this skullcap often shows up en masse.

Stachys albens

Whitestem Hedgenettle

Lamiaceae (Mint family)

Erect perennial herb 0.5–2.5 m, usually branched; stem and herbage covered with dense, tangled, cobwebby white hairs. Strongly aromatic, with a smell not entirely pleasant.

LEAVES: Opposite, felty-hairy; blades 3–15 cm, bases cordate, margins scalloped or toothed.

FLOWERS: Inflorescences 10–30 cm, with sometimes distinct clusters of 10–12 flowers separated along axis; calyx 4–5.5 mm, lobes triangular and shorter than tube. Corollas 12–17 mm with tube 6–9 mm; 2-lipped, with the upper lip 3.5–5.5 mm and the lower 6–8 mm, perpendicular to corolla tube.

FRUITS: Ovoid or oblong, brown to black.

SIERRA DISTRIBUTION: Most counties.

ELEVATION: < 3000 m

HABITAT: Moist or wet areas, seeps, springs, streamsides; foothill, montane, subalpine, and sagebrush zones.

FLOWERING TIME: May–Oct

NOTES: A highly variable plant, but distinguished by felty, tangled, cobwebby white hairs, especially on the stems, and the absence of glandular hairs. Stems on larger plants are much thicker than any other *Stachys* spp.

Stachys rigida var. *rigida*

Rough Hedgenettle

Lamiaceae (Mint family)

Erect to decumbent perennial herb up to 1 m, with soft or stiff hairs as well as glandular hairs.

LEAVES: Opposite; blades 5–9 cm, oblong to lanceolate with bases rounded to cordate (not tapered gradually), margins minutely toothed or crenate.

FLOWERS: Inflorescence with clusters of 6–10 flowers subtended by leaflike bracts. Calyx radial, 5-lobed; corollas 2-lipped with corolla tubes 6–10 mm; upper lip entire, and lower lip 3-lobed with central lobe longest. Ring of hairs inside the corolla tube strongly oblique (magnification required).

FRUITS: Nutlets 4, ovoid, brown to black.

SIERRA DISTRIBUTION: West slope, Fresno County north.

ELEVATION: 400–2000 m

HABITAT: Moist to somewhat dry sites; foothill and montane zones.

FLOWERING TIME: Jul–Aug

NOTES: Most abundant in the northern SN, along with bugle hedgenettle, *S. ajugoides*: < 1000 m, stems < 0.6 m, soft-hairy leaves are silky hairy and wedge-shaped (tapered) at their bases, and with clusters of 3–6 white (or pink) flowers.

Trichostema lanceolatum
Vinegar Weed
Lamiaceae (Mint family)

Tall, bushy annual herb < 1 m; hairy (some hairs are glandular). Strong, vinegarlike smell.
LEAVES: Opposite, 2–7 cm, lanceolate, entire, with distinctive veins.
FLOWERS: Racemes in axils; calyx lobes 5; corolla lobes 5, with upper 2 lobes spreading and longer than others. Stamens 4, 13–20 mm, and arched above corolla.
FRUITS: Nutlets 4, ridged, hairy.
SIERRA DISTRIBUTION: West slope.
ELEVATION: < 1300 m
HABITAT: Dry, open, often disturbed areas; foothill and lower montane zones.
FLOWERING TIME: Jun–Sep
NOTES: Blooms in summer and fall when most other species have finished. The long, arched anthers of vinegar weed illustrate especially well why the genus is called "blue curls." Pollinated by bees.

Trichostema oblongum
Mountain Bluecurls
Lamiaceae (Mint family)

Annual herb < 5 dm; hairy (some hairs are glandular). Strongly aromatic.
LEAVES: Opposite, 2–5 cm, broadly elliptic, entire.
FLOWERS: In axils; calyx lobes 5; corolla lobes 5, with lower lip longer (2–3.5 mm) than others. Stamens 4, 3–6 mm, and arched above corolla.
FRUITS: Nutlets 4, ridged, hairy.
SIERRA DISTRIBUTION: All, except Inyo County.
ELEVATION: 100–3000 m
HABITAT: Seeps and drying vernally moist areas; foothill through montane zones.
FLOWERING TIME: Jun–Sep
NOTES: The very strong, vinegarlike smell of this plant is often the first indication of its presence. Take a moment to look at the tiny flowers with a 10x hand lens; they are exquisite! Self-pollinated.

Utricularia intermedia

Flat-Leaved Bladderwort

Lentibulariaceae (Bladderwort family)

Semiaquatic carnivorous plant with 2 types of stems: Some float on the water's surface and primarily have highly dissected, flattened green leaves and few insect-capturing "bladders"; other stems are rooted, white, lack green leaves, and have bladders (modified leaves) that are triggered by sensitive hairs to open a flap and suck small organisms into the bladderlike container.

LEAVES: Green leaves are feathery looking, 3–15 mm, 3-parted from base, each segment further dissected into linear ± flattened lobes, with 5–12 bristles along margins.

FLOWERS: Peduncles < 25 cm, bearing 1–5 flowers; corollas 10–16 mm, yellow, with lower lip twice as long as upper, and back of corolla with cylindric spur (not visible in photo).

FRUITS: Capsules circumscissile.

SIERRA DISTRIBUTION: Relatively uncommon; currently documented from Tulare, Fresno, Mariposa, Plumas, and Butte Counties.

ELEVATION: 1200–2700 m

HABITAT: Shallow pools and streamlets, wet meadows; montane and subalpine zones.

FLOWERING TIME: Jul–Sep

NOTES: Three other *Utricularia* spp. occur in the SN. One of these is *U. macrorhiza*, which has leaves 5–90 mm, inflorescences of 5–20 flowers, the corollas 1–2 cm, with the lower lip larger than the upper, and a spur below the lower lip with an upward hook at the end. The other 2 spp. are *U. gibba*, with inflorescences of 1–4 flowers, the corollas 6–15 mm, and the upper lip the same size as the lower lip; and *U. minor*, with 2–6 flowers, the corollas 6–8 mm, and the lower lip twice the size of the upper. Proper ID requires dissecting and a compound microscope, specimens with both kinds of stems if applicable, and patience with unique morphology and terminology. Excellent photos of all features for each species are available on CalPhotos, posted by the genus expert.

Limnanthes alba ssp. *alba*

White Meadowfoam

Limnanthaceae (Meadowfoam family)

Erect or ascending annual herb 8–40 cm; hairy.

LEAVES: Alternate, pinnately compound, 2–10 cm; leaflets 5–11, entire or 3-lobed.

FLOWERS: Sepals densely hairy, and petals curve inward as they finish flowering, turning pink when dry. Style 3–6 mm; anthers 2 mm.

FRUITS: Nutlets wide-ridged.

SIERRA DISTRIBUTION: West slope, Fresno County north.

ELEVATION: < 1400 m

HABITAT: Wet sites in grassland, vernal pool margins; foothill zone.

FLOWERING TIME: Apr–Jun

NOTES: See notes for *L. douglasii*. Dried petals fold inward to cover nutlets in fruiting plants. Another ssp., *versicolor*, has hairless or sparsely hairy sepals, smooth or turbercled nutlets, and anthers 1 mm, and occurs 150–2100 m from Tuolumne County north.

Limnanthes douglasii ssp. *rosea*

Douglas' Meadowfoam

Limnanthaceae (Meadowfoam family)

Erect or ascending annual herb 3–35 cm; hairless.

LEAVES: Alternate, pinnately compound, 3–7 cm; leaflets 5–13, entire or deeply pinnately lobed.

FLOWERS: Bell- or cup-shaped; sepals 4–17 mm; petals 10–18 mm, with veins pink, purple or cream; petals reflex away from center as they finish flowering, revealing ov ules. Style 3–8 mm; anthers cream, pink, reddish, or black (before release of pollen).

FRUITS: Nutlets ± 4; surface sculpted by protuberances (need a 10x lens).

SIERRA DISTRIBUTION: West slope, Tulare to Butte Counties.

ELEVATION: < 800 m

HABITAT: Wet sites, vernal pool margins; foothill zone.

FLOWERING TIME: Mar–May

NOTES: An important feature separates the 2 main groups of *Limnanthes*: whether the petals move away from or toward the developing fruit. *L. douglasii* is in the former group, as seen in the lower right of the photo where petals are gone, revealing the nutlets (compare with *L. alba* and *L. montana*). Other ssp. of *L. douglasii* in the Sierra are ssp. *douglasii,* which has petals with yellow bases, found in Butte County; and ssp. *striata*, which has the corolla funnel-shaped, anthers cream-colored, and veins dark, found on west slope < 800 m. ID to ssp. can be difficult, even impossible without fruiting plants. Refer to *The Jepson Manual.*

Limnanthes montana

Mountain Meadowfoam

Limnanthaceae (Meadowfoam family)

Erect or ascending annual herb 10–40 cm; hairless or with sparse, long hairs.

LEAVES: Alternate, pinnately compound, 3–15 cm; leaflets 7–11, entire or lobed.

FLOWERS: Funnel-shaped; sepals 3–6 mm; petals 7–12 mm, with veins purple, and petals curving inward as they finish flowering. Style 2.5–4 mm; filaments 2.5–4 mm, anthers 0.5–1 mm and cream-colored.

FRUITS: Nutlets with small, cone-shaped protuberances at top (need a 10x lens).

SIERRA DISTRIBUTION: West slope, Tulare to Butte Counties.

ELEVATION: 200–1800 m

HABITAT: Wet sites, streamsides; foothill and montane zones.

FLOWERING TIME: Mar–May

NOTES: Compare with *L. alba*.

Linum lewisii var. *lewisii*

Prairie Flax

Linaceae (Flax family)

Hairless perennial herb 5–80 cm.

LEAVES: Alternate, simple, 10–20 mm, linear, narrowly lanceolate or oblanceolate.

FLOWERS: Sepals 4–6 mm, inner 2 covered by the outer 3; petals 6–15 mm. Styles 2–5, 2–12 mm, at least as long as the stamens, separate; stigmas headlike.

FRUITS: Capsule 10-valved. Seeds usually 10.

SIERRA DISTRIBUTION: Both slopes, sporadic; in Fresno County upper montane zone in Kaiser Wilderness, associated with limestone soils.

ELEVATION: < 3660 m

HABITAT: Open, often rocky or sandy areas; montane through alpine zones.

FLOWERING TIME: Apr–Jul

NOTES: Related to *L. perenne*, the European species from which linen fabric is manufactured (from stem fibers). *L. perenne* may be spreading in wild areas from birdseed mixes, and it can be challenging to tell it apart from *L. lewisii*; *L. perenne* is heterostylous: The styles are either much shorter or much longer than stamens.

Linnaea borealis var. *longiflora*

Twin Flower

Linnaeaceae (Twin Flower family)

Creeping, evergreen subshrub 15–20 cm.
LEAVES: Opposite, simple, 12–18 mm, serrate above middle; with 1 main vein from base.
FLOWERS: Nodding in pairs at tips of erect peduncles; corollas 10–13 mm, white or pink.
FRUITS: Capsule 3-valved.
SIERRA DISTRIBUTION: El Dorado County north.
ELEVATION: 200–1500 m
HABITAT: Moist, shady sites in conifer forest; montane zone.
FLOWERING TIME: Jun–Aug
NOTES: Formerly included in the Honeysuckle family, Caprifoliaceae. The genus, and more recently the family, are named after Carl Linnaeus, a Swedish physician and biologist who in the 1700s created the system of binomial nomenclature that is still used today to classify living organisms. Linnaeus was especially fond of this particular wildflower.

Mentzelia affinis

Yellow Comet

Loasaceae (Loasa family)

Erect, hairy annual herb 5–47 cm.
LEAVES: Alternate, 1–17 cm; lower leaves toothed to pinnately lobed, and upper entire to lobed.
FLOWERS: Cymes, with green inflorescence bracts, lanceolate to obovate, entire to toothed; sepals 5, 2–7 mm, persistent; petals 5, 3–10 mm, ± ovate, and yellow, with bases yellow or orange. Ovary inferior, style 3–6.5 mm; stamens numerous, < 6.5 mm; outer filaments threadlike.
FRUITS: 12–28 mm x 1–3 mm wide, narrowly cylindric, erect to curved < 90°. Seeds 1–2 mm x ± 1 mm wide, ± prism-shaped with sharp angles.
SIERRA DISTRIBUTION: Tulare and Kern Counties.
ELEVATION: < 1800 m
HABITAT: Sandy or rocky soils; foothill zone.
FLOWERING TIME: Apr–May
NOTES: More common in southern CA, the Mojave Desert, the Coast Ranges, and the Central Valley; this species is found in scattered locations in the southern Sierra, mostly in Kern County. Intergrades with *M. dispersa*.

Mentzelia crocea

Sierra Blazing Star

Loasaceae (Loasa family)

Erect annual herb 34–100 cm; hairy.

LEAVES: Basal in rosettes; cauline alternate, sessile; 1–40 cm, pinnately toothed or lobed.

FLOWERS: Inflorescence bracts green, 6- to 10-toothed; sepals 7–20 mm; petals 5, 21–36 mm x 8–17 mm wide, yellow, usually with orange bases. Ovary inferior, style 2–3.5 cm; stamens numerous, 11–28 mm, the outer 5 wider (up to 2 mm) than the inner.

FRUITS: Capsule 20–35 mm x 3–5 mm wide, obconic, erect or curved 45°. Seed 1.5–2 mm, irregularly rounded or angular, surface rough, dark mottled.

SIERRA DISTRIBUTION: West slope, Tulare to Tuolumne Counties.

ELEVATION: < 1700 m

HABITAT: Sandy or rocky slopes, roadcuts; foothill and montane zones.

FLOWERING TIME: Mar–Jun

NOTES: One of the showiest blazing stars, along with *M. laevicaulis*, but *M. crocea* has a narrower range, shorter petals, and is annual rather than perennial.

Mentzelia dispersa

Bushy Blazing Star

Loasaceae (Loasa family)

Erect annual herb 7–48 cm; whitish stems, branched from base, varying from essentially a single-stemmed plant to an extremely bushy, vigorous one (especially after fire).

LEAVES: Basal in rosettes, cauline sessile and smaller above; alternate, < 10 cm, pinnately toothed or entire.

FLOWERS: Cymes; calyx lobes 1–3.5 mm, persistent; petals 5, 2–8 mm, yellow (sometimes with orange base). Ovary inferior, style 1–3.5 mm; stamens numerous, < 4.5 mm.

FRUITS: Capsule 7–25 mm x 1–2.5 mm wide, cylindrical, sometimes curved, hairy.

SIERRA DISTRIBUTION: All counties.

ELEVATION: < 3100 m

HABITAT: Open, often disturbed areas in full sun; foothill through subalpine zones.

FLOWERING TIME: May–Aug

NOTES: Occurs in great numbers after foothill chaparral and montane zone fires. Seeds are eaten by California Indians.

Mentzelia laevicaulis var. *laevicaulis*

Smooth-Stemmed Blazing Star

Loasaceae (Loasa family)

Erect perennial herb 22–100 cm; hairless to hairy.

LEAVES: Basal in rosettes, and along with lowest cauline leaves they are petioled, ≤ 24 cm; pinnately lobed. Distal cauline leaves alternate, sessile, 2–10 cm, pinnately toothed.

FLOWERS: Inflorescence bracts linear to lanceolate, entire or 4- to 5-lobed, leaflike; sepals 15–46 mm; petals 40–80 mm x 8–18 mm wide, yellow above, white to paler yellow below; tips acute. Ovary inferior, style 2.5–7 cm; stamens numerous, threadlike, 15–55 mm; outermost 5 linear, filaments 40–55 mm, 1–2 mm wide.

FRUITS: Capsule 15–44 mm x 8–13 mm wide, cylindric, erect. Seed 3–4 mm x 2.5 mm wide, lenticular (shaped like a lentil), winged.

SIERRA DISTRIBUTION: Both slopes, Nevada County south, scattered locations.

ELEVATION: < 2900 m

HABITAT: Sandy or rocky slopes, roadcuts; foothill and montane zones.

FLOWERING TIME: May–Oct

NOTES: One of the showiest blazing stars; compared to *M. crocea*, this plant is perennial rather than annual and has longer, more slender-looking petals with lighter yellow upper surfaces and white to even paler yellow lower surfaces.

Fremontodendron californicum

California Flannelbush

Malvaceae (Mallow family)

Erect evergreen shrub 2–5 m, plant stellate hairy.

LEAVES: Alternate, petioles 0.4–3 cm; blades 1–7 cm, palmately lobed, dark green above and paler below, stipules 2 mm.

FLOWERS: Borne singly, 3.5–6 cm wide, with 5 petal-like sepals, no true petals.

FRUITS: Capsule 2–4 cm, ovoid, bristly; seeds 3.5–5.5 mm, brown to black. Most have a large elaiosome, a structure containing oils and fats that attracts ants; ants feed the elaiosome to their broods and discard the seed, aiding in dispersal. Small squirrels and deer mice also consume the elaiosomes, then cache seeds.

SIERRA DISTRIBUTION: West slope; east slope, Kern County.

ELEVATION: 400–2350 m

HABITAT: Dry slopes and rocky areas; foothill and sagebrush/pinyon-juniper zones.

FLOWERING TIME: Apr–Jul

NOTES: The Western Mono people (and others) use the strong, fibrous stems for cordage; stems are collected in the spring, especially after a fire, when they grow strong and straight. There are many uses for the fiber: tying things together, making nets, and combining with milkweed to make a tumpline (a means of carrying a load with weight borne on the head). "Slippery elm" is the English name used in the central Sierra by Mono people, but it is not an elm.

Malacothamnus fremontii var. *fremontii*

Fremont's Bush Mallow

Malvaceae (Mallow family)

Shrub or erect, woody perennial herb to 3 m; white-woolly to shaggy hairy, with stalked hairs.

LEAVES: Blades < 9 cm, round, 5-lobed.

FLOWERS: Spikes or panicles; calyx < 1.2 cm, bractlets < ½ length of calyx; petals < 2 cm; fused column of filaments ± 1 cm.

FRUITS: Segments 5–14, 2–4 mm, notched, brown.

SIERRA DISTRIBUTION: West slope, Amador County south; east slope, Inyo County.

ELEVATION: < 2500 m

HABITAT: Dry, open, often rocky sites in chaparral, coniferous forest; often found on nongranitics such as limestone; foothill and montane zones.

FLOWERING TIME: May–Jul

NOTES: The only other species of *Malacothamnus* in the Sierra is *M. marrubioides*; it is < 2 m, has yellowish hairs, bractlets the same size as the calyx, and is mostly in Kern County but also probably occurs in Fresno County.

Sidalcea asprella ssp. *asprella*

Sierra Foothills Checkerbloom

Malvaceae (Mallow family)

Decumbent or ascending perennial herb 3–10 dm; from short rhizomes. Stems are stellate-hairy, especially at base; hairs mostly have 4 rays. Stipules ± 3 mm.

LEAVES: Stem and basal leaves usually lobed similarly (lower may be more shallowly lobed); upper surfaces stellate-puberulent.

FLOWERS: Open racemes, sometimes 1-sided; pedicels 2–5 mm; calyx 7–10 mm, densely stellate-puberulent. Female flowers have petals 1–2 cm; bisexual flowers have petals 2–2.5 cm.

FRUITS: Segments 6–8, 3–3.5 mm, inflated, pitted, and glandular on back.

SIERRA DISTRIBUTION: West slope, Madera County north.

ELEVATION: 1500–3000 m

HABITAT: Dry to moist areas in meadows, open grassy areas; foothill and montane zones.

FLOWERING TIME: Jun–Aug

NOTES: Formerly *Sidalcea malviflora* ssp. *asprella*.

Sidalcea glaucescens

Waxy Checkerbloom

Malvaceae (Mallow family)

Decumbent or ascending perennial herb 2–5 dm (up to 7 dm), glaucous; from thick, woody taproots; not spreading via rooting at nodes. Stems are clustered, the base usually with 7- to 10-rayed stellate hairs; upper stem hairless.

LEAVES: Basal with petioles 6–14 cm, blades 2–6 cm x 2–6 cm wide, deeply palmately 5- to 7-lobed; note that leaves are mostly cauline; upper leaves often fewer lobed or simple; blades 1.5–4 cm wide, hairless to stellate-puberulent.

FLOWERS: Open racemes, mostly 1-sided, zigzagging with age; pedicels 3–10 mm; calyx 6–10 mm (larger in fruit), stellate-puberulent; petals 9–12 mm, pink to pinkish-purple. Flowers may be bisexual or pistillate.

FRUITS: Segments 6–8, 3–3.5 mm, inflated; deeply pitted, net-veined on sides and back, glandular puberulent on back; beak of each segment very short.

SIERRA DISTRIBUTION: West slope, Tulare County north; both slopes Alpine County north.

ELEVATION: 1500–3000 m

HABITAT: Dry to moist areas, meadow edges, mixed conifer and red fir forest; montane and sagebrush/pinyon-juniper zones.

FLOWERING TIME: Jun–Aug

NOTES: *Sidalcea* spp. are notoriously difficult to tell apart! The similar-looking cut-leaf checkerbloom, *S. multifida*, is a rare plant mostly confined to Mono and Tulare Counties 2000–2800 m, the many stems are 1–4 dm, ascending to erect, gray-glaucous, stellate-hairy, leaves are mostly basal, and more deeply lobed than *S. glaucescens*, flowers have larger petals (10–25 mm), flowering is May-July.

Sidalcea hirsuta

Hairy Checkerbloom

Malvaceae (Mallow family)

Erect annual herb 3–8 dm; hairless below and generally densely bristly-hairy above.

LEAVES: Cauline, petioles 2–7 cm; longer on lower than upper leaves. Lower leaves unlobed, crenate, blades 1–2.5 cm x 1–2.5 cm; upper deeply lobed to base, lobes 7–9, linear, blades 3–8 x 3–8 cm.

FLOWERS: Dense, spikelike inflorescences, 2–5 cm; calyx 8–10 mm, densely stellate-hairy and bristly; petals 13–25 mm, pink. Outer filaments fused into a continuous tube reaching just below anthers.

FRUITS: Inflorescence elongated up to 20 cm in fruit; segments 3–4 mm, net-veined and pitted; top and back usually puberulent; beak 1 mm, tip bristled.

SIERRA DISTRIBUTION: West slope, Tulare County north, patchy.

ELEVATION: < 1100 m

HABITAT: Vernally wet areas, vernal pools, grassland; foothill zone.

FLOWERING TIME: Apr–Jun

NOTES: This annual checkerbloom has no basal leaves and puberulent fruits (segments of all *Sidalcea* fruits look somewhat like wedges in a wheel of cheese).

Sidalcea oregana ssp. *spicata*

Oregon Checkerbloom

Malvaceae (Mallow family)

Perennial herb 3–8 dm; taprooted, covered with soft hairs or nearly hairless.

LEAVES: Basal and lower leaves scalloped to deeply lobed, and upper leaves more deeply 5- to 7-lobed; blades 3–10 cm wide.

FLOWERS: Dense spike or panicle, pedicels 1–3 mm; calyx 3.5–7 mm, densely stellate-hairy; petals 10–15 mm, light to dark pink.

FRUITS: Each segment 2.5–3 mm, smooth or roughened, with tiny beak 0.1–0.3 mm.

SIERRA DISTRIBUTION: Kern and Inyo Counties north, sparse to absent in many west slope counties.

ELEVATION: 1250–3050 m

HABITAT: Meadows, streamsides, wet areas; montane and subalpine zones.

FLOWERING TIME: Jun–Aug

NOTES: Distinguished from most other checkerblooms by the dense inflorescence (but compare with *S. ranunculacea* in the southern Sierra). Highly variable; ssp. *oregana* also occurs in the Sierra, and the 2 interbreed.

Sidalcea ranunculacea
Marsh Checkerbloom
Malvaceae (Mallow family)

Perennial herb 2–5 dm; rhizomatous, with stem ascending, base with bristly hairs, and top with stellate hairs.

LEAVES: Basal and lower leaves shallowly 5-lobed, and upper leaves deeply 5- to 7-lobed; blades 2.5–6 cm wide; with stellate hairs or sparse bristles.

FLOWERS: Spikelike; lower flower clusters separated by visible length of inflorescence stalk; pedicels 1–3 mm; calyx 5–9 mm, stellate-puberulent; petals 5–15 mm, pink, sometimes white-veined.

FRUITS: Each segment is 2.5 mm, roughened, sparsely stellate-puberulent.

SIERRA DISTRIBUTION: Mostly Tulare and Kern Counties, a few records for Inyo County.

ELEVATION: 1850–3050 m

HABITAT: Meadows, streamsides, moist areas often near giant sequoia groves; montane and subalpine zones.

FLOWERING TIME: Jun–Aug

NOTES: Uncommon. Compare with *S. oregana.* They overlap in the southern High Sierra, but *S. ranunculacea* generally has smaller leaves, smaller stature, and broader petals.

Sidalcea reptans ✿
Sierra Checkerbloom
Malvaceae (Mallow Family)

Perennial herb 2–5 dm; stem prostrate, creeping along ground rooting at nodes; lower stem has long, dense 2–3 mm hairs, often based on swollen pads; upper stem at least somewhat stellate-puberulent.

LEAVES: Basal and lower leaves roundish, with long petioles, blades 2.5–5 cm wide, unlobed but deeply crenate, hairless or bristly with unbranched hairs; upper stem blades deeply 5- to 7-lobed.

FLOWERS: Inflorescence open, ascending to erect, axis < 20 cm, stellate-puberulent and ± glandular; calyx 8–10 mm, stellate-puberulent with longer, glandular marginal hairs; petals 12–20 mm, purple or dark pink, white-veined.

FRUITS: Segments ± 3 mm, deeply pitted and net-veined, stellate-puberulent, beak ± 1 mm, also stellate-puberulent.

SIERRA DISTRIBUTION: West slope, Tulare to Nevada Counties.

ELEVATION: 1120–2500 m

HABITAT: Meadows and understory of conifer forest; montane zone.

FLOWERING TIME: Jun–Aug

NOTES: Distinguished from other perennial *Sidalcea* spp. by prostrate stems spreading horizontally by rooting freely rooting at nodes, in contrast to inflorescences that are ascending to erect, and stellate hairs on the fruit segments, including the beaks.

Menyanthes trifoliata

Buckbean

Menyanthaceae (Buckbean family)

Perennial herb, with stout, ascending stem and creeping rhizomes covered with previous years' leaf bases.
LEAVES: Basal and alternate, leaflets 3, each 2–12 cm.
FLOWERS: Racemes on peduncles 20–40 cm. Corollas 9–15 mm, funnel-shaped, with lobes 5–8 mm, and long-hairy on upper surfaces. Style 1, persistent; stamens 5, filaments slender, anthers arrow-shaped.
FRUITS: Capsule 6–8 mm, 2-valved.
SIERRA DISTRIBUTION: Mostly west slope, Tulare County north.
ELEVATION: 1250–3200 m
HABITAT: Fens, wet meadows, and shallow water at margins of small lakes; montane and subalpine zones.
FLOWERING TIME: May–Aug
NOTES: Finding buckbean in bloom is a treat! When it grows in fens, it is often found with other acid-loving species, such as sundew and western blueberry, as well as peat moss, *Sphagnum*.

Calandrinia menziesii

Red Maids

Montiaceae (Miner's Lettuce family)

Hairless, low-growing annual herb with stems spreading, 3–45 cm.
LEAVES: Alternate, 1–10 cm, linear or narrowly oblanceolate.
FLOWERS: Inflorescence a leafy raceme; sepals 2, hairless or ciliate; petals 5, 5–14 mm. Stamens 3–15.
FRUITS: Capsule 3-valved, longer than the calyx by < 3 mm (if > 3 mm, then it is *C. breweri*). Seeds have netted surfaces.
SIERRA DISTRIBUTION: West slope, all.
ELEVATION: < 2200 m
HABITAT: Open grasslands, sandy areas, roadsides, disturbed areas; foothill and montane zones.
FLOWERING TIME: Feb–May
NOTES: Formerly *C. ciliata*. One of the most common early blooming wildflowers of the foothills. The related *C. breweri* is found from Mariposa to Tulare Counties; it has capsules ≥ 3 mm than the calyx, smaller (3–5 mm) petals, and fewer (3–6) stamens. *C. breweri* is especially likely to be found after wildfires and other ground disturbance < 1200 m.

Calyptridium monospermum

Oneseed Pussypaws

Montiaceae (Miner's Lettuce family)

Fleshy, hairless perennial herb; from stout, unbranched taproot. Multiple stems < 50 cm radiate from 1 (sometimes dense) basal rosette. Stems often change position from horizontal to ± erect at midday.

LEAVES: Basal in rosettes with leaves 1.5–6 cm; smaller stem leaves also usually present, alternate, 0.8–3 cm.

FLOWERS: Inflorescence is ± headlike umbels (in larger plants, coiled cymes); each flower has 2 roundish bracts enclosing the sepals and petals. Sepals 2, 3–8 mm, round, scarious; petals 4, 3–7 mm, pink or white. Style 2–4 mm, threadlike; stamens 3, anthers pink.

FRUITS: Capsule 2–3.5 mm, ovate to roundish, 2-valved. Seeds 1–4.

SIERRA DISTRIBUTION: West slope, all.

ELEVATION: 500–3700 m

HABITAT: Open sandy and gravelly areas; foothill zone (barely), primarily montane zone.

FLOWERING TIME: Apr–Sep

NOTES: Common above the foothill zone, though occasionally plants are found this low. Can be confused with *C. umbellatum*, but the look of *C. monospermum* is of multiple flowering stems radiating from each basal rosette. The mechanism for movement of stems from prostrate to upright is not understood, but it may help the plant avoid extreme heat on the ground at midday.

Calyptridium umbellatum

Pussypaws

Montiaceae (Miner's Lettuce family)

Fleshy, hairless perennial herb; from stout taproot branched at soil surface, usually with > 1 rosette produced from each branch. Stem 1 per rosette, < 60 cm.

LEAVES: Flowering stem 1 per rosette; rosettes have leaves 1.5–7 cm, and stem leaves ± absent.

FLOWERS: Terminal, headlike umbels 1–7 cm wide; each flower has 2 roundish bracts enclosing the sepals and petals. Sepals 2, 3–8 mm round, scarious; petals 4, 3–7 mm, pink or white. Style 2–4 mm, threadlike; stamens 3, anthers yellow to red.

FRUITS: Capsule 2–3 mm, ovate to roundish, 2-valved. Seeds 1–8.

SIERRA DISTRIBUTION: Tulare and Inyo Counties north.

ELEVATION: 500–4300 m

HABITAT: Open sandy and gravelly areas, rocky sites; all zones, but primarily above foothill zone.

FLOWERING TIME: May–Oct

NOTES: Compare carefully with *C. monospermum*. Much confusion exists regarding ID of these 2 spp., with many plants that are actually *C. monospermum* being called *C. umbellatum*, even by experienced botanists. At this writing, many of the photos on CalPhotos are actually *C. monospermum*. DNA research on the *Calyptridium umbellatum* complex, including these 2 spp., is underway and may clarify taxonomy.

Claytonia lanceolata

Western Springbeauty

Montiaceae (Miner's Lettuce family)

Hairless perennial herb with stem 5–15 cm; erect, from stout, rounded tubers.

LEAVES: Basal leaves absent, or when present, petioled, 5–8 cm, oblanceolate; **s**tem leaves opposite, sessile, 1–7 cm, linear to narrowly ovate, free (not fused).

FLOWERS: Short-stalked or in a sessile raceme, flowers 3–15 per raceme, with 1 small bract below flowers. Sepals 2, 3–7 mm; petals 5, 5–12 mm, white or light pink, bases sometimes yellow.

FRUITS: Capsule 3.5–4.5 mm, 3-valved. Seed 2–2.5 mm, round, shiny.

SIERRA DISTRIBUTION: Mostly west slope, Tulare County north.

ELEVATION: 1500–2600 m

HABITAT: Gravelly sites, meadows; montane zone.

FLOWERING TIME: May–Jul

NOTES: Sporadic in its distribution. Compare with *Lewisia triphylla*—petal number is always 5 for *C. lanceolata*, and 5–9 for *L. triphylla*. If fruits are visible, note that *Claytonia* fruits open along 3 valves.

Claytonia nevadensis

Sierra Springbeauty

Montiaceae (Miner's Lettuce family)

Perennial herb < 15 cm; rhizomatous.

LEAVES: Basal 2–15 mm; blades 10–50 cm x 0.2–3 cm wide, widely ovate or spatulate, and wedge-shaped at base, with the linear petiole often buried. Cauline leaves sessile, < 2 cm, blades ovate.

FLOWERS: Inflorescences sessile or with short peduncles. Flowers 2–8 per inflorescence, 12–16 mm diameter, with 1 bract below lowest flower. Sepals 2, 3–8 mm; petals 5, 6–10 mm, pink.

FRUITS: Capsule 3–4 mm, 3-valved. Seed 1.5–2 mm, round, black, and shiny.

SIERRA DISTRIBUTION: Both slopes where elevation is high enough, Tulare County north.

ELEVATION: 2200–3500 m

HABITAT: Moist to wet sites in sand or gravel near streams, springs, snowmelt; subalpine and alpine zones.

FLOWERING TIME: Jul–Sep

NOTES: Found along streamlets formed by melting snow on unglaciated surfaces. A large population still occurs where the species was first discovered, on the northwest shoulder of Mount Dana.

Claytonia parviflora ssp. *parviflora*

Streambank Springbeauty

Montiaceae (Miner's Lettuce family)

Hairless annual herb with stem 1–15 cm; from taproots.

LEAVES: Basal in rosettes, 1–8 cm; blades ± diamond-shaped or triangular with bases truncate. Stem leaves ± fused into a disk or partially free on 1 side.

FLOWERS: Dense raceme of 3–30 flowers, with 1 small bract below flowers, and pedicels reflexed. Sepals 2, 1.5–3 mm; petals 5, 2–3.5 mm, white or light pink.

FRUITS: Capsule 2–3 mm, 3-valved. Seed 1–2 mm, elliptic, smooth, black, and shiny.

SIERRA DISTRIBUTION: West slope, all.

ELEVATION: < 2300 m

HABITAT: Moist, often disturbed sites; foothill and montane zones.

FLOWERING TIME: Jan–May

NOTES: This plant is the unsung hero of burned areas, proliferating rapidly to protect burned soil from erosion during rainfall, and filling the air with a sweet fragrance from countless tiny, white flowers abuzz with insect visitors. The reddish-orange dead leaves continue to provide soil cover long into the winter. Compare with the very similar-looking *C. perfoliata* ssp. *perfoliata*—the key is the shape of the basal leaves, which are linear in this species; however, intermediates are encountered frequently.

Claytonia perfoliata ssp. *perfoliata*

Miner's Lettuce

Montiaceae (Miner's Lettuce family)

Hairless, fleshy annual herb with stem 1–40 cm; from taproots.

LEAVES: Basal in rosettes; blades < 4 cm, < 3x longer than wide, and elliptic to deltate; upper stem leaves fused into a disk 1–10 cm diameter, round, square, or asymmetric.

FLOWERS: Racemes 1-sided, of 5–40 flowers, with 1 small bract below flowers, and pedicels reflexed. Sepals 2; petals 5, 2–5 mm, white or light pink.

FRUITS: Capsule 1.5–4 mm, 3-valved. Seeds 1.2–2.7 mm, smooth, black, and shiny; elaiosome 1–3 mm.

SIERRA DISTRIBUTION: Mostly west slope, all.

ELEVATION: < 2200 m

HABITAT: Moist, often disturbed sites; foothill and montane zones.

FLOWERING TIME: Jan–May

NOTES: Compare with the very similar-looking *C. parviflora* ssp. *parviflora*. The shape of the basal leaves are ± linear in *C. parviflora* and have distinctly expanded blades in *C. perfoliata*. However, the full gamut of leaf shapes can be found. This plant grows mixed with *C. parviflora* on burned slopes.

Claytonia rubra ssp. *rubra*

Redstem Springbeauty

Montiaceae (Miner's Lettuce family)

Hairless annual herb < 15 cm, with overall form often moundlike.

LEAVES: Basal in rosettes, each leaf 1–8 cm, usually deltate to diamond-shaped, and purple underneath. Stem leaves < 4 cm wide, ± fused (round), and sometimes free along 1 side, in which case free corners are square.

FLOWERS: Inflorescence of 3–30 flowers, with 1 small bract below all the flowers, above fused stem leaves. Sepals 2, 1.3–3 mm; petals 5, 2–3.5 mm, white or light pink.

FRUITS: Capsule 2–3 mm, 3-valved. Seeds 1–2 mm, elliptic, smooth, black and shiny.

SIERRA DISTRIBUTION: All.

ELEVATION: < 2500 m

HABITAT: Partial shade/openings in foothill chaparral, foothill woodland, and conifer forest; foothill and montane zones.

FLOWERING TIME: Apr–Jul

NOTES: Proliferates after fires, along with *C. parviflora* and *C. perfoliata*. In years 1–2 after fire, *C. rubra* plants sometimes form geometrically perfect domes, which is probably the result of having unlimited sun, water, and nutrients.

Lewisia cantelovii

Cantelow's Lewisia

Montiaceae (Miner's Lettuce family)

Perennial herb with stem 10–60 cm; procumbent to suberect, from ramified taproots.

LEAVES: Rosettes of numerous fleshy leaves 2.5–8 cm, spoon-shaped, finely toothed, tapering to wide petioles.

FLOWERS: Panicles of few flowers, open, with small, gland-toothed bracts; flowers rise on several stalks from each rosette and held well above leaves; pedicels 0.3–4 mm. Calyx of 2 sepals ± ½ the size of the corolla, with margins gland-toothed; petals 5–7, 6–9 mm, white to light pink, with 5–7 darker veins. Stigmas 3, stamens 5–6.

FRUITS: Capsule ± 3 mm, spheric to ovoid, circumscissile. Seeds 1–3, 1.2–1.5 mm, shiny, dark.

SIERRA DISTRIBUTION: West slope, El Dorado to Plumas Counties.

ELEVATION: 385–1370 m

HABITAT: Granitic cliff faces, rocky outcrops, serpentine seeps; foothill and montane zones.

FLOWERING TIME: May–Oct

NOTES: A rare plant. Locations on public websites are often obscured to prevent illegal poaching, which has impacted this species. The similar-looking saw-toothed lewisia, *L. serrata*, has longer (3–8 mm) pedicels and smaller flowers (petals 5–6 mm), and is even rarer.

Lewisia nevadensis

Nevada Lewisia

Montiaceae (Miner's Lettuce family)

Fleshy perennial herb; from stout taproots.

LEAVES: 5–15 per rosette, 3–13 cm, linear or narrowly lanceolate.

FLOWERS: Inflorescence of 1–3 flowers, with 2 bracts toward base of stem (obscured in photo); pedicels 10–40 mm; sepals 2, ovate, entire or jagged but not glandular, with sharp tips; petals 5–10, 9–20 mm, tips blunt or rounded. Stigmas 3–6, stamens 6–15.

FRUITS: Capsule 5–10 mm, circumscissile.

SIERRA DISTRIBUTION: All, except Kern County.

ELEVATION: 1300–3000 m

HABITAT: Streamsides, moist gravel flats, forest openings, meadows; montane, subalpine, and alpine zones.

FLOWERING TIME: May–Aug

NOTES: Notice that the 2 petals attached where the sepals connect are the widest, contributing to an overall oval look unique to this species. Compare with *L. pygmaea*, which has smaller petals and glandular sepals.

Lewisia pygmaea

Dwarf Lewisia

Montiaceae (Miner's Lettuce family)

Fleshy perennial herb 1–5 cm; from stout taproots.

LEAVES: Rosette, each leaf 2–9 cm, linear or linear-lanceolate.

FLOWERS: Inflorescence of 1–3 flowers, with 2 bracts toward base of stem (obscured in photo); pedicels 2–5 mm; sepals 2, rounded or ovate (not sharp-tipped), toothed or with pale glands; petals 5–9, 4–10 mm, obovate, white, light to dark pink, often striped. Stigmas 3–6, stamens 4–8.

FRUITS: Capsule 4–5 mm, circumscissile.

SIERRA DISTRIBUTION: Tulare and Inyo Counties to Nevada County.

ELEVATION: 1700–3750 m

HABITAT: Moist gravel flats, streamlets draining snowbanks, rocky sites, meadows; subalpine and alpine zones.

FLOWERING TIME: Jul–Aug

NOTES: Compare with *L. nevadensis*. Another similar plant is *L. glandulosa*, an alpine plant < 10 cm with leaves withered by flowering time, sepal margins with dark glands (usually red), 6–8 white to pink petals < 8 mm, and exactly 6 stamens and 4 styles.

Lewisia rediviva var. *rediviva*

Bitterroot

Montiaceae (Miner's Lettuce family)

Low-growing, fleshy perennial herb; from branched taproots.
LEAVES: Many per rosette, each leaf 0.5–5 cm, linear.
FLOWERS: Inflorescence of 1 flower, with 2–7 slender, paperlike bracts below center of stem; pedicels 10–30 mm; sepals 4–9, somewhat petal-like; petals 10–19, 18–35 mm, white, pink, or lavender. Stigmas 4–9, stamens 30–50.
FRUITS: Capsule 5–6 mm, circumscissile.
SIERRA DISTRIBUTION: West slope, scattered locations < 1000 m; east slope, Inyo and Mono Counties < 2500 m.
ELEVATION: 60–2500 m
HABITAT: Streamsides, moist gravel flats, forest openings, meadows; foothill, montane, subalpine, and alpine zones.
FLOWERING TIME: Mar–Jun
NOTES: Unmistakable! The number of petals, stamens, and sepals distinguishes this species from others.

Lewisia triphylla

Three-Leaved Lewisia

Montiaceae (Miner's Lettuce family)

Perennial herb 2–7 cm; from round tubers.
LEAVES: Basal leaves withered when flowers open; stem leaves opposite if there are 2, whorled if there are 3–5. Leaves 1–6 cm, linear, tapered to base, tips obtuse.
FLOWERS: Inflorescence of 1 to many flowers; bracts below flowers lanceolate, smooth-margined; pedicels 5–15 mm; sepals 2, ovate, smooth-margined; petals 5–9, 4–7 mm, white or pink with darker veins. Stigmas and stamens 3–5.
FRUITS: Capsule 3–4 mm, circumscissile.
SIERRA DISTRIBUTION: Mostly west slope, Tulare and Mono Counties north.
ELEVATION: 1300–3400 m
HABITAT: Moist to wet gravelly or sandy flats, rocky areas; montane, subalpine and alpine zones.
FLOWERING TIME: May–Aug
NOTES: Despite the species name meaning "3-leaved," plants do not always have 3 leaves. Forms showy patches of pink on wet gravel or sand, next to or on rocky outcrops in the early spring (which may be August in the alpine zone!). Compare with *Claytonia lanceolata*, which grows in drier sites and has lanceolate leaves with acute tips, as well as fruits that split open longitudinally rather than via a horizontal line of weakness as in *Lewisia.*

Montia chamissoi

Toad Lily

Montiaceae (Miner's Lettuce family)

Fleshy perennial herb 2–30 cm; prostrate to erect, often matted or sometimes floating in its wet habitat; forming leafy stolons that trail along the ground with pink bulblets (allowing it to overwinter).

LEAVES: Opposite, 5–50 mm, oblanceolate, not reduced toward top of stem.

FLOWERS: Inflorescence of 2–8 flowers (some replaced by bulblets); sepals 2, 1.5–2 mm; petals 5, 5–9 mm, white or pink; stamens 5 with pink or lavender stamens.

FRUITS: 1–1.5 mm, 3-valved, valves rolling in, forcing expulsion of seeds.

SIERRA DISTRIBUTION: All.

ELEVATION: 1300–3700 m

HABITAT: Streamsides, moist gravel flats, forest openings, meadows; montane, subalpine, and alpine zones.

FLOWERING TIME: May–Aug

NOTES: Native bees and flies pollinate toad lily. The bumpy black seeds have small, white structures called elaiosomes that are rich in lipids and proteins. Ants (most likely) transport the seeds to their nests, larvae eat the elaiosomes, then the rest of the seed goes to the ants' garbage dump to germinate.

Myrica hartwegii

Sierra Sweet Bay

Myricaceae (Wax Myrtle family)

Deciduous shrub 1–2 m; dioecious, twigs are glandular-hairy, sticky. Strongly aromatic.

LEAVES: Alternate, 4–8 cm, toothed, especially toward tips.

FLOWERS: Plants have either male or female flowers. Male inflorescences 1–2 cm, ± spheric, and each flower has 3–6 stamens subtended by 2 light brown bracts. Female inflorescences with 1 bract and 2 bractlets subtending each pistil, becoming spongy as fruit matures. No petals.

FRUITS: Nutlets with winglike bractlets still attached, hairless.

SIERRA DISTRIBUTION: West slope, Madera to Nevada Counties.

ELEVATION: 300–1800 m

HABITAT: Streamsides; foothill and montane zones.

FLOWERING TIME: May–Jun

NOTES: The photo shows a female plant with fruits maturing; it can be seen along the Merced River in the Wawona Campground in Yosemite. The same plant can change genders and produce all male flowers one year, then all female flowers the next.

Lysimachia latifolia

Pacific Starflower

Myrsinaceae (Myrsine family)

Perennial herb 5–30 cm; hairless.

LEAVES: 25–90 mm, 10–50 mm wide, ovate to obovate, located mostly near stem tips.

FLOWERS: Petals 5–7, 8–15 mm wide, pink, fused.

FRUITS: Spherical capsules, 5-valved.

SIERRA DISTRIBUTION: West slope, Tuolumne County north.

ELEVATION: < 1400 m

HABITAT: Dense, shaded woodland and forest, often near streams; foothill and montane zones.

FLOWERING TIME: Apr–Jul

NOTES: Formerly included in Primulaceae. At first glance this species can be mistaken for a monocot, especially if you see an individual with 6 petals. However, the leaf venation shows it to be a eudicot.

Eriodictyon californicum

California Yerba Santa

Namaceae (Nama family)

Shrub 1–3 m; twigs hairless or with a few hairs, sticky, sprouting back promptly after fire or other disturbance.

LEAVES: Alternate; blades ≤ 15 cm, hairless, shiny, and sticky above, and hairy below; margins toothed or sometimes smooth.

FLOWERS: Corollas 8–12 mm, funnel-shaped, light lavender to purple (occasionally white). Ovary hairy.

FRUITS: Capsule 2–3 mm, sticky. Seeds 2–20.

SIERRA DISTRIBUTION: West slope, all.

ELEVATION: < 1830 m

HABITAT: Open areas primarily in chaparral; foothill and lower montane zones.

FLOWERING TIME: Apr–Jul

NOTES: Formerly in Boraginaceae, then Hydrophyllaceae. A reliable source of abundant nectar for native pollinators and especially attractive to bees. Indians make a tea with it.

Eriodictyon parryi

Poodle-Dog Bush

Namaceae (Nama family)

Erect subshrub 1–3 m; glandular, sticky. Strong odor.
LEAVES: Alternate, densely arranged; blades 4–30 cm, lanceolate, usually toothed, and densely glandular.
FLOWERS: Inflorescences branched and dense; calyx lobes 3–6 mm, glandular, long-hairy; corollas 10–20 mm, funnel-shaped, glandular.
FRUITS: Capsule 3–4 mm, glandular-hairy. Seeds numerous, shiny, black, angled with netlike sculpturing (magnification required).
SIERRA DISTRIBUTION: West slope, Fresno County (south of the Kings River) south.
ELEVATION: < 2440 m
HABITAT: Open, dry areas, especially after fires; foothill, montane, and sagebrush/pinyon-juniper zones.
FLOWERING TIME: May–Aug
NOTES: Formerly *Turricula parryi*; and had been placed in Boraginaceae and Hydrophyllaceae in the past. Touching poodle-dog bush can cause severe dermatitis (itching, painful blisters) and an anaphylactic reaction in some people.

Nama demissa var. *demissa*

Purple Mat

Namaceae (Nama family)

Annual herb with ± prostrate stems 3–20 cm; herbage has fine-strigose hairs < 1 mm.
LEAVES: Cauline and alternate, sessile, 1–4 cm, linear to spoon-shaped, and green rather than gray-green.
FLOWERS: Terminal clusters; calyx lobes 3–8 mm, lance-linear; corollas 7–14 mm, purple-pink to blue purple, with lobes 2–3 mm, limb 5–9 mm diameter. Style 3–6 mm, stamens 4–7 mm.
FRUITS: Capsule 3–4 mm, ovoid or elliptic. Seeds 0.5 mm, elliptic-ovoid.
SIERRA DISTRIBUTION: Kern and Tulare Counties.
ELEVATION: < 1780 m
HABITAT: Open sandy or gravelly flats and slopes; montane and sagebrush/pinyon-juniper zones.
FLOWERING TIME: Mar–May
NOTES: Formerly in Boraginaceae and before that, Hydrophyllaceae. A plant primarily of the desert that reaches into the southern and eastern Sierra.

Fraxinus dipetala

California Ash, Two-Petaled Ash

Oleaceae (Olive family)

Deciduous shrub 1.5–3 m (sometimes small tree to 6 m); multistemmed.

LEAVES: Opposite, pinnately compound; leaflets 3–9, 2–4.5 cm.

FLOWERS: Inflorescences 8–15 cm; flowers bisexual; calyx minute; petals 2, 3.5–6.5 mm, white. Stamens 2.

FRUITS: Achenes 20–32 mm, 5–9 mm wide, winged.

SIERRA DISTRIBUTION: West slope, Tulare County north.

ELEVATION: < 1300 m

HABITAT: Chaparral; foothill zone.

FLOWERING TIME: Apr–Jun

NOTES: In contrast to the related tree, Oregon ash, *Fraxinus latifolia*, which is restricted to wet or moist areas, *F. dipetala* is found on dry slopes in foothill chaparral.

Camissonia campestris ssp. *campestris*

Mojave Suncup

Onagraceae (Evening Primrose family)

Hairless annual herb 0.5–2.5 dm; ± erect (sometimes decumbent), peeling at base.

LEAVES: 5–30 mm, linear to narrowly elliptic, with small, widely spaced marginal teeth.

FLOWERS: Hypanthium 1.5–5 mm; sepals 3–8 mm, fused in pairs and reflexed when flowers are open; petals 3.5–15 mm, fading red with age, with 1–2 red spots at base. Stigma longer than anthers, headlike.

FRUITS: Capsule 20–43 mm x 0.7–2 mm wide, sessile, straight or wavy, lumpy where seeds present.

SIERRA DISTRIBUTION: West slope, Mariposa County south.

ELEVATION: < 1525 m

HABITAT: Open sandy flats and slopes; foothill zone.

FLOWERING TIME: Mar–May

NOTES: Compare with *C. sierrae* ssp. *sierrae*, which has much broader leaves covered with soft hairs.

Camissonia sierrae ssp. *sierrae*

Sierra Suncup

Onagraceae (Evening Primrose family)

Finely hairy annual herb < 15 cm; decumbent to ascending, stem peeling.

LEAVES: Alternate, 5–18 mm, elliptic or narrowly ovate, minutely toothed.

FLOWERS: Racemes, with flowers nodding in bud and erect in bloom; sepals 3–4.2 mm, fused in pairs and reflexed in flower; petals 4–7 mm, fading red, with 2 red spots at base. Ovary inferior, style 4.5–7 mm; stigma at least as long as anthers, headlike.

FRUITS: Capsule 20–30 mm, sessile, often wavy or curved, and lumpy where seeds present.

SIERRA DISTRIBUTION: West slope, Tulare to Calaveras Counties.

ELEVATION: 500–1300 m (probably higher)

HABITAT: Open sandy or gravelly flats and rocky outcrops; foothill and montane zones.

FLOWERING TIME: Apr–Jun

NOTES: SN endemic, along with the rarer, self-pollinated Mono hot springs evening primrose, ssp. *alticola*, which has sepals 1.2–3 mm and petals 2.2–4 mm without red dots, and grows in similar habitats in the montane zone in the central SN, 2000–2400 m.

Camissoniopsis hirtella

Hairy Suncup

Onagraceae (Evening Primrose family)

Annual herb < 60 cm; stems ascending, with dense, short, spreading hairs.

LEAVES: Basal in rosettes, alternate, 10–80 mm, ovate, finely toothed, and irregularly "ruffled."

FLOWERS: Open at dawn; hypanthium 1–3 mm, sepals 2.5–6 mm, fused in pairs or separating when flowers open; petals 2–9 mm, with no orange spots. Ovary inferior, stigma at least as long as anthers, headlike.

FRUITS: Capsule 13–20 m, cylindric, coiled when fully mature.

SIERRA DISTRIBUTION: West slope, El Dorado County south.

ELEVATION: < 2300 m

HABITAT: Open slopes and gravelly areas; especially abundant after wildfire; foothill and montane zones.

FLOWERING TIME: Mar–Aug

NOTES: Formerly *Camissonia hirtella*.

Chamerion angustifolium ssp. *circumvagum*

Fireweed

Onagraceae (Evening Primrose family)

Tall perennial herb 0.5–3 m; hairless or strigose toward top of plant. Often forms large patches via rhizomes.

LEAVES: Alternate; petioles 2–7 mm; blades 1.5–20 cm, lanceolate, midvein hairy below.

FLOWERS: Racemes of showy pink to magenta flowers; sepals 7–16 mm; petals 10–25 mm. Ovary inferior, stigma longer than stamens, pollen blue.

FRUITS: 4–10 cm, grayish, with pedicel elongated to 7–20 mm. Seeds have white hair tufts.

SIERRA DISTRIBUTION: Tulare and Inyo Counties north.

ELEVATION: 900–3500 m

HABITAT: Common along roadsides, disturbed areas, after fires, rocky areas; upper montane, subalpine, and alpine zones.

FLOWERING TIME: Jul–Sep

NOTES: Formerly *Epilobium angustifolium* ssp. *circumvagum*. Circumboreal.

Chamerion latifolium

Dwarf Fireweed

Onagraceae (Evening Primrose family)

Clumped perennial herb < 7 dm; glaucous or strigose.

LEAVES: Alternate; petioles 0–3 mm; blades 1–10 cm, elliptic to lanceolate, hairless to hairy on veins and margins.

FLOWERS: Racemes of light pink to purple flowers; sepals 10–24 mm; petals 10–24 mm. Ovary inferior, stigma at least as long as stamens, pollen gray.

FRUITS: 3–10.5 cm, grayish-hairy or hairless, with pedicel elongated to 5–18 mm.

SIERRA DISTRIBUTION: Tuolumne, Mono, and Inyo Counties.

ELEVATION: 2500–3100 m

HABITAT: Fellfields, talus, rock cracks, glacial moraines; subalpine and alpine zones.

FLOWERING TIME: Jul–Aug

NOTES: Formerly *Epilobium latifolium*. Circumboreal. A species of arctic and alpine habitats that reaches its southern limit in the SN in the Convict Creek Basin in Inyo County and around Mount Whitney.

Circaea alpina ssp. *pacifica*

Enchanter's Nightshade

Onagraceae (Evening Primrose family)

Erect perennial herb with slender stems 1–5 dm; rhizomatous, forming small clumps to large patches in moist sites.
LEAVES: Opposite, 3–11 cm, ovate to round.
FLOWERS: Panicle of racemes with glandular hairs, and flowers small; sepals 2, 2 mm, white (reflexed), and petals 2, 1–1.5 mm, erect, white. Stigma 2-lobed; stamens 2, pollen yellow.
FRUITS: Roundish, burlike, indehiscent. Seed 1.
SIERRA DISTRIBUTION: Kern County north.
ELEVATION: 700–2400 m
HABITAT: Shaded, moist forest floor and riparian areas; montane zone.
FLOWERING TIME: May–Aug
NOTES: A distinctive, delicate plant.

Clarkia biloba ssp. *biloba* ✿

Bilobed Clarkia

Onagraceae (Evening Primrose family)

Erect annual herb sometimes reaching heights of 1 m, but usually 20–80 cm.
LEAVES: Alternate; petioles < 1.5 cm; blades 2–6 cm, linear to linear-lanceolate.
FLOWERS: Inflorescence is recurved in bud; buds are pendulous; sepals have reddish cast and remain united when flowers open; petals narrowly heart-shaped, lavender to magenta; lobes $\frac{1}{5}$–$\frac{1}{2}$ petal length. Stigma longer than anthers; anthers outer 4 lavender, inner 4 much paler to white, and shorter.
FRUITS: 8-grooved.
SIERRA DISTRIBUTION: Mariposa to El Dorado Counties.
ELEVATION: < 1500 m
HABITAT: Foothill woodland, chaparral of mother lode, sometimes on serpentine; foothill zone.
FLOWERING TIME: May–Jun
NOTES: Intergrades with and difficult to distinguish from ssp. *australis*, which is only found in Mariposa and Tuolumne Counties to the south, and ssp. *brandegeeae*, which occurs from El Dorado County north. All 3 sspp. are SN endemics. The very rare Merced clarkia, *C. lingulata*, known from only 2 populations in the Merced River canyon, has been clearly shown to be derived from ssp. *australis* in a relatively recent speciation event. It looks like ssp. *biloba* aside from having tongue-shaped petals, and no notch.

Clarkia cylindrica ssp. *clavicarpa* ✿

Speckled Clarkia

Onagraceae (Evening Primrose family)

Erect annual herb < 6 dm.

LEAVES: Alternate; petioles < 5 mm; blades 1–6 cm, narrowly lanceolate or linear.

FLOWERS: Inflorescence is recurved at apex in buds; buds are pendent. Sepals red-purple, remaining united and off to one side; petals 10–35 mm, fan-shaped, with base bright red-purple, then white blending into pink on outer petals, often flecked with darker pink spots. Stigma longer than anthers; outer 4 anthers lavender with filaments wider, and inner 4 shorter and lighter-colored.

FRUITS: Capsule 2–5 cm, overall cylindric but wider distally (like a club), 4-grooved.

SIERRA DISTRIBUTION: West slope foothills, Tuolumne County south.

ELEVATION: < 1000 m

HABITAT: Grassy slopes, roadsides, openings; foothill and montane zones.

FLOWERING TIME: Apr–Jun

NOTES: The ssp. name *clavicarpa* means "club + fruit," referring to the club-shaped capsules. Compare with *C. dudleyana*.

Clarkia dudleyana

Dudley's Clarkia

Onagraceae (Evening Primrose family)

Erect annual herb < 7 dm.

LEAVES: Alternate; petioles < 1 cm; blades 1.5–7 cm, linear to oblong.

FLOWERS: Inflorescence is recurved in buds; buds are pendulous. Corollas bowl-shaped; sepals pink to maroon, all 4 remaining united and off to 1 side; petals fan-shaped, lavender to pink, usually with streaks of white, sometimes flecked with darker pink spots. Stigma longer than anthers; outer 4 anthers lavender, inner 4 shorter and lighter-colored.

FRUITS: 8-grooved.

SIERRA DISTRIBUTION: West slope, Nevada County south.

ELEVATION: < 1500 m

HABITAT: Grassy slopes, roadsides, openings; foothill and montane zones.

FLOWERING TIME: May–Jul

NOTES: The widely fan-shaped petals with white streaks are characteristic of this common clarkia.

Clarkia heterandra

Mountain Clarkia

Onagraceae (Evening Primrose family)

Erect annual herb < 6 dm glandular-puberulent.

LEAVES: Alternate; petioles 5–20 mm; blades 2–8 cm, ovate to lanceolate.

FLOWERS: Inflorescence axis straight in bud; buds are drooping. Sepals fused in 4s, sometimes pairs; petals 5–6 mm, uniformly pink, often with darker speckles. Stigma not projecting beyond anthers; stamens 8, the inner 4 smaller.

FRUITS: 2–3 mm, roundish, indehiscent.

SIERRA DISTRIBUTION: West slope, Tulare to Mariposa Counties, Butte County.

ELEVATION: 500–1700 m

HABITAT: Shady areas; foothill and lower montane zones.

FLOWERING TIME: May–Jul

NOTES: Distinguished from all other clarkias by the small, nutlike fruit.

Clarkia lassenensis

Mount Lassen Clarkia

Onagraceae (Evening Primrose family)

Erect annual herb < 90 cm; puberulent.

LEAVES: Alternate; petioles < 1 cm; blades 2–5 cm, linear to linear-lanceolate.

FLOWERS: Inflorescence is recurved in bud; buds are pendent. Sepals united when flowers open; petals 8–16 mm, pink/lavender, lighter to white toward red-purple base; corollas bowl-shaped. Stigma not longer than anthers; anthers 8, alike.

FRUITS: 2.5–4 cm, 8-grooved, hairy, beak 2 mm.

SIERRA DISTRIBUTION: El Dorado and Alpine Counties north.

ELEVATION: 500–2000 m

HABITAT: Open areas; foothill and montane zones.

FLOWERING TIME: May–Jun

NOTES: Limited to the northern Sierra, where it is distinguished from other clarkias with similar petal shape by the stigma not being exserted beyond the stamens and the 8-grooved hairy ovary/fruit.

Clarkia modesta

Modest Clarkia

Onagraceae (Evening Primrose family)

Erect annual herb < 7 dm; puberulent.

LEAVES: Alternate; petioles < 15 mm; blades 2–4 cm, linear to lanceolate or elliptic.

FLOWERS: Inflorescence axis is recurved in bud; buds are drooping. Sepals remain fused in 4s; petals 8–12 mm, diamond-shaped, uniformly pink, often with darker speckles. Stigma not projecting beyond anthers; anthers 8, inner ones smaller and paler than outer 4.

FRUITS: 8-grooved.

SIERRA DISTRIBUTION: West slope, Tulare to Mariposa Counties, Butte County.

ELEVATION: < 1100 m

HABITAT: Shade or partial shade; foothill and lower montane zones.

FLOWERING TIME: Apr–Jun

NOTES: Distinguished from other clarkias with similar petal shape by the uniformly pale pink petals, which are smaller and more delicate.

Clarkia purpurea ssp. *quadrivulnera*

Four-Spot, Winecup

Onagraceae (Evening Primrose family)

Erect to decumbent annual herb < 1 m; hairless to puberulent.

LEAVES: Alternate; petioles 0.2 mm; blades 1.5–7 cm, linear or slightly wider.

FLOWERS: Inflorescence is erect in bud; buds are erect. Sepals separate completely or in pairs; petals 10–15 mm, fan-shaped, pink with a dark spot toward tip or entirely dark wine purple. Stigma not longer than anthers; anthers 8, all alike; pollen blue.

FRUITS: 1–3 cm, 8-grooved, beak 0–2 mm.

SIERRA DISTRIBUTION: West slope.

ELEVATION: < 1500 m

HABITAT: Open, sunny areas, grassy or sandy; foothill and montane zones.

FLOWERING TIME: Apr–Aug

NOTES: This one is tricky because the same ssp. encompasses pink petals with purple spots or entirely dark purple petals, and sometimes intermediates. *Clarkia purpurea* ssp. *viminea* could be mistaken for *C. williamsonii*, but petals are slightly smaller on average, and if buds are still present the sepals would be fused to the tips in ssp. *viminea* and separate at the tips in *C. williamsonii*.

Clarkia rhomboidea

Diamond Clarkia

Onagraceae (Evening Primrose family)

Tall, erect annual herb < 1 m.

LEAVES: Alternate; petioles 5–25 mm; blades 2–6 cm, lanceolate to ovate.

FLOWERS: Inflorescence is recurved in bud; buds are pendent. Sepals separate completely; petals 7–15 mm, pink with darker flecks, with claw 2-lobed. Stigma not longer than anthers; anthers 8, all alike; pollen blue.

FRUITS: 4-grooved.

SIERRA DISTRIBUTION: East slope, all.

ELEVATION: < 2500 m

HABITAT: Openings and shaded areas; foothill and montane zones.

FLOWERING TIME: May–Jul

NOTES: Diamond clarkia generally occurs at the highest elevations of all the SN clarkias, in mixed conifer forest. It tends to grow in small, sparse patches but can be exceptionally tall, robust, and abundant after wildfire.

Clarkia unguiculata

Elegant Clarkia

Onagraceae (Evening Primrose family)

Erect annual herb < 1 m; hairless, glaucus.

LEAVES: Alternate; petioles 0–1 cm; blades 1–6 cm, lanceolate to ovate.

FLOWERS: Inflorescence is erect in bud; buds are somewhat downward pointing, hairy, with some hairs up to 3 mm. Hypanthium 2–5 mm; sepals green, pink, or maroon, remaining united and off to one side or reflexed individually; petals 1–2.5 cm, with a narrow claw at least as large as the blade. Blade varies from ± triangular to ± round, sometimes ruffled. Stigma longer than anthers; outer 4 anthers red, and inner 4 shorter and lighter-colored.

FRUITS: 8-grooved, hairy like sepals.

SIERRA DISTRIBUTION: West slope, all.

ELEVATION: < 1500 m

HABITAT: Grassy slopes, roadsides, openings; foothill and lower montane zones.

FLOWERING TIME: May–Jul

NOTES: In the southern Sierra (Tulare County) at about 500 m, the similar-looking Springville clarkia, *C. springvillensis*, may be encountered. This very rare plant is listed as state endangered and federally threatened. It differs in having puberulent buds without long, straight hairs, petal blades that are ± diamond-shaped, usually with purple spots at bases, very narrow, linear claws, and fruits with hairs like sepals (no long hairs up to 3 mm, as in *C. unguiculata*).

Clarkia williamsonii
Williamson's Clarkia
Onagraceae (Evening Primrose family)

Tall, erect annual herb < 1 m.

LEAVES: Alternate; petioles < 1 cm; blades 2–7 cm, linear to lanceolate.

FLOWERS: Inflorescence is erect in bud; buds are erect. Sepals remain united in pairs or separate completely, free at tips; petals 1–3 cm, fan-shaped, lavender with a white area in center and a large dark spot above the middle (very rarely solid wine red). Stigma longer than anthers; anthers 8, all alike.

FRUITS: 8-grooved.

SIERRA DISTRIBUTION: West slope, Sierra County south.

ELEVATION: 400–2000 m

HABITAT: Grassy slopes, roadsides, openings; foothill and montane zones.

FLOWERING TIME: Apr–Sep

NOTES: In a good rain year and often after wildfires, the plant puts on a "superbloom" all on its own!

Epilobium anagallidifolium
Alpine Willowherb
Onagraceae (Evening Primrose family)

Mat-forming perennial herb < 20 cm; stoloniferous, stems S-shaped, prostrate or decumbent.

LEAVES: Opposite, 8–25 mm.

FLOWERS: Inflorescence is nodding, sometimes glandular. Flowers nod in bud; sepals 2–5 mm; petals 3–9 mm, pink to lavender, deeply notched.

FRUITS: 2–4 cm, purplish.

SIERRA DISTRIBUTION: Both slopes, Tulare County north.

ELEVATION: 2100–3700 m

HABITAT: Moist, usually rocky areas in areas of recent snow melt; subalpine and alpine zones.

FLOWERING TIME: Jul–Sep

NOTES: Found in alpine areas throughout the northern hemisphere.

Epilobium brachycarpum

Tall Annual Willowherb

Onagraceae (Evening Primrose family)

Erect annual herb 2–20 dm; hairless below, stem peeling at base; glandular-hairy above.

LEAVES: Opposite below, alternate above, usually early deciduous, 10–55 mm, linear to narrowly elliptic, folded at least somewhat along midrib, and hairless, with petioles 0–4 mm.

FLOWERS: Inflorescence is a panicle or raceme; hypanthium 1.5–8 mm; sepals 1.2–8 mm; petals 2–20 mm, white to pink-purple. Stigma may be 4-lobed; stamens no longer than the pistil.

FRUITS: Capsule 15–32 mm, hairless or hairy. Seeds brown, vertically ridged, with hair tufts.

SIERRA DISTRIBUTION: Both slopes, all.

ELEVATION: < 3300 m

HABITAT: Abundant in disturbed sites like roadsides, recently burned, logged, or cleared areas; foothill, montane, and subalpine zones.

FLOWERING TIME: Jun–Sep

NOTES: This species blooming late summer or fall is remarkably variable. Sometimes plants look scruffy, with small flowers that do not particularly catch the eye. Other populations have larger, pink flowers that are much showier than in the smaller-flowered populations. All flower sizes and degrees of aesthetic appeal between these 2 extremes exist as well. The peeling lower stems are diagnostic.

Epilobium canum ssp. *latifolium*

California Fuchsia, Zauschneria

Onagraceae (Evening Primrose family)

Perennial herb or subshrub 1–5 dm; hairy, with some hairs glandular.

LEAVES: Opposite, 8–70 mm, lanceolate or ovate, with margins entire or toothed, and green rather than grayish.

FLOWERS: Bilaterally symmetrical; hypanthium 17–34 mm; sepals 4–10 mm, red-orange; petals 7–15 mm. Ovary inferior; stigma exserted, 4-lobed, and longer than stamens.

FRUITS: Capsule 15–30 mm, sessile, beaked.

SIERRA DISTRIBUTION: All.

ELEVATION: < 3300 m

HABITAT: Rocky places, often where there is some available moisture; foothill and montane zones.

FLOWERING TIME: Jun–Dec

NOTES: A wonderful native plant for gardens, providing late summer to fall color and attracting hummingbirds and native insect pollinators. Available from native plant nurseries. Co-occurring in much of the range of ssp. *latifolium* < 2200 m is ssp. *canum*, which has gray foliage and ± linear leaves.

Epilobium ciliatum ssp. *ciliatum*

Fringed Willowherb

Onagraceae (Evening Primrose family)

Erect, clumped perennial herb 5–12 dm, with lines of tiny hairs arranged vertically on stem, glandular hairy toward top. Vegetative reproduction is by fleshy basal rosettes or fleshy shoots.

LEAVES: Opposite (except below inflorescence), 1–12 cm, lanceolate or ovate, finely toothed, with veins visible, petioles 0–8 mm.

FLOWERS: Inflorescence not leafy, densely strigose, and spreading hairy, usually glandular; flowers radially symmetrical; hypanthium 0.5–2.6 mm; sepals 2–7.5 mm, white to light pink; petals 2–9 mm. Ovary inferior; stigma exserted, not lobed, and same length as stamens.

FRUITS: Capsule 15–100 mm, hairy. Seeds brown, vertically ridged, with hair tufts.

SIERRA DISTRIBUTION: All.

ELEVATION: < 4100 m

HABITAT: Meadows, streamsides, moist disturbed areas; montane and subalpine zones.

FLOWERING TIME: Jun–Oct

NOTES: The ridged seeds (need a 10x lens) are the trick to identifying this species. *E. ciliatum* ssp. *glandulosum* also occurs throughout the Sierra < 3500 m; it has larger (4–14 mm), pink-purple petals, and elongated, leafy, glandular inflorescences.

Epilobium densiflorum

Denseflower Willowherb

Onagraceae (Evening Primrose family)

Erect annual herb 0.3–1.5 m; sometimes branched toward top, stem peeling at base, hairy.

LEAVES: Opposite at base, alternate above, 18–85 mm, narrowly lanceolate to linear.

FLOWERS: Inflorescence densely leafy, often glandular; hypanthium 1.3–4 mm; sepals 2–8 mm; petals 3–10 mm. Ovary inferior; stigma headlike or with 4 small lobes.

FRUITS: Capsule 4–11 mm, splitting open to base, not beaked. Seeds lack hair tufts.

SIERRA DISTRIBUTION: All, but no records for Inyo County yet.

ELEVATION: < 2600 m

HABITAT: Streamsides, meadows, moist sites, seeps, ditches, swales; foothill and montane zones.

FLOWERING TIME: May–Oct

NOTES: This summer-blooming annual often grows with *E. torreyi*.

Epilobium glaberrimum ssp. *glaberrimum*

Glaucous Willowherb

Onagraceae (Evening Primrose family)

Clumped perennial herb 2–8.5 cm; from scaly, branching rootstocks, short scaly stolons, with no turions; stem is ascending, and branched distally if branched; glaucous.

LEAVES: Opposite, sessile, 20–70 mm, narrowly lanceolate or oblong lanceolate, ascending, clasping.

FLOWERS: Bracted raceme; hypanthium 0.7–2.3 mm; sepals 4, 3–7.5 mm, erect; petals 4, 4–12 mm, notched, and white to pink to purplish-pink. Ovary inferior; stigma clublike, not lobed; stamens 8.

FRUITS: Capsule 45–75 mm, erect or ascending, sometimes sparsely hairy, with fruiting pedicel 5–25 mm. Seeds 0.7–1.0 mm, papillate, without hair tuft.

SIERRA DISTRIBUTION: Both slopes, Tulare and Inyo Counties north.

ELEVATION: < 3000 m

HABITAT: Streamsides, springs, meadows, other wet sites; sometimes on dry, rocky slopes; montane and subalpine zones.

FLOWERING TIME: Jun–Sep

NOTES: *Epilobium glaberrimum* ssp. *fastigiatum* occurs 1200–3800 m in similar habitats, with stem < 3.5 dm, leaves 10–25 mm and lanceolate to narrowly ovate, sepals 2.5–5.2 mm, petals 3.4–8 mm, and shorter (20–55 mm) fruits.

Epilobium hallianum

Hall's Willowherb

Onagraceae (Evening Primrose family)

Erect perennial herb 1–6 dm; with nodding inflorescence, strigose in lines along stem, with turions underground (inset).

LEAVES: Opposite, mostly sessile, 5–45 mm; lower leaves ovate, and upper lance-linear, held ± upright, and may be slightly toothed.

FLOWERS: Inflorescence is drooping, glandular; hypanthium 0.5–1.7 mm; sepals 1.2–2.8 mm; petals 1.6–5.5 mm, white to light pink. Ovary inferior, stigma not lobed.

FRUITS: Capsule 24–60 mm, with fruiting pedicel 10–40 mm. Seeds 1.1–1.6 mm, minutely bumpy, with hair tufts.

SIERRA DISTRIBUTION: All.

ELEVATION: 1500–3700 m

HABITAT: Meadows, streamsides; montane, subalpine, and alpine zones.

FLOWERING TIME: Jun–Oct

NOTES: Another herbaceous, white-flowered willowherb with a drooping inflorescence is *E. lactiflorum*. It has capsules twice as long, no turions, and seeds with netted rather than bumpy surfaces.

Epilobium minutum

Minute Willowherb

Onagraceae (Evening Primrose family)

Annual herb < 4 dm; taprooted, with stem peeling at base, hairless to strigose, branched or simple, often with a reddish cast.

LEAVES: Opposite near base, alternate toward top of stem, sessile or short-petioled, 9–25 mm, lanceolate to oblanceolate to elliptic, ± hairless.

FLOWERS: In upper leaf axils; hypanthium 1.1–1.5 mm; sepals 1.3–2.5 mm; petals 2–5 mm, white to pink, notched. Ovary inferior; stigma clublike.

FRUITS: Capsule 9–28 mm, curved, ± hairless, with fruiting pedicel 3–10 mm. Seed 0.9–1.2 mm, netted, with hair tuft not persistent.

SIERRA DISTRIBUTION: West slope, Mariposa County north; both slopes, Lake Tahoe north.

ELEVATION: < 2320 m

HABITAT: Dry to moist sites, often disturbed, can be abundant after fire; montane zone.

FLOWERING TIME: Apr–Sep

NOTES: Compare with *E. brachycarpum*.

Epilobium obcordatum

Rock Fringe

Onagraceae (Evening Primrose family)

Clumped or matted perennial herb or subshrub < 1.5 dm.

LEAVES: Opposite, lance-elliptic or oval, often glaucous.

FLOWERS: Inflorescence is glandular; hypanthium 3–5.5 mm; sepals 6–15 mm; petals 11–24 mm. Ovary inferior; stigma exserted, 4-lobed.

FRUITS: Capsule 20–45 mm, glandular. Seeds have white hair tufts.

SIERRA DISTRIBUTION: Tulare and Inyo Counties north.

ELEVATION: 1700–4000 m

HABITAT: Rocky areas; subalpine and alpine zones.

FLOWERING TIME: Jul–Sep

NOTES: The common name "rock fringe" is appropriate, as these plants are frequently seen decoratively growing along the edge of rocks.

Epilobium torreyi

Torrey's Willowherb

Onagraceae (Evening Primrose family)

Erect annual herb 4–65 cm; stem peeling at base, grayish-hairy.
LEAVES: Opposite at base, alternate above, 5–45 mm, linear to linear-lanceolate, hairy except at base.
FLOWERS: Inflorescence has small, leaf-like bracts, glandular; hypanthium 0.4–1 mm; sepals 0.7–2 mm; petals 1–3.2 mm. Ovary inferior; stigma headlike or with 4 small lobes.
FRUITS: Capsule 8–14 mm, splitting open to base, not beaked. Seeds without hair tufts.
SIERRA DISTRIBUTION: West slope, all.
ELEVATION: < 2600 m
HABITAT: Streamsides, meadows, moist sites, seeps, ditches, swales; foothill and montane zones.
FLOWERING TIME: May–Aug
NOTES: This summer-blooming annual often grows with *E. densiflorum*, which tends to make this less showy plant easy to overlook.

Gayophytum diffusum ssp. *parviflorum*

Small Flowered Groundsmoke

Onagraceae (Evening Primrose family)

Erect annual herb < 6 dm; hairless, strigillose (sometimes villous), usually many-branched from midstem rather than near base, with branching forked toward top of plant.
LEAVES: Alternate, 1–6 cm, narrow; upper leaves smaller than lower.
FLOWERS: Lowermost (first) flowers at 1–20 nodes from base of plant; sepals 4, 0.9–2 mm, reflexed singly or in pairs; petals 4, 1.2–3 mm, white. Stigma spheric and as long as stamens, stamens 0.9–2 mm, and all are shorter than petals.
FRUITS: Capsule 3–15 mm, longer than the pedicel, ± cylindric to somewhat flattened, and constricted weakly to strongly between seeds (knobby); valves 4, all separating, and edges wavy. Seeds 3–18, hairless or puberulent.
SIERRA DISTRIBUTION: Both slopes, all.
ELEVATION: 800–3700 m
HABITAT: Open areas; montane, subalpine, sagebrush/pinyon-juniper zones.
FLOWERING TIME: May–Sep
NOTES: Capsules are knobby their full length because of constrictions between seeds, which is different from the irregularly knobby look of *G. heterozygum* resulting from up to half the ovules aborting. The plants form masses, often with other *Gayophytum* spp., especially after wildfire, exemplifying the common name "groundsmoke."

Gayophytum eriospermum

Coville's Groundsmoke

Onagraceae (Evening Primrose family)

Erect annual herb < 1 m, branched toward top of plant.
LEAVES: Alternate, 2–7.5 cm, narrow; upper leaves smaller than lower.
FLOWERS: Inflorescences axillary, flowers white with yellow bases; petals 4–8 mm, longer than stamens. Ovary inferior, stigma much longer than stamens; stamens 3–7 mm.
FRUITS: 4–16 mm, appearing bumpy. Seeds 4–12.
SIERRA DISTRIBUTION: El Dorado County south, including Inyo County (no records for Mono County but likely present).
ELEVATION: 100–3200 m
HABITAT: Openings; montane and subalpine zones.
FLOWERING TIME: Jun–Oct
NOTES: This is the largest flowered species of any SN *Gayophytum.* The relatively large flowers with yellow at bases of corollas (pollinator signal) and the long style separating the stigma from self-pollen are indicators of insect-mediated pollination. Another larger-flowered taxon, *G. diffusum* ssp. *diffusum*, could be encountered within the range of *G. eriospermum* in Alpine, Tulare, and Kern Counties. It has petals 3–7 mm, stems < 60 cm, and blooms May–Sep.

Gayophytum racemosum

Blackfoot Groundsmoke

Onagraceae (Evening Primrose family)

Erect annual herb < 4 dm; branched from base of plant, with many secondary branches.
LEAVES: Alternate, 1–2.5 cm, narrow; upper leaves similar in size to lower.
FLOWERS: Lowermost (first) flowers at 1–3 nodes from base of plant; petals 1.3–1.8 mm, longer than stamens, white, fading reddish. Stigma not exserted beyond anthers; stamens 0.8–1.5 mm.
FRUITS: 10–15 mm, larger than the pedicel, flattened, ± grooved along center. Seeds 10–35, hairless or puberulent.
SIERRA DISTRIBUTION: Both slopes, Kern and Inyo Counties north.
ELEVATION: 1000–4000 m
HABITAT: Gravelly or sandy sites, drying flats or edges of meadows; montane, subalpine, and sagebrush zones.
FLOWERING TIME: May–Oct
NOTES: Common in subalpine and alpine zones. *G. humile* looks similar but is hairless. Zigzag groundsmoke, *G. heterozygum* (inset), looks similar and overlaps with *G. racemosum* from Tulare County north on both slopes up to 3000 m, but has irregularly lumpy fruits due to about 50% of seeds aborting.

Oenothera elata ssp. *hirsutissima*

Tall Evening Primrose

Onagraceae (Evening Primrose family)

Erect biennial herb 10–25 dm; densely minutely strigose and glandular, mixed with longer hairs.

LEAVES: Alternate, sessile, 4–25 cm, oblanceolate to lanceolate or elliptic, entire or dentate.

FLOWERS: Spikes; hypanthium 20–48 mm; sepals 4, 27–48 mm, green or reddish, tips free in bud for 3–6 mm, reflexed singly or remaining connected in 2s or 3s; petals 25–52 mm, yellow, fading red-orange. Ovary inferior, style 1, stigma deeply 4-lobed; stamens 8, anthers 8–15 mm, attached to filaments at middle.

FRUITS: Capsule 20–65 mm x 4–7 mm wide, cylindric to narrowly lanceolate, not coiled. Seeds 1–1.8 mm, angled, irregularly pitted.

SIERRA DISTRIBUTION: East slope, Inyo to Plumas Counties.

ELEVATION: < 2800 m

HABITAT: Moist sites, roadsides, meadows, streams; montane, subalpine, and sagebrush zones.

FLOWERING TIME: Jun–Sep

NOTES: The large, fragrant flowers open at dusk, attracting a variety of moth pollinators.

Taraxia subacaulis

Northern Suncup

Onagraceae (Evening Primrose family)

Stemless perennial herb; from stout taproot, hairless, fleshy.

LEAVES: Basal, simple, ± entire to slightly pinnately lobed or toothed; blades 2–22 cm.

FLOWERS: Hypanthium 1.5–3 mm; sepals 4.1–13 mm, petals 5–16 mm. Ovary inferior, with sterile tip 15–80 mm.

FRUITS: Capsule 4-angled, not obviously swollen by seeds, straight or curved.

SIERRA DISTRIBUTION: Mostly east slope, Tulare, Mono and Alpine Counties; west slope from Placer County north.

ELEVATION: 700–2500 m

HABITAT: Open, moist sites, meadows, often in clay soils; montane and sagebrush zones.

FLOWERING TIME: May–Jul

NOTES: Formerly *Camissonia subacaulis*.

Taraxia tanacetifolia

Tansyleaf Suncup

Onagraceae (Evening Primrose family)

Stemless, hairy perennial herb; from stout taproot.

LEAVES: Basal, simple, 6.5–32 cm, deeply and irregularly pinnately lobed, densely hairy.

FLOWERS: Hypanthium 4–6.5 mm; sepals 5.5–13 mm; petals 8–23 mm. Ovary inferior, has sterile tip 14–55 mm.

FRUITS: Capsule 7–25 mm, long-tapered, straight or curved, lumpy where seeds present.

SIERRA DISTRIBUTION: Nevada County north; east slope, Inyo and Mono Counties.

ELEVATION: 700–2500 m

HABITAT: Open, exposed sites, especially clay soils; montane and sagebrush zones.

FLOWERING TIME: May–Jul

NOTES: Formerly *Camissonia tanacetifolia*.

Aphyllon franciscanum

Franciscan Broomrape

Orobanchaceae (Broomrape family)

Perennial herb 5–20 cm; stems are underground, 1 or clustered; roots are modified into structures that absorb nutrients from the host plant, frequently on *Eriodictyon californicum*, and also other hosts, but never *Artemisia*.

LEAVES: Reduced to fleshy, nongreen scales.

FLOWERS: Racemes; pedicels emerge from the ground (looking like stems at first) of unequal lengths, glandular-puberulent; flowers 4–15, bracts > 6, pedicels 3–15 cm; bractlets absent. Calyx lobes 5, 3–7 mm, triangular, acuminate; corollas 15–30 mm, curved, ultimately becoming erect, yellow or pinkish; lobes rounded to acute. Stigma 2-lobed; anthers hairy.

FRUITS: Capsule 4-valved. Seeds < 0.7 mm.

SIERRA DISTRIBUTION: Mono and Tulare Counties south.

ELEVATION: < 2300 m

HABITAT: Dry slopes and flats, foothill and montane zones.

FLOWERING TIME: Apr–Jul

NOTES: Formerly included with *A. fasciculatum*, which occurs in the southeastern SN on *Artemisia* and has pink, rounded corolla lobes.

Aphyllon corymbosum

Flat-Top Broomrape

Orobanchaceae (Broomrape family)

Perennial herb 3–17 cm; parasitic, with stem branches clustered, thickened at base.

LEAVES: Reduced to fleshy, nongreen scales.

FLOWERS: Inflorescences 3–4 cm, branched, few-flowered; upper pedicels sessile, lower 5–30 mm, and 2 bractlets below pedicels. Calyx 12–18 mm, lobes 5 and longer than tube; corollas 20–30 mm, 2-lipped, sparsely glandular, with lips 5–8 mm, pink to purple with darker veins. Stigma 2-lobed; anthers densely hairy.

FRUITS: Capsule 4-valved. Seeds < 0.7 mm.

SIERRA DISTRIBUTION: Mono and Tulare Counties south.

ELEVATION: 1200–2800 m

HABITAT: Openings, usually on *Artemisia tridentata;* montane, subalpine, and sagebrush-pinyon juniper zones.

FLOWERING TIME: Jun–Aug

NOTES: Previously known as *Orobanche californica* var. *corymbosa*. California *Orobanche* are now *Aphyllon*. California broomrape, *A. californicum* ssp. *feudgei*, and 2 other sspp. of *A. californicum* occur in the Sierra: Ssp. *feudgei* occurs in the southern Sierra and east slope < 2500 m, mostly on *Artemisia tridentata*, and blooms May–Jul; ssp. *grayanum* occurs sporadically from Fresno County north < 2600 m, and blooms May–Aug; it occurs primarily on *Symphyotrichum* and *Erigeron* in moist areas. Ssp. *jepsonii* occurs on perennial and woody Asteraceae < 2200 m in Butte and Plumas Counties in dry places; the corolla is 25–40 mm, with dark purple veins; bloom time is Jul–Sep.

Aphyllon purpureum

Purple Broomrape

Orobanchaceae (Broomrape family)

Perennial herb < 5 cm; parasitic.

LEAVES: Reduced to fleshy, nongreen scales.

FLOWERS: Flowers 1–2; bracts < 6; pedicels (aboveground stalks) 3–12 cm, bractlets absent. Calyx lobes 4–8 mm, threadlike; corollas 12–35 mm, ± horizontal, pale purple to yellowish or white; lobes rounded, ciliate. Stigma 2-lobed; anthers hairy.

FRUITS: 2-valved; style persistent. Seeds < 1.0 mm.

SIERRA DISTRIBUTION: Mostly west slope.

ELEVATION: < 3100 m

HABITAT: Primarily in moist places; on herbs, especially *Senecio aronicoides*, *Antennaria*, *Sedum*, Apiaceae, Saxifragaceae; foothill and montane zones.

FLOWERING TIME: Mar–Jul

NOTES: Pollinated by bees.

Castilleja attenuata

Valley Tassels

Orobanchaceae (Broomrape family)

Erect annual herb 10–50 cm; densely hairy, nonglandular.

LEAVES: Alternate, sessile, 20–80 mm, with 0–3 linear lobes.

FLOWERS: Inflorescences 3–30 cm x 1–2 cm wide; bracts below calyces are 15–35 mm, white- or yellow-tipped, with 3 lobes. Calyx 10–20 mm, lobed ½ its depth on upper side, ¾ on lower side, ⅓ on sides, and colored similarly to bracts; corollas 12–30 mm, with beak 4–5 mm, lower lip 3–4 mm with pouches 2 mm wide x 1–1.5 mm deep. Stigma included, beak puberulent, straight; filaments hairless.

FRUITS: Capsule 7–11 mm. Seed < 1 mm, and coat has shallowly netted pattern.

SIERRA DISTRIBUTION: West slope, all.

ELEVATION: < 1600 m

HABITAT: Open, sunny sites; foothill and montane zones.

FLOWERING TIME: Mar–May

NOTES: Compare with *C. lineariloba*.

Castilleja applegatei ssp. *pinetorum*

Wavy-Leaved Indian Paintbrush

Orobanchaceae (Broomrape family)

Erect, often bushy perennial herb 30–60 cm; green, with glandular hairs, sticky.

LEAVES: Alternate, sessile, 20–70 mm, lobes 0–3, margins wavy.

FLOWERS: Inflorescences 5–20 cm; bracts below calyx, 15–25 mm, linear-lanceolate, red or sometimes yellow, lobes 0–7. Calyx 16–25 mm, divided > ⅓ on top and bottom, < ⅕ on sides, lobes lanceolate, acute. Corollas 25–35 mm, yellowish; upper lip (beak) 11–16 mm, the same size as the corolla tube; lower lip reduced to green bulges 1–3 mm. Stigma ± 2-lobed, inserted or exserted.

FRUITS: 8–15 mm.

SIERRA DISTRIBUTION: All.

ELEVATION: 300–3600 m

HABITAT: Open rocky areas; foothill, montane, and sagebrush zones.

FLOWERING TIME: Jun–Aug

NOTES: Three other sspp. of *C. applegatei* occur in the SN. One is ssp. *pallida*, with stems 10–25 cm, leaf lobes 3, calyx 13–15 mm, divided no more than ¼ on sides; corollas 16–22 mm, beak 7–10 mm; it is found on both slopes, Tulare and Inyo Counties north, in upper montane to alpine zones. A second is ssp. *martinii*, with stems 30–80 cm, leaf lobes 0 (or 3), calyx 15–25 mm, divided ⅕–⅙ on sides, lobes ovate, obtuse, with the corolla 25–40 mm, beak 12–18 mm; it can be found in Inyo, Kern, and Tulare Counties 300–2800 m. A third is ssp. *disticha*, with stems 30–80 cm, leaf lobes 0 (or 3), calyx 12–18 mm, divided ⅙ on sides, lobes lanceolate, corollas 25–35 mm, beak 14–18 mm; it is on both slopes 2000–3000 m, Kern and Inyo to El Dorado Counties (intergrades with ssp. *pinetorum*).

Castilleja densiflora ssp. *densiflora*

Dense Flower Owl's Clover

Orobanchaceae (Broomrape family)

Erect annual herb 10–40 cm; branched from middle, if branched at all.

LEAVES: Alternate, sessile, 20–80 mm, with 0–3 linear lobes.

FLOWERS: Inflorescences 3–25 cm, 2.5–4 cm wide; bracts below calyces are 10–25 mm, linear, tipped with purple or white; lobed ⅔ depth on upper surface, ½ below and on the sides, with lobes colored similarly to bracts. Corollas 12–30 mm, beak 5–6 mm, puberulent, straight; lower lip 4–5 mm, with pouches 4–6 mm wide x 3–4 mm deep; teeth on pouches, 2 mm. Stigma 2-lobed, exserted; filaments hairless.

FRUITS: Capsule 7–10 mm. Seed 0.5 mm; coat has shallow, netted pattern.

SIERRA DISTRIBUTION: West slope, Placer County south.

ELEVATION: < 1600 m

HABITAT: Open, sunny sites; foothill and montane zones.

FLOWERING TIME: Mar–May

NOTES: Resembles purple owl's clover but does not have the fuzzy, curved beak above the stigma and has a 2-lobed stigma, in contrast to the headlike stigma of *C. exserta*.

Castilleja exserta ssp. *exserta*

Purple Owl's Clover

Orobanchaceae (Broomrape family)

Erect annual herb 10–45 cm; densely hairy, with some hairs glandular.

LEAVES: Alternate, sessile, 10–50 mm, with 5–9 linear lobes.

FLOWERS: Inflorescences 2–20 cm, 2–4 cm wide; bracts below calyces are 10–25 mm, purple or white, and linear with wider tips. Calyx 10–22 mm, lobed ½ its depth on the upper side, ⅔ on lower side, and colored similarly to bracts. Corollas 12–30 mm, beak 6–7 mm, densely hairy with a hooked tip; lower lip 4–6 mm, with pouches 3–8 mm wide x 3–4 mm deep. Stigma included but globelike and visible below hooked, hairy beak; filaments hairy.

FRUITS: Capsule 10–15 mm. Seed 1–2 mm; coat has a deeply netted pattern.

SIERRA DISTRIBUTION: West slope, Butte and Plumas Counties south.

ELEVATION: < 1400 m

HABITAT: Open, sunny sites; foothill and montane zones.

FLOWERING TIME: Mar–May

NOTES: Purple owl's clover forms enormous blankets of pinkish-purple visible from miles away in a good rain year. This is one of our most common and beautiful annual wildflowers; it is also extremely variable and can hybridize with *C. attenuata*, *C. densiflora*, and *C. lineariloba*. The curved, hairy beak is diagnostic—compare with *C. densiflora*.

Castilleja foliolosa
Woolly Indian Paintbrush
Orobanchaceae (Broomrape family)

Bushy perennial herb or subshrub 30–60 cm; white or gray woolly, with hairs feltlike and branched (need a 10x lens).
LEAVES: Alternate, sessile, 10–50 mm, linear, gray, with lobes 0–3.
FLOWERS: Inflorescences 3–20 cm; bracts below calyx, 15–25 mm; lobes 0–3, obtuse or squared off. Calyx 15–18 mm, divided $1/3$–$2/5$ on upper and lower surfaces, and not divided (or barely notched) on the sides; corollas 18–25 mm, with upper lip (beak) the same size as the corolla tube, and the lower lip 2 mm, dark green, included in calyx. Stigma somewhat 2-lobed.
FRUITS: Capsule 10–15 mm. Seed 1–2 mm, deeply netted.
SIERRA DISTRIBUTION: West slope, Mariposa County north.
ELEVATION: < 1800 m
HABITAT: Open rocky areas; foothill zone.
FLOWERING TIME: Mar–Jun
NOTES: A similar species with feltlike foliage and branched hairs is *C. pruinosa*. It differs by having acute rather than obtuse calyx lobes, with the sides as well as the upper and lower surfaces of the calyx divided ($1/4$ on sides in *C. pruinosa*, compared to notched at most in *C. foliolosa*).

Castilleja lacera
Cutleaf Indian Paintbrush
Orobanchaceae (Broomrape family)

Erect annual herb 10–40 cm, with both short glandular hairs and longer-spreading hairs.
LEAVES: Simple to 5-lobed, 10–50 mm.
FLOWERS: Inflorescences 3–15 cm, 2–3 cm wide; bracts 10–20 mm, entirely green (no colored tips); lobes 3–7, linear-lanceolate. Calyx 8–13 mm; corollas 13–22 mm, dark yellow, with upper lip (beak) 4–6 mm, and lower lip 3–5 mm with 3 distinctive, balloonlike pouches 4–6 mm wide (across all 3 pouches) and 3–6 mm deep (side view). Stigma usually visible (exserted), 2-lobed.
FRUITS: 5–8 mm.
SIERRA DISTRIBUTION: All.
ELEVATION: < 2800 m
HABITAT: Open areas, often sites that are vernally moist but dry out later in the summer; foothill through montane zones.
FLOWERING TIME: May–Jul
NOTES: Compare with *C. tenuis*, which has shallower and less balloonlike pouches on the lower lip, and is generally a lighter shade of yellow.

Castilleja lemmonii

Lemmon's Indian Paintbrush

Orobanchaceae (Broomrape family)

Erect, mostly unbranched perennial herb 10–20 cm, with glandular and spreading hairs.

LEAVES: Alternate, sessile, 20–40 mm, with 0–3 linear lobes.

FLOWERS: Inflorescences 3–12 cm; bracts below calyces are 10–15 mm, purple to reddish, with 3–5 lobes. Calyx 16–18 mm, divided ½–⅔ on upper and lower surfaces, about ⅛ on sides; corollas 16–20 mm, light yellow, with lower lip 5–7 mm, pouches shallow white or purple, and small, triangular teeth. Stigma exserted, 2-lobed, beak 7–9 mm.

FRUITS: Capsule 7–9 mm. Seed 1–1.5 mm; coat has a shallow, netted pattern.

SIERRA DISTRIBUTION: Both slopes, Tulare and Inyo Counties north.

ELEVATION: 1550–3700 m

HABITAT: Moist to wet meadows; montane through alpine zones.

FLOWERING TIME: Jul–Aug

NOTES: This is *the* bright pink (not orange or red) paintbrush in high mountain meadows. Pollinated by bees. Includes plants formerly classified as *C. culbertsonii.*

Castilleja linariifolia

Wyoming Indian Paintbrush

Orobanchaceae (Broomrape family)

Perennial herb 30–100 cm; plant is yellowish to gray-green, becoming purple in age; hairless or puberulent.

LEAVES: Alternate, sessile, 20–80 mm, linear, with margins folded upward, lobes 0–3; leaves smaller than calyx.

FLOWERS: Inflorescences 5–20 cm, open below and dense above; bracts 15–30 mm, red to yellow, narrowly 3-lobed, and smaller than calyx. Calyx 20–35 mm, puberulent, divided ⅙–⅓ on upper surface, approximately ⅔ on lower surface, and approximately ⅛ on sides, with lobes curved upward; corollas 25–45 mm, beak yellow-green, with beak and tube approximately equal in length; lower lip 2–3 mm, dark green. Stigma indistinctly 2-lobed.

FRUITS: Capsule 10–15 mm. Seeds 1.5–2 mm; coat loose-fitting and shallowly netted.

SIERRA DISTRIBUTION: Mostly east slope; on west slope in Plumas, El Dorado, Tulare, and Kern Counties.

ELEVATION: 1000–3350 m

HABITAT: Open, often rocky areas, more common on east slope; montane, subalpine, sagebrush/pinyon-juniper zones.

FLOWERING TIME: Jun–Sep

NOTES: This tall paintbrush is distinguished by its linear leaves with margins folding slightly to strongly upward, and bracts with 3 narrow lobes.

Castilleja lineariloba

Valley Tassels

Orobanchaceae (Broomrape family)

Erect annual herb 15–45 cm; glandular, herbage has stiff, spreading hairs.

LEAVES: Alternate, sessile, 20–75 mm, lance-linear, with 0–7 linear lobes.

FLOWERS: Inflorescences 2–5 cm x 2.5–4 cm wide; bracts below calyces are 12–25 mm, cream, yellow, or purple-tipped, with 5–7 lobes. Calyx 15–25 mm, divided $\frac{1}{3}$ on all 4 sides; corollas 15–30 mm, white, pink, or yellow, with beak 4–5 mm, and lower lip 3–4 mm with pouches 4–5 mm wide x 2 mm deep, with purple dots at base. Stigma exserted, beak puberulent, and straight; filaments hairless.

FRUITS: Capsule 7–9 mm. Seed ± 1 mm; coat has deeply netted pattern.

SIERRA DISTRIBUTION: West slope, all.

ELEVATION: < 1800 m

HABITAT: Open, sunny sites; foothill and montane zones.

FLOWERING TIME: Mar–May

NOTES: Compare with *C. attenuata*.

Castilleja miniata ssp. *miniata*

Giant Red Indian Paintbrush

Orobanchaceae (Broomrape family)

Erect, stout perennial herb 40–80 cm, with long, soft hairs toward the top of the plant.

LEAVES: Alternate, sessile, 30–60 mm, lanceolate, and entire.

FLOWERS: Inflorescences 3–15 cm; bracts below calyces are 15–35 mm, bright red (sometimes yellow), with 0–5 lobes. Calyx 15–30 mm, divided $\frac{1}{2}$–$\frac{2}{3}$ on upper surface, $\frac{2}{3}$ on lower surface, about $\frac{1}{4}$ on sides; corollas 20–40 mm, yellow-green on upper surface with red margins, and lower lip 1–2 mm with dark green, incurved teeth. Stigma barely exserted, slightly 2-lobed; beak 5 mm, about to the same length as the tube.

FRUITS: Capsule 10–12 mm. Seed 1.5–2 mm; coat has shallow netted pattern.

SIERRA DISTRIBUTION: Tulare and Inyo Counties north.

ELEVATION: 1500–3500 m

HABITAT: Wet to moist meadows and streamsides; montane, subalpine, and alpine zones.

FLOWERING TIME: May–Sep

NOTES: This is the tallest paintbrush, reliably found in wet sites, and sometimes almost appearing bushy.

Castilleja nana

Alpine Paintbrush

Orobanchaceae (Broomrape family)

Perennial herb 5–25 cm; stems clustered, herbage soft-hairy, not glandular; green or sometimes purplish.

LEAVES: Alternate, sessile, 10–35 mm, lanceolate to linear, with 0–5 lobes.

FLOWERS: Inflorescences 3–13 cm; bracts below calyx are 15–30 mm, lobes 3–5, acuminate, yellow-green or purplish with green margins. Calyx 12–20 mm, with 4 roughly equal lobes, each lobe 3.5–7 mm (about ⅓–½ the length of the calyx), roughly equal, lance-linear. Corollas 15–20 mm, with upper lip (beak) 4–6 mm, pale yellow with purple blotches, and lower lip 3–5 mm; pouches 3, shallow. Stigma slightly exserted, dark (black).

FRUITS: Capsule 8–12 mm. Seed 1–1.5 mm; coat loose-fitting and shallowly netted, with ladderlike thickenings (need a 10x lens).

SIERRA DISTRIBUTION: Both slopes, Tulare and Inyo Counties north to Sierra County.

ELEVATION: 2400–4200 m

HABITAT: Dry, rocky and gravelly sites, fellfields; alpine zone.

FLOWERING TIME: Jul–Aug

NOTES: Can be confused with *C. pilosa*, with which it interbreeds. Plants from the eastern Sierra have a pinkish cast, as shown in the inset.

Castilleja peirsonii

Peirson's Paintbrush

Orobanchaceae (Broomrape family)

Perennial herb 15–40 cm; herbage varies from hairless to having long nonglandular hairs mixed with short, glandular hairs.

LEAVES: Alternate, sessile, 10–50 mm, lanceolate or oblong, with 0–5 lobes.

FLOWERS: Inflorescences 3–15 cm; bracts below calyx are 15–25 mm, lobes 3–5, bright red or orange. Calyx 12–20 mm, divided ½ on upper and lower surfaces, ¼–⅓ on sides, with lobes acute, sometimes notched; corollas 15–28 mm, with upper lip (beak) about ½ length of corolla tube, and lower lip 1–2 mm, dark green, included in the calyx. Stigma exserted, 2-lobed, black.

FRUITS: Capsule 6–9 mm. Seed 1–1.5 mm, with coat shallowly netted.

SIERRA DISTRIBUTION: Fresno and Inyo Counties north.

ELEVATION: 1500–3400 m

HABITAT: Moist to wet meadows; montane, subalpine, and alpine zones.

FLOWERING TIME: Jul–Aug

NOTES: Formerly referred to as *C. parviflora*. However, *C. parviflora* does not actually occur in CA, and plants formerly called *C. parviflora* belong in *C. peirsonii*.

Castilleja pilosa

Hairy Paintbrush

Orobanchaceae (Broomrape family)

Perennial herb 8–35 cm; erect or decumbent, with stiff, spreading hairs; not glandular.

LEAVES: Alternate, sessile, 10–50 mm, lanceolate to linear, with 0–5 lobes.

FLOWERS: Inflorescences 3–20 cm; bracts below calyx, 10–30 mm, lobes 3–5, pale green or purplish, margins white, and central lobe rounded or truncate, sometimes acute. Calyx 10–20 mm, divided ½ their length into 4 subequal lobes, which are lance-linear; corollas 13–22 mm, with upper lip (beak) 4–6 mm, pale yellowish-green, and lower lip 3–5 mm; pouches 3, shallow. Stigma ± exserted, 2-lobed, dark (black).

FRUITS: Capsule 6–9 mm. Seed 1–1.5 mm, with coat shallowly netted and walls ladderlike.

SIERRA DISTRIBUTION: Both slopes, Tulare and Inyo Counties north.

ELEVATION: 1200–3400 m

HABITAT: Dry, rocky and gravelly sites; subalpine, alpine, and sagebrush zones.

FLOWERING TIME: Jun–Aug

NOTES: Interbreeds with *C. nana*.

Castilleja praeterita ✿

Salmon Creek Paintbrush

Orobanchaceae (Broomrape family)

Erect, branched perennial herb 10–45 cm; with stiff hairs; not glandular.

LEAVES: Alternate, sessile, 30–50 mm, with 0–3 linear lobes.

FLOWERS: Inflorescences 8–14 cm; bracts below calyces are 15–25 mm, pale green with yellow tips (sometimes pale red), with 3–5 lobes. Calyx 14–18 mm, divided > ½ on upper and lower surfaces, ⅛ on sides; corollas 13–16 mm, with margins yellow or purple, and lower lip 2 mm with narrow pouches and dark green, incurved teeth. Stigma exserted, 2-lobed, beak 5 mm, puberulent.

FRUITS: Capsule 8–10 mm. Seed 1–1.5 mm; coat has shallow netted pattern.

SIERRA DISTRIBUTION: Tulare and Kern Counties.

ELEVATION: 2200–3400 m

HABITAT: Dry meadows, especially where Rothrock sagebrush is present; montane, subalpine, and sagebrush/pinyon-juniper zones.

FLOWERING TIME: Jul–Aug

NOTES: Inflorescences may be reddish rather than yellow (not shown).

Castilleja tenuis

Hairy Indian Paintbrush

Orobanchaceae (Broomrape family)

Erect annual herb 10–45 cm tall, with long, spreading hairs (glandular hairs toward the top).

LEAVES: Simple to lobed (3–5 lobes), 10–40 mm.

FLOWERS: Inflorescences 5–25 cm, 1–3 cm wide, with bracts 10–30 mm, entirely green (no colored tips), and lanceolate lobes 3–7. Calyx 8–12 mm; corollas 12–20 mm, white or light yellow, with upper lip (beak) 4–5 mm, and lower lip 3–4 mm with 3 distinctive, balloon-like pouches, which measure 2–4 mm wide (across all 3 pouches), 2 mm deep (side view), balloonlike. Stigma not easily visible (included).

FRUITS: Capsule 6–9 mm.

SIERRA DISTRIBUTION: All.

ELEVATION: < 2800 m

HABITAT: Moist openings and meadows from; foothill and montane zones.

FLOWERING TIME: May–Aug

NOTES: Compare with *C. lacera*, which has much deeper, more balloonlike pouches on the lower lip, and is generally a deeper yellow than *C. tenuis.*

Cordylanthus rigidus ssp. *rigidus*

Stiffbranch Bird's-Beak

Orobanchaceae (Broomrape family)

Erect, many-branched, yellowish-green annual herb to 1.5 m; sometimes with a reddish cast; bristly, generally not glandular.

LEAVES: Alternate, 1–4 cm, linear, and entire or with 3 lobes.

FLOWERS: Dense clusters of 5–15 flowers, each cluster up to 3 cm wide. Corollas 12–20 mm, mostly hidden by calyx of about the same length and of similar appearance. Flower subtended by ≥ 1, 3-lobed bracts. Lower corolla lip has U-shaped maroon mark.

FRUITS: Capsule 8–10 mm. Seeds 1.5–2 mm.

SIERRA DISTRIBUTION: West slope, Kern to Mariposa Counties.

ELEVATION: < 2700 m

HABITAT: Open areas; foothill and montane zones.

FLOWERING TIME: Aug–Oct

NOTES: Distinguished from *C. tenuis* ssp. *tenuis* by its yellowish-green, bristly foliage and larger flower clusters, each with 5+ flowers.

Cordylanthus tenuis ssp. *tenuis*

Slender Bird's-Beak

Orobanchaceae (Broomrape family)

Erect, branched annual herb with wiry stem to 1.2 m; hairless to glandular-hairy with long, soft hairs.

LEAVES: Alternate, 1–6 cm, linear, entire.

FLOWERS: Loose clusters of 1–3 flowers, each cluster up to 3 cm wide. Corollas 10–20 mm, mostly hidden by calyx. Lower corolla lip pouched, with maroon streaks or blotches.

FRUITS: Capsule 6–8 mm. Seeds 1.5–2.5 mm.

SIERRA DISTRIBUTION: West slope, Fresno County north.

ELEVATION: < 2600 m

HABITAT: Open areas; foothill and montane zones.

FLOWERING TIME: Jul–Sep

NOTES: Distinguished from *C. rigidus* by gray-green rather than yellowish-green foliage, hairs that are soft rather than bristly, and flower clusters with no more than 3 flowers each. Fresno County bird's-beak, ssp. *barbatus*, usually has 3-lobed outer inflorescence bracts, but distinguishing between them is tricky.

Orthocarpus cuspidatus ssp. *cryptanthus*

Short-Flowered Owl's Clover

Orobanchaceae (Broomrape family)

Annual herb 10–40 cm; puberulent, sparsely glandular, with herbage green, then tinged with purple in age; stem is simple or highly branched.

LEAVES: Alternate, sessile, 10–50 mm, ± lanceolate.

FLOWERS: Inflorescence is dense, 2–10 cm; bracts below each flower, 2–20 mm, purple/pink distally and with 2 narrow lobes on either side at base, mostly hiding flowers. Calyx 7–10 mm, 4-lobed, divided ⅔ to base on upper surface, ½ on lower, ¼ on sides. Corollas 9–14 mm, lips purple/pink, densely puberulent, with upper lip folded lengthwise, forming a hood (beak) that contains the 4 anthers and the style; beak 3–4 mm, only 0–1 mm longer than lower lip, straight; lower lip has 3 pouches 1–2 mm deep.

FRUITS: 8–15 mm.

SIERRA DISTRIBUTION: Both slopes, Madera and Mono Counties north.

ELEVATION: 1500–3200 m

HABITAT: Open, rocky/pebbly areas, drying meadows; montane, subalpine, and sagebrush zones.

FLOWERING TIME: Jun–Aug

NOTES: One way that *Orthocarpus* differs from *Castilleja* and *Triphysaria* is that the upper 2 corolla lobes fold at the tip, enclosing the anthers and style, with an opening that faces down (rather than the tips facing straight forward). Another *Orthocarpus* found 1500–3000 m in the SN is *O. luteus*, yellow owl's clover, with bright yellow flowers well exserted from green calyces, and leaves similar to floral bracts. Yellow owl's clover occurs sporadically in moist sites from Kern to Plumas Counties, on both slopes, but primarily in the eastern Sierra; it is much more common in the Great Basin.

Pedicularis attollens

Little Elephant's Heads

Orobanchaceae (Broomrape family)

Erect perennial herb 0.6–6 dm; hairless below the woolly inflorescence.

LEAVES: Basal and alternate, pinnately lobed, 3–20 cm, with total number of lobes 13–25, serrate.

FLOWERS: Inflorescences 2–30 cm, woolly, with bracts smaller than flowers. Flowers have upper lips (beaks) modified to resemble curved elephant's trunks; lower lips 3-lobed, with 2 outer lobes somewhat resembling elephant ears. Corollas 4.5–7 mm, beak 3–5 mm.

FRUITS: Capsule 6–10 mm, 2.5–4 mm wide, with surface netlike.

SIERRA DISTRIBUTION: Tulare and Inyo Counties north.

ELEVATION: 1200–4000 m

HABITAT: Meadows, streamsides; montane through alpine zones.

FLOWERING TIME: Jun–Sep

NOTES: Compare with *P. groenlandica*, which has larger flowers with much longer beaks ("trunks"), more distinct "ears," and hairless inflorescences.

Pedicularis densiflora

Indian Warrior

Orobanchaceae (Broomrape family)

Leafy perennial herb 1–5.5 dm.

LEAVES: 5–28 cm, bipinnately lobed with 13–41 main segments, each of these also lobed, giving leaves a fernlike appearance.

FLOWERS: Inflorescences 4–12 cm; bracts reddish-orange and engulfing flowers. Calyx 8–15 mm, hairy, also red-orange. Corollas 23–36 mm, cylindric, orange, red, or yellowish.

FRUITS: Capsule 10–20 mm.

SIERRA DISTRIBUTION: West slope, all.

ELEVATION: 1200–3600 m

HABITAT: Dry slopes, rocky sites; foothill and lower montane zones.

FLOWERING TIME: Mar–May

NOTES: Early to emerge in spring; a partial root parasite on oak and manzanita.

Pedicularis groenlandica
Elephant's Head
Orobanchaceae (Broomrape family)

Erect hairless perennial herb 0.8–8 dm.

LEAVES: Basal and alternate, 3–25 cm, pinnately lobed, lobes 25–51, serrate.

FLOWERS: Inflorescences 1–30 cm, mostly hairless, with bracts = flowers; flowers have upper lips (beaks) modified, resembling curved elephant's trunks, and lower lips 3-lobed with 2 outer lobes somewhat resembling elephant's ears; corollas 8–15 mm, beak 6–13 mm.

FRUITS: Capsule 6–9 mm, 2.5–4 mm, with surface netlike.

SIERRA DISTRIBUTION: Both slopes, Tulare and Inyo Counties north.

ELEVATION: 1200–3600 m

HABITAT: Meadows, streamsides; montane, subalpine, and alpine zones.

FLOWERING TIME: Jun–Sep

NOTES: Compare with the generally smaller *P. attollens*, which has a much shorter beak ("trunk") and a hairy inflorescence. If these 2 similar spp. tend to confuse you, remember that *g* stands for *groenlandica,* which has a glabrous (hairless) inflorescence, and a *great*, big elephant's trunk.

Pedicularis racemosa
Leafy Lousewort
Orobanchaceae (Broomrape family)

Perennial herb 1.2–8 dm; decumbent, branched toward top, mostly hairless.

LEAVES: Cauline and alternate, simple, 2–10 cm.

FLOWERS: Inflorescences 1–5 cm, with lower bracts at least as large as flowers. Calyx 4.5–8 cm, 2-lobed; corollas 10–16 mm, 5–7.5 mm, white, yellow, or purple, forming a curved beak; lower lip 5–9 mm, 3-lobed.

FRUITS: Capsule 10–16 mm. Seed 1.5–3 mm, with smooth surface.

SIERRA DISTRIBUTION: Placer County north.

ELEVATION: 900–2300 m

HABITAT: Dry forest floor; montane zone.

FLOWERING TIME: Jun–Aug

NOTES: Although this plant somewhat resembles elephant's heads, it differs by growing in dry habitats and by the fact that the leaves are simple (not bipinnate). This lousewort is found only in the northern Sierra.

Pedicularis semibarbata

Lousewort

Orobanchaceae (Broomrape family)

Ground-hugging, leafy perennial herb with a "ferny" look.

LEAVES: Basal 5–22 cm, overtopping inflorescence, bipinnately lobed with 11–25 main segments, each of these also lobed.

FLOWERS: Inflorescences 3–12 cm; bracts hairy; flowers often hidden by leaves. Calyx 8–10 mm, ciliate; corollas 15–24 mm, cylindrical, yellow, with orange or maroon tips.

FRUITS: Capsule 6.5–10 mm.

SIERRA DISTRIBUTION: West slope, all; east slope, north of Tahoe.

ELEVATION: 1200–3600 m

HABITAT: Dry floor of coniferous forest; montane through alpine zones.

FLOWERING TIME: May–Jul

NOTES: *P. semibarbata* is a root parasite, tapping into roots of ponderosa pine and *Lupinus fulcratus*, and likely other hosts. Interestingly, Yosemite flowers were pollinated exclusively by the mason bee *Osmia tristella*. This bee was also observed nesting in nearby ponderosa pines that were likely providing nutrients via their roots to *P. semibarbata*.

Triphysaria eriantha ssp. *eriantha*

Johnny-Tuck, Butter-and-Eggs

Orobanchaceae (Broomrape family)

Erect annual herb 10–35 cm; plant is ± purple, puberulent, with both glandular and nonglandular hairs present.

LEAVES: Alternate, 10–50 mm, 3- to 7-lobed.

FLOWERS: Inflorescences 2–15 cm, dense distally; bracts 5–18 mm, 3- to 5-lobed. Calyx 8–13 mm, divided ¼–½; corollas 10–25 mm, yellow, or white fading rose-pink, and beak dark purple; tube densely puberulent, with lower lip about the same length as the beak; pouches 3–4 mm deep.

FRUITS: Capsule 5–8 mm, oblong. Seeds 30–50, dark brown.

SIERRA DISTRIBUTION: West slope foothills, all counties.

ELEVATION: < 1600 m

HABITAT: Grasslands, oak/pine woodland, chaparral; foothill and lower montane zones.

FLOWERING TIME: Mar–May

NOTES: Formerly *Orthocarpus erianthus*.

Paeonia brownii

Brown's Peony

Paeoniaceae (Peony family)

Perennial herb 20–40 cm; from fleshy root clusters, stems 1 to several, simple.

LEAVES: Basal and cauline, 5–8 per stem, glaucous below; deeply ternately dissected, with primary divisions 3–6 cm x 2–5 cm wide, and ultimate segments elliptic with tapered bases.

FLOWERS: Single or few in terminal inflorescences; sepals 5, leathery and persistent; petals 5–10, 8–13 mm, round, smaller than inner sepals, maroon to bronze, with margins greenish or yellow. Stamens numerous, filaments 3–5 mm, anthers 2–4 mm.

FRUITS: Follicles 2–5.

SIERRA DISTRIBUTION: Both slopes, Mono and Tuolumne Counties north.

ELEVATION: 200–3000 m

HABITAT: Open, dry conifer forest, chaparral; foothill, montane, and sagebrush zones.

FLOWERING TIME: Apr–Jun

NOTES: *P. brownii* produces copious nectar from a lobed disc at the base of the pistils (hidden from view during flowering). In an Oregon study, a variety of insects visited flowers throughout the blooming season due to consistent nectar production. Primary pollinators were wasp queens (Vespidae) and large flower flies (Syrphidae, bumblebee mimics), and other insect taxa likely also transferred pollen, e.g., sweat bees (Halictidae).

Argemone munita

Chicalote, Prickly Poppy

Papaveraceae (Poppy family)

Prickly annual or perennial herb 6–15 dm. Sap yellow.

LEAVES: Alternate, 5–15 cm; upper and lower surfaces prickly, and lower leaves lobed at least halfway to midrib.

FLOWERS: Corollas 5–13 cm diameter, petals 2.5–5 cm, white, crinkled. Stamens 150–250.

FRUITS: 35–55 mm, ovate, prickly.

SIERRA DISTRIBUTION: Mostly east of crest, but found sparingly on the west slope.

ELEVATION: 700–3000 m

HABITAT: Dry, open areas in full sun; foothill, montane, and sagebrush/pinyon-juniper zones.

FLOWERING TIME: Jun–Sep

NOTES: A similar species, *A. corymbosa*, also occurs in the eastern Sierra. It differs in having fewer (100–120) stamens, orange rather than yellow sap, and lower leaves lobed less than halfway to midrib, as well as being generally shorter with shorter (< 3.5 cm) petals.

Dendromecon rigida

Bush Poppy

Papaveraceae (Poppy family)

Evergreen shrub 1–3 m. Sap clear.

LEAVES: Alternate, 2.5–10 cm, 7–25 mm wide, with finely toothed margins.

FLOWERS: Solitary, bright yellow; petals 2–3 cm, satiny, falling immediately after pollination.

FRUITS: Capsule 5–10 cm, splitting open from base.

SIERRA DISTRIBUTION: West slope, all.

ELEVATION: < 1900 m

HABITAT: Dry slopes, especially in chaparral; foothill zone.

FLOWERING TIME: Apr–Jun

NOTES: Seeds are adapted to fire, and flushes of bush poppy seedlings are commonly seen in burned areas in foothill chaparral.

Dicentra formosa

Pacific Bleeding Heart

Papaveraceae (Poppy family)

Perennial herb 20–45 cm; rhizomatous, sometimes forming large patches.

LEAVES: Dissected ternately, each division ternately dissected again (or in other words, much divided and fernlike), 20–50 cm, glaucous at least underneath.

FLOWERS: Inflorescence of 2–30 flowers, each a nodding pink "heart." Sepals 2, shedding once flowers open; petals 4, pink, with the outer 2 petals 14–19 mm, varying in color from rose to cream or yellow.

FRUITS: Capsule 14–20 mm, oblong.

SIERRA DISTRIBUTION: All.

ELEVATION: < 2300 m

HABITAT: Moist areas in coniferous forest; montane zone.

FLOWERING TIME: May–Jul

NOTES: A beautiful native plant for garden use in shady, moist, well-drained sites. There is abundant information on the internet about this species and the many cultivars that have been bred from it. If you live in a wildland setting, it's best to attempt to find plants from locally native stock.

Dicentra pauciflora

Shorthorn Steer's Head

Papaveraceae (Poppy family)

Perennial herb 3–7 cm (< 1 dm); from tubers.
LEAVES: Basal leaves 1–3, 4–7 cm, 2–3 times dissected.
FLOWERS: Inflorescences of 1–3 flowers; sepals 2, 5–6 mm, lance-oblong; petals 4, the outer 2 recurved < ½ the length of the corolla; outer petals vary in color from white to pink to lavender, and inner petals purple; corollas ± 15 mm.
FRUITS: Capsule 10–15 mm, ovate.
SIERRA DISTRIBUTION: This species has an interesting distribution: Tulare and Kern Counties in the south, Plumas and Butte Counties in the north, and only 1 site in Yosemite National Park.
ELEVATION: 1200–3000 m
HABITAT: Open gravelly areas; montane through alpine zones.
FLOWERING TIME: Jun–Jul
NOTES: Compare with *D. uniflora*.

Dicentra uniflora

Steer's Head

Papaveraceae (Poppy family)

Perennial herb 3–7 cm (< 1 dm); from tubers.
LEAVES: Basal leaves 1–3, 4–7 cm, 2–3 times dissected.
FLOWERS: Solitary; sepals 2, 5–6 mm, lance-oblong; petals 4, pink, the outer 2 recurved about ½ the length of the corolla, varying in color from white to pink to lavender; corollas ± 15 mm.
FRUITS: Capsule 9–14 mm, cone-shaped.
SIERRA DISTRIBUTION: Tulare and Mono Counties north.
ELEVATION: 1000–3700 m
HABITAT: Moist gravelly areas; montane through alpine zones.
FLOWERING TIME: May–Jul
NOTES: This diminutive plant emerges just as the snow melts and is often long gone by the time most people are in its habitat. Compare with *D. pauciflora*.

Ehrendorferia chrysantha

Golden Ear-Drops

Papaveraceae (Poppy family)

Tall perennial herb 50–160 cm, with leafy stems.

LEAVES: Basal and stem leaves 15–30 cm, 2–3 times pinnately dissected.

FLOWERS: Inflorescence of many golden-yellow flowers, biradial; sepals 2, falling after flowering; petals 4, the outer 2 free, with pouched base, and the inner 2 joined at tips, with crests on backs. Stamens 6.

FRUITS: Capsule 15–25 mm, long-conic, splitting open from tip.

SIERRA DISTRIBUTION: Amador County south.

ELEVATION: < 1900 m

HABITAT: Dry slopes, especially burned areas in chaparral; foothill zone.

FLOWERING TIME: Apr–Sep

NOTES: Formerly classified as *Dicentra chrysantha*. Abundant and spectacular in chaparral burn areas.

Eschscholzia caespitosa

Tufted Poppy

Papaveraceae (Poppy family)

Erect to branched, tufted annual herb 5–30 cm tall; hairless (sometime glaucous).

LEAVES: Basal and alternate, finely dissected into linear lobes.

FLOWERS: Buds erect and hairless with tips long, pointed, > ¼ the length of the bud. Petals 10–25 mm, yellow to orange; no red rim at base of petals as in *E. californica*.

FRUITS: 4–8 cm.

SIERRA DISTRIBUTION: West slope, all.

ELEVATION: < 1500 m

HABITAT: Open areas in chaparral and foothill woodland, especially in river canyons; foothill zone.

FLOWERING TIME: Feb–Jun

NOTES: *E. caespitosa* is by far the most common orange poppy in much of the Sierra; in good water years and especially after wildfires, the plants blanket hillsides with brilliant, almost blinding orange.

Eschscholzia californica

California Poppy

Papaveraceae (Poppy family)

Freely branched annual (sometimes perennial) herb 5–60 cm; tall, hairless (sometimes glaucous).

LEAVES: Basal and alternate, finely dissected into linear lobes.

FLOWERS: Buds erect; petals 20–60 mm, fan-shaped, yellow to orange, often with darker orange spots at base. Receptacle has 2 rims, the outer one expanded into a red or purple, saucerlike structure.

FRUITS: Capsule 3–9 cm.

SIERRA DISTRIBUTION: All.

ELEVATION: < 2000 m

HABITAT: Open grassy or rocky areas; many zones.

FLOWERING TIME: Mar–Jun

NOTES: The CA state flower. The red, saucerlike rim at base of the petals is diagnostic and quickly distinguishes this species from the similar *E. caespitosa*, with which it grows intermixed in many sites. The California poppy occurs in every one of the 58 counties of CA, and when seen along highways and near urban areas, it can be difficult to determine whether a given population is native to that site or persisting from a "wildflower" seed mix or highway roadside beautification project. The populations in Mono County at 2500 m elevation are doubtless introduced, perhaps above the natural upper elevation limit for the species.

Eschscholzia lobbii

Frying Pans

Papaveraceae (Poppy family)

Erect, hairless annual herb 5–15 cm; tall, with leafless stems.

LEAVES: Basal only, finely dissected into 2-ternate linear lobes.

FLOWERS: 1 per stem, buds erect; petals 7–12 mm, yellow.

FRUITS: Capsule 4–7 cm. Seeds 1.4–2 mm, brown, with raised ridges appearing burlike.

SIERRA DISTRIBUTION: West slope.

ELEVATION: < 830 m

HABITAT: Open, sunny areas; foothill zone.

FLOWERING TIME: Mar–Apr

NOTES: Frying pans are not as common as *E. caespitosa* or *E. californica*, and they are easily distinguished by their bright yellow petals and leafless stems.

Hesperomecon linearis

Narrow-Leaved Meconella

Papaveraceae (Poppy family)

Annual herb 3–40 cm; herbage long-hairy.
LEAVES: Basal 5–85 mm, linear, densely hairy.
FLOWERS: 1 per inflorescence, nodding in bud, with stalk longer than leaves; petals 6, 3–30 mm, cream-colored with yellow base or outer yellow and inner cream. Stigmas 3; stamens numerous.
FRUITS: Capsule 10–15 mm, ovoid, breaking open by 3 longitudinal valves. Seeds 0.4 mm, dark and shiny.
SIERRA DISTRIBUTION: West slope, Tulare and Kern Counties.
ELEVATION: < 1000 m
HABITAT: Open, sunny areas in grasslands, chaparral, and foothill woodland; foothill zone
FLOWERING TIME: Mar–Jun
NOTES: Formerly *Meconella linearis*. Compare with *Meconella californica*, which has wider, hairless leaves.

Meconella californica

California Fairypoppy

Papaveraceae (Poppy family)

Delicate, hairless annual herb 3–20 cm, with slender, erect to ascending stems.
LEAVES: 3–25 mm; basal oblanceolate, and cauline more linear.
FLOWERS: Solitary, both terminal and in leaf axils; sepals 3; petals 6, 5–10 mm, white or cream-colored (sometimes with outer 3 yellow). Stigma lobes 3; stamens 12, in 2 series, with outer ones shorter.
FRUITS: Capsule 25–28, cylindric, dehiscent, splitting open from tip. Seeds numerous, 0.5–1 mm, shiny brown or black.
SIERRA DISTRIBUTION: West slope.
ELEVATION: < 1000 m
HABITAT: Open, sunny areas; foothill zone.
FLOWERING TIME: Mar–May
NOTES: The outer petals of some plants are bright yellow. The yellow fades at night and reappears in daylight.

Papaver heterophyllum
Wind Poppy
Papaveraceae (Poppy family)

Erect annual herb 30–60 cm. Sap yellow.
LEAVES: Basal and cauline, 2–12 cm, deeply pinnately lobed, with lobes also pinnately lobed (or toothed); hairless or sparsely hairy.
FLOWERS: Solitary, in axils; buds nodding; sepals 2, falling when flowers open; petals 4, 10–20 mm, red-orange with purple bases. Stigma headlike on a slender style; stamens numerous, filaments purple, and anthers yellow.
FRUITS: Capsule 1–2 cm, dehiscing by pores below stigma. Seeds numerous, 0.7 mm, and kidney-shaped; the surface has netlike ridges, brown or black.
SIERRA DISTRIBUTION: West slope, Calaveras County south.
ELEVATION: < 1000 m
HABITAT: Open, sunny areas in chaparral and foothill woodlands; foothill zone.
FLOWERING TIME: Apr–May
NOTES: Formerly *Stylomecon heterophylla*. Wind poppy appears in greatest abundance after fires, sometimes turning slopes orange 1 year post-fire.

Platystemon californicus
Cream Cups
Papaveraceae (Poppy family)

Hairy annual herb, with many erect stems from base.
LEAVES: Opposite, sessile, 2–5 cm, entire, with long, soft hairs.
FLOWERS: Solitary; buds nodding; sepals 3, with long, soft hairs; petals 6, 8–16 mm, cream-colored with varying degrees of yellow toward tips. Stamens > 12.
FRUITS: Capsule 10–16 mm, breaking into segments 1-seeded.
SIERRA DISTRIBUTION: West slope, all.
ELEVATION: < 1000 m
HABITAT: Open, sunny areas in chaparral and foothill woodland; foothill and montane zones.
FLOWERING TIME: Mar–May
NOTES: Unmistakable; nothing else looks like this beautiful poppy.

Parnassia palustris

Marsh Grass-Of-Parnassus

Parnassiaceae (Grass-of-Parnassus family)

Erect, hairless perennial herb to 50 cm, with basal leaves.
LEAVES: Petioles long; blades 2–5 cm, ovate to lanceolate, with 5 distinct veins.
FLOWERS: "Stem" actually the inflorescence stalk, varying in length 5–50 cm, and the stem "leaf" a bract. Flowers borne singly; calyx lobes 3–7 mm. Petals 5, white with distinct veins; gland-tipped clusters at the base of each petal are staminodes (sterile stamens) that help attract pollinators by appearing to have nectar (nectar is produced at the base of the staminodes, however). Pistil 1; stamens 5.
FRUITS: Capsule 8–12 mm.
SIERRA DISTRIBUTION: All.
ELEVATION: 900–3400 m
HABITAT: Wet streambanks and meadows; montane, subalpine, and alpine zones.
FLOWERING TIME: Jul–Oct
NOTES: Formerly *P. californica*; formerly in Saxifragaceae. In the northern Sierra, you might encounter *P. fimbriata*, which looks similar but has fringed petal bases.

Diplacus aurantiacus

Orange Bush Monkeyflower

Phrymaceae (Lopseed family)

Shrub or subshrub 0.5–1.2 m; hairless or minutely hairy and minutely stalked-glandular.
LEAVES: Opposite, sessile, 16–60 mm x $2\text{–}20$ mm wide, narrowly oblong to narrowly elliptic-lanceolate, with margins entire or shallowly toothed, plane or revolute. Upper leaf surface hairless, and undersurface sparsely to densely hairy.
FLOWERS: Inflorescences axillary, with 2–4 flowers per node. Calyx 18–30 mm, hairless, sometimes minutely hairy and minutely stalked-glandular, and not inflated in fruit. Corollas yellow-orange to orange, not marked with spots or stripes; corolla tube + throat 25–30 mm; limb 20–30 mm diameter, 2-lipped to nearly rotate; lobes 5, oblong, with the upper 2 each shallowly, asymmetrically incised. Styles glandular, stigmas exserted with 2 equal lobes; anthers included, hairless.
FRUITS: Capsule 18–32 mm.
SIERRA DISTRIBUTION: West slope, except Madera County.
ELEVATION: < 800 m
HABITAT: Cliffs, canyons, disturbed areas; foothill zone.
FLOWERING TIME: Mar–Jun
NOTES: Formerly *Mimulus aurantiacus*. Compare with *D. grandiflorus*, another orange-flowered bushy monkeyflower: it has 2 flowers per node, pale orange corollas with tube + throat 38–57 mm, and 2-lobed corolla lobes.

Diplacus bicolor

Harlequin Monkeyflower

Phrymaceae (Lopseed family)

Annual herb 1–14 cm; puberulent.
LEAVES: 7–23 mm, linear to elliptical, entire.
FLOWERS: Pedicels 1–3 mm, shorter than calyx; calyx 4–8 mm with lobes 1–3 mm, acute, and calyx ribs dark, puberulent; corollas magenta or yellow, persisting after blooming, tube + throat 13–18 mm; corolla throat has 3–5 dark stripes and 2 red-brown blotches that are hard to see in magenta flowers.
FRUITS: Capsule 6–10 mm.
SIERRA DISTRIBUTION: West slope, Madera County south.
ELEVATION: 1500–3270 m
HABITAT: Open sandy or gravelly areas; montane and subalpine zones.
FLOWERING TIME: May–Sep
NOTES: Formerly *Mimulus whitneyi*. Hybridizes with *D. mephiticus* and *D. leptaleus* at higher elevations, and with *D. constrictus* at lower elevations.

Diplacus bolanderi

Bolander's Monkeyflower

Phrymaceae (Lopseed family)

Erect annual herb 2–90 cm; densely glandular-hairy.
LEAVES: Sessile, 0.5–7 cm long, narrowly ovate or oblanceolate. Lower leaves lack hairs, and upper leaves densely hairy.
FLOWERS: Pedicels 2–5 mm, much shorter than calyx; calyx 7–27 mm, with wide ribs, swollen in fruit, with unequal lobes 1–9 mm. Corolla persists in fruit, tube + throat 12–30 mm, magenta with 2 white ventral ridges.
FRUITS: 7–20 mm.
SIERRA DISTRIBUTION: El Dorado County south (may occur north but needs documentation).
ELEVATION: < 1900 m
HABITAT: Open, usually disturbed or burned areas; foothill and montane zones.
FLOWERING TIME: May–Jul
NOTES: Recognized by the magenta flowers with white ridges on the floor of the corolla, frequently arranged along the stem in pairs at each node. Forms large masses of plants of varying heights after fires or other types of clearing. Flower size can be quite variable within a population.

Diplacus congdonii

Congdon's Monkeyflower

Phrymaceae (Lopseed family)

Small annual herb < 12 cm; puberulent to hairy.
LEAVES: Blades 8–32 mm, oblanceolate or elliptic, purple underneath, and green with purple veins above; bases of blades and petioles ciliate.
FLOWERS: Pedicels 2–5 mm, shorter than calyx; calyx 5–14 mm with unequal lobes and hairy ribs. Cleistogamous flowers may be present. In open flowers, corolla tube + throat are 8–30 mm, with limb pink to magenta, lower lobe longest.
FRUITS: 4–8.5 mm, oblique at base, laterally compressed.
SIERRA DISTRIBUTION: West slope.
ELEVATION: < 1300 m
HABITAT: Flat to sloping areas, granitic or serpentine soils, often disturbed areas; foothill and montane zones.
FLOWERING TIME: Mar–May
NOTES: Easily overlooked due to small stature, but flowers are beautiful and unique.

Diplacus douglasii

Purple Mouse Ears

Phrymaceae (Lopseed family)

Small annual herb 0.3–4 cm; usually hairy.
LEAVES: Blades 5–28 mm, ovate to obovate, shiny green.
FLOWERS: Pedicels 2–4 mm, shorter than calyx; calyx 8–14 mm with unequal lobes. Corolla tube + throat are 20–41 mm, with limb magenta. Upper lobes 4–5 mm, obvious, forming the mouse's "ears"; lower lobes < 1 mm (almost nonexistent).
FRUITS: Capsule 3–7 mm, oblique or ovoid.
SIERRA DISTRIBUTION: West slope.
ELEVATION: < 1200 m
HABITAT: Flat to gently sloping areas, rocky ledges, often vernally moist, in many soil types; foothill and montane zones.
FLOWERING TIME: Mar–Apr
NOTES: Relatively uncommon in the SN, possibly overlooked as it blooms early and becomes difficult to see by early summer.

Diplacus grandiflorus

Sticky Monkeyflower

Phrymaceae (Lopseed family)

Shrub or subshrub to 150 cm.

LEAVES: Opposite, sessile, 20–80 mm, linear or narrowly elliptic, with upper surface glossy green and often sticky, and lower surface hairless; both sides of leaf uniformly green, with edges rolled under.

FLOWERS: Inflorescence has 2 flowers per node; pedicels 4–25 mm; calyx 23–33 mm, corolla tube + throat 38–57 mm, and 5 lobes are 2-lobed.

FRUITS: Capsule 16–25 mm, splitting open along upper capsule.

SIERRA DISTRIBUTION: West slope, El Dorado County north to Butte and Plumas Counties.

ELEVATION: < 1530 m

HABITAT: Rocky areas or hillsides in chaparral, lower coniferous forest; foothill and montane zones.

FLOWERING TIME: Mar–Jun

NOTES: Formerly *Mimulus aurantiacus* var. *grandiflorus*. Compare with orange bush monkeyflower, *D. aurantiacus*, which is similar-looking, but flowers are 2–4 per node, corollas are smaller (tube + throat 25–30 mm), and leaf surfaces are hairy below.

Diplacus graniticola

Granite-Crack Monkeyflower

Phrymaceae (Lopseed family)

Annual herb 6–15 cm; plants are compact (internodes usually shorter than leaves); herbage has gland-tipped hairs 1.0–1.6 mm.

LEAVES: Blades 20–40 mm, lanceolate to ovate.

FLOWERS: 2 per node; pedicels 1–3 mm; calyx 8–12 mm, dark-ribbed, glandular-hairy, with lobes 1–3 mm and ± equal, acute to acuminate. Corollas pink to magenta, tube + throat 15–20 mm, limb 10–16 mm wide, lips ± equal, with lower lip bent ± downward; yellow ridges on floor of throat near mouth, partially or fully bordered with magenta or purple. Stigma and anthers included, anthers ciliate.

FRUITS: Capsule 6–10 mm.

SIERRA DISTRIBUTION: West slope, Tuolumne County south.

ELEVATION: 300–2100 m

HABITAT: Cracks and crevices in granitic outcrops; foothill and montane zones.

FLOWERING TIME: Apr–Sep

NOTES: Described as a new species in 2017, this plant was separated from the more widespread *D. layneae* based on its restriction to granitic cracks and crevices, internodes being shorter than leaves, more numerous flowering nodes (4–15 vs. 3–6 in *D. layneae*), calyx ribs more narrowly darkened, tissue between ribs green or purple rather than white/membranous, corollas with 2 blotches at base of upper and lower lips, distinct yellow ridges on floor of throat, and a dark line usually extending to the tip of the corolla lobe.

Diplacus kelloggii

Kellogg's Monkeyflower

Phrymaceae (Lopseed family)

Glandular-hairy annual herb 3–30 cm.
LEAVES: 6–40 cm, with lower surface purplish in many plants.
FLOWERS: Pedicels 2–6 mm, much shorter than calyx. Corollas magenta to purplish-red, deeper toward throat; tube + throat 20–45 mm, more than twice the length of the calyx; upper corolla lobes 4–5 mm, lower 2–3 mm. Stigma and anthers exserted.
FRUITS: Capsule 6–12 mm long, somewhat curved, indehiscent.
SIERRA DISTRIBUTION: West slope, Tulare County north.
ELEVATION: 50–1600 m
HABITAT: Open rocky or gravelly areas, often on steep slopes in talus or soil; foothill and montane zones.
FLOWERING TIME: Mar–Jun
NOTES: This gorgeous monkeyflower is worth a special trip. In April, look for it along the Moss Creek trail north of El Portal in the Merced River Canyon.

Diplacus leptaleus

Sierra Monkeyflower

Phrymaceae (Lopseed family)

Small, erect annual herb 1–14 cm; usually glandular-hairy.
LEAVES: Sessile or with short petioles, 5–25 mm, linear to oblanceolate, usually clustered toward top of stem.
FLOWERS: Pedicels 1–2 mm, calyx 3–6 mm, with glandular-hairy ridges, lobes 1–2 mm, equal and acute. Corollas 5–9 mm, purple or white, with throat yellowish; corolla lobes lack notches and often have a darker line down center, with lower lobes paler.
FRUITS: Capsule 4–6 mm.
SIERRA DISTRIBUTION: All.
ELEVATION: 2000–3400 m
HABITAT: Moist areas, sandy or gravelly areas; montane through alpine zones.
FLOWERING TIME: Jun–Aug
NOTES: Compare with *Erythranthe breweri*, which also has very small flowers, but has pedicels longer than the calyx.

Diplacus mephiticus
Skunky Monkeyflower
Phrymaceae (Lopseed family)

Annual herb 1–15 cm; glandular-hairy. Strong, skunky odor.
LEAVES: 6–30 mm, linear to lanceolate, often crowded together toward top of stem.
FLOWERS: Pedicels 2–7 mm, shorter than calyx; calyx 7–15 mm with lobes 1–3 mm, unequal. Corollas pink or yellow (plants with both colors often grow together, as in this photo), persisting after blooming; tube + throat 11–19 mm, well exserted from calyx.
FRUITS: Capsule 6–11 mm, dehiscent.
SIERRA DISTRIBUTION: Tulare County north.
ELEVATION: 1850–3650 m
HABITAT: Open sandy or gravelly areas often on or near rocky outcrops; montane through subalpine zones.
FLOWERING TIME: May–Aug
NOTES: In the southern Sierra, compare with *D. bicolor*; be prepared to surrender and just call it *Diplacus* and enjoy its beauty whether or not you can conclusively name it (these 2 spp. hybridize where their ranges overlap).

Diplacus pictus
Calico Monkeyflower
Phrymaceae (Lopseed family)

Hairy annual herb 2–38 cm, with 4-angled stems.
LEAVES: 7–45 mm, oblanceolate or obovate.
FLOWERS: Pedicels 2–6 mm, shorter than calyx; calyx 6–13 mm. Corolla tube + throat 6.5–18 mm, radially symmetrical; corolla lobes white with maroon veins.
FRUITS: Capsule 7–17 mm.
SIERRA DISTRIBUTION: Southern Sierra, Tulare and Kern Counties.
ELEVATION: < 1300 m
HABITAT: Open, sunny areas in foothill woodland; foothill zone.
FLOWERING TIME: Mar–May
NOTES: Rare and exquisite.

Diplacus torreyi

Torrey's Monkeyflower

Phrymaceae (Lopseed family)

Erect, delicate to robust, glandular-hairy annual herb 5–35+ cm; often branched.

LEAVES: Petioles short (0–10 mm) or absent; blades 6–40 mm, oblanceolate or elliptic.

FLOWERS: Pedicels 1–3 mm, shorter than calyx; calyx 5–10 mm with lobes 0.5–2 mm, not swollen in fruit, puberulent. Corolla tube + throat 9–18 mm; lower lip projects forward, and the upper reflexes backward; lower lip base has 2 gold stripes edged with magenta. Stigma and anthers no longer than corolla throat; anthers hairless.

FRUITS: Capsule 6–11 mm.

SIERRA DISTRIBUTION: West slope, Fresno County north.

ELEVATION: 600–2000 m

HABITAT: Dry or moist, sunny or partially shaded areas in many plant communities; up through the mixed conifer zone.

FLOWERING TIME: May–Aug

NOTES: Formerly *Mimulus torreyi*. Flourishes and grows > 35 cm and many-branched ("bushy") after fire and other disturbances. The gold stripes edged with darker pink and the architecture of the upper and lower lip help to distinguish this species from similar species (*D. mephiticus* and *D. bicolor*).

Erythranthe acutidens

Kings River Monkeyflower

Phrymaceae (Lopseed family)

Hairless annual herb 3–30 cm.

LEAVES: Basal in rosettes, ± petioled, 7–20 cm; stem leaves sessile, ovate with rounded or cordate bases.

FLOWERS: Pedicels 5–15 mm; calyx 5–9 mm, hairless, with lobes 0.5–1 mm, subequal. Corollas rose to lavender; tube + throat 8–12 mm, limb 7–12 mm wide; corolla lobes deeply notched.

FRUITS: 4–9 mm, within a swollen calyx elongated to 6–11 mm.

SIERRA DISTRIBUTION: West slope, Tulare to Calaveras Counties.

ELEVATION: < 2000 m

HABITAT: Partially shady openings; soils that are vernally moist or near a water source; foothill through montane zones.

FLOWERING TIME: Apr–Jun

NOTES: Previously combined with *Erythranthe inconspicua* in *Mimulus inconspicuus*. Both species have wide, sessile stem leaves, but *E. acutidens* is showier due to its wider corolla diameter. *E. inconspicua*, small-flowered monkeyflower, is found in moist sites 200–2100 m in most west slope counties, with corolla tube + throat 5–9 mm, and limb 5–6 mm. In recent years I have observed *E. acutidens* in great numbers in burned areas, and it is more common than previously thought in vernally moist sites at the end of spring.

Erythranthe arenaria

Sand-Loving Monkeyflower

Phrymaceae (Lopseed family)

Hairy annual herb 5–20 cm (taller in good water years); slimy to the touch.

LEAVES: Opposite; petioles 0–3 mm; blades 5–12 mm, elliptic to lanceolate, bases rounded to tapered.

FLOWERS: Pedicels 10–23 mm; calyx 5–7 mm, hairy; with lobes 0.5–1 mm, about equal in length, acute. Corollas yellow, tube + throat 9–12 mm, limb up to 14 mm wide, throat red-spotted.

FRUITS: Capsule 4–7 mm.

SIERRA DISTRIBUTION: West slope, Kern to Amador Counties.

ELEVATION: 500–2600 m

HABITAT: Sandy washes, flats, wet areas by ponds and lakes, seasonal streams; foothill and montane zones.

FLOWERING TIME: May–Sep

NOTES: Formerly *Mimulus arenarius*. A helpful character not mentioned in the eFlora is the presence of white patches or what looks like a brushing of white on the lower corolla throat. If it is difficult to determine whether annual or perennial, one could mistake *E. moschata* (perennial) for *E. arenaria*. The former is also slimy but has slender rhizomes, and often the red/purple throat markings look like stippled stripes rather than spots.

Erythranthe bicolor

Bicolored Monkeyflower

Phrymaceae (Lopseed family)

Annual herb 4–27 cm; puberulent.

LEAVES: 4–30 cm, often wider than 3 mm, linear to oblanceolate, sometimes irregularly toothed.

FLOWERS: Pedicels 6–30 mm; calyx 5.5–12 mm, red-dotted. Calyx lobes 1–3.5 mm, acute, all about the same length. Corollas 15–25 mm, with the upper lip white or yellow, and the lower yellow. Corolla lobes 5, all 2-lobed.

FRUITS: Calyx corky in fruit; capsule 4–8 mm.

SIERRA DISTRIBUTION: West slope, Tulare County north.

ELEVATION: 360–2100 m

HABITAT: Moist flats and seeps; foothill through montane zones.

FLOWERING TIME: Apr–Jun

NOTES: Formerly *Mimulus bicolor*. When this species is bicolored, there is no mistaking it as *the* yellow-and-white monkeyflower. However, some populations are comprised of yellow-flowered plants with no white upper lips, in which case carefully compare with *E. montioides* (smaller plants, corolla lobes notched rather than 2-lobed, leaves < 3 mm wide, and plants generally smaller and occurring at higher elevations than *E. bicolor*).

Erythranthe breweri

Brewer's Monkeyflower

Phrymaceae (Lopseed family)

Small annual herb 2–15 cm (up to 20 cm); with small flowers, usually glandular hairy.

LEAVES: Sessile, 3–35 mm, linear.

FLOWERS: Pedicels 3–15 mm (and elongating in fruit); calyx 3–9 mm, with dark ridges, lobes 1 mm, equal. Corollas 4–10 mm, pink to magenta, with 2 yellow ventral ridges with all 5 lobes approximately equal (often not strongly 2-lipped), each with a shallow notch.

FRUITS: 3.5–6.5 mm.

SIERRA DISTRIBUTION: All.

ELEVATION: 1200–3400 m

HABITAT: Moist areas; foothill through alpine zone.

FLOWERING TIME: May–Aug

NOTES: Plants are typically just a few centimeters tall, but following burns they can reach the upper height limit. Compare with *Diplacus leptaleus*, which has pedicels shorter than the calyces.

Erythranthe cardinalis

Scarlet Monkeyflower

Phrymaceae (Lopseed family)

Glandular-hairy perennial herb with stems 25–80 mm; rhizomatous, decumbent or ascending, tending to occur in clumps, with bright red-orange flowers.

LEAVES: Opposite, 20–80 cm, with 3–5 primary veins originating from the base of the leaf (not pinnately veined); margins toothed, especially toward tips.

FLOWERS: Pedicels 50–80 mm. Calyx 20–30 mm, symmetrical, hairy, with lobes 4–5 mm, acute; corollas scarlet, tube + throat 40–50 mm.

FRUITS: Capsule 16–18 mm.

SIERRA DISTRIBUTION: Both slopes, all.

ELEVATION: 1000–2600 m

HABITAT: Riparian areas and moist areas with seepage; foothill and montane zones.

FLOWERING TIME: Apr–Oct

NOTES: Hummingbird-pollinated. A large patch of scarlet monkeyflower is a great place to hummingbird-watch.

Erythranthe erubescens

California Blushing Monkeyflower

Phrymaceae (Lopseed family)

Stout, hairy perennial herb with stem 30–90 cm; rhizomatous.

LEAVES: Opposite, 30–90 mm, ovate to lanceolate, often clasping, with 3–5 primary veins originating from the base of the leaf (not pinnately veined).

FLOWERS: Racemes, with pedicels 45–90 mm; calyx 15–22 cm, symmetrical, hairy, and lobes 4–6 mm; corollas pink with purple stripes, and tube + throat 20–53 mm.

FRUITS: Capsule 13–15 mm.

SIERRA DISTRIBUTION: West slope except around Lake Tahoe, Tulare County north.

ELEVATION: 1400–3500 m

HABITAT: Riparian areas and moist areas with seepage; mixed conifer through subalpine zones.

FLOWERING TIME: Jun–Aug

NOTES: Formerly included in *E. lewisii*. Easy to recognize; the only tall, perennial, pink monkeyflower found in wet areas in the montane zone and above. The only other species with the same distinctively veined leaves is *E. cardinalis*, which is also very easy to recognize by its bright red-orange flowers (hybrids do occur occasionally).

Erythranthe floribunda

Many-Flowered Monkeyflower

Phrymaceae (Lopseed family)

Hairy annual herb 5–20 cm (up to 40 cm or longer in good water years); slimy to the touch. Stems are erect to decumbent, or sometimes sprawling or climbing.

LEAVES: Opposite, petioles 1–12 mm; blades 8–25 mm, lanceolate to ovate, bases squared off or cordate, veins pinnate.

FLOWERS: Pedicels 5–20 mm; calyx hairy, 4–8 mm; lobes 0.8–2 mm, about equal in length, acute. Corolla yellow, tube + throat 5–12 mm, limb 3–4 mm wide, throat may or may not have red spots.

FRUITS: 4–7 mm.

SIERRA DISTRIBUTION: All, but until the records reflect only true *E. floribunda* with its tiny corollas, the distribution will not be clear (many records show plants with much larger corollas that cannot be *E. floribunda*).

ELEVATION: < 2500 m

HABITAT: Sandy washes, flats, wet areas by ponds and lakes, seasonal streams; foothill and montane zones, likely others.

FLOWERING TIME: Apr–Aug

NOTES: Formerly *Mimulus floribundus*. This plant has the smallest corolla of any yellow-flowered *Erythranthe* in the SN. The corolla is just downright tiny (4–8 mm wide looking down upon the limb)!

Erythranthe guttata

Seep-Spring Monkeyflower

Phrymaceae (Lopseed family)

Annual or rhizomatous perennial herb ranging in height 2–150 cm; extremely variable, hairless or hairy.

LEAVES: 0.5–12 cm, round to ovate with entire or toothed margins, sometimes pinnatifid. Upper leaves sessile or clasping stem, and lower leaves have long petioles.

FLOWERS: Inflorescence is a bracted raceme, usually with > 5 flowers. Pedicels 10–80 mm; calyx 6–30 mm, often red-dotted, and bell-shaped but becoming asymmetric when mature, with the 2 lower lobes curved upward. Corollas 10–40 mm, yellow, strongly 2-lipped, and lower lip usually has several smaller spots and a large, red spot.

FRUITS: Capsule 5–10 mm, indehiscent.

SIERRA DISTRIBUTION: All.

ELEVATION: < 4100 m

HABITAT: Wet to moist places in many plant communities, sometimes semiaquatic, forming floating mats; all zones.

FLOWERING TIME: May–Aug

NOTES: By far the most common and the most variable of the monkeyflowers.

Erythranthe laciniata ✿

Cut-Leaved Monkeyflower

Phrymaceae (Lopseed family)

Erect annual herb 3–38 cm; usually hairless, sometimes≈branched.

LEAVES: Petioles 0–35 mm; blades 3–55 mm; most leaves pinnately lobed with linear or broader lobes, and these sometimes also have lobes.

FLOWERS: 2–8 per raceme; pedicels 5–25 mm. Calyx 8–10 mm, swollen in fruit with lower 2 calyx lobes curved upward; corollas 7–15 mm, yellow, and lower lip usually with a large red spot.

FRUITS: Capsule 3–8 mm.

SIERRA DISTRIBUTION: Tulare to Butte Counties.

ELEVATION: > 900 m

HABITAT: Moist to wet, sandy or gravelly areas, often on or near rocky outcrops; foothill chaparral through alpine zones.

FLOWERING TIME: Apr–Aug

NOTES: In Plumas and Butte Counties, *E. filicifolia* also has pinnately lobed leaves, but there are at least 8 lobes on each side of the leaf. *Laciniate* means cut into deep, narrow, irregular lobes.

Erythranthe montioides

Montia-Like Monkeyflower

Phrymaceae (Lopseed family)

Delicate annual herb ranging in height 1–18 cm; puberulent.
LEAVES: Opposite, 4–25 mm, rarely wider than 2 mm.
FLOWERS: Pedicels 5–20 mm, calyx 5–7 mm, puberulent, lobes equal. Corollas 12–18 mm, yellow.
FRUITS: Capsules, 3–5 mm.
SIERRA DISTRIBUTION: West slope, Nevada County south.
ELEVATION: 1600–2900 m
HABITAT: Open areas, often along streams in coniferous forest; montane through subalpine zones.
FLOWERING TIME: Jun–Aug, just after snowmelt
NOTES: This plant is quite abundant in the Huntington Lake area of the Sierra National Forest, and in Sequoia & Kings Canyon National Parks.

Erythranthe moschata

Musk Monkeyflower

Phrymaceae (Lopseed family)

Erect or ascending perennial herb with stems 5–30 cm; rhizomatous, usually slimy to the touch due to dense, soft, glandular hairs (but may be hairless). The musky smell may be a myth.
LEAVES: Appearing crowded on stems, and longer than internodes; petioles 0–15 mm; blades 1–6 cm, with margins irregularly toothed or nearly entire.
FLOWERS: Pedicels 10–15 mm, longer than the calyx; calyx 8–12 mm, and calyx lobes 2–5 mm, equal. Corollas yellow; tube 2–4 mm wide, except at opening, mostly cylindric; tube + throat 15–26 mm, and floor deeply grooved. Many corollas have lines of stippled purple on the throat, as in the photo; this may not apply to all plants, but when you see it, it's a good character.
FRUITS: Capsule approximately 4–9 mm.
SIERRA DISTRIBUTION: Mostly west slope, Fresno County north.
ELEVATION: 1000–3200 m
HABITAT: Moist to wet areas; lower coniferous forest through subalpine zones.
FLOWERING TIME: Jul–Aug
NOTES: Common in wet areas above the foothill woodland/chaparral zone and below the alpine zone. Often forms sprawling masses in wet areas.

Erythranthe primuloides

Primrose Monkeyflower

Phrymaceae (Lopseed family)

Low-growing, mat-forming perennial herb with stem 0.5–12 cm; rhizomatous, forms bulblets in autumn.

LEAVES: Usually in distinct rosettes, 7–35 mm, oblong or obovate, usually with very long hairs on upper surface.

FLOWERS: Pedicels 10–120 mm, erect, and much longer than calyx; calyx 5–9 mm, hairless. Corollas yellow, tube + throat 8–20 mm, and there is a reddish spot at the base of each lower lip lobe.

FRUITS: Capsule 6–7 mm.

SIERRA DISTRIBUTION: All.

ELEVATION: 1200–3660 m

HABITAT: Wet areas along streams and in meadows; montane to alpine zones.

FLOWERING TIME: Jun–Aug

NOTES: One of the most common species of Sierra meadows, a pioneering species that quickly colonizes disturbed areas.

Erythranthe tilingii

Mountain Monkeyflower

Phrymaceae (Lopseed family)

Perennial herb 2–40 cm; erect or decumbent stems, hairless to somewhat hairy; spreads by rhizomes or stolons.

LEAVES: Blades 5–30 cm, broadly elliptical to round; upper leaves sessile, and lower leaves with short petioles.

FLOWERS: Pedicels 10–90 mm; calyx 7–25 mm, often red-dotted, with upper 2 lobes curving upward once fruits ripe. Corollas yellow, tube + throat 17–45 mm, and lower throat folds densely hairy and nearly closing mouth; lower lip has faint reddish spots or not.

FRUITS: Capsule 7–8 mm, indehiscent.

SIERRA DISTRIBUTION: All.

ELEVATION: 1400–3700 m

HABITAT: Wet areas; montane, subalpine, and alpine zones.

FLOWERING TIME: Jul–Sep

NOTES: Where ranges overlap, may intergrade with *E. guttata*. *E. tilingii* usually has ≤ 5 flowers and a solid yellow or more faintly red-dotted lower lip than *E. guttata*.

Mimetanthe pilosa

Shaggy False Monkeyflower

Phrymaceae (Lopseed family)

Densely shaggy-hairy annual herb 2–35 cm; erect to ascending.

LEAVES: Sessile, 10–30 mm, lanceolate to oblong.

FLOWERS: Pedicels 10–35 mm, elongating when plant is in fruit; calyx 6–7 mm, bell-shaped, with lobes 3–4 mm, unequal. Corollas yellow, tube + throat 7–8 mm.

FRUITS: 4–7 mm.

SIERRA DISTRIBUTION: All.

ELEVATION: < 2900 m

HABITAT: Moist, sandy or gravelly areas; all zones but alpine.

FLOWERING TIME: May–Sep

NOTES: Formerly *Mimulus pilosus*. The small, yellow flowers, bell-shaped calyx with extremely long calyx lobes, and long, soft, wavy white hairs distinguish this monkeyflower from all others. Sometimes seen in great numbers after fires.

Phytolacca americana var. *americana* (non-native)

American Pokeweed

Phytolaccaceae (Pokeweed family)

Erect to leaning perennial herb up to 3 m (and occasionally taller); robust with pink stems and large, irregularly shaped tubers.

LEAVES: Petioles 1–6 cm; blades 8–35 cm, 3–18 cm wide (sometimes larger), lanceolate or ovate.

FLOWERS: Open racemes, 10–30 cm, often drooping, with peduncle < 15 cm; pedicels 3–13 mm. Sepals 5, petal-like, white to pink; petals absent. Styles 6–12 (equal to number of carpels), slender; stamens 10.

FRUITS: Berry 6–11 mm wide, purple-black. Seed 1 per carpel (chamber), 2.5–3 mm, lenticular, black, shiny.

SIERRA DISTRIBUTION: West slope, Tulare County north.

ELEVATION: < 1200 m

HABITAT: Variable, from wet (growing in the middle of a flowing stream) to dry slopes and flats. Primarily in disturbed areas to date.

FLOWERING TIME: Aug–Dec

NOTES: American pokeweed is native to the eastern US but recently has been an aggressive invader in the SN along streams and disturbed sites. Pokeweed populations spread rapidly after tree mortality in the 2014–17 drought resulted in heavy equipment and crews moving contaminated soil from site to site. Birds and bears eat and spread the seeds as well. All parts of this plant are highly toxic, though in its native range very young leaves are cooked and eaten as poke salad (also polk salad, poke sallet). This plant inspired the song "Polk Salad Annie," by Tony Joe White.

A. letaleum -calyx lobes approx. equal
corolla base sac-like
upper calyx
lobe >> others
A. vexillocalyculatum ssp. intermedium

Antirrhinum leptaleum

Spurred Snapdragon

Plantaginaceae (Plantain family)

Erect, glandular-hairy annual herb 8–60 cm.

LEAVES: Lower leaves opposite, upper alternate; petioles 1–7 mm; blades 7–40 mm, tips obtuse.

FLOWERS: 1 per axil; pedicels 1–2 mm; calyx has glandular and nonglandular hairs, lobes 5, with the upper longest but not markedly unequal; corollas 7–14 mm, white to pale pink or lavender, with no contrasting veins.

FRUITS: Capsule spheric or ovoid, with 1–2 holes at tips of lower chambers, which allow seeds to exit. Seeds numerous, ± 1 mm, black with a netlike pattern.

SIERRA DISTRIBUTION: West slope, Kern to Butte Counties.

ELEVATION: < 2100 m

HABITAT: Burned areas, disturbed sites, roadsides; foothill and montane zones.

FLOWERING TIME: Jun–Aug

NOTES: Wiry snapdragon, *A. vexillocalyculatum* ssp. *intermedium* (inset), is 1–10 cm, with unequal calyx lobes, and pink corollas 10–14 mm, often with darker veins; the plant has large bulges on the lower lip closing the throat, and is found < 1400 m from Mariposa County north on the west slope, in loose gravelly areas, serpentine, and burned areas. Both are SN endemics.

Collinsia childii

Child's Blue-Eyed Mary

Plantaginaceae (Plantain family)

Glandular-hairy annual herb 8–35 cm.

LEAVES: Opposite, 1–4 cm, entire or toothed.

FLOWERS: Inflorescences open, not whorled, and densely glandular; pedicels longer than the calyx, with calyx lobes blunt-tipped and glandular; corollas 6–11 mm, and upper and lower lips equal, white or lavender. Filaments hairless.

FRUITS: Pedicels spreading horizontally or ascending. Capsule < 5 mm wide, with calyx lobes longer. Seeds 2.

SIERRA DISTRIBUTION: West slope, Mariposa County south.

ELEVATION: 1000–2200 m

HABITAT: Shaded areas in forests; mostly montane zone.

FLOWERING TIME: Apr–Jul

NOTES: The sap of *C. childii* produces stains on skin even more dramatically than *C. tinctoria*, and the color is a darker rusty hue than the iodine orange of *C. tinctoria*.

Collinsia heterophylla

Chinese Houses

Plantaginaceae (Plantain family)

Erect annual herb ranging in height 10–50 cm.

LEAVES: Opposite, lanceolate, with toothed margins; plain green without mottling.

FLOWERS: Inflorescences distinct whorls of purple-and-white flowers. Pedicels shorter than calyx, which is 5-lobed.

FRUITS: Capsule 5 mm.

SIERRA DISTRIBUTION: West slope, Butte and Nevada Counties south.

ELEVATION: < 1800 m

HABITAT: Shaded areas; foothill and montane zones.

FLOWERING TIME: Apr–Jun

NOTES: Distinguished from the similar tincture plant by the longer upper corolla lip, unmottled leaves, and earlier blooming time, and by having a short spur 1–2 mm at the base of the upper filaments. Does not stain skin orange-brown when crushed.

Collinsia sparsiflora var. *collina*

Few-Flowered Blue-Eyed Mary

Plantaginaceae (Plantain family)

Annual herb 5–20 cm.

LEAVES: Opposite, linear to oblong, with margins entire.

FLOWERS: Inflorescences open, not whorled, and hairless or slightly hairy, not glandular; flowers 1–3 per node; calyx lobes sharp-tipped. Corollas 5–8 mm, lavender to purple (sometimes white), and lowest lobe has sparse, long hairs near tip, with throat barely angled to the tube, and pouch at base of upper tube hidden by calyx.

FRUITS: Capsule spheric and red-blotched; fruiting calyx lobes much larger than capsule. Seeds 4–12, flat, round, and winged.

SIERRA DISTRIBUTION: All.

ELEVATION: < 1200 m

HABITAT: Dry slopes, roadsides; foothill chaparral and woodland; ponderosa pine forest; foothill and montane zones.

FLOWERING TIME: Mar–Apr

NOTES: *Collinsia sparsiflora* var. *sparsiflora* is also found in the SN. It has larger (7–20 mm) corollas, with the pouch at base of upper corolla tube visible beyond the calyx, and the corolla throat strongly angled to the tube.

Collinsia tinctoria

Tincture Plant

Plantaginaceae (Plantain family)

Erect annual herb ranging in height 20–60 cm.

LEAVES: Opposite, lanceolate-deltate, entire to toothed, with white mottling.

FLOWERS: Inflorescence is densely glandular-hairy; flowers in distinct, dense whorls; corollas 12–20 mm, with upper lip much shorter than lower lip.

FRUITS: Capsule 4 mm.

SIERRA DISTRIBUTION: West slope.

ELEVATION: < 1800 m

HABITAT: Shaded or partially shaded areas; foothill and montane zones.

FLOWERING TIME: Apr–Jun

NOTES: When the inflorescence is crushed, the sap stains skin orange-brown (iodinelike), hence the name "tincture plant." The mottled leaves, with later flowering time, and much shorter upper corolla lip distinguish this species from the similar *C. heterophylla*.

Collinsia torreyi var. *wrightii*

Wright's Blue-Eyed Mary

Plantaginaceae (Plantain family)

Glandular-hairy annual herb 5–25 cm, carpeting the ground in good years.

LEAVES: Opposite, 5x longer than wide, linear or elliptic.

FLOWERS: Inflorescence is open, not whorled, and densely glandular; pedicels longer than the calyx, and calyx lobes blunt-tipped and glandular. Corollas 4–6 mm, with upper lip shorter, and white; the lower lip has 3 lobes, and the center lobe is much smaller than the lateral lobes. Filaments hairless (upper pair may be hairy at base).

FRUITS: At least some pedicels reflexed (sometimes pedicel forms S-shape) in fruit. Capsule 3 mm. Seeds 2.

SIERRA DISTRIBUTION: Both slopes, all.

ELEVATION: 800–4000 m

HABITAT: Sandy or gravelly soil in conifer forest; montane, subalpine, and alpine zones.

FLOWERING TIME: May–Aug

NOTES: If the 3 lower corolla lobes are ± equal and the leaves are linear and > 6x longer than wide, with corollas 6–9 mm, you're looking at *C. torreyi* var. *torreyi*, which co-occurs and is also common throughout much of the range of *C. torreyi* var. *wrightii* up to 3000 m. Another similar plant occurring within the same geographical and elevation range is blue-eyed Mary, *C. parviflora*; plants are < 40 cm, generally hairless (though sometimes sparsely glandular), and, in contrast to *C. torreyi*, leaf margins are rolled under, calyx lobes are sharp-tipped (acute) rather than blunt-tipped, and corollas are 4–8 mm with white tube/throats rather than blue/purple.

Keckiella breviflora var. *breviflora*

Bush Penstemon

Plantaginaceae (Plantain family)

Drought-deciduous shrub 5–20 dm, hairless, glaucous.
LEAVES: Opposite, 10–40 mm, oblanceolate or lanceolate, serrate (sometimes entire).
FLOWERS: Panicles of white, cream, or pink-tinged flowers; calyx 4–8 mm; lobes 5, ± equal in size, glandular-hairy; corollas 12–18 mm, bilateral, with upper 2 lobes forming a hood, and lower 3 lobes reflexed, and rounded. Stamens 4, filaments densely hairy at base, anthers hairless, and staminode hairless at tip.
FRUITS: Capsule ovoid. Seeds many, irregularly angled.
SIERRA DISTRIBUTION: West slope, Butte and Sierra Counties south.
ELEVATION: < 2000 m
HABITAT: Chaparral, lower conifer forest; foothill and montane zones.
FLOWERING TIME: May–Jul
NOTES: *K. breviflora* var. *glabrisepala* has hairless calyces, grows up to 2800 m, and occurs in all SN counties, both slopes.

Linaria vulgaris (non-native, invasive)

Butter-and-Eggs, Yellow Toadflax

Plantaginaceae (Plantain family)

Erect to ascending perennial herb 3–10 dm; hairless.
LEAVES: Whorled or opposite, crowded, 2.5–5 cm, linear, entire, hairless or occasionally sparsely hairy.
FLOWERS: Inflorescence is a dense, glandular-hairy raceme; pedicels 1–4 mm; calyx lobes ± 3 mm, unequal; corollas 18–32 mm, yellow with orange-hairy lower lip.
FRUITS: Capsule 9–12 mm. Seed 1.5 mm, flat, winged.
SIERRA DISTRIBUTION: Fresno and Mono Counties north, but expected elsewhere.
ELEVATION: < 2150 m
HABITAT: Disturbed areas; montane and sagebrush zones, but expected in other vegetation types.
FLOWERING TIME: Jun–Aug
NOTES: *L. vulgaris* is listed by the Cal-IPC as a weed of "moderate" ecological impact at the state level. Its deep and aggressive root system, ability to spread from root fragments, and high production of seeds that remain viable up to 8 years make this plant a threat to wildlands. It is officially listed as a noxious weed in 46 states. The similar but even more damaging Dalmatian toadflax, *L. dalmatica* ssp. *dalmatica*, is taller (< 12 dm) and has wider leaves and larger (2–5 cm) flowers. Dalmatian toadflax is an A-rated noxious weed in CA and should be reported to the local county agriculture department.

Penstemon azureus var. *azureus*

Azure Penstemon

Plantaginaceae (Plantain family)

Hairless perennial herb 20–70 cm; woody at base, often glaucous.

LEAVES: Mostly cauline and opposite, 1–6.5 cm x 5–17 mm wide, lanceolate to oblanceolate; uppermost leaves widest at base and usually clasping stem.

FLOWERS: Inflorescence is hairless; calyx 3–6.5 mm, and lobes have short, pointed tips 0.2–1.5 mm; corollas 20–35 mm, blue to lavender, and hairless. Anther sacs 1.8–3.5 mm, dehiscing ½–⅔ from point of attachment to the filament; staminode hairless.

FRUITS: Capsule.

SIERRA DISTRIBUTION: Patchy distribution on west slope, from Tulare to Butte Counties; east slope, Nevada and Sierra Counties.

ELEVATION: 500–3100 m

HABITAT: Dry to moist open areas, often rocky or sandy; foothill and montane zones.

FLOWERING TIME: May–Aug

NOTES: Leaves in the photos look short and wide, but leaf shape and size are variable, and longer thinner leaves are common as well. Regardless, the uppermost (distal) leaves are sessile and clasping the stem. Var. *angustissimus* occurs < 800 m and has linear leaves 4–9 cm x 2–7.5 mm wide, longer (5–7.5 mm) calyx lobes with long, tapered tips 1.5–3.2 mm.

Penstemon davidsonii var. *davidsonii*

Davidson's Penstemon

Plantaginaceae (Plantain family)

Mat-forming shrub with stems < 1 dm.

LEAVES: Mostly basal, 5–30 mm, with stem leaves much smaller; leaves entire, hairless, green (not glaucous).

FLOWERS: Inflorescences glandular; calyx 7–13 mm; corollas 20–36 mm, blue-violet, with dense, white hairs on the inner floor of the corolla tube. Staminode included in tube, densely covered with pale yellow hairs.

FRUITS: Capsule.

SIERRA DISTRIBUTION: All.

ELEVATION: 2000–3750 m

HABITAT: Rocky areas; montane through subalpine zones.

FLOWERING TIME: Jul–Aug

NOTES: Think: huge flowers on a tiny, little shrub.

Penstemon deustus var. *suffrutescens*

Hot-Rock Penstemon

Plantaginaceae (Plantain family)

Erect subshrub < 4 dm; with stems bent horizontally at base, nearly hairless or glandular-hairy.

LEAVES: Stem leaves 8–50 mm, 7–28 mm wide, ovate to round, toothed.

FLOWERS: Inflorescence is glandular, of about 10 clusters of flowers; calyx 3.5–6.5 mm, with lobes slender and long-tapered; corollas 9–11 mm, white with a brown upper lip, which is shorter than the lower lip. Staminode bearded.

FRUITS: Capsule.

SIERRA DISTRIBUTION: Alpine County north.

ELEVATION: 600–2200 m

HABITAT: Forest floor, montane zone.

FLOWERING TIME: May–Jul

NOTES: The only mostly white-flowered penstemon; all others are pink, purple, or red. Two other varieties occur in the Sierra: Var. *deustus* has a white upper corolla lip, and var. *pedicellatus* has shorter calyx lobes and lanceolate rather than ovate-round leaves (see the Jepson eFlora for details).

Penstemon gracilentus

Slender Penstemon

Plantaginaceae (Plantain family)

Hairless perennial herb 25–65 cm; base is woody.

LEAVES: Basal and cauline, opposite, 40–100 mm, linear to narrowly lanceolate, entire.

FLOWERS: Inflorescences glandular; calyx 4–6 mm, with lobes lanceolate; corollas 15–20 mm, blue-purple, floor varies from hairless to shaggy-hairy. Anther sacs 1–1.3 mm, dehiscing ½ their length at proximal end (from point of attachment to filament); staminode yellow-hairy.

FRUITS: Capsule 5–7 mm.

SIERRA DISTRIBUTION: Both slopes, Calaveras County north.

ELEVATION: 900–2800 m

HABITAT: Dry areas; montane, subalpine, and sagebrush/pinyon-juniper zones.

FLOWERING TIME: Jun–Aug

NOTES: Compare with *P. heterodoxus*.

Penstemon grinnellii var. *scrophularioides*

Grinnell's Beardtongue

Plantaginaceae (Plantain family)

Erect perennial herb 4.5–8.5 dm; new growth is glaucous.
LEAVES: Thick, fleshy, 50–90 mm, lanceolate, tending to arch and to fold lengthwise; margins at least somewhat toothed.
FLOWERS: Inflorescences glandular; calyx 4.7–8.5 mm, lobes ovate; corollas 26–35 mm, blue-violet, with shape unique due to abrupt expansion of tube into wide throat, which measures 10–18 mm across when flattened. Corollas glandular inside and out. Staminode exserted, bearded with dense, yellow hairs.
FRUITS: Capsule 10–14 mm x 5–8 mm.
SIERRA DISTRIBUTION: Tulare and Inyo Counties south.
ELEVATION: 500–2850 m
HABITAT: Dry sites, rocky areas; foothill and montane zones.
FLOWERING TIME: Apr–Jul
NOTES: Another variety, var. *grinnellii*, also occurs in the southern Sierra. Plants are shorter (< 6 dm), leaves are not glaucous, and flowers are white, light pink, or lavender.

Penstemon heterodoxus var. *heterodoxus*

Sierra Penstemon

Plantaginaceae (Plantain family)

Erect perennial herb 5–20 cm; mat-forming.
LEAVES: Basal and cauline, 5–50 mm, 2–10 mm wide, entire, sometimes folding lengthwise, hairless.
FLOWERS: Inflorescence of 1–2 clusters; corollas 10–16 mm, glandular-hairy on outer surface, blue to purple. Staminode moderately covered with yellow-brown hairs.
FRUITS: Capsule.
SIERRA DISTRIBUTION: Fresno and Inyo Counties north.
ELEVATION: 1900–3200 m
HABITAT: Meadows, slopes, rocky areas; montane and subalpine zones.
FLOWERING TIME: Jul–Aug
NOTES: Compare with *P. rydbergii*. *P. heterodoxus* var. *cephalophorus* also occurs in the SN, a taller (> 20 dm) plant with 2–6 flower clusters.

Penstemon incertus

Mojave Beardtongue

Plantaginaceae (Plantain family)

Rounded shrub 20–100 cm, with young stems hairless and glaucous.

LEAVES: 40–70 mm, linear (sometimes narrowly lanceolate), with blades rolled inward; thick (leathery), longest at midstem.

FLOWERS: Inflorescences glandular; calyx 4.5–7.3 mm, with lobes ovate, glandular or hairless; corollas 23–32 mm, throat 8–12 mm wide when flattened, violet to purple with blue lobes; corollas glandular outside and hairless inside (but with small patch of hairs on floor). Anther sacs 1.8–2.4 mm, dehiscing full length; staminode densely hairy, not exserted.

FRUITS: Capsule 10–17 mm.

SIERRA DISTRIBUTION: Both slopes, Inyo, Kern, and Tulare Counties.

ELEVATION: 900–2300 m

HABITAT: Dry, open, sandy to rocky sites; sagebrush/pinyon-juniper zone.

FLOWERING TIME: May–Jun

NOTES: A shrub of the southern and eastern SN and the Mojave Desert.

Penstemon laetus var. *laetus*

Mountain Blue Penstemon

Plantaginaceae (Plantain family)

Subshrub 2–8 dm; herbage is hairy, and inflorescence is densely glandular.

LEAVES: Mostly on the stems (cauline); upper leaves 1.5–10 cm, 2–11 mm wide, linear to lanceolate, entire.

FLOWERS: Inflorescence hairs glandular, flowers 21–25 mm, calyx 4.7–9 mm. Anther sacs split open $\frac{2}{3}$–$\frac{3}{5}$ their length.

FRUITS: Capsule.

SIERRA DISTRIBUTION: All.

ELEVATION: < 2500 m

HABITAT: Dry soils or rocky areas; foothills and montane forests.

FLOWERING TIME: May–Jul

NOTES: This species is distinguished from similar-appearing blue/purple penstemons by its glandular-hairy inflorescence. The glands appear as tiny globules of moisture that could be thought of as luscious. Remembering "Luscious *laetus*" may help you to tell it apart from *P. azureus* and *P. speciosus*, neither of which have glandular inflorescences.

Penstemon newberryi var. *newberryi*

Mountain Pride

Plantaginaceae (Plantain family)

Mat-forming subshrub 1.2–3 dm.

LEAVES: Mostly basal, 10–40 cm; stem leaves smaller, obovate, finely serrate.

FLOWERS: Inflorescence hairs glandular; corollas 20–30 mm.

FRUITS: Capsule.

SIERRA DISTRIBUTION: All.

ELEVATION: 1000–3700 m

HABITAT: Rocky outcrops, cliffs, talus; montane through alpine zones.

FLOWERING TIME: Jun–Aug

NOTES: Unmistakable, this plant is easy to identify by simply looking at the photo. One of the characteristic beauties of the upper montane zone and above.

Penstemon parvulus

Small Penstemon

Plantaginaceae (Plantain family)

Loosely matted perennial herb 15–30 cm; woody at base, with herbage hairless and glaucous.

LEAVES: Opposite, 10–45 mm, lanceolate to ovate, sometimes barely serrate; uppermost leaves sessile with clasping bases.

FLOWERS: Inflorescences hairless; calyx 3–5 mm, lobes lanceolate to obovate; corollas 14–20 mm, blue or blue-violet, hairless. Anther sacs 1.4–2 mm, with inner margins hairy, dehiscing ½–⅔ their length from the point of attachment to the filament; staminode hairless.

FRUITS: Capsule.

SIERRA DISTRIBUTION: West slope, patchy distribution from Tulare to Nevada Counties.

ELEVATION: 400–3100 m

HABITAT: Open, rocky areas; foothill, montane, and subalpine zones.

FLOWERING TIME: Jun–Aug

NOTES: *P. parvulus* looks similar to *P. azureus* var. *azureus*, but it's a shorter plant with smaller flowers. In Northern CA, these taxa intergrade.

Penstemon rostriflorus

Bridge's Penstemon

Plantaginaceae (Plantain family)

Subshrub or robust perennial herb 30–100 cm; hairless or with stem that is finely hairy at base.

LEAVES: Opposite, 20–70 mm, linear to lanceolate, entire.

FLOWERS: Inflorescences glandular; calyx 4–8 mm, with lobes lanceolate; corollas 22–33 mm, lips long, with lower lip strongly reflexed and upper lip forming a hood over anthers. Anther sacs 1.8–2.5 mm.

FRUITS: Capsule.

SIERRA DISTRIBUTION: Both slopes, Alpine County south.

ELEVATION: 500–3500 m

HABITAT: Dry areas; montane zone, sagebrush up through subalpine zones.

FLOWERING TIME: Jun–Aug

NOTES: Formerly *Penstemon bridgesii*. Unmistakable, bright red penstemon.

Penstemon rydbergii var. *oreocharis*

Mountain Blue Penstemon

Plantaginaceae (Plantain family)

Erect perennial herb 2–6 dm, mostly hairless.

LEAVES: Basal and cauline, oblanceolate, entire, with upper stem leaves smaller.

FLOWERS: Inflorescences of 1–6 many-flowered clusters, mostly hairless but when hairs present, nonglandular. Calyx 3–5 mm; corollas 9–14 mm, hairless except inner floor of tube, blue to purple. Anther sacs dehisce full length; staminode densely covered with yellow hairs.

FRUITS: Capsule.

SIERRA DISTRIBUTION: Fresno and Inyo Counties north.

ELEVATION: 1000–3600 m

HABITAT: Dry edges and fully moist areas of meadows and streambanks, rocky areas; montane through alpine zones.

FLOWERING TIME: May–Aug

NOTES: Compare with the similar *P. heterodoxus*, which has glandular-hairy inflorescences including outer corolla surfaces. It may help to remember that the *r* in *rydbergii* is "rid" of hairs, (mostly hairless), and the *h* in *heterodoxus* is much more "hairy."

Penstemon speciosus

Showy Penstemon

Plantaginaceae (Plantain family)

Perennial herb 0.5–6 dm; stems with short hairs.

LEAVES: Basal leaves form a mat, upper stem leaves 35–90 mm, lanceolate, entire, tending to fold lengthwise.

FLOWERS: Inflorescences hairless or with short hairs (occasionally glandular). Calyx 6–13 mm, with lobes lanceolate or ovate; corollas 25–37 mm, 7–13 mm wide (when pressed flat), white inside, and hairless. Anther sacs dehisce full length; staminode hairless or with yellow or orange hairs along 1–7 mm of tip.

FRUITS: Capsule.

SIERRA DISTRIBUTION: All.

ELEVATION: 850–3800 m

HABITAT: Dry areas in many habitats; montane, sagebrush/pinyon-juniper, and subalpine zones.

FLOWERING TIME: May–Aug

NOTES: Flowers large and extra showy, often dwarfing the foliage.

Plantago erecta

California Dwarf Plantain

Plantaginaceae (Plantain family)

Hairy annual herb 3–30 cm.

LEAVES: Basal 3–13 cm, narrowly oblanceolate to linear.

FLOWERS: Peduncle (inflorescence stalk in this case appearing to be the stem) 3–30 cm, tipped by a spike 0.5–3 cm. Calyx deeply 4-lobed; corollas tubular, with 4 membranous corolla lobes 2–2.7 mm, spreading or reflexed. Stigma hairy; stamens not conspicuously exserted.

FRUITS: Capsule circumscissile. Seeds 2.

SIERRA DISTRIBUTION: West slope foothills, all counties.

ELEVATION: < 1200 m

HABITAT: Sandy, gravelly, clay or serpentine areas, grassy openings; foothill and montane zones.

FLOWERING TIME: Mar–May

NOTES: This little native plantain is often overlooked because it blends in so well with its background, but it is quite exquisite upon close examination with a 10x hand lens.

Plantago lanceolata (non-native)

English Plantain, Buckhorn Plantain

Plantaginaceae (Plantain family)

Erect perennial herb < 80 cm; from fibrous-haired caudexes or taproots.

LEAVES: Basal 5–25 cm, lanceolate to lance-oblong, parallel-veined.

FLOWERS: Peduncle (inflorescence stalk in this case appearing to be the stem) 20–80 cm, grooved, and tipped by a dense spike 2–8 cm. Calyx deeply 4-lobed, with outer lobes united more than halfway up; corollas tubular, with 4 membranous corolla lobes 2–2.5 mm. Stigma exserted, hairy; stamens exserted and obvious when shedding pollen.

FRUITS: Capsules circumscissile. Seeds 1–2.

SIERRA DISTRIBUTION: All.

ELEVATION: < 2600 m

HABITAT: Disturbed areas, sometimes in undisturbed sites; all zones.

FLOWERING TIME: Apr–Aug

NOTES: This tall plantain, originally from Europe, is a common garden and orchard weed, as well as a weed of wild areas. Being wind-pollinated, plants produce copious amounts of pollen, which for many people causes allergic reactions.

Plantago major (non-native)

Common Plantain

Plantaginaceae (Plantain family)

Perennial (sometimes annual) herb < 60 cm; from fibrous-haired caudexes.

LEAVES: Basal 5–18 cm, heart-shaped or widely elliptic, parallel-veined.

FLOWERS: Peduncle (inflorescence stalk in this case appearing to be the stem) 5–60 cm, tipped by an elongated, dense spike 3–20 cm. Calyx deeply 4-lobed; corollas tubular, with 4 membranous corolla lobes 0.5 mm. Stigma exserted, hairy; stamens exserted.

FRUITS: Capsules circumscissile. Seeds 5–16.

SIERRA DISTRIBUTION: All.

ELEVATION: < 2600 m

HABITAT: Moist, disturbed areas; all zones.

FLOWERING TIME: Apr–Sep

NOTES: *Plantago major* is originally from Europe. As with *P. lanceolata*, its pollen is a moderate allergen. The leaves and other parts of the plant have been used in herbal healing for centuries.

Veronica americana

American Brooklime

Plantaginaceae (Plantain family)

Hairless perennial herb 5–60 cm; from rhizomes, with stem decumbent, branched, rooting at nodes in wet soil.
LEAVES: Opposite; blades petioled, 5–50 mm, lanceolate to ovate, toothed.
FLOWERS: Racemes of 10–25 flowers each in leaf axils; bracts linear, pedicels 5–10 mm; sepals 4, ± 3 mm, oblanceolate; corollas 7–10 mm, slightly bilateral, with upper lobe the largest, purplish-blue with dark lines. Style 1.5–3 mm, unlobed; stamens 2.
FRUITS: 3–4 mm, ± round, slightly if at all notched.
SIERRA DISTRIBUTION: All.
ELEVATION: < 3000 m
HABITAT: Moist and wet sites, streamsides, meadows, lakeshores; montane, subalpine, and sagebrush zones.
FLOWERING TIME: May–Aug
NOTES: Submerged plants are sometimes observed fully underwater and with no flowers. This aquatic morph can be quite confusing; the leaves are much larger than the partially submerged or terrestrial plants, and they can withstand swiftly flowing water (e.g., below Friant Dam in Fresno County).

Veronica serpyllifolia ssp. *humifusa*

Thyme-Leaved Veronica

Plantaginaceae (Plantain family)

Perennial herb 5–30 cm; arising from rhizomes, with stem decumbent, ± hairy.
LEAVES: Opposite, sessile, 12–25 mm, elliptic to ovate, entire (sometimes crenate).
FLOWERS: Terminal raceme, glandular-hairy, with pedicels 5–10 mm; sepals 4, 2.5–4 mm, oblanceolate; corollas 6–7 mm, slightly bilateral, with the upper lobe the largest, bright blue with dark lines. Style 2–3 mm; stamens 2.
FRUITS: 2.8–3.7 mm, ± heart-shaped, wider than long.
SIERRA DISTRIBUTION: All.
ELEVATION: < 3200 m
HABITAT: Moist and wet sites, streamsides, meadows, lakeshores; montane, subalpine, and sagebrush zones.
FLOWERING TIME: Apr–Aug
NOTES: The glandular-hairy terminal (not axillary) racemes help to distinguish this species.

Veronica wormskjoldii

American Alpine Speedwell

Plantaginaceae (Plantain family)

Perennial herb 10–25 cm; from rhizomes, with stem mostly erect but rooting at nodes, with long, wavy hairs.

LEAVES: Opposite (upper leaves sometimes alternate), sessile, 20–40 mm, lanceolate to elliptic, entire or crenate.

FLOWERS: Glandular-hairy, racemes terminal, with pedicels 2–6 mm; sepals 4, 3.5–5.5 mm, lanceolate; corollas 6–10 mm, slightly bilateral, with upper lobe the largest, deep purple or blue. Style 0.8–1.3 mm, unlobed; stamens 2.

FRUITS: 4.5–6.5 mm, ± heart-shaped, longer than wide.

SIERRA DISTRIBUTION: Tulare and Inyo Counties north.

ELEVATION: 1500–3500 m

HABITAT: Moist sites, streamsides, meadows, lakeshores; subalpine and alpine zones.

FLOWERING TIME: Jun–Aug

NOTES: This plant of the high subalpine and low alpine zones is characterized by its long, wavy hairs and the deep blue or violet color of the flowers. The glands of the inflorescence can be dense enough to give the plant a sticky feel.

Allophyllum divaricatum

Purple False Gily-Flower

Polemoniaceae (Phlox family)

Erect to sprawling annual herb < 60 cm, with glandular hairs. Skunky odor.

LEAVES: Alternate; lower leaves pinnately 3- to 13-lobed, and stem leaves deeply 3-lobed, with the middle lobe the longest; widest leaflets or lobes > 4 mm wide.

FLOWERS: Inflorescences clusters of 2–8 flowers; corollas 8–22 mm, with tube red-violet; lobes 2–4 mm, pink. Stamens unequal, and often the longest one exserted.

FRUITS: Pedicels elongate as fruits mature, and the calyx membrane splits as the spherical fruit expands. Seeds 1–3 per chamber, brown.

SIERRA DISTRIBUTION: All.

ELEVATION: < 2200 m

HABITAT: Dry, open areas, frequently in disturbed areas including newly burned sites; foothill and montane zones.

FLOWERING TIME: Apr–Jun

NOTES: The glandular stem hairs are long-stalked. *A. gilioides* looks similar but has short-stalked stem glands and leaves or leaf lobes < 4 mm wide.

Allophyllum integrifolium

White False Gily-Flower

Polemoniaceae (Phlox family)

Annual herb 5–25 cm; with stems conspicuously soft glandular-hairy.

LEAVES: Lower leaves entire to toothed or irregularly lobed, and the upper 3-lobed, with the widest lobes 5–15 mm wide.

FLOWERS: Inflorescences open, with flowers in clusters of 2–4; corollas 6–11 mm, white or pale blue.

FRUITS: Pedicels elongate as fruits mature. Seeds 1–2 per chamber, brown.

SIERRA DISTRIBUTION: All.

ELEVATION: 1200–2700 m

HABITAT: Dry to moist openings; montane and subalpine zones.

FLOWERING TIME: Jun–Aug

NOTES: The only white *Allophyllum* in the Sierra.

Collomia grandiflora

Grand Collomia

Polemoniaceae (Phlox family)

Erect annual herb 1–10 dm; sometimes branched toward top of plant.

LEAVES: Basal leaves toothed; stem leaves lanceolate to linear with entire margins. Upper leaf surface hairless, and lower surface glaucous; glandular.

FLOWERS: In terminal heads of sessile, cream- to salmon-colored flowers. Calyx 7–10 mm, corollas 15–30 mm. Pollen blue.

FRUITS: Capsule ellipsoid. Seeds 2–3 per chamber.

SIERRA DISTRIBUTION: All.

ELEVATION: < 2900 m

HABITAT: Dry, sunny or shady areas, roadsides; foothill and montane zones.

FLOWERING TIME: May–Jul

NOTES: A distinctive *Collomia*, with its salmon-colored corollas; not easily confused with any other species.

Collomia heterophylla
Variable-Leaf Collomia
Polemoniaceae (Phlox family)

Erect to decumbent, many-branched annual herb < 20 cm; glandular-hairy, often clammy to the touch.

LEAVES: Alternate (occasionally opposite below); lower leaves lobed to pinnately dissected, with lobes variable in shape and depth, and upper leaves entire or have some lobes.

FLOWERS: Inflorescences terminal and in leaf axils, flowers sessile (no pedicels). Calyx 5–8 mm; corollas 10–14 mm, pink or white with yellow tube. Filaments unequal.

FRUITS: Capsule ellipsoid. Seeds 2–3 per chamber.

SIERRA DISTRIBUTION: West slope, Tulare County north.

ELEVATION: < 2000 m

HABITAT: Open and partially shaded areas, abundant after wildfire; foothill and montane zones.

FLOWERING TIME: Apr–Jun

NOTES: This species has highly variable, lobed leaves, often on the same plant. Most likely to be confused with *C. tinctoria* or *C. linearis*, but both of these have entire leaves. Also compare with *Microsteris gracilis*.

Collomia linearis
Narrow-Leaved Collomia
Polemoniaceae (Phlox family)

Erect, simple or branched annual herb 1–6 dm; internodes congested, and glandular-hairy toward top of plant.

LEAVES: Basal leaves entire or toothed, and cauline leaves sessile, 15–60 mm, linear to lanceolate, entire. Lower leaves on stem glandular, upper hairless.

FLOWERS: Sessile in bracted head-like clusters of 7–20 at top of plant. Calyx 4–7 mm; corollas 8–15 mm, white or pink. Stigma and stamens included; pollen white.

FRUITS: Capsule approximately 5 mm.

SIERRA DISTRIBUTION: Both slopes, Tulare and Inyo Counties north.

ELEVATION: 1000–3700 m

HABITAT: Open areas; all zones except alpine.

FLOWERING TIME: May–Jul

NOTES: Compare with *C. heterophylla*, which is much more branched and has at least some lobed stem leaves, as well as having a shorter stature. Also compare with *C. tinctoria*, which has blue pollen and axillary flowers.

Collomia rawsoniana

Rawson's Flaming Trumpet

Polemoniaceae (Phlox family)

Perennial herb with leafy stems to 4 dm; rhizomatous, forming large, sometimes clonal patches; herbage is glandular-hairy. Skunky odor.

LEAVES: Alternate, 3–8 cm, ovate to elliptic, toothed.

FLOWERS: Terminal clusters of 3–7, with calyx 8–12 mm; corollas 25–40 mm, red-orange. Stamens and stigma exserted; pollen blue.

FRUITS: Capsule ellipsoid. Seed 1 seed per chamber, released explosively.

SIERRA DISTRIBUTION: Madera County.

ELEVATION: 750–2200 m

HABITAT: Moist areas along streams and meadow edges in coniferous forest; montane zone.

FLOWERING TIME: Jun–Sep

NOTES: Endemic to the San Joaquin River watershed upslope of Bass Lake and Mammoth Pool in Madera County. However, within its small range it can be found growing abundantly along streams. Recent field studies indicate that genetic variability is relatively high for a rare plant, and that sexual reproduction is occurring adequately. Conservation measures have been in place since the late 1970s, and the Sierra National Forest manages *C. rawsoniana* as a Species of Conservation Concern. Primarily hummingbird-pollinated.

Collomia tinctoria

Staining Collomia

Polemoniaceae (Phlox family)

Many-branched annual herb 2–8 cm; branches ± equal; stems have long hairs tipped by translucent glands.

LEAVES: Basal and stem leaves sessile, 15–60 mm, linear to lanceolate, entire, glandular; upper leaves longer than upper flowers.

FLOWERS: Clusters of 2–3 flowers in leaf axils. Calyx 5–7 mm, with lobes awnlike; corollas 8–14 mm, narrow, with tube maroon to purple, lobes pink. Stigma and stamens exserted; pollen blue.

FRUITS: Capsule. Seed 1 per chamber.

SIERRA DISTRIBUTION: Both sides of crest, Tulare County north.

ELEVATION: 600–3000 m

HABITAT: Dry, open gravelly and rocky areas; foothill through subalpine zones.

FLOWERING TIME: Jun–Jul

NOTES: Thrives after low- to moderate-intensity wildfire.

Eriastrum densifolium ssp. *austromontanum*

Southern Mountain Eriastrum

Polemoniaceae (Phlox family)

Erect or spreading perennial herb; herbage is nearly hairless to woolly.

LEAVES: Alternate, 10–50 mm, with 2–15 linear lobes.

FLOWERS: Inflorescences terminal (occasionally axillary as well), of 10–20 flowers, with bracts 5- to 9-lobed. Calyx lobes unequal, white-woolly; corollas 16–18 mm, funnel-shaped, blue, with tube puberulent, and lobes 5–7 mm. Style with 3 stigmas, exserted; stamens 5–6 mm, white, exserted, with anthers sagittate.

FRUITS: Capsule 3-chambered. Seeds 1 to several per chamber.

SIERRA DISTRIBUTION: Inyo, Kern, and Tulare Counties.

ELEVATION: 1300–3000 m

HABITAT: Dry, open areas; foothill woodland, chaparral, lower coniferous forest; foothill and montane zones.

FLOWERING TIME: Jun–Aug

NOTES: The genus name *Eriastrum* derives from Greek and means "woolly star."

Eriastrum pluriflorum ssp. *pluriflorum*

Many-Flowered Eriastrum

Polemoniaceae (Phlox family)

Erect annual herb 2–32 cm, not usually glandular.

LEAVES: Alternate, 20–50 mm, entire or with up to 10 threadlike lobes; hairless to woolly.

FLOWERS: Inflorescences woolly; corollas 16–24 mm, salverform, with tube 8–16 mm, white, yellow, or blue; lobes 5–6 mm, with tube + throat generally > 2x the lobes, bright blue or lavender. Stamens 3–5 mm, exserted, usually equal in length and attached at or below sinuses of corolla lobes.

FRUITS: Capsule.

SIERRA DISTRIBUTION: West slope, Madera County south.

ELEVATION: < 2000 m

HABITAT: Open, grassy areas; foothill and montane zones.

FLOWERING TIME: May–Jul

NOTES: *Pluriflorum* means "many-flowered"; and heads can have up to 50 flowers.

Gilia capitata ssp. *pedemontana*

Blue Gilia, Ball-Headed Gilia

Polemoniaceae (Phlox family)

Erect, branched annual herb 3–9 dm; hairless or glandular-hairy.
LEAVES: Lower leaves 1–2 times pinnately lobed, with lobes 5–20 mm x < 1 mm wide; upper leaves smaller.
FLOWERS: Inflorescence is a dense spherical head of 25–100 white to blue flowers; calyx 3–5.2 mm (larger in fruit), densely hairy, with lobes recurved; corollas 7–11 mm, with tube exserted from calyx, light blue-violet, lobes ± 2 mm wide.
FRUITS: Capsule 3–4.5 mm, ovoid.
SIERRA DISTRIBUTION: West slope, Tulare County north.
ELEVATION: < 2000 m
HABITAT: Open rocky areas; foothill and lower montane zones.
FLOWERING TIME: Mar–May
NOTES: There are several other sspp. of *Gilia capitata* in the SN. To hone your skills at distinguishing them, I recommend using the *Gilia* key in the Jepson eFlora, with a dissecting microscope.

Gilia tricolor ssp. *diffusa*

Bird's-Eye Gilia

Polemoniaceae (Phlox family)

Branched annual herb 1.3–8 dm.
LEAVES: Basal, 1–2 times pinnately lobed, and upper leaves palmately lobed, with lobes linear.
FLOWERS: Inflorescences in open clusters with long pedicels 10–40 mm, with black small-stalked glands. Calyx 3–4 mm; corollas 7–13 mm, with tube yellow, throat with 2 purple spots per lobe, lobes 3–6 mm wide, blue-violet toward tips, paler at base.
FRUITS: Capsule 3–6 mm, ovoid, smaller than the calyx.
SIERRA DISTRIBUTION: West slope, all.
ELEVATION: < 2000 m
HABITAT: Open grassy areas; foothill and lower montane zones.
FLOWERING TIME: Mar–Jul
NOTES: Intergrades with and difficult to distinguish from ssp. *tricolor*, which has dense clusters of flowers on pedicels 1–5 mm and larger (10–19 mm) corollas. In Tulare and Kern Counties on the west slope 700–1700 m grows the similar-looking inland gilia, *G. interior*; it is shorter (6–13 cm), with tufted-woolly leaves and corollas 6–11 mm.

Ipomopsis aggregata ssp. *aggregata*

Scarlet Gilia

Polemoniaceae (Phlox family)

Erect biennial or short-lived perennial 3–8 dm; hairless or glandular-hairy; each plant flowers only once.

LEAVES: Alternate; basal 3–5 cm with 9–11 pinnate lobes, withered by flowering time; stem leaves 5–7 times pinnately lobed, and lobes have acute tips.

FLOWERS: Lateral clusters of 3–7; calyx lobes 3–4 mm, and corolla tube 20–30 mm; lobes red or pink with yellow mottling on throat. Pollen yellow or white (occasionally light blue).

FRUITS: Capsule.

SIERRA DISTRIBUTION: Tulare and Inyo Counties north.

ELEVATION: 1100–3500 m

HABITAT: Open, sunny areas; montane through subalpine zones.

FLOWERING TIME: Jun–Sep

NOTES: Similar to and found growing with *I. aggregata* ssp. *bridgesii*, which has leaves with blunt-tipped lobes, flowers 1–5 per cluster, calyx lobes 1–2 mm, pollen always blue. Ssp. *bridgesii* is also found further to the south in Kern County.

Ipomopsis congesta ssp. *montana*

Globe Gilia

Polemoniaceae (Phlox family)

Clump-forming perennial herb 1–2 dm.

LEAVES: Alternate, 1–4 cm, palmately 3- to 5-lobed.

FLOWERS: Terminal, headlike inflorescences. Calyx 3–5 mm, corolla tube < 10 mm, yellowish; lobes 5–10 mm, white.

FRUITS: Capsule.

SIERRA DISTRIBUTION: West slope, Tuolumne County north; east slope, Inyo County north.

ELEVATION: 1500–3700 m

HABITAT: Open sandy or rocky areas; montane through alpine zones.

FLOWERING TIME: May–Sep

NOTES: Similar to *I. congesta* ssp. *congesta*, which has pinnately lobed leaves.

Leptosiphon bicolor

Baby Stars

Polemoniaceae (Phlox family)

Erect or ascending annual herb 2–21 cm; sparsely hairy.
LEAVES: Opposite, 3–13 mm, with 5–7 linear lobes.
FLOWERS: Inflorescences in dense heads, flowers 1–2 open per head, opening midmorning and closing by evening. Flower total length 15–30 mm; calyx 6–11 mm; corolla tube 12–32 mm, hairy, with yellow throat; corolla lobes < 4 mm, white to pink, usually with no basal spot.
FRUITS: Capsule.
SIERRA DISTRIBUTION: West slope.
ELEVATION: < 1700 m
HABITAT: Open grassy areas; foothill and lower montane zones.
FLOWERING TIME: Mar–Jun
NOTES: Compare with *L. ciliatus*. The corolla tube of *L. bicolor* is extraordinarily long in comparison.

Leptosiphon chrysanthus ssp. *chrysanthus*

Golden Desert-Trumpet

Polemoniaceae (Phlox family)

Annual herb; with threadlike stem, hairless or hairy, sometimes glandular.
LEAVES: Opposite, lobes 3–6 mm, linear.
FLOWERS: 1 per inflorescence, with peduncle 5–15 mm, threadlike. Sepals 5, 4–6 mm, fused at base, with membrane wider than lobes (ribs); corollas funnel-shaped, tube 3–5 mm, with hairy ring inside and out; corolla lobes 5, 5–7 mm, bright yellow, with throat brighter yellow or sometimes maroon. Stigma lobes 3, 3–4 mm, exserted; filaments hairless, attached above hairy ring.
FRUITS: Capsule.
SIERRA DISTRIBUTION: Inyo and Kern Counties.
ELEVATION: < 2000 m
HABITAT: Dry slopes and flats in pinyon pine-juniper woodlands; pinyon-juniper zone.
FLOWERING TIME: Mar–Jun
NOTES: Formerly *Leptosiphon aureus* ssp. *aureus*. Primarily a plant of the desert, but reaches into the southern SN.

Leptosiphon ciliatus
Whisker Brush
Polemoniaceae (Phlox family)

Diminutive, hairy annual herb 2–30 cm; simple or branched at base, hairy.

LEAVES: Opposite, with 5–11 linear lobes 5–20 mm.

FLOWERS: Inflorescences in dense heads with deeply lobed bracts edged with white hairs; flower total length 25–35 mm. Calyx 7–10 mm; corolla tube 12–25 mm, hairy, with yellow throat; corolla lobes 2–4 mm, pink with spot at base, rounded or squared off.

FRUITS: Capsule.

SIERRA DISTRIBUTION: All.

ELEVATION: < 3000 m

HABITAT: Sandy, gravelly, and rocky areas, as well as grassy areas; foothill through subalpine zones.

FLOWERING TIME: Mar–Aug

NOTES: Large specimens could be confused with small-flowered variants of *L. montanus,* but the corolla tube should always be < 25 mm (usually much less) in *L. ciliatus*, and the corolla lobes will be shorter (2–4 mm rather than 5–8 mm).

Leptosiphon filipes
Thread Linanthus
Polemoniaceae (Phlox family)

Intricately branched annual herb 5–20 cm; often with reddish stems.

LEAVES: Opposite, with linear lobes 3–7 mm.

FLOWERS: Peduncle 4–12 mm, slender, with 1 flower. Calyx lobes hairy, membrane the same size as the lobes; corollas funnel-shaped, with the tube hairy inside, and throat yellow with white or pink lobes, each 2–3 mm. Stigmas exserted; stamens hairy at base.

FRUITS: Capsule.

SIERRA DISTRIBUTION: West slope.

ELEVATION: < 1300 m

HABITAT: Dry, open areas; foothill and lower montane zones.

FLOWERING TIME: Apr–Jun

NOTES: The species name *filipes* derives from the Latin root *fili-*, meaning "thread," and refers to the extremely slender, threadlike stems on this delicate plant.

Leptosiphon montanus

Mustang Clover

Polemoniaceae (Phlox family)

Erect, hairy annual herb 10–60 cm; exceedingly variable in flower color and size.

LEAVES: Opposite, with 5–11 linear lobes 20–30 mm.

FLOWERS: Inflorescences in dense heads, with deeply lobed bracts edged with white hairs; flower total length 25–35 mm. Calyx 10 mm; corolla tube 25–30 mm, glandular hairy, throat yellow; corolla lobes 5–8 mm, white to pink, often with purple spot at base.

FRUITS: Capsule.

SIERRA DISTRIBUTION: West slope, Placer County south.

ELEVATION: < 1700 m

HABITAT: Dry slopes and flats; foothill and montane zones.

FLOWERING TIME: Apr–Jun

NOTES: One of the primary showy wildflowers of roadsides and foothill slopes in the spring, forming large masses of white or pink, or even sometimes magenta.

Leptosiphon pachyphyllus

Bush-Gilia

Polemoniaceae (Phlox family)

Hairy perennial herb or subshrub 1–2 dm.

LEAVES: Opposite, with 5–9 flattened linear to oblanceolate lobes 10–15 mm.

FLOWERS: In clusters. Calyx 9–10 mm; corolla tube 12–15 mm, longer than calyx, hairy, with throat white-yellow; corolla lobes 6–9 mm.

FRUITS: Capsule. Seeds > 2 mm.

SIERRA DISTRIBUTION: Both slopes, Kern and Inyo to Sierra Counties.

ELEVATION: 1700–3100 m

HABITAT: Open areas; upper montane, subalpine, and sagebrush zones.

FLOWERING TIME: Jun–Sep

NOTES: Similar to *L. nuttallii* ssp. *pubescens*, which has corolla tubes shorter than the calyx, linear leaves, and is mostly in the high desert east of the Sierra.

Linanthus dichotomus ssp. *dichotomus*

Evening Snow

Polemoniaceae (Phlox family)

Hairless annual herb 5–20 cm; glaucous.
LEAVES: 10–22 mm, linear, with 3–7 lobes.
FLOWERS: Corollas funnel-shaped, with tube 7–10 mm, purple; throat white with lobes 10–16 mm, white within, and purplish shading on outer margins.
FRUITS: Capsule obovoid, smaller than the calyx.
SIERRA DISTRIBUTION: All.
ELEVATION: < 1800 m
HABITAT: Open rocky, gravelly or sandy areas; foothill and lower montane zones.
FLOWERING TIME: Apr–Jun
NOTES: Flowers remain closed during the day and open at dusk. The plants are amazingly camouflaged during the day, then showy and unmissable at dusk.

Linanthus parryae

Sandblossoms

Polemoniaceae (Phlox family)

Decumbent or erect annual herb with stem 2–10 cm; hidden by leaves, glandular-hairy.
LEAVES: Opposite, crowded, with linear lobes 5–15 mm.
FLOWERS: Sessile, with inflorescence bracts crowded; calyx 6–8 mm; corollas funnel-shaped, white or blue-purple, with tube 1 mm, throat 1 mm, and lobes 8–12 mm, with 1–2 purple spots at base.
FRUITS: Capsule. Seeds 18–36, gelatinous when wet.
SIERRA DISTRIBUTION: Both slopes, Kern County; a few sites in Inyo County reach the SN, but most are farther to the east.
ELEVATION: < 2000 m
HABITAT: Open sandy flats; foothill and pinyon-juniper zones.
FLOWERING TIME: Mar–May
NOTES: Sometimes found in mixed populations of white-and-blue-flowered plants.

Linanthus pungens ssp. *pulchriflorus*

Granite Gilia

Polemoniaceae (Phlox family)

Low-growing, perennial herb to subshrub 1–3 dm; woody at base, hairs both glandular and simple.

LEAVES: Alternate (lower leaves sometimes opposite), palmate, lobes 3–13, spine-tipped.

FLOWERS: Corollas 15–25 mm, funnel-shaped, tube + throat 7–15 mm, lobes 7–10 mm, rounded, white to pink.

FRUITS: Capsule obovoid and slender.

SIERRA DISTRIBUTION: All, distribution scattered.

ELEVATION: 1700–4000 m

HABITAT: Open rocky, gravelly or sandy areas; montane through alpine zones.

FLOWERING TIME: Apr–Jun

NOTES: Formerly *Leptodactylon pungens*. Could be confused with *Phlox diffusa*, but *P. diffusa* has simple, opposite leaves. Flowers open in the evening.

Loeseliastrum matthewsii

Desert Calico

Polemoniaceae (Phlox family)

Soft-hairy annual herb; stem erect.

LEAVES: Alternate, simple, 1–4 cm, toothed, and the teeth have bristles; longest blades oblanceolate.

FLOWERS: Inflorescence bracts leaflike; pedicels glandular; calyx $\frac{1}{2}$–$\frac{3}{4}$ the length of the corolla tube, and calyx lobes bristle-tipped. Corollas 11–21 mm, bilaterally symmetrical, white to rose-purple, with upper lip 5–11 mm, 3-lobed, $\frac{3}{4}$ the length of the corolla tube or longer than the tube; lobes have bright maroon arches and a white blotch, and inward-pointing projections on either side of the middle lobe; lower lip 4–7 mm, 2-lobed, lobes truncate, 3-toothed.

FRUITS: Capsule ovoid.

SIERRA DISTRIBUTION: East slope, Kern and Inyo Counties.

ELEVATION: < 1800 m

HABITAT: Open rocky, gravelly or sandy areas; montane through alpine zones.

FLOWERING TIME: Mar–Jul

NOTES: Formerly *Langloisia matthewsii*. A plant of the desert that sneaks up the east slope of the Sierra in Kern and Inyo Counties.

Microsteris gracilis
Slender Phlox
Polemoniaceae (Phlox family)

Erect, glandular-hairy annual herb < 20 cm; varying from 1 simple stem to a bushy, many-branched plant.
LEAVES: Opposite, sometimes alternate above, 10–30 mm.
FLOWERS: Occur in pairs in upper leaf axils. Calyx 5–10 mm; corolla tube 8–12 mm, yellowish; lobes 1–2 mm, squared or notched, bright pink or whitish. Style included in tube.
FRUITS: Capsule. Seeds gelatinous when wet.
SIERRA DISTRIBUTION: All.
ELEVATION: < 3300 m
HABITAT: Open, often damp, areas; foothill through subalpine zones.
FLOWERING TIME: Mar–Aug
NOTES: Formerly *Phlox gracilis*.

Navarretia intertexta
Needleleaf Navarretia
Polemoniaceae (Phlox family)

Prickly, branched annual herb with stem 8–28 cm; often wider than tall, with branches ascending, white-hairy.
LEAVES: Lower leaves opposite, upper alternate, usually deeply 1–2 times pinnately lobed.
FLOWERS: Inflorescence is headlike, with bracts 10–20 mm, pinnately lobed, longer than the flowers, white-hairy like the calyx; calyx lobes 4–5, entire or 3-pronged; corollas 6–8 mm, white or light blue, with lobes ovate. Style and stamens exserted.
FRUITS: Capsule, dehiscing when wet. Seeds brown with pitted surface.
SIERRA DISTRIBUTION: West slope, most counties.
ELEVATION: < 2100 m
HABITAT: Open, vernally wet areas, vernal pools, meadows, grassy sites; foothill and montane zones.
FLOWERING TIME: May–Jul
NOTES: Formerly *N. intertexta* ssp. *intertexta*. The taxon formerly known as *N. intertexta* ssp. *propinqua* has now been separated out as *N. propinqua*; it has stamens that are level with the corolla lobes rather than exserted from them.

Navarretia leptalea ssp. *bicolor*

Bridges' Gilia

Polemoniaceae (Phlox family)

Erect annual herb 4–26 cm; with primary branches spreading, ± hairless to glandular (glands black).

LEAVES: Alternate, linear to narrowly elliptic, and entire (lower leaves sometimes pinnately lobed).

FLOWERS: Open, with pedicels long and threadlike; calyx 5-lobed, tips acuminate; corollas 8–15 mm, throat 2–5 mm, < 2x lobes, tube and throat yellow. Stamens 5, the longest exserted from throat along style.

FRUITS: Capsule, dehiscing from tip.

SIERRA DISTRIBUTION: Both slopes, Inyo and Fresno Counties north.

ELEVATION: 1500–3100 m

HABITAT: Open rocky or gravelly areas; montane and subalpine zones.

FLOWERING TIME: Jun–Sep

NOTES: *N. leptalea* ssp. *leptalea* co-occurs at elevations 900–2100 m (both slopes, Tulare and Inyo Counties north) and differs by having longer (13–21 mm) corollas with purple throats 6–8 mm, and throat 2–3x lobes; the bloom time is Jun–Aug. Difficult to see in the photos, but present in both sspp. are minute, stalked glands on the outside of the corolla tube and throat.

Navarretia pubescens

Downy Pincushion Plant

Polemoniaceae (Phlox family)

Prickly, erect annual herb 14–33 cm; 1 primary axis (not branched from base); stem is tan to reddish-brown, and internodes are usually shorter than leaves.

LEAVES: Lower leaves opposite, upper alternate, and bipinnately lobed, with lobes linear and lower lobes hairless. Upper leaves puberulent to hairy, with tips linear or wider and many-toothed.

FLOWERS: Inflorescences headlike; bracts below each flower pinnately lobed, the outer bracts longer (1.5–2x head), and inner bracts shorter with clasping bases, tip toothed. Bracts and calyx glandular-hairy; calyx lobes 2-toothed; corollas 10–16 mm, with throat reddish-purple, lobes blue-purple, and undersides of lobes glandular. Style and stamens exserted.

FRUITS: Capsule ovoid, 1-chambered, dehiscing below middle.

SIERRA DISTRIBUTION: West slope, most counties.

ELEVATION: < 1850 m

HABITAT: Open areas on gravel, sand, clay; foothill and montane zones.

FLOWERING TIME: May–Jul

NOTES: Look for the glandular hairs on the calyces and inflorescence bracts, positioned at middle with nonglandular hairs toward the tips (a 10x hand lens helps for this).

Phlox condensata

Cushion Phlox

Polemoniaceae (Phlox family)

Cushion-forming, alpine perennial herb; woody at base.

LEAVES: Opposite, 3–5 mm, lanceolate, unlobed, ciliate.

FLOWERS: Calyx 5–6 mm, with glandular hairs; corollas trumpet-shaped, white or light pink; tube 8–10 mm, lobes 4–6 mm.

FRUITS: Capsule.

SIERRA DISTRIBUTION: Both slopes, Tulare and Inyo to El Dorado Counties.

ELEVATION: 2000–4000 m

HABITAT: Dry, exposed rocky areas, fellfields, talus slopes; mostly alpine zone.

FLOWERING TIME: Jun–Aug

NOTES: Formerly *Phlox caespitosa* var. *condensata*. A great example of an alpine cushion plant.

Phlox diffusa

Spreading Phlox

Polemoniaceae (Phlox family)

Mat-forming perennial herb 1–3 dm; rhizomatous, with woody base.

LEAVES: Opposite, 10–15 mm, linear, unlobed, hairless except for woolly base.

FLOWERS: Calyx 8–10 mm, hairs not glandular; corollas trumpet-shaped, white, blue, or pink; tube 9–13 mm, lobes 6–7 mm. Style 3–5 mm.

FRUITS: Capsule.

SIERRA DISTRIBUTION: All.

ELEVATION: 1100–3600 m

HABITAT: Open rocky, gravelly or sandy areas; montane through subalpine (occasionally alpine) zones.

FLOWERING TIME: May–Aug

NOTES: Compare with *Linanthus pungens*, which has lobed leaves and funnel-shaped rather than trumpet-shaped corollas.

Phlox speciosa

Showy Phlox

Polemoniaceae (Phlox family)

Erect, branched perennial herb 2–4 dm.

LEAVES: Opposite, 30–50 mm, narrowly lanceolate; upper leaves have glandular hairs.

FLOWERS: Calyx 7–10 mm, glandular-hairy, with short, sharp lobes; corollas trumpet-shaped, pink (sometimes white); tube 10–15 mm, white; lobes pink, notched. Style 2–4 mm, included in tube.

FRUITS: Capsule.

SIERRA DISTRIBUTION: Tulare County north (not in Mono or Inyo Counties).

ELEVATION: < 1850 m

HABITAT: Forested areas, roadsides, rocky areas; foothill, montane, and sagebrush zones.

FLOWERING TIME: Apr–Jun

NOTES: This unique, bubblegum-pink flower is difficult to confuse with any other species.

Polemonium californicum

Low Polemonium

Polemoniaceae (Phlox family)

Glandular-pubescent perennial herb 1–2.5 dm; rhizomatous, with stems solitary to clumped.

LEAVES: Alternate, pinnately compound, all < 15 cm, sometimes larger at base of plant. Leaflets 9–25, 10–25 mm, 5–15 mm wide, entire; terminal 3 sometimes fused at base.

FLOWERS: Open inflorescence of 5–25 flowers, ± same level as the tallest leaves. Calyx 6–8 mm; corollas 7–15 mm diameter, bell-shaped.

FRUITS: 2–4 mm, 2–3 mm wide. Seeds 6–10, brown.

SIERRA DISTRIBUTION: Fresno County north.

ELEVATION: 1600–3100 m

HABITAT: Open or shaded sites; montane and subalpine zones.

FLOWERING TIME: Jun–Aug

NOTES: Can interbreed with *P. pulcherrimum* in the High Sierra, so if you find yourself confused it could be because you have encountered a hybrid.

Polemonium eximium

Sky Pilot

Polemoniaceae (Phlox family)

Tufted, glandular-hairy perennial herb 1–4 dm; rhizomatous. Musky odor.

LEAVES: Mostly basal, 4–13 cm, with cauline smaller and alternate. Leaves pinnately compound with 20–23 leaflets, each with 3–5 deep divisions.

FLOWERS: Inflorescences headlike and composed of many flowers; calyx 5–10 mm, corollas 12–15 m.

FRUITS: Capsule 5 mm x 3 mm wide. Seeds < 6, brown.

SIERRA DISTRIBUTION: Tulare and Inyo Counties to Tuolumne County.

ELEVATION: 2564–3640 m

HABITAT: Dry rocky slopes and ridges; alpine zone.

FLOWERING TIME: Jul–Aug

NOTES: Sky pilot is unmistakable, one of the endemic gems of the central High Sierra, and must be earned by hiking to some of the higher elevations of the range. An old tradition was to pick a stem of sky pilot as a souvenir; this is not ecologically sustainable nor permissible on public lands. A photograph is far preferable!

Polemonium occidentale

Western Polemonium

Polemoniaceae (Phlox family)

Tall, erect perennial herb 4–10 dm; from short rhizomes; hairless or glandular-pubescent above, with stems usually solitary.

LEAVES: Cauline only (not basal), alternate, 6–40 cm, 1–9 cm wide, hairless, with larger leaves toward base of stem. Leaves pinnately compound, usually with 20+ leaflets 8–35 mm, lanceolate, entire.

FLOWERS: Inflorescences of 10–35 flowers; calyx 4–5 mm, lobes larger than the tube; corollas 8–15 mm, bell-shaped, purple or light blue; tube blue.

FRUITS: Capsule 3–5 mm.

SIERRA DISTRIBUTION: Fresno County north.

ELEVATION: 900–3300 m

HABITAT: Moist areas; montane and subalpine zones.

FLOWERING TIME: Jun–Aug

NOTES: The tallest *Polemonium*, and the only SN species with hairless leaves.

Polemonium pulcherrimum var. *pulcherrimum*

Showy Polemonium

Polemoniaceae (Phlox family)

Clumped, glandular-hairy perennial herb 1–2 dm; from short rhizomes.

LEAVES: Alternate, pinnately compound, with basal leaves < 6 cm, 5–10 mm wide, and cauline much smaller. Leaflets 9–22, 3–4 mm, ovate-round, and terminal 3 not fused at base.

FLOWERS: Inflorescences dense, of 4–5 flowers, above the level of the tallest leaves; calyx 4–5 mm, with lobes nearly the same size as the tube; corollas 5–8 mm, 5–11 mm wide, rotate to bell-shaped, lobes blue-purple.

FRUITS: Capsule 3–4 mm, 2 mm wide. Seeds ≤ 3, brown or black.

SIERRA DISTRIBUTION: Sierra crest, Fresno County north.

ELEVATION: 2400–2700 m

HABITAT: Dry rocky areas; subalpine and alpine zones.

FLOWERING TIME: Jun–Aug

NOTES: Compare with *P. californicum*.

Rhinotropis cornuta var. *cornuta*

Sierra Milkwort

Polygalaceae (Milkwort family)

Subshrub 6–9 dm; rhizomatous, with many slender stems.

LEAVES: Alternate, 1–6.5 cm, ovate to nearly linear, with no stipules, margins entire.

FLOWERS: Racemes of pealike flowers 8.5–14 mm, with sepals and petals all pinkish; sepals 5, the inner (lateral) 2 enlarged and petal-like; petals 3, ± united and enclosing the 3 stamens, the lower ones boatlike, with a short beak.

FRUITS: Capsule 5.9–10 mm. Seed 4.5–7.3 mm, black, hairy, with a hairy white aril (a projection on the seed, often a reward for ants).

SIERRA DISTRIBUTION: Fresno and Inyo Counties north.

ELEVATION: 150–1830 m

HABITAT: Open, dry or moist areas; foothill and montane zones.

FLOWERING TIME: May–Sep

NOTES: Formerly *Polygala cornuta* var. *cornuta*. South of Tuolumne County, this species is only known from a few scattered localities; it is much more common in the northern Sierra.

Aconogonon davisiae

Davis' Knotweed

Polygonaceae (Buckwheat family)

Perennial herb 15–40 cm; ascending or erect, branching from base, with herbage hairless to scabrous.

LEAVES: Alternate, 0.5–3 cm, with ocrea brown, hairless. Petioles short (0.2–0.6 mm); blades 2.2–5 cm x 1.0–2.2 cm wide, varying in shape from lanceolate to ovate, with truncate or cordate bases; glaucous.

FLOWERS: Axillary inflorescences of 1–5 flowers. Perianth 2–4 mm; lobes 5, bell-shaped, petal-like, greenish-yellow or pinkish-white, the outer 2 smaller than the inner 3. Styles 3, stigmas headlike; stamens 8.

FRUITS: 3.5–8 mm, ovoid, 3-angled, yellow-brown, shiny.

SIERRA DISTRIBUTION: Both slopes, primarily west slope, from Madera County north.

ELEVATION: 1500–3000 m

HABITAT: Talus slopes and meadows, streamsides, wet areas of snow accumulation; upper montane through alpine zones, also sagebrush zone.

FLOWERING TIME: Jun–Aug

NOTES: Formerly *Polygonum davisiae*.

Aconogonon phytolaccifolium

Pokeweed Knotweed

Polygonaceae (Buckwheat family)

Erect perennial herb 70–150 cm; hairless or hairy.

LEAVES: Alternate, 1–3 cm, ocrea brown. Petioles 5–20 mm; blades 5–15 cm, 1.4–7.5 cm wide, lanceolate.

FLOWERS: Primary inflorescences terminal and paniclelike, with some smaller inflorescences in upper axils. Peduncles 0.5–10 cm; flowers in groups of 1–3, with pedicels 1–4 mm, perianth 2.5–3.5 mm, white to greenish.

FRUITS: 4–7.5 mm x 2.5–4.5 mm wide, ovoid, exserted, yellow-brown, and shiny.

SIERRA DISTRIBUTION: Primarily west slope, Fresno County north.

ELEVATION: 1500–2800 m

HABITAT: Meadows, streamsides, moist rocky areas; montane through subalpine zones.

FLOWERING TIME: Jun–Aug

NOTES: Formerly *Polygonum phytolaccifolium*. Ironically, the non-native pokeweed, *Phytolacca americana* var. *americana*, whose similarity of leaf shape and arrangement inspired the species name of this plant, is spreading in CA.

Bistorta bistortoides

Western Bistort

Polygonaceae (Buckwheat family)

Erect perennial herb 20–70 cm; from contorted rhizomes, with stems unbranched.

LEAVES: Mostly basal, with brownish ocrea (sheath composed of fused stipules) forming a cylinder 10–25 mm on the stem at the leaf base. Blades 4–30 cm x 1–4.8 cm wide, elliptic to oblanceolate, hairless or sparsely hairy and glaucous below, hairless above, with prominent midvein. Stem leaves 2–4, smaller and sessile toward top, larger and petioled toward base.

FLOWERS: Inflorescences 20–40 mm x 10–25 mm wide, ovoid or short-cylindrical; flowers bisexual, on short pedicels 2–10 mm. Perianth parts 5, fused at base, with perianth 4–5 mm, bell-shaped, white or light pink. Stamens 5–8, exserted.

FRUITS: Achenes 3–4 mm, ovoid, shiny, brownish.

SIERRA DISTRIBUTION: Both slopes, all.

ELEVATION: 1500–3000 m

HABITAT: Meadows, streamsides, wet areas; montane through alpine zones, and sagebrush zone.

FLOWERING TIME: Apr–Jul

NOTES: Common and distinctive, easy to recognize. The species name refers to the twisted rhizome; *bi-* means "two," and *tort* refers to the contorted shape. It is a host plant for many species of butterflies and moths.

Chorizanthe membranacea

Pink Spineflower

Polygonaceae (Buckwheat family)

Erect hairy annual herb 1–6 dm.
LEAVES: Basal and alternate, 1.5–5 cm, ± linear, woolly below.
FLOWERS: Inflorescence bracts in 2s or whorls of 3–5, with awns 0.5–1 mm; involucre 3–4 mm, 6-ribbed, with 5 membranous-margined teeth 0.5–1 mm; awns 0.7–1.5 mm at tips of teeth and hooked. Flowers 1–2 per involucre, 2.5–3 mm, with 6-lobed perianth of petal-like sepals that are hairy, and white or pink. Stamens 9.
FRUITS: 2.5–3 mm, elliptic.
SIERRA DISTRIBUTION: All west slope foothill counties.
ELEVATION: < 1400 m
HABITAT: Open rocky, sandy, grassy sites; foothill zone.
FLOWERING TIME: Apr–Jul
NOTES: The gray and light pink of this spineflower might initially cause it to blend into the background of drying grasses or the sandy soil of its habitat, but upon close inspection this is a delicate, beautiful plant blooming in late spring and early summer.

Chorizanthe xanti var. *xanti*

Pinyon Spineflower

Polygonaceae (Buckwheat family)

Erect, branched annual herb 0.5–2.5 dm; hairy.
LEAVES: Basal and alternate, 0.3–1.5 cm, 0.3–0.8 mm wide, woolly below.
FLOWERS: Inflorescence bracts 2, with awns 0.5–1 mm; involucre 3–4.5 mm, 6-ribbed, with 6 teeth, each 0.7–1.5 mm; awns at tip of teeth and hooked. Flower 1 per involucre, with involucre 4.5–6 mm, sparsely hairy; perianth 6-lobed; sepals petal-like, hairy, rose-pink to reddish. Stamens 9.
FRUITS: 4–4.5 mm, elliptic.
SIERRA DISTRIBUTION: Tulare and Kern Counties.
ELEVATION: < 1600 m
HABITAT: Sandy and gravelly sites; foothill, montane, and sagebrush zones.
FLOWERING TIME: Apr–Jul
NOTES: Look for this plant in the spring along Highway 178 through the Kern River canyon, and in the vicinity of Lake Isabella.

Eriogonum fasciculatum var. *polifolium*

California Buckwheat

Polygonaceae (Buckwheat family)

Subshrub to shrub 2–5 dm x 2–20 dm wide; spreading to rounded in overall outline; stem tomentose or hairy, rarely hairless.

LEAVES: Cauline, in alternate clusters (fasciculate); petioles 0.1–0.3 cm; blades 0.6–1.5 mm x 0.05–0.4 mm wide, oblanceolate, white tomentose to canescent on both surfaces. Upper leaf surface sometimes hairless or nearly so, and green; margins usually plane but sometimes rolled under.

FLOWERS: Inflorescences cymose to cymose-umbellate, compact to open; involucres 2.5–3.5 mm x 0.2–0.6 mm wide, top-shaped to bell-shaped, and canescent; perianths without stipes, canescent.

FRUITS: Achenes 1.8–2.5 mm, hairless.

SIERRA DISTRIBUTION: Both slopes, Kern and Tulare Counties; occasionally east slope, Inyo County (most Inyo County locations are just east of the Sierra).

ELEVATION: 300–2500 m

HABITAT: Dry, often sandy sites; foothill, sagebrush, and pinyon-juniper zones.

FLOWERING TIME: Year-round

NOTES: *E. fasciculatum* var. *fasciculatum*, a coastal variety, has been planted outside its native range along many federal highways in the SN in an attempt to use "native" plant species for erosion control. This variety differs by its hairless upper leaf surfaces, involucres, and corollas. Within its native range, primarily Southern CA coast and desert, it occurs < 300 m, but planted swaths along highways occur higher.

Eriogonum incanum

Frosted Wild Buckwheat

Polygonaceae (Buckwheat family)

Mat-forming perennial herb < 2 dm; mat 0.5–4 dm wide, and woolly; female and male flowers are on separate plants (dioecious).

LEAVES: Basal, with blades 0.5–1.5 cm, 0.3–0.7 cm wide, woolly.

FLOWERS: Inflorescences 0.5–2 cm wide, headlike when young, becoming umbellate in age; involucres 2.5–3 mm, woolly, with 5–8 teeth. Flowers hairless, bright yellow, narrowed at base into a stipe 0.5–1 mm; female flower 4–6 mm, often reddish in fruit, perianth lobes oblanceolate, and male flowers 2–3 mm, lobes ovate.

FRUITS: 3–3.5 mm, hairless with pubescent tips.

SIERRA DISTRIBUTION: Inyo and Fresno Counties to Sierra County.

ELEVATION: 2300–4000 m

HABITAT: Dry sandy, gravelly, or rocky sites; subalpine, alpine, and sagebrush zones.

FLOWERING TIME: Jun–Sep

NOTES: Extremely common in rocky and gravelly flats of subalpine zone. A close look helps to determine that the plants are different sexes. The green hairstreak, Pacific dotted blue, and Gorgon copper butterflies all use this as a food plant.

Eriogonum lobbii

Lobb's Wild Buckwheat

Polygonaceae (Buckwheat family)

Decumbent perennial herb 0.3–3 dm x 1–2.5 dm diameter; sometimes mat-forming, tomentose.

LEAVES: Basal, with blades 1–4 cm, round, tomentose on lower surface, and less so or sometimes hairless on upper surface.

FLOWERS: Inflorescences 1–4 cm x 1.4 cm wide, round, headlike, branches tomentose; involucres 5–10 mm, generally tomentose.

FRUITS: 3–3.5 mm, hairless with pubescent tips.

SIERRA DISTRIBUTION: Inyo and Fresno Counties to Sierra County.

ELEVATION: 2300–4000 m

HABITAT: Dry sandy, gravelly, or rocky sites; subalpine, alpine, and sagebrush zones.

FLOWERING TIME: Jun–Sep

NOTES: Extremely common in rocky and gravelly flats of subalpine zone. A close look helps to determine that the plants are different sexes. The green hairstreak, Pacific dotted blue, and Gorgon copper butterflies all use this as a food plant.

Eriogonum nudum var. *nudum*

Nude Wild Buckwheat

Polygonaceae (Buckwheat family)

Erect perennial herb 3–10 dm; usually hairless.

LEAVES: Basal, with blades 1–5 cm, 0.5–2 cm wide, green and hairless or sparsely hairy above, woolly below.

FLOWERS: Inflorescences 20–50 cm, with branches hairless; involucres 2–5 per cluster, 3–5 mm, hairless or slightly hairy. Flowers 2–3 mm, hairless, white (rarely yellow).

FRUITS: 1.5–3.5 mm, hairless.

SIERRA DISTRIBUTION: All.

ELEVATION: < 3000 m

HABITAT: Dry soil, sandy or gravelly sites, grassland, rocky sites; foothill, montane, subalpine, alpine, and sagebrush zones.

FLOWERING TIME: Jun–Sep

NOTES: Nude wild buckwheat is represented by 8 varieties in the SN. Several are very rare and would be unlikely to be encountered unless you were looking specifically for them. Var. *deductum* is more or less the high-elevation replacement of var. *nudum* found in the alpine zone; however, var. *nudum* has been documented as high as 3000 m.

Eriogonum ovalifolium var. *nivale*

Sierra Cushion Wild Buckwheat

Polygonaceae (Buckwheat family)

Mat-forming, cushionlike perennial herb < 5 cm, 0.5–3 dm wide; woolly.

LEAVES: Basal, with blades 0.2–0.8 mm, narrow to roundish, woolly, and edges sometimes brownish.

FLOWERS: Inflorescences 1–2.5 cm wide, headlike; involucres 2–4, 3–4.5 mm, woolly to hairy; flowers hairless, white or pinkish, with 3 outer lobes, 2–4 mm, and inner lobes 3–7 mm, thinner; bright yellow to reddish, lobes equal, bases not narrowed to stipes.

FRUITS: 2–3 mm, hairless.

SIERRA DISTRIBUTION: Both slopes, Tulare and Inyo Counties to Nevada County.

ELEVATION: 1700–4200 m

HABITAT: Dry sandy, gravelly sites; subalpine, alpine, and sagebrush zones.

FLOWERING TIME: Jun–Sep

NOTES: A common cushion plant of the alpine zone, look for it on windswept rides, talus slopes, and alpine fellfields. A good place to see it is the Dana Plateau.

Eriogonum rosense var. *rosense*

Mount Rose Wild Buckwheat

Polygonaceae (Buckwheat family)

Mat-forming perennial herb < 1 dm, 0.5–5 dm wide; woolly, glandular.

LEAVES: Basal, with blades 0.4–1.5 cm, 0.25–0.6 cm wide, woolly, glandular.

FLOWERS: Inflorescences < 1.5 cm wide, headlike; involucres 3–6, 3–5 mm, rigid, with teeth 5–8, and glandular. Flowers 2–3 mm, hairless or glandular, bright yellow; lobes equal, bases not narrowed to stipes.

FRUITS: 1.5–2.5 mm, hairless.

SIERRA DISTRIBUTION: Inyo and Fresno Counties to Sierra County.

ELEVATION: 2300–4000 m

HABITAT: Dry sandy, gravelly, or rocky sites; subalpine, alpine, and sagebrush zones.

FLOWERING TIME: Jul–Sep

NOTES: This species can be grown as a rock garden plant (it is an alpine plant, so it requires special conditions). Investigate local native plant sales or contact CNPS to learn more.

Eriogonum roseum

Wand Wild Buckwheat

Polygonaceae (Buckwheat family)

Erect, branched annual herb 1–8 dm; stems sparsely to densely woolly, grayish to brownish or reddish.

LEAVES: Basal and cauline, 1–3 cm x 0.5–1 cm wide, obovate or oblong, and short-woolly.

FLOWERS: Inflorescences 10–70 cm, flowers in globelike clusters directly on stalk; hairless. Perianth parts 6, 1.5–2 mm, white, pink, or yellow.

FRUITS: 1.8–2 mm, 3-angled, hairless.

SIERRA DISTRIBUTION: Primarily west slope, Butte County south.

ELEVATION: < 2200 m

HABITAT: Dry sandy and gravelly sites; foothill, montane, and sagebrush zones.

FLOWERING TIME: May–Nov

NOTES: Wand wild buckwheat is appropriately named, as the inflorescence branches are usually long, straight, and wandlike. This is a late-summer- and fall-blooming plant, often forming massive fields or solid roadside displays. The Kawaiisu people of the southern SN's Tehachapi region use the seeds of this buckwheat for food, as do other Native Californians.

Eriogonum saxatile

Rock Buckwheat

Polygonaceae (Buckwheat family)

Perennial herb 2–4 dm; forming mats 0.5–2 dm diameter, with stem hairy to tomentose.

LEAVES: Mostly basal, with blades 1–2.5 cm, round to oblanceolate, and white-tomentose.

FLOWERS: Inflorescences 20–25 cm, with branches hairy to tomentose; bracts 3–4, 1.5–7 mm, triangular; no peduncles; involucres 3–4 mm, tomentose; teeth 5–6, 0.8–1.5 mm, erect. Flowers 5–7 mm, with base narrowing to a winged, stipelike base above the joint with the pedicels. Perianth light yellow, white, or pink and hairless; the outer 3 perianth parts 3–5 mm, oblanceolate to lanceolate, and narrower, and the inner 3 parts 4–6 mm, obovate, and wider. Stamens 2.5–5 mm, filaments pilose at base.

FRUITS: 1.5–3.5 mm, hairless.

SIERRA DISTRIBUTION: Southern Sierra, Fresno, Inyo, Tulare, and Kern Counties.

ELEVATION: 800–3400 m

HABITAT: Rocky outcrops and cliffs, volcanic flats, sandy or gravelly soil; foothill, montane, and sagebrush zones.

FLOWERING TIME: May–Oct

NOTES: The striking, white-woolly leaves are bound to attract attention to this handsome plant. A 10x magnifier should show the pleated, stipelike corolla base and the pilose filament bases.

Eriogonum spergulinum var. *spergulinum*

Spurry Wild Buckwheat

Polygonaceae (Buckwheat family)

Erect, branched annual herb 1–4 dm; internodes glandular.

LEAVES: Basal and cauline, 1–3 cm, linear, short-hairy.

FLOWERS: 1 per involucre, hairless; sepals 6, 2.5–2.5 mm wide, petal-like (no petals), white or pinkish, often with pink midrib.

FRUITS: 1.5–2.3 mm, elliptic, hairless.

SIERRA DISTRIBUTION: Tulare and Inyo Counties north.

ELEVATION: 1200–3000 m

HABITAT: Dry sandy or rocky sites; montane, subalpine, and sagebrush zones.

FLOWERING TIME: Jun–Sep

NOTES: *E. spergulinum* var. *reddingianum* looks nearly identical and is more abundant than var. *spergulinum*. It ranges 1300–3400 m in the same habitat. These 2 are difficult to distinguish, but var. *reddingianum* flowers can be hairless or hairy, and are slightly smaller (1.5–2.5 mm). All Kern, Butte, and Lassen County plants should be var. *reddingianum*, based on current distribution information, but this is subject to change if new verified collections of var. *spergulinum* are confirmed in these counties. Var. *pratense* is an uncommon plant restricted to high elevations > 2800 m in Tulare and Inyo Counties; it is prostrate to ascending, with hairy flowers < 2 mm.

Eriogonum umbellatum var. *furcosum*

Sierra Nevada Sulphur Flower

Polygonaceae (Buckwheat family)

Subshrub 3–6 dm x 3–8 dm wide; stem sparsely hairy to hairless.

LEAVES: Basal, with blades 1–2.5 cm, 0.3–0.8 cm wide, hairless or somewhat woolly on upper surface and tomentose below, margins flat (not wavy).

FLOWERS: Compound umbels, ± hairless; involucre tubes 2–3 mm; teeth (lobes) 1.5–3 mm; flowers 6–8 mm, hairless, bright yellow, with a long narrowed basal tube (stipe) ± 2 mm.

FRUITS: 2–7 mm, hairless with hairy tip.

SIERRA DISTRIBUTION: All (no records yet for Mono County).

ELEVATION: 1200–3000 m

HABITAT: Dry sandy or rocky sites; montane, subalpine, and sagebrush zones.

FLOWERING TIME: Jun–Sep

NOTES: *E. umbellatum* is extremely variable, evidenced by its classification into 41 varieties across western North America. Of the 7+ varieties of *E. umbellatum* in the SN, var. *furcosum* is the most common compound umbellate taxon. The distinguishing features for the species, regardless of variety: It is a perennial or subshrub with umbellate or compound-umbellate inflorescences with leafy bracts at base, hairless bisexual flowers with basal stipes, and involucres with reflexed lobes at least as large as the tube.

Eriogonum ursinum var. *ursinum*

Bear Valley Wild Buckwheat

Polygonaceae (Buckwheat family)

Mat-forming perennial herb 0.4–4 dm; stems thinly tomentose to nearly hairless.

LEAVES: Basal, with blades 0.8–1.4 cm x 0.5–1.2 cm wide, ovate, densely white-tomentose below, and greenish and sparsely tomentose to nearly hairless above.

FLOWERS: Compound umbels, with bracts at base of primary 1° umbel (proximal) 3–8, 0.5–1.5 cm x 0.3–1 cm wide, lanceolate; bracts of 2° umbels (distal) are smaller, 1–5 mm, positioned directly below involucres (not halfway down stalk). Involucres turbinate, hairy, with teeth much smaller than the tube, erect. Flowers 4–6 mm, including narrowed, 0.5–0.8 mm stipe above the joint with the pedicel; perianth hairless, pale yellow or cream, not tinged red or pink. Stamens 3–5 mm, exserted; filaments hairy proximally.

FRUITS: 3–3.5 mm, 3-angled, hairless except pubescent beak.

SIERRA DISTRIBUTION: West slope, northern El Dorado County north.

ELEVATION: 900–2500 m

HABITAT: Dry sandy, gravelly, or rocky sites; foothill, montane, and sagebrush zones.

FLOWERING TIME: May–Sep

NOTES: Common in the northern Sierra, forming large mats with showy inflorescences on leafless stalks. May intergrade with the rare var. *erubescens*, which has been observed in the northern Sierra occasionally. Var. *erubescens* differs most obviously in that the 2° umbel bracts are about halfway down the stalk from the 2° umbels, and they are linear and brown or tan, not green and leaflike.

Eriogonum wrightii var. *subscaposum*

Short-Stemmed Bastard Sage

Polygonaceae (Buckwheat family)

Mat-forming subshrub 0.5–3 dm, up to 3 dm wide; usually woolly but sometimes hairless.

LEAVES: Mostly cauline; blades 0.5–1 cm, oblanceolate to elliptic, woolly to glabrous, margins sometimes curled under.

FLOWERS: Inflorescences 5–20 cm, branched; flowers 2–3 mm, white-pink, held within an involucre 1.5–4 mm, varying from woolly to hairless.

FRUITS: 2–2.5 mm.

SIERRA DISTRIBUTION: All.

ELEVATION: 200–3400 m

HABITAT: Dry sandy, gravelly, or rocky sites; montane, subalpine, and sagebrush/pinyon-juniper zones.

FLOWERING TIME: Jun–Sep

NOTES: Generally a fall-blooming species, the light pink flowers are delicate and plants are showier than many other buckwheats.

Oxyria digyna

Alpine Mountainsorrel

Polygonaceae (Buckwheat family)

Erect, hairless, evergreen perennial herb 0.5–5 dm; from thick caudexes, with roots fibrous, rarely rhizomatous.

LEAVES: Basal and alternate near base of stem, with stipules fused, translucent, deciduous or persistent; blades 0.5–6.5 cm, kidney-shaped, ± fleshy, and petioles much longer than blade.

FLOWERS: Panicles 2–20 cm, erect; flowers 2–6 per node, bisexual, with 4 perianth parts, the outer 2 parts 1.2–2.5 mm. Styles 2; stamens 6.

FRUITS: Achenes 3–4.5 mm x 2.5–5 mm wide, 2 wings red or pink, veiny.

SIERRA DISTRIBUTION: Both slopes, Tulare and Inyo Counties north.

ELEVATION: 1800–4000 m

HABITAT: Rock crevices and ledges, talus; alpine zone.

FLOWERING TIME: Jul–Sep

NOTES: Alpine mountainsorrel occurs in arctic tundra and throughout the northern hemisphere. Pollen has been identified in 12,600-year-old peat bogs, suggesting that plants recolonized soon after ice age glaciers retreated or possibly that they survived glaciation.

Polygonum shastense

Shasta Knotweed

Polygonaceae (Buckwheat family)

Prostrate subshrub 5–40 cm; gnarled.

LEAVES: Alternate, often crowded distally, with ocreae (fused stipules) 3–5 mm, hairless, and deciduous distally; petioles 0–0.5 mm; blades 5–25 mm x 2–5 mm wide, lanceolate to elliptic, 1-veined, with leathery texture and margins rolled under.

FLOWERS: Inflorescences of 2–6 flowers in distal axils, with pedicels 2–4 mm; flowers bisexual of 5 perianth parts fused at base, 7%–15% of corolla length; corollas 5–9 mm, white or pink. Stamens 8.

FRUITS: Achenes 3–4 mm, narrowly ovate, shiny brown.

SIERRA DISTRIBUTION: Both slopes, Tulare and Mono Counties north.

ELEVATION: 2100–3400 m

HABITAT: Gravelly and rocky areas; montane through alpine zones.

FLOWERING TIME: Jul–Sep

NOTES: The only woody *Polygonum* in the SN.

Pterostegia drymarioides

Woodland Threadstem

Polygonaceae (Buckwheat family)

Sprawling, diffusely branching annual herb with stems 10–40 cm (up to 1 m); hairy, often reddish.

LEAVES: Opposite, 0.3–2 cm, heart-shaped.

FLOWERS: Cymelike inflorescences, with 2–3 flowers per involucre; flowers 0.9–1.2 mm, hairy; perianth has 6 pale yellow or pink lobes. Male and female flowers are separate on the same plant (monoecious).

FRUITS: 1.2–1.5 mm, obconic, winged.

SIERRA DISTRIBUTION: West slope (east slope in Kern County).

ELEVATION: 1200–3000 m

HABITAT: Shady, sandy, gravelly, and rocky areas; foothill, montane, and sagebrush/pinyon-juniper zones.

FLOWERING TIME: Mar–Jul

NOTES: While easy to overlook, this unique little plant is worth a look with your hand lens; you may become quite fond of it once you make its acquaintance.

Rumex acetosella (non-native)

Sheep Sorrel

Polygonaceae (Buckwheat family)

Erect to ascending perennial herb 10–45 cm; from creeping rhizomes or vertical taproots; herbage is hairless; plant is dioecious.

LEAVES: Basal and alternate, with ocreae (fused stipules) brown at base and the distal half silvery to white and lacerated; leaf blades 2–6 cm x 0.3–2 cm wide, variously shaped but usually hastate or sagittate (with distinct basal lobes).

FLOWERS: Plants either male or female; inflorescences terminal (generally occupying the upper ⅔ of the stem), flowers 5–8 per whorl; pedicels 1–3 mm. Perianth parts 6, the outer 3 small, and the inner 3 the same size or slightly larger (1.2–1.7 mm x 0.5–1.3 mm wide). Male flowers have 6 stamens, with filaments distinct and hairless, anthers yellow, ovate to elongate; female flowers have 3 styles.

FRUITS: Achenes 0.9–1.5 mm x 0.6–0.9 mm wide, brown, enveloped in perianth lobes.

SIERRA DISTRIBUTION: Both slopes, all.

ELEVATION: < 3400 m

HABITAT: Usually disturbed sites, often moist, often acidic, but found increasingly spreading in undisturbed sites; all zones.

FLOWERING TIME: Apr–Jul

NOTES: From Eurasia and introduced nearly worldwide by now, this plant is part of an extremely variable polyploid complex. The plant is weedy in gardens and wildlands, and listed as an invasive weed by Cal-IPC. Plants contain oxalic acid, which in large enough quantities can be poisonous; it has little forage value for livestock (nor, presumably, for native wildlife).

Rumex paucifolius

Alpine Sheep Sorrel

Polygonaceae (Buckwheat family)

Erect to ascending perennial herb 10–40 cm; stems densely clustered, from creeping rhizomes, with hairless herbage; plant is ± dioecious (some flowers are bisexual).

LEAVES: Basal and alternate; ocreae present; leaf blades 3–7 cm x 1–3 cm wide, lanceolate, with base tapered to petiole (without basal lobes).

FLOWERS: Plants either male or female; inflorescences terminal (generally occupying the upper ⅔ of the stem), in whorls of 3–10 flowers, with pedicels 1–3 mm. Perianth lobes 6, the inner 3 lobes 2.8–3.8 mm x 2.7–3.6 mm wide, larger than the outer 3, and larger than the fruit. Male flowers have 6 stamens, with filaments distinct and hairless, and anthers ovate to elongate; female flowers have 3 styles.

FRUITS: Achenes 1.2–1.8 mm x 0.8–1 mm wide, brown, enveloped in inner perianth lobes.

SIERRA DISTRIBUTION: Both slopes, Tulare and Inyo Counties north.

ELEVATION: 2000–4000 m

HABITAT: Lakeshores, streambanks, other moist sites; montane through alpine zones.

FLOWERING TIME: May–Aug

NOTES: *R. paucifolius* is found in higher-elevation, more pristine sites than *R. acetosella*, but where they overlap, leaf shape should quickly distinguish the 2 spp.

Primula fragrans

Scented Shooting Star

Primulaceae (Primrose family)

Perennial herb 15–60 cm; densely glandular-hairy.

LEAVES: Basal 20–40 cm, oblanceolate, tapered to petioles, entire.

FLOWERS: Inflorescence with 5–10 flowers. Petals 5, fused, magenta or lavender (sometimes white); corolla lobes 15–25 mm. Stigma headlike, wider than style; corolla tube fused and usually covering the anther bases, and anther tips pointed.

FRUITS: Capsule 5-valved.

SIERRA DISTRIBUTION: Tulare, Inyo, and Fresno Counties.

ELEVATION: 2400–3600 m

HABITAT: Moist sites; subalpine and alpine zones.

FLOWERING TIME: Jun–Jul

NOTES: Formerly *Dodecatheon redolens*. Could be confused with *P. jeffreyi*, but always has 5 corolla lobes, and the yellow, bulging apex of the corolla tube hides the anther bases slightly. Grows along the trail to Mount Whitney.

Primula hendersonii

Mosquito Bills

Primulaceae (Primrose family)

Perennial herb 12–48 cm; hairless to glandular-hairy.
LEAVES: Basal 2–16 cm, elliptic to ovate or obovate, narrowed abruptly to petiole, entire to ± toothed, length less than twice the width.
FLOWERS: 3–17 per umbel; petals 4–5, fused (flowers with 4–5 corolla lobes can be found on the same plant), pink to lavender (sometimes white), with corolla lobes 6–23 mm. Stigma about the same width as style; filament tube < 3 mm wide, wrinkled.
FRUITS: Capsule circumscissile.
SIERRA DISTRIBUTION: West slope, all.
ELEVATION: < 1900 m
HABITAT: Shady to partially shady areas; foothill and lower montane zones.
FLOWERING TIME: Feb–May
NOTES: Formerly *Dodecatheon hendersonii*. This is the most common low-elevation shooting star. Also found in foothills < 1000 m is Padre's shooting star, *Primula clevelandii* var. *patula*, with white (or light pink) corollas about the same length as *P. hendersonii*, and fat-looking filament tubes 3–4 mm wide with a yellow or white spot below each purple anther.

Primula jeffreyi

Jeffrey's Shooting Star, Sierra Shooting Star

Primulaceae (Primrose family)

Perennial herb 15–60 cm; hairless except stem and inflorescences, which are glandular-hairy.
LEAVES: Basal 9–50 cm, generally oblanceolate, narrowed gradually to petioles, and entire or crenate.
FLOWERS: 3–18 per umbel; flowers 10–25 mm, with petals 4–5, fused, magenta, lavender or white. Corolla tube does not hide base of anthers.
FRUITS: Capsule circumscissile or valved, sometimes both on the same plant.
SIERRA DISTRIBUTION: Throughout, east and west of the crest.
ELEVATION: 600–3000 m
HABITAT: Moist to dry meadows; montane through subalpine zones.
FLOWERING TIME: Jun–Aug
NOTES: Formerly *Dodecatheon jeffreyi*. The most common shooting star in montane through subalpine zones; the plant can have 4–5 corolla lobes.

Primula suffrutescens

Sierra Primrose

Primulaceae (Primrose family)

Subshrub up to 20 cm tall; creeping along the ground.
LEAVES: Basal in rosettes, 15–35 mm, spatulate, hairless, with bases tapered and tips generally rounded, dentate.
FLOWERS: Petals 5, 7–10 mm, fused, glandular-hairy, magenta with yellow throat.
FRUITS: 5-valved capsules, about the same length as the calyx.
SIERRA DISTRIBUTION: Both slopes, Tulare and Inyo Counties north.
ELEVATION: 2000–4200 m
HABITAT: Rock crevices in areas with some soil moisture; subalpine and alpine zones.
FLOWERING TIME: Jun–Aug
NOTES: One of the smash hits of the Sierra high country! Look for it on the domes around Yosemite Valley and in the Convict Creek basin on the east slope.

Primula tetrandra

Alpine Shooting Star

Primulaceae (Primrose family)

Perennial herb 4–30 cm tall; hairless.
LEAVES: Basal in rosettes, 2–20 cm, linear to linear-oblanceolate, narrowed gradually to petiole.
FLOWERS: 1–10 per umbel; petals 4, 8–16 mm, fused, lavender to magenta. Stigma larger in diameter than style; anthers 4, and filaments not fused.
FRUITS: Capsule 5-valved.
SIERRA DISTRIBUTION: Throughout, east and west of the crest.
ELEVATION: 1900–3700 m
HABITAT: Wet meadows, streambanks; montane through alpine zones.
FLOWERING TIME: Jun–Aug
NOTES: Formerly *Dodecatheon alpinum*. Alpine shooting star is not limited to the alpine zone. These delicate plants can cover entire subalpine meadows a few weeks after snowmelt and are often associated with large mosquito populations!

Aconitum columbianum ssp. *columbianum*

Monkshood

Ranunculaceae (Buttercup family)

Erect perennial herb 30 cm to 2 m tall.

LEAVES: Stem leaves alternate, 5–15 cm wide, deeply 3- to 7-divided, with lobes or divisions themselves variously toothed or lobed.

FLOWERS: Upper sepal hooded, enclosing upper 2 petals and stamens, with no spur.

FRUITS: Follicles 10–15 mm.

SIERRA DISTRIBUTION: Both slopes, Tulare and Inyo Counties north.

ELEVATION: 600–3500 m

HABITAT: Wet meadows, spring-fed areas, streamsides; coniferous forest; montane through alpine zones.

FLOWERING TIME: Jul–Sep

NOTES: This is the only *Aconitum* in the SN. Members of the genus are known for containing diterpene alkaloids that can be highly poisonous, even fatal.

Actaea rubra

Baneberry

Ranunculaceae (Buttercup family)

Perennial herb 2–8 dm; branched toward top of plant, sparsely puberulent.

LEAVES: Cauline, mostly opposite, with petioles; lower 2–3 leaves ternate, upper 1–2 ternate. Leaflets toothed to variously cut.

FLOWERS: Pyramidal racemes of ± 25 radial flowers; sepals 3–5, 2–3 mm, petal-like, early deciduous; petals 4–10, 2–3 mm, spoon-shaped, white. The showiest aspect of the flowers are the numerous white stamens (4–7 mm).

FRUITS: Berry to 1 cm wide, red (rarely white).

SIERRA DISTRIBUTION: All.

ELEVATION: < 2800 m

HABITAT: Deep, dry to moist soils in shaded or open sites, often stream banks; montane zone.

FLOWERING TIME: Jun–Aug

NOTES: The fruits are toxic to humans.

Anemone drummondii var. *drummondii*

Drummond's Windflower

Ranunculaceae (Buttercup family)

Perennial herb 10–25 cm; from branched caudexes.
LEAVES: Basal leaves 5–15; blades 2-ternate, sparsely soft-hairy, with the ultimate divisions < 1.5 mm wide. Stem leaves are actually 1–3 involucre bracts that appear leaflike.
FLOWERS: 1–3 per stem, with showy sepals and no petals. Sepals 8–20 mm, 6–10 mm wide, white to bluish. Anthers > 80.
FRUITS: Achenes 2–4 mm, woolly but not covered with long, feathery hairs, with beak 2–4 mm, straight, hairless.
SIERRA DISTRIBUTION: Inyo and Tuolumne Counties north.
ELEVATION: 1200–3350 m
HABITAT: Rocky and gravelly slopes, areas of snow accumulation; upper montane, subalpine, and alpine zones.
FLOWERING TIME: May–Aug
NOTES: Blooms early.

Aquilegia formosa

Western Columbine

Ranunculaceae (Buttercup family)

Erect perennial herb 20–80 cm; hairless, and glaucous above.
LEAVES: Basal and lower cauline, 2-ternate, with leaflets 7–45 mm; upper leaves simple to 3-lobed.
FLOWERS: Nodding, with 5 red, petal-like sepals alternating with 5 petals with a yellow blade and a red tubular spur 12–23 mm.
FRUITS: Follicles 15–28 mm, beak 9–12 mm.
SIERRA DISTRIBUTION: All.
ELEVATION: < 3600 m
HABITAT: Streamsides, wet or moist areas; foothill, montane, and subalpine zones.
FLOWERING TIME: Apr–Sep
NOTES: Pollinated primarily by hummingbirds. In cultivation, this is a nice addition to a native plant garden (it requires shade and moisture).

Aquilegia pubescens

Sierra Columbine

Ranunculaceae (Buttercup family)

Erect perennial herb 15–50 cm; from caudexes, glandular-hairy above.

LEAVES: Basal and cauline, deeply 3-lobed to 1–3 times ternate, with leaflets 10–25 mm.

FLOWERS: Erect, with 5 petal-like sepals alternating with 5 petals with an expanded blade and a long, slender spur ± 25 mm.

FRUITS: Follicles 19–25 mm, beak 10–12 mm.

SIERRA DISTRIBUTION: Inyo and Tulare Counties north to Mono and Tuolumne Counties.

ELEVATION: 2600–3700 m

HABITAT: Rocky and gravelly slopes; subalpine and alpine zones.

FLOWERING TIME: Jul–Aug

NOTES: Pollinated primarily by hawkmoths. Sometimes hybridizes with *A. formosa*, resulting in spectacular patches of multicolored flowers with intermediate traits (inset).

Caltha leptosepala

Marsh Marigold

Ranunculaceae (Buttercup family)

Hairless, fleshy perennial herb < 50 cm.

LEAVES: Petioles < 25 cm; blades to 9 cm wide, kidney-shaped to round, with margins scalloped.

FLOWERS: 1–4 per peduncle; peduncles 4–32 cm, longer than leaves. Sepals 5–11, white, petal-like (no petals).

FRUITS: Follicles 4–15, 7–18 mm, erect, with straight or curved beak.

SIERRA DISTRIBUTION: West slope, Tulare County north; both slopes Tuolumne and Mono Counties north.

ELEVATION: 900–3300 m

HABITAT: Wet meadows and along streams and lakes in coniferous forest; montane and subalpine zones.

FLOWERING TIME: May–Aug

NOTES: A good place to see this plant, which blooms just after snowmelt, is in Yosemite National Park in the meadows along Glacier Point Road.

Clematis lasiantha

Chaparral Clematis, Pipestem Clematis

Ranunculaceae (Buttercup family)

Woody vine 4–5 m, climbing on other vegetation.

LEAVES: Opposite, with 3–5 leaflets, 3+ times lobed, with the largest leaflet 1.5–6 cm.

FLOWERS: Unisexual inflorescences have 3–5 flowers; sepals 10–21 mm, white to yellowish, hairy on both sides, petal-like (no petals). Male flowers have 50–100 stamens that are much shorter than sepals; female flowers have 75–100 pistils.

FRUITS: Achenes with feathery, elongated styles, in fluffy heads.

SIERRA DISTRIBUTION: West slope, all.

ELEVATION: < 2000 m

HABITAT: Canyons, slopes, abundant in foothill chaparral; foothill and lower montane zones.

FLOWERING TIME: Jan–Jun

NOTES: Creek clematis, *C. ligusticifolia*, occurs on the west slope < 2400 m, and flowers Jun–Sep; it has vines up to 6 m, leaves have 5–15 leaflets, the petal-like sepals are shorter (< 10 mm), and stamens are fewer (25–50) than *C. lasiantha*. Creek clematis occurs in moister areas than *C. lasiantha*.

Clematis ligusticifolia

Creek Clematis

Ranunculaceae (Buttercup family)

Woody vine climbing to 6 m (occasionally as high as 20 m).

LEAVES: Opposite, with 5–15 leaflets that are lobed or toothed, with the largest leaflet 2–9 cm.

FLOWERS: Unisexual inflorescences of 5 to many flowers; petals absent. Calyx of sepals 4, 6–10 mm, white to cream, petal-like, hairy on undersides only. Male flowers have 25–50 stamens, 5–9 mm; female flowers have 25–65 pistils.

FRUITS: Achenes hairy, with feathery, elongated style, in fluffy heads.

SIERRA DISTRIBUTION: Scattered distribution on both slopes, all.

ELEVATION: < 2400 m

HABITAT: Streamsides, wet places; below subalpine zone.

FLOWERING TIME: Jun–Sep

NOTES: Native American uses include treatment of skin infections; extractions of plants or leaves in hot water are used to treat eczema and chest pain, sores, and boils.

Delphinium glaucum

Giant Mountain Larkspur

Ranunculaceae (Buttercup family)

Hairless perennial herb l–2 m tall; stems whitish due to glaucous "bloom," especially toward tips; stems are hollow.

LEAVES: Basal and on lower ⅔ of stem, mostly hairless; blades 8–15 cm wide, with 5–7 major lobes, each themselves lobed.

FLOWERS: Inflorescence is usually branched, flowers 50+ per raceme, often with leaflike bracts throughout the inflorescence and sometimes just at base. Flowers purple or blue, with spurs 10–19 mm.

FRUITS: Follicles 9–20 mm, smooth or slightly hairy.

SIERRA DISTRIBUTION: Both slopes, Tulare and Inyo Counties north, absent from Fresno County.

ELEVATION: 1500–3200 m

HABITAT: Moist to wet meadows, streamsides; montane, subalpine, and sagebrush zones.

FLOWERING TIME: Jul–Sep

NOTES: *The* biggest larkspur in the SN!

Delphinium gracilentum

Greene's Larkspur

Ranunculaceae (Buttercup family)

Perennial herb 15–80 cm; the base of stem is hairless, often reddish, and narrowed to a weak attachment with root.

LEAVES: Basal leaves 0–3 appear at flowering time; the outline is pentagonal to round, with ± 5 primary lobes, wedge-shaped, 3- to 7-lobed, and the terminal lobe wider above the middle. Cauline leaves 2–5 at flowering time, on the lower ⅓ of stem, with narrower lobes; petioles 3–15 cm; blades 1.5–4 cm x 3–7 cm wide, and hairless or slightly glandular-pubescent,

FLOWERS: Inflorescences of 5–20+ flowers, with pedicels 1.5–4.5 cm, held ± 90° away from stem, hairless or glandular; sepals dark blue or purple (sometimes pink or white), with reflexed lateral sepals 7–13 mm, spur 8–12 mm, often angled ± 30° above horizontal.

FRUITS: Follicles 8–16 mm, hairless to glandular, ± curved when fully mature. Seeds have "collars" (tiny crowns of clear tissue at tips) and bumpy surfaces.

SIERRA DISTRIBUTION: West slope, Kern County north; both slopes, northern Mono County north.

ELEVATION: 150–2700 m

HABITAT: Dry to moist, partially shaded sites usually where soil is well developed, often in coniferous forest; montane zone.

FLOWERING TIME: Mar–Jun

NOTES: *D. patens* ssp. *patens* is similar-looking and hybridizes with *D. gracilentum* in the northern Sierra, making ID rather impossible. Theoretically, *C. patens* differs in that the terminal leaf lobes are wider at the middle, seeds are smooth and shiny not bumpy, and habitat is grassland and woodland < 1100 m, with patchy distribution on west slope.

Delphinium hansenii ssp. *hansenii*

Hansen's Larkspur

Ranunculaceae (Buttercup family)

Stout, hairy perennial herb to 1.8 m tall.
LEAVES: Mostly basal, hairy, deeply lobed, with lobes 3–18; leaves are withered by flowering time.
FLOWERS: Long racemes 6–40 cm, of > 12 white or light pink flowers, tinged with darker pink or green; spur 8–13 mm, and lower petals usually hairy on upper surfaces.
FRUITS: Follicles 8–20 mm.
SIERRA DISTRIBUTION: Butte County south.
ELEVATION: 150–3000 m
HABITAT: Rocky and gravelly places; foothill and montane zones.
FLOWERING TIME: May–Jul
NOTES: The uncommon ssp. *ewanianum* is found < 600 m in the foothills from Tulare to El Dorado Counties; it is a striking plant with maroon to dark purple flowers.

Delphinium nudicaule

Red Larkspur

Ranunculaceae (Buttercup family)

Hairless perennial herb 2–8 dm; base of stem tapered to a slender attachment to root.
LEAVES: Basal and on lower stem, hairless; blades 3–10 cm, with upper leaves much smaller; lobes 3–10.
FLOWERS: Racemes of < 10 flowers; sepals point forward, spur 12–34 mm, and lower petals 2–3 mm.
FRUITS: Follicles 13–26 mm, curved.
SIERRA DISTRIBUTION: Primarily the northern Sierra, Butte and Sierra Counties north, but also documented from Tuolumne and Tulare Counties.
ELEVATION: < 2600 m
HABITAT: Rocky slopes; foothill and montane zones.
FLOWERING TIME: Mar–Jun
NOTES: Pollinated by hummingbirds.

Delphinium nuttallianum

Nuttall's Larkspur

Ranunculaceae (Buttercup family)

Perennial herb 5–50 cm; base of stem is hairless or puberulent, and narrowed to attachment with root.

LEAVES: Basal leaves 0–3 at flowering time; cauline 2–5 at flowering, on lower ⅓ of stem, with petioles 3–15 cm. Blades 1.5–4 cm x 3–7 cm wide, usually with ≤ 5 wedge-shaped lobes, incised > 50% to base, hairless or slightly glandular-pubescent.

FLOWERS: Inflorescences of < 12 flowers, with pedicels 1.5–4.5 cm, held ± 90° away from stem, about 1 cm apart, hairless or glandular; sepals dark blue or purple (sometimes pink or white), and reflexed lateral sepals 7–13 mm; spur 7–15 mm, often held slightly above the horizontal (± 30°).

FRUITS: Follicles 7–17 mm, curved when fully mature. Seeds smooth, with "collars" (low rim of tissue at tips).

SIERRA DISTRIBUTION: Tulare and Inyo Counties north.

ELEVATION: 300–3300 m

HABITAT: Open, dry to moist sites; montane, subalpine, and sagebrush zones.

FLOWERING TIME: May–Jul

NOTES: Hybridizes with *D. depauperatum*, *D. polycladon*, and others, forming a difficult complex with many intermediates that are better admired than identified!

Delphinium purpusii

Rose-Flowered Larkspur

Ranunculaceae (Buttercup family)

Tall, hairless perennial herb 3–12 dm; the base of stem is tapered to a slender attachment to root, branched at top.

LEAVES: Basal and on lower half of stem, 1.5–6 cm x 2–10 cm wide, round in outline, ± puberulent.

FLOWERS: Racemes of 12–20 flowers; pedicels have yellow glands; sepals rose to magenta, petal-like, with the top sepal forming a spur 10–19 mm; petals white, inconspicuous.

FRUITS: Follicles 11–29 mm, erect.

SIERRA DISTRIBUTION: Tulare and Kern Counties.

ELEVATION: 300–1300 m

HABITAT: Rocky slopes, cliffs; foothill zone.

FLOWERING TIME: Mar–May

NOTES: Rare, CRPR 1B.3. Not likely to be confused with any other larkspur.

Delphinium variegatum ssp. *variegatum*

Royal Larkspur

Ranunculaceae (Buttercup family)

Erect, hairy perennial herb usually < 0.5 m tall.
LEAVES: Basal and on lower stem; petioles have long hairs, and blades deeply 3- to 15-lobed, ± overlapping.
FLOWERS: Inflorescences often branched, flowers noticeably large, with sepals deep royal blue or purple, spur 10–19 mm, straight or curved downward last 3 mm of tip; lateral sepals 10–25 mm, and lower petals 4–11 mm, hairy on upper margins.
FRUITS: Follicles 9–19 mm, often have colored veins.
SIERRA DISTRIBUTION: Tulare County north.
ELEVATION: < 800 m
HABITAT: Grassland, chaparral, oak woodland; foothill zone.
FLOWERING TIME: Mar–May
NOTES: One of the most striking larkspurs of the foothill zone; in a good year, a large patch can be seen from quite a distance.

Enemion occidentale

Western Rue-Anemone

Ranunculaceae (Buttercup family)

Erect perennial herb ranging 8–25 cm tall.
LEAVES: 3–12 cm, and leaflet segment lobes 2–3 cm and shallow (< ½ the length of the segment); segment lobes 2–9 mm wide.
FLOWERS: Borne singly; sepals 5–11 mm, white or pink, petals absent. Stamens > 20.
FRUITS: Follicles 8–10.5 mm, hairless, with no stalk at base.
SIERRA DISTRIBUTION: All, except east side.
ELEVATION: < 1700 m
HABITAT: Shaded, dry to moist areas; foothill and lower montane zones.
FLOWERING TIME: Mar–Jun
NOTES: Formerly *Isopyrum occidentale*. Blooms before other spring wildflowers in a given area. From El Dorado County north, *E. stipitatum* may be encountered as well; it has fruits with stalklike bases (0.5–2 mm), stamens < 15, sepals shorter (3.5–6 mm), and leaf lobe segments narrower (< 4 mm wide).

Ranunculus alismifolius var. *alismellus*

Plantainleaf Buttercup

Ranunculaceae (Buttercup family)

Perennial herb 6–21 cm; erect to ascending, hairless, with stem 1–3 mm wide.

LEAVES: Basal and cauline, 2–5 cm, elliptic to ovate, tapered at base, with margins smooth.

FLOWERS: Sepals 5, 2–6 mm, 1–4 mm wide, green, reflexed, falling early; petals 4–8, 5–8 mm x 2–6 mm wide, yellow, shiny.

FRUITS: Achenes in spherical clusters, 1.6–2.8 mm x 1.2–2 mm wide, green, smooth-walled, with beaks 0.4–1.2 mm and straight or curved.

SIERRA DISTRIBUTION: Tulare and Inyo Counties north.

ELEVATION: 1400–3400 m

HABITAT: Damp to wet sites; montane and subalpine zones.

FLOWERING TIME: May–Aug

NOTES: Also occurring in the same range is the generally taller var. *alismifolius*, which has slightly longer (6–14 cm) leaves with toothed margins, and longer (7–11 mm) petals.

Ranunculus aquatilis var. *aquatilis*

White Water Buttercup

Ranunculaceae (Buttercup family)

Perennial herb 20–80 cm; aquatic or prostrate on mud, rooting at nodes.

LEAVES: Floating and submersed (both shown in photo); floating leaves 0.4–1.1 cm x 0.7–2.3 cm wide, 3-parted, parts fan-shaped or obovate, and crenate; submersed leaves divided into 3–6 linear, thread-like segments, with connate stipules.

FLOWERS: Sepals 5, 2–4 mm, 1–2 mm wide, spreading or reflexed, falling early; petals 5, 4–7 mm x 1–5 mm wide, white or white with yellow claw.

FRUITS: Achenes 15+, 1.6–2 mm x 1–1.4 mm, lenticular, with walls transversely wrinkled, beaks 1–1.4 mm, threadlike, and persistent.

SIERRA DISTRIBUTION: All.

ELEVATION: < 2000 m

HABITAT: Streams, ponds, lakes, mudflats; below subalpine zone.

FLOWERING TIME: Feb–Jun

NOTES: *Ranunculus aquatilis* var. *diffusus* occurs in all SN counties in the same habitats, but occurs up to 3200 m elevation and differs in that both floating and submersed leaves are finely dissected into threadlike segments; the flowering time is Mar–Sep, and achene bodies are slightly smaller (1–1.8 mm x 0.8–1.2 mm wide), with beaks 0.2–1.2 mm.

Ranunculus californicus var. *californicus*

California Buttercup

Ranunculaceae (Buttercup family)

Erect to decumbent perennial herb 18–70 cm, with spreading hairs or hairless.

LEAVES: Basal and lower stem leaves 2.8–5.8 cm x 4–6 cm wide, 3-lobed to compound with 3–5 leaflets; upper stem leaves smaller and deeply lobed or compound.

FLOWERS: Sepals 5, reflexed a short distance up from base, and falling early; petals 9–17, longer than twice their width (7–14 mm x 2–6 mm wide).

FRUITS: Achenes disklike, head with smooth walls, beak 0.2–0.8 mm, curved.

SIERRA DISTRIBUTION: All west slope counties.

ELEVATION: < 2000 m

HABITAT: Dry and vernally moist, grassy areas, woodlands; foothill and lower montane zones.

FLOWERING TIME: Mar–Aug

NOTES: Compare with *R. occidentalis* var. *occidentalis*, which co-occurs and has fewer petals, which are wider in relation to their length.

Ranunculus eschscholtzii var. *oxynotus*

Alpine Buttercup

Ranunculaceae (Buttercup family)

Erect to decumbent perennial herb 5–25 cm; hairless, with a noticeable cluster of previous years' leaf bases.

LEAVES: Basal 1.5–4 cm wide, roundish in outline, and 3- to 9-lobed; stem leaves smaller, more deeply lobed.

FLOWERS: Sepals 5, yellowish; petals 5, 6–13 mm, waxy yellow.

FRUITS: Achenes 1.4–3 mm, disklike, head ovoid, with smooth walls, beak 0.6–1.8 mm, straight.

SIERRA DISTRIBUTION: Tulare and Inyo Counties north to Nevada County.

ELEVATION: 2700–4300 m

HABITAT: Dry to moist talus slopes and meadows, often near late-melting snowbanks; subalpine and alpine zones.

FLOWERING TIME: Jul–Sep

NOTES: *R. eschscholtzii* var. *eschscholtzii* occurs in the High Sierra as well, but isn't associated with snowbanks and doesn't have the cluster of dead leaves from previous years.

Ranunculus hystriculus

Waterfall Buttercup

Ranunculaceae (Buttercup family)

Hairless perennial herb; arising from caudexes.

LEAVES: Basal, bright green, roundish, with scalloped or lobed margins.

FLOWERS: Flowering stalks 8–25 cm. Inflorescences of 1–3 flowers are on leafless stalks, with 5 white, petal-like sepals 6–10 mm surrounding small, green petals 2–3 mm.

FRUITS: Spherical clusters; fruit 3–4 mm, green, sparsely hairy, with hook at tip of beak when mature.

SIERRA DISTRIBUTION: West slope, Tulare County north.

ELEVATION: < 1700 m

HABITAT: Wet places along streams or on seeping rock faces; foothill and montane zones.

FLOWERING TIME: Feb–Jul (depending on elevation)

NOTES: Formerly *Kumlienia hystricula*.

Ranunculus occidentalis var. *occidentalis*

Western Buttercup

Ranunculaceae (Buttercup family)

Erect to decumbent, hairy perennial herb 10–60 cm.

LEAVES: Basal and lower stem leaves 1.5–5.3 cm x 2.2–8 cm wide, 3-lobed to compound, with 3 leaflets; lobes or leaflets ± wedge-shaped; upper stem leaves much smaller, and deeply lobed or compound.

FLOWERS: Sepals 5, reflexed a short distance up from base, falling early; petals 5–6, 5–10 mm x 3–6 mm wide.

FRUITS: Heads of disklike achenes, with smooth or bristly walls, beak 1–1.4 mm, curved.

SIERRA DISTRIBUTION: All west slope counties.

ELEVATION: < 1500 m

HABITAT: Grassy areas, woodlands; foothill and lower montane zones.

FLOWERING TIME: Mar–Jul

NOTES: Compare with *R. californicus* var. *californicus*, which co-occurs and has 9–17 petals with a more elongate shape.

Ranunculus orthorhynchus var. *orthorhynchus*

Straight Beaked Buttercup

Ranunculaceae (Buttercup family)

Erect to decumbent perennial herb 15–50 cm; not rooting at nodes; herbage is hairy.

LEAVES: Basal and lower stem leaves 2.8–12.5 cm x 2.5–14 cm wide; basal leaves compound, ternate or pinnate, with leaflets 3–5, 1- to 2-lobed or parted, and entire to dentate; upper stem leaves smaller.

FLOWERS: Sepals 5, 5–11 mm x 2–4 mm wide, reflexed 1–2 mm from base, falling early; petals 5–6, 8–18 mm x 4–6 mm wide, rounded, often reddish below.

FRUITS: Head of disklike achenes 2.8–4.5 mm, beak 3–3.8 mm, straight.

SIERRA DISTRIBUTION: West slope, Tulare County north.

ELEVATION: < 2200 m

HABITAT: Wet or marshy areas such as montane meadows, streamsides; montane zone.

FLOWERING TIME: Mar–Aug

NOTES: *Ortho* means "straight," and *rhynchus* means "snout, bill (or beak)," together referring to the long, straight beaks on the fruits.

Thalictrum fendleri var. *fendleri*

Fendler's Meadow-Rue

Ranunculaceae (Buttercup family)

Erect perennial herb to 2 m; male and female plants are separate.

LEAVES: Basal and cauline, compound (1–4 times ternate), 7–46 cm, with leaflets lobed; leaves hairless except the upper surface of upper leaves have small glandular hairs (need a 20x lens).

FLOWERS: Unisexual; sepals 4, 2–5 mm, green; no petals. Male flowers have 15–28 pendulous stamens, with slender filaments; female flowers have 7–20 pistils.

FRUITS: Achenes 7–20, 4–8 mm, compressed laterally; beak 1.5–4 mm, 2- to 3-ribbed.

SIERRA DISTRIBUTION: Fresno and Inyo Counties north.

ELEVATION: 500–3200 m

HABITAT: Wet areas such as streamsides and meadows; foothill through alpine zones.

FLOWERING TIME: May–Aug

NOTES: Another variety of *T. fendleri*, var. *polycarpum*, occurs throughout the Sierra as well and is documented from Tulare and Kern Counties (both may occur in these counties). *T. fendleri* var. *polycarpum* has hairless, glaucous leaves (no glandular hairs on upper leaves are visible at 20x), and less compressed fruits with netlike patterns.

Thalictrum sparsiflorum

Few-Flowered Meadow-Rue

Ranunculaceae (Buttercup family)

Erect perennial herb 0.6–1.8 m; plants have bisexual flowers.
LEAVES: Basal and cauline, compound (1–4 times ternate), 4–30 cm, with leaflets lobed and minutely glandular-hairy.
FLOWERS: Bisexual; sepals 5, 2.5–4 mm, white to greenish, petal-like; no petals. Stamens 10–20, with filaments wider toward tips.
FRUITS: Achenes 6–22, 4–6 mm, pointing downward; beak 1–1.5 mm.
SIERRA DISTRIBUTION: Both slopes, Tulare and Inyo Counties north.
ELEVATION: 1400–3500 m
HABITAT: Moist to wet areas, streamsides, meadows; montane and subalpine zones.
FLOWERING TIME: Jul–Aug
NOTES: In the alpine zone of Inyo and Mono Counties, *T. alpinum* is another bisexual meadow-rue. Plants are < 15 cm, leaves are mostly basal (1 stem leaf only is present), and fruiting pedicels are strongly recurved.

Ceanothus cordulatus

Mountain Whitethorn

Rhamnaceae (Buckthorn family)

Spiny, many-branched shrub < 1.5 m; grayish-white bark, glaucous.
LEAVES: Alternate, < 3 cm, and elliptic or ovate, evergreen; upper surface grayish-green and glaucous, paler below, and 3-veined from base, with petioles 3–6 mm.
FLOWERS: Numerous in clusters < 4 cm, white.
FRUITS: Capsule 5 mm x 4–5 mm wide, triangular, 3-lobed, lobes crested.
SIERRA DISTRIBUTION: All.
ELEVATION: 400–2900 m
HABITAT: Chaparral and forests; montane and subalpine zones.
FLOWERING TIME: May–Jul
NOTES: One of the most common shrubs of the montane zone, forming dense, impenetrable (spiny) stands, sometimes codominant with *Arctostaphylos patula*, *Quercus vacciniifolia*, and *Chrysolepis sempervirens*. Sprouts back after fire.

Ceanothus cuneatus var. *cuneatus*

Buck Brush, Wedge-Leaf Ceanothus

Rhamnaceae (Buckthorn family)

Erect shrub to 3 m; grayish bark, often with an overall funnel-shaped form.

LEAVES: Opposite, 8–30 mm, wedge-shaped, thick, evergreen, with margins entire; upper surfaces grayish-green and hairless, and lower surfaces have fine, white hairs; 1-veined from base.

FLOWERS: Dense panicles 1–2.5 cm; flowers creamy white.

FRUITS: Capsule 5 mm wide, 3-lobed, with horns on top.

SIERRA DISTRIBUTION: West slope, all.

ELEVATION: < 1800 m

HABITAT: Dry, open slopes and flats in chaparral and forests; foothill and montane zones.

FLOWERING TIME: Mar–May

NOTES: One of the most common shrubs in CA. Used in basketry and to make seed-beaters for harvesting seeds. A major food source for larvae of the California tortoiseshell butterfly (*Nymphalis californica*), which are collected and roasted by the Western Mono. The pupae attach to branches and wiggle vigorously in response to noise or vibration, facilitating their harvest. The elders I learned from told me they shout, "Huya!" to start the wiggling, and that if it's not wiggling, they don't collect it because it's deaf (said with much laughter). Possibly the nonwiggling pupae are nearing metamorphosis and wouldn't be as tasty. This behavior may have evolved to avoid predation. Huya, as they are called, have boom years, often after above normal precipitation, their arrival marked by defoliation of buck brush across a small watershed.

Ceanothus diversifolius

Pine Mat Ceanothus

Rhamnaceae (Buckthorn family)

Mat-forming shrub < 30 cm; reddish or greenish, soft-hairy branches.

LEAVES: Alternate, < 5.5 cm, evergreen; variable in size and shape, but generally ovate or obovate with toothed margins, both surfaces densely soft-hairy, 1-veined.

FLOWERS: Sparse racemes < 3 cm; flowers pale blue.

FRUITS: Capsule < 4 mm, smooth with crests.

SIERRA DISTRIBUTION: West slope, all.

ELEVATION: 900–2200 m

HABITAT: Open forest floor; montane zone.

FLOWERING TIME: May–Jun

NOTES: Distinctive among mat-forming *Ceanothus* for the soft, hairy, variably sized and shaped leaves (compare with *C. prostratus*). Jepson's dodder, *Cuscuta jepsonii*, a cute and at least temporally rare, parasitic plant, only attaches to *Ceanothus diversifolius* and *C. prostratus*.

Ceanothus integerrimus

Deerbrush

Rhamnaceae (Buckthorn family)

Erect, deciduous shrub to 4 m.

LEAVES: Alternate, < 8 cm, evergreen; light green above, paler underneath, with margins gland-toothed, 3-veined from base.

FLOWERS: Inflorescence < 15 cm, flowers white, blue, or pink.

FRUITS: Capsule 4–5 mm, sticky, 3-lobed, with small crests.

SIERRA DISTRIBUTION: West slope, all.

ELEVATION: < 2100 m

HABITAT: Many plant communities; foothill and montane zones.

FLOWERING TIME: May–Jul

NOTES: An extremely common shrub of the montane zone that sprouts and seeds after fires and other disturbance, temporarily dominating formerly forested areas until trees return.

Ceanothus lemmonii

Lemmon's Ceanothus

Rhamnaceae (Buckthorn family)

Ascending, evergreen shrub < 1 m, with grayish twigs.

LEAVES: Alternate, < 3 cm, evergreen; dark green and hairless to hairy above, densely hairy below, with margins gland-toothed, 1- to 3-veined from base.

FLOWERS: Inflorescences < 5 cm, dense; flowers pale blue.

FRUITS: Capsule < 4 mm, 3-lobed, with lobes crested.

SIERRA DISTRIBUTION: West slope, Tuolumne to Plumas Counties, Kern County.

ELEVATION: 350–1300 m

HABITAT: Open slopes; foothill and montane zones.

FLOWERING TIME: Apr–Jul

NOTES: Most common in the northern Sierra, but documented from 2 localities in Kern County, with no records for Mariposa through Tulare Counties.

Ceanothus leucodermis

Chaparral Whitethorn

Rhamnaceae (Buckthorn family)

Spiny, erect shrub < 4 m; bark is smooth and white to greenish white, glaucous.

LEAVES: Alternate, < 4 cm, elliptic or ovate, evergreen; light green above, paler below, with both surfaces glaucous; 3-veined from base, petioles 2–3 mm.

FLOWERS: Inflorescences < 11 cm, flowers white to light blue.

FRUITS: < 6 mm, 3-lobed, no crests on top, sticky.

SIERRA DISTRIBUTION: West slope, Nevada County south.

ELEVATION: < 1800 m

HABITAT: Chaparral; foothill zone.

FLOWERING TIME: Apr–Jun

NOTES: A common shrub of the foothill chaparral. Somewhat similar in appearance to mountain whitethorn, but chaparral whitethorn is found at lower elevations than *C. cordulatus*. Chaparral whitethorn is generally taller and has longer inflorescences and more striking white bark than *C. cordulatus*.

Ceanothus parvifolius

Littleleaf Ceanothus

Rhamnaceae (Buckthorn family)

Low-growing, deciduous shrub < 1.2 m; hairless, with yellowish to reddish stems.

LEAVES: Alternate, 1–2 cm, 3-ribbed from base, dark green above, and paler beneath; mostly entire.

FLOWERS: Inflorescences < 7 cm, flowers light to dark blue.

FRUITS: 4–5 mm, 3-lobed, no crests on top.

SIERRA DISTRIBUTION: West slope, Tulare to Plumas Counties.

ELEVATION: 1350–2100 m

HABITAT: Openings and roadsides, mostly in mixed conifer forest; lower montane zone.

FLOWERING TIME: May–Jul

NOTES: This shrub is common in giant sequoia groves in Yosemite. It is easily distinguished from other *Ceanothus* spp. by its low-growing habit and small, thin, alternate leaves with shiny upper surfaces.

Ceanothus prostratus var. *prostratus*

Mahala Mat

Rhamnaceae (Buckthorn family)

Mat-forming, evergreen shrub sprawling up to 2.5 m wide.

LEAVES: Opposite, < 2.5 cm, evergreen; dark green, densely hairy below, 3- to 9-toothed toward tip of leaf; 1-veined from base.

FLOWERS: Inflorescences < 2 cm, flowers blue, lavender, or purple.

FRUITS: Capsule 3–9 mm, with 3 crested lobes and prominent horns on sides.

SIERRA DISTRIBUTION: Calaveras County north.

ELEVATION: 900–2200 m

HABITAT: Open flats in coniferous forest; montane zone.

FLOWERING TIME: Apr–Jun

NOTES: Common in Calaveras County and northward, but intriguing records exist in CCH for Mariposa, Tulare, and Inyo Counties; field verification is needed to clarify whether these are indeed *C. prostratus*.

Ceanothus tomentosus

Woolly Leaf Ceanothus

Rhamnaceae (Buckthorn family)

Erect, woolly, evergreen shrub to 3 m.

LEAVES: Alternate, < 3 cm, evergreen; dark green and finely hairy above, and very woolly below, with margin gland-toothed; 3-veined from base.

FLOWERS: Inflorescences < 9 cm, flowers blue or white.

FRUITS: Capsule 4 mm, sticky, shallowly 3-lobed, with crests present.

SIERRA DISTRIBUTION: Mariposa County north.

ELEVATION: < 1500 m

HABITAT: Dry slopes; foothill and montane zones.

FLOWERING TIME: May–Jul

NOTES: In Mariposa County, the rare *Horkelia parryi* is closely associated with this shrub. Look for both on Feliciana Ridge, and for *C. tomentosus* in similar habitats around Midpines.

Ceanothus velutinus

Snowbrush Ceanothus, Tobacco Brush

Rhamnaceae (Buckthorn family)

Erect, evergreen shrub to 2 m.

LEAVES: Alternate, 2.5–8 cm, evergreen; shiny, dark green, and pale underneath, with margins gland-toothed; 3-veined from base.

FLOWERS: Inflorescences < 12 cm, flowers white.

FRUITS: Capsule 3–4 mm, 3-lobed, without crests.

SIERRA DISTRIBUTION: All.

ELEVATION: < 3000 m

HABITAT: Open wooded areas; montane and subalpine zones.

FLOWERING TIME: Apr–Jul

NOTES: There are no records for several central Sierra counties (Mariposa, Madera, Calaveras, Amador); this shrub is most common in the northern Sierra and on the east slope.

Frangula californica ssp. *tomentella*

Chaparral Coffeeberry

Rhamnaceae (Buckthorn family)

Erect, evergreen shrub to 5 m; reddish-brown twigs, terminal buds lacking scales.

LEAVES: Alternate, 3–7 cm, oval to elliptical with serrate margins, evergreen, thick, leathery; leaves have bright to dull green upper surfaces, and are woolly-hairy below with prominent veins.

FLOWERS: Dense umbels of 6–50 flowers, with peduncles 4–18 mm. Flowers 2–3 mm, greenish-yellow, with 5 sepals, 5 inconspicuous petals, and pedicels < 20 mm.

FRUITS: 10–15 mm, black. Seeds 3.

SIERRA DISTRIBUTION: Tulare County north.

ELEVATION: < 1500 m

HABITAT: Chaparral and woodland; mostly foothill zone.

FLOWERING TIME: Jun–Jul

NOTES: Previously known as *Rhamnus tomentella* ssp. *tomentella*. Another ssp. occurring in the SN is *F. californica* ssp. *cuspidata*, which has leaves 2–6 cm with abrupt points.

Frangula rubra ssp. *rubra*

Sierra Coffeeberry

Rhamnaceae (Buckthorn family)

Erect to spreading, deciduous shrub to 2 m; red-gray twigs, terminal buds lacking scales, and hairy.
LEAVES: Alternate, 1.5–7 cm, thin, extremely variable, hairless.
FLOWERS: Inflorescences of 6–15 flowers with pedicels 1–12 mm. Sepals 5, petals 5, inconspicuous.
FRUITS: < 10 mm, black. Seeds 2.
SIERRA DISTRIBUTION: West slope only, all.
ELEVATION: < 2200 m
HABITAT: Dry areas; foothill and montane zones.
FLOWERING TIME: Mar–Jun
NOTES: Formerly *Rhamnus rubra*. Distinguished from *F. californica* by the thinner (deciduous) leaves, smaller fruits with 2 rather than 3 seeds, and smaller stature. Yosemite buckthorn, *F. rubra* ssp. *yosemitana*, occurs in the western central SN at 1000–2200 m; its leaves are finely hairy.

Rhamnus ilicifolia

Holly-Leaf Redberry

Rhamnaceae (Buckthorn family)

Rigidly branched, evergreen shrub to 4 m; with bark gray, branches often spiny, and terminal buds with densely hairy scales ± 3 mm.
LEAVES: Alternate, 2–4 cm, round to ovate with spiny margins, surfaces hairless to hairy; overall architecture not plane (lower surfaces ± concave), and veins on lower surface not prominent.
FLOWERS: Inflorescence of 1–6 unisexual flowers, unisexual. Sepals 4, no petals. Style extruding.
FRUITS: 8 mm, bright red. Seeds 2.
SIERRA DISTRIBUTION: West slope.
ELEVATION: < 2000 m
HABITAT: Dry slopes and flats, especially in chaparral; foothill and montane zones.
FLOWERING TIME: Mar–Jun
NOTES: One of the more distinctive chaparral shrubs when the bright red berries are ripe. The leaves might resemble a live oak but are distinguished by the fact that they are concave as seen from below, or irregularly curved such that they are not plane (live oak leaves are not always plane, but there's a look to the redberry leaves that you get good at with experience).

Adenostoma fasciculatum var. *fasciculatum*

Chamise, Greasewood

Rosaceae (Rose family)

Shrub to 4 m; with burl, much branched with overall outline inversely conical.

LEAVES: Alternate, ≤ 1 cm, stiff, evergreen, in clusters, with tiny stipules.

FLOWERS: Inflorescences 4–12 cm, elongate; flowers cream to white, with corolla lobes roundish. Ovary superior; stamens 10–15.

FRUITS: Achenes.

SIERRA DISTRIBUTION: West slope, patchy distribution.

ELEVATION: < 1600 m

HABITAT: Dry ridges and slopes, often forming dense, impenetrable cover; foothill zone.

FLOWERING TIME: May–Jun

NOTES: Chamise is a common, often dominant species in the chaparral vegetation type across most of CA; it sprouts from a burl as well as producing a flush of seedlings post-fire. Researchers have demonstrated that seedlings are stimulated to germinate by smoke.

Amelanchier utahensis

Utah Service-Berry

Rosaceae (Rose family)

Deciduous shrub < 5 m; with grayish or reddish bark.

LEAVES: Blades 1.3–4.5 cm, 1–4.5 cm wide, hairy below (finely hairy in fruit), and toothed from above center of margin to tip.

FLOWERS: Inflorescences in racemes or clusters of 3–6 flowers; sepals 5, with long, soft hairs; petals 5, 6–11 mm. Styles 2–4.

FRUITS: Pome 4–10 mm wide, purple; sepals present on ripening fruit.

SIERRA DISTRIBUTION: All.

ELEVATION: < 3400 m

HABITAT: Open rocky slopes, canyons, forests; all zones.

FLOWERING TIME: Apr–Jun

NOTES: One other species of *Amelanchier* occurs in the SN, *A. alnifolia* var. *pumila*. It is a smaller (< 3 m) shrub with hairless lower leaf surfaces by the time the plant is in fruit, with 4–5 styles, and fruits 8–9 mm wide.

Cercocarpus betuloides var. *betuloides*

Birch-Leaf Mountain Mahogany

Rosaceae (Rose family)

Evergreen shrub 1–3 m.

LEAVES: Blades 1–4 cm, round to obovate, toothed on upper half, darker above, and paler and hairy below.

FLOWERS: Inflorescences of 1–6 flowers. Hypanthium 5–8 mm (7–13 mm when fruit is ripe), funnel-shaped, densely hairy; sepals 5 and triangular; petals absent. Style 1; stamens 25–45, anthers hairy.

FRUITS: Achenes > 9 cm, long, curved, featherlike.

SIERRA DISTRIBUTION: West slope, all.

ELEVATION: < 2500 m

HABITAT: Dry slopes and rocky areas; foothill and montane zones.

FLOWERING TIME: Mar–May

NOTES: The common name derives from the hardness and color of the wood, though it is not related to mahogany, a tropical tree in the Meliaceae. The hardness of the wood makes it ideal for digging sticks, fishing spears, and arrow tips made by Native Americans.

Cercocarpus ledifolius var. *intermontanus*

Curl-Leaf Mountain Mahogany

Rosaceae (Rose family)

Evergreen shrub or small tree 2–7 m.

LEAVES: Blades 1–3 cm, ± 9 mm wide, elliptic or oblanceolate, smooth-margined, and hairless, or if hairy below, veins still visible.

FLOWERS: Inflorescences of 4–10 flowers. Hypanthium 2.5–6 mm (6–10 mm when fruit is ripe), funnel-shaped, densely hairy; sepals 5, and triangular; petals absent. Style 1; stamens 15–25, anthers hairless.

FRUITS: Achenes 3–7 cm, with long, curved, featherlike style.

SIERRA DISTRIBUTION: Mostly east of the crest, but not always, Tulare and Inyo Counties north.

ELEVATION: 1000–3256 m

HABITAT: Dry rocky slopes; subalpine and sagebrush/pinyon-juniper zones.

FLOWERING TIME: Apr–Aug

NOTES: Occasionally sprouts after fire, but mostly comes back from seed. Nitrogen-fixing bacteria are found in root nodules and have been demonstrated to increase soil nitrogen.

Chamaebatia foliolosa

Bear Clover, Mountain Misery

Rosaceae (Rose family)

Low-growing, mat-forming, evergreen shrub 2–6 dm. Highly aromatic.

LEAVES: 2–3 times pinnately compound, 2–10 cm, fernlike, glandular-sticky.

FLOWERS: Inflorescences terminal, few-flowered; flowers have 5 white petals, each 6–9 mm, sometimes notched at tips. Pistil 1, ovule 1.

FRUITS: Achenes.

SIERRA DISTRIBUTION: All.

ELEVATION: 600–2350 m

HABITAT: Shaded, dry to moist areas in coniferous forest and meadow edges; montane zone.

FLOWERING TIME: May–Jul

NOTES: Bear clover carpets the forest floor in many parts of the Sierra, holding the soil in place and resprouting rapidly after fire or logging. Roots can extend to 1.5 m (5 ft) below the soil surface, and a single rhizome can be 25 m (82 ft) long!

Chamaebatiaria millefolium

Desertsweet, Fern Bush

Rosaceae (Rose family)

Many-branched, evergreen shrub 6–20 dm; stellate-hairy, and glandular. Strongly aromatic.

LEAVES: Alternate, 1–2 times pinnately compound, 2–8 cm, fernlike, glandular-sticky.

FLOWERS: Inflorescences 3–15 cm, of 20–400 flowers; flowers have 5 white petals, ± 5 mm, sometimes notched at tips. Pistils 4–5, ovules > 2.

FRUITS: Follicles 3–5 mm, red-brown.

SIERRA DISTRIBUTION: East slope.

ELEVATION: 900–3400 m

HABITAT: Dry rocky areas; sagebrush zone.

FLOWERING TIME: Jun–Aug

NOTES: Native Americans east of the Sierra make a tea from the leaves to treat cramps and stomach ailments.

Coleogyne ramosissima

Blackbrush

Rosaceae (Rose family)

Evergreen shrub 2–20 dm; with an overall rounded shape, many-branched, and thorny; drought deciduous. Aromatic.

LEAVES: In opposite clusters. Simple, 5–15 mm, linear to oblanceolate, with stipules persistent.

FLOWERS: Single at the tips of branches; pedicel with 2 bractlets is immediately below flower; petals absent; sepals 4 (sometimes 5), yellow, often tinged red below. Ovary superior; pistil 1; stamens 20–25.

FRUITS: Achenes 4–6 mm, crescent-shaped, hairless.

SIERRA DISTRIBUTION: East slope, Inyo, Kern, and Mono Counties.

ELEVATION: 1000–2000 m

HABITAT: Dry, open areas; sagebrush and pinyon-juniper zones.

FLOWERING TIME: Apr–Jun

NOTES: Primarily a plant of the desert, blackbrush extends up the east slope of the Sierra. Look for it in the Rock Creek drainage, Whitney Portal area, and Short Canyon.

Comarum palustre

Marsh Cinquefoil

Rosaceae (Rose family)

Loosely matted perennial herb; vegetative stems are creeping or floating, and they root at nodes; flowering stems 2–5 dm, erect.

LEAVES: Alternate, pinnate or nearly palmately compound, 5–20 cm; leaflets 15–70 mm, toothed.

FLOWERS: Inflorescences of < 10 flowers. Sepals reddish, triangular, > petals; petals 2–6 mm, ovate or elliptic, and darker red. Pistils many; stamens 20–25.

FRUITS: Achenes 1–1.5 mm, hairless, with a spongy receptacle.

SIERRA DISTRIBUTION: Tuolumne County north, montane zone.

ELEVATION: < 2400 m

HABITAT: Marshes, boggy areas, fens.

FLOWERING TIME: May–Aug

NOTES: Formerly *Potentilla palustris*.

Dasiphora fruticosa
Shrubby Cinquefoil
Rosaceae (Rose family)

Low-growing shrub < 1 m; nonglandular.
LEAVES: Alternate, 1–3 cm, leathery; pinnately compound, with 2–3 leaflets per side or subpalmate.
FLOWERS: 1–6 at branch ends; petals 5, 5–10 mm, longer than sepals. Pistils numerous; stamens 20–25.
FRUITS: Achenes hairy.
SIERRA DISTRIBUTION: Fresno and Inyo Counties north.
ELEVATION: 2000–3600 m
HABITAT: Variable; from moist meadows and streambanks to open rocky sites, talus, scree; subalpine and alpine zones.
FLOWERING TIME: Jun–Sep
NOTES: Formerly *Potentilla fruticosa*. This is the only bright yellow, potentilla-like shrub in the High Sierra. Circumboreal.

Drymocallis glandulosa var. *glandulosa*
Sticky Cinquefoil
Rosaceae (Rose family)

Tufted perennial herb 20–60 cm; usually erect, hairy, with glandular hairs abundant at base of stem.
LEAVES: Basal 10–25 cm, pinnately compound, with 2–3 pairs of lateral leaflets; terminal leaflet 25–55 mm, with teeth 7–15 per side, each tooth double-toothed. Cauline leaves alternate, smaller.
FLOWERS: Inflorescences leafy, with widely spreading branches (angles 30°–55°); pedicels 1–5 mm, the lowermost ones longer (up to 15 mm), glandular; bractlets 5 below sepals (hypanthium bractlets) are 2.5–6 mm x 0.5–2 mm wide. Sepals 4.5–7 mm, spreading, and widely obtuse; petals 3.5–5 mm, pale yellow to cream-colored.
FRUITS: 1–1.5 mm, brown or reddish.
SIERRA DISTRIBUTION: Both slopes, all.
ELEVATION: 400–2000 m
HABITAT: Moist, shaded sites; montane and subalpine zones.
FLOWERING TIME: May–Jul
NOTES: Formerly *Potentilla glandulosa* ssp. *glandulosa*. Co-occurring in the same general range is *D. glandulosa* ssp. *reflexa*, which differs by having leafless inflorescences, bright yellow petals, and sepals that are reflexed rather than spreading.

Drymocallis lactea var. *lactea*

Sierran Woodbeauty

Rosaceae (Rose family)

Many-stemmed, usually erect perennial herb 10–60 cm; tufted, hairy, with hairs at base of stem < 1 mm, and glandular hairs absent or few.

LEAVES: Basal 5–20 cm, pinnately compound, with leaflets in 3–4 pairs, the terminal leaflet 20–50 mm, and teeth 6–11 per side. Cauline leaves alternate, smaller.

FLOWERS: Inflorescences narrow, and branches angled close to the main stem; bractlets 5 below sepals are 2–5 mm. Pedicel hairs dense, mostly nonglandular. Sepals 3–8 mm; petals 4–8 mm, cream-colored.

FRUITS: 1 mm, brown.

SIERRA DISTRIBUTION: Tulare and Inyo Counties north.

ELEVATION: 1800–3700 m

HABITAT: Moist, often rocky sites; montane, subalpine, and alpine zones.

FLOWERING TIME: Jun–Sep

NOTES: Includes cream-petaled plants formerly included in *Potentilla glandulosa* ssp. *nevadensis.* The similar-looking *Drymocallis hansenii* has a taller (30–90 cm) single stem with abundant long (2–3 mm) hairs at the base, most of them glandular. *D. hansenii* (formerly *Potentilla glandulosa* ssp. *hansenii*) ranges from Fresno County north on the west slope (1200–2200 m).

Drymocallis pseudorupestris var. *crumiana*

Crum's Woodbeauty

Rosaceae (Rose family)

Tufted to matted, alpine perennial herb 5–25 cm; dense, glandular hairs at base.

LEAVES: Basal 6–9 cm, pinnately compound, with 4 pairs of leaflets, the terminal leaflet 5–20 mm; cauline leaves alternate, smaller.

FLOWERS: Inflorescence branches held 20°–40° from axis. Pedicels 3–10 mm, with dense hairs, many glandular. Sepals 4–6 mm with hairs both glandular and long (1–1.5 mm), and nonglandular; petals 4–8 mm, cream to light yellow.

FRUITS: 1 mm, light brown.

SIERRA DISTRIBUTION: Both slopes and crest, Tulare, Inyo, and Fresno Counties.

ELEVATION: 2300–3900 m

HABITAT: Rocky areas; subalpine and alpine zones.

FLOWERING TIME: Jul–Sep

NOTES: Formerly *Potentilla glandulosa* ssp. *pseudorupestris*. Sparingly distributed along the SN crest, var. *saxicola*, cliff woodbeauty, has 3 leaflets per side, calyx hairs glandular and usually also with short, nonglandular hairs < 1 mm (sometimes no nonglandular hairs), and styles are golden-brown.

Fragaria vesca
Wood Strawberry
Rosaceae (Rose family)

Low-growing perennial herb 3–15 cm.

LEAVES: Thin, petioles 3–25 cm, and stalk of the central leaflet < 2 mm (leaflets appear sessile). Leaflets 3, 1.5–7 cm, densely silky-hairy below, and hairless or sparsely hairy above; toothed from below the middle, with the central leaflet having 12–21 teeth, and the central tooth at tip of leaflet a little bigger or smaller than adjacent teeth, but not markedly smaller.

FLOWERS: Inflorescences often longer than leaves; flowers ± 15 mm wide, with sepals 4–8 mm, and petals 5–8 mm, white, not much longer than sepals.

FRUITS: Receptacle 5–10 mm, enlarged, fleshy, incompletely covered with red fruits.

SIERRA DISTRIBUTION: West slope, Tulare County north.

ELEVATION: < 2150 m

HABITAT: Shaded, dry to moist areas in coniferous forest and meadow edges; montane zone.

FLOWERING TIME: Mar–Jun

NOTES: Can be difficult to distinguish from *F. virginiana* where the 2 spp. overlap, though at lower elevations this species predominates, and at higher elevations (subalpine zone up) *F. virginiana* takes over. Pay attention to leaflet toothing and leaflet stalks, hairiness, and length of petals in relation to sepals.

Fragaria virginiana
Mountain Strawberry
Rosaceae (Rose family)

Low-growing perennial herb 2–12 cm; stoloniferous, forming large mats.

LEAVES: Thin, petioles 1–25 cm, and the stalk of the central leaflet 1–3 mm; all leaflets have at least short stalks. Leaflets 1.5–6 cm, usually hairless below, and usually hairless and glaucous above; toothed from above middle, with the central leaflet having 7–13 teeth, and the central tooth at tip of leaflet much shorter than adjacent teeth.

FLOWERS: Inflorescences shorter than leaves; flowers 10–20 mm wide, sepals 3–6 mm, and petals 4–9 mm, white, longer than sepals.

FRUITS: Receptacle 10 mm wide, enlarged, fleshy, covered with red fruits, each fruit 1.5 mm.

SIERRA DISTRIBUTION: Both slopes, Fresno and Inyo Counties north.

ELEVATION: 1200–3300 m

HABITAT: Shaded areas in forests and in and around meadows; montane and high subalpine zones.

FLOWERING TIME: May–Aug

NOTES: See notes under *F. vesca*.

Geum macrophyllum var. *perincisum*

Large-Leaved Avens

Rosaceae (Rose family)

Erect, hairy perennial herb 3–10 dm.
LEAVES: Mostly basal, pinnate, with a disproportionately large, rounded, terminal leaflet. Stem leaves smaller.
FLOWERS: Open cymes, with glandular hairs on pedicels. Petals 5, yellow, sepals 5, reflexed with age. Pistils and stamens many.
FRUITS: Achenes with elongated styles, combining to form burlike heads.
SIERRA DISTRIBUTION: All.
ELEVATION: 1000–3300 m
HABITAT: Moist to wet areas; montane and subalpine zones.
FLOWERING TIME: May–Aug
NOTES: The enormous, fanlike terminal leaflet of the basal leaf is the key to identifying large-leaved avens. The burlike head of fruits is also distinctive.

Geum triflorum var. *ciliatum*

Prairie Smoke

Rosaceae (Rose family)

Gray-green, woolly perennial herb 2–5 dm; rhizomatous, forming patches.
LEAVES: Compound, 4–30 cm, with leaflets wedge-shaped, deeply 2- to 3-lobed, and the lobes deeply toothed.
FLOWERS: Inflorescences of 1–3 flowers, cream to pale green, tinged with pink, purple, or maroon. Sepals shorter than petals; bracts 5, slender, subtending the sepals.
FRUITS: Achenes have styles 15–40 mm, elongated, plumose, combining to form woolly heads.
SIERRA DISTRIBUTION: Both slopes, Tuolumne and Mono Counties north.
ELEVATION: 1300–3200 m
HABITAT: Open coniferous forest, edges of meadows, sagebrush flats; montane, subalpine, and sagebrush/pinyon-juniper zones.
FLOWERING TIME: May–Jul
NOTES: The fruiting head is as attractive as the flower in this species.

Heteromeles arbutifolia

Toyon, Christmas Berry

Rosaceae (Rose family)

Evergreen shrub or small tree < 10 m; bark gray, no spines.
LEAVES: Alternate; blades 10–50 mm, elliptical, with margins toothed, and veins reaching teeth; no stipules.
FLOWERS: Panicles dome-shaped, many-flowered; flowers small, white, with short, triangular sepals and 5 petals < 4 mm. Ovary inferior, styles 2–3; stamens 10.
FRUITS: Pome 5–10 mm wide, bright red. Seeds 2–3, large, smooth.
SIERRA DISTRIBUTION: West slope.
ELEVATION: < 1005 m
HABITAT: Mostly in chaparral; foothill zone.
FLOWERING TIME: Apr–Jul
NOTES: This attractive native shrub has been used extensively in landscapes and gardens and can be grown from seed or from stem cuttings. The red, berrylike pomes persist cheerfully throughout the winter. It is a good drought-resistant plant that can be trained to various shapes. The flowers are visited by bees, butterflies, and other insects, and the berries are consumed by birds, including mockingbirds, American robins, and cedar waxwings. Mammals including coyotes and bears also eat and disperse the berries.

Holodiscus discolor var. *microphyllus*

Small-Leaved Cream Bush, Rock Spiraea

Rosaceae (Rose family)

Low-growing shrub 0.3–1 m; twigs are soft-hairy, lateral branchlets 0.3–2 cm.
LEAVES: Alternate, ± sessile; blades 0.3–3 cm, ± wedge-shaped, densely hairy on both surfaces, and toothed above middle of margin; the teeth themselves entire (leaf margins not double-toothed); no stipules present.
FLOWERS: Inflorescences 2–8 cm x 1.5–5 cm wide, unbranched or with few branches, and the axis densely hairy; sepals 5, 1–2 mm; petals 5, 1.5–2 mm, white. Style 1 mm.
FRUITS: Achenes 1–1.5 mm, often glandular.
SIERRA DISTRIBUTION: Both slopes, Plumas County south.
ELEVATION: 1100–4000 m
HABITAT: Rocky sites, cliffs; upper montane, subalpine, and alpine zones.
FLOWERING TIME: Jun–Sep
NOTES: Intergrades with *H. discolor* var. *discolor*, which co-occurs in the montane and subalpine zones < 3200 m. Technically, var. *discolor* is a taller (1.5–6 m) shrub with longer (1.5–8 cm) leaves, and petioles 2–15 mm rather than sessile, margins double-toothed (teeth are themselves toothed), and larger (2–25 cm x 5–25 cm wide), many-branched inflorescences.

Horkelia fusca var. *parviflora*

Dusky Horkelia

Rosaceae (Rose family)

Tufted to mat-forming perennial herb 10–60 cm; herbage green.
LEAVES: Basal and alternate, pinnately compound, 4–15 cm; leaflets 4–8 pairs, 5–15 mm, ± 5-toothed, and teeth ¼–½ to base of leaflet.
FLOWERS: Headlike clusters of 5–20 flowers; sepals 2–3 mm; petals 2–4 mm, wedge-shaped. Pistils 10–20, styles 1 mm; stamens 10.
FRUITS: 1.2 mm.
SIERRA DISTRIBUTION: Tulare and Mono Counties north.
ELEVATION: 1400–3300 m
HABITAT: Open sandy flats, meadow edges, damp places in forests; montane and subalpine zones.
FLOWERING TIME: Jun–Sep
NOTES: Compare with *H. tridentata* var. *tridentata*, especially the number of leaflet teeth and the petal shape. *H. californica* var. *elata* could superficially be confused with *H. fusca*, but it is a taller (< 1.2 m) plant, bright green with roundish leaflets with 20+ teeth. It occurs in moist areas or formerly wet areas in the foothill and lower montane zones from Tulare to Amador Counties < 1830 m.

Horkelia parryi

Parry's Horkelia

Rosaceae (Rose family)

Mat-forming perennial herb; stem 10–30 cm, ascending to erect; herbage is bright green, and hairless to sparsely hairy.
LEAVES: Basal and alternate, pinnately compound, 5–10 cm, with 3–6 leaflets per side, 5- to 10-toothed.
FLOWERS: Open cymes of 5–10 flowers; pedicels 5–15 mm, erect in flower; bractlets 5, lanceolate, at base of and alternate to the sepals; sepals 5, 4–6 mm; petals 4–7 mm, white. Pistils 20–50, styles 2 mm; stamens 10.
FRUITS: Pedicels recurved in fruit. Achenes 10–50, enclosed by the calyx appearing to be nodding.
SIERRA DISTRIBUTION: West slope, Mariposa to El Dorado Counties.
ELEVATION: < 1100 m
HABITAT: Openings in chaparral, lower-elevation ponderosa pine or knobcone pine forest; foothill and montane zones. Associated with the Ione formation to the north, and on phyllite in Mariposa County.
FLOWERING TIME: Apr–Sep
NOTES: CRPR 1B.2, Parry's horkelia can form large, vigorous, continuous patches in openings on ridgetops in Mariposa County after fire and even on dozer lines. Recent repeated bulldozing of the same fire lines every few years may be detrimental, however. The bright green foliage and nodding fruits are diagnostic.

Horkelia tridentata var. *tridentata*

Three-Toothed Horkelia

Rosaceae (Rose family)

Tufted, decumbent to ascending perennial herb < 45 cm; grayish.

LEAVES: Basal and alternate, pinnately compound, 3–12 cm, with 2–5 pairs leaflets, ± 3-toothed < ¼ to base of leaflet.

FLOWERS: Clusters of 3–40 flowers; sepals 1.5–3 mm; petals 1.5–4 mm, narrowly oblanceolate. Pistils 5–15, styles 1–2.5 mm; stamens 10.

FRUITS: 1.5–2 mm.

SIERRA DISTRIBUTION: Tulare County north.

ELEVATION: 300–2500 m

HABITAT: Open sandy flats, meadow edges, damp places in forests; montane zone.

FLOWERING TIME: Jun–Sep

NOTES: Compare with *H. fusca* var. *parviflora*; check the number of leaflet teeth and the petal shape. Another variety, *H. tridentata* var. *flavescens*, is decumbent, petals are widely oblanceolate and shorter than sepals, and it occurs mostly on serpentine from El Dorado County north.

Horkeliella purpurascens

Purple False Horkelia

Rosaceae (Rose family)

Tufted, grayish perennial herb 15–90 cm; ascending to erect, glandular and nonglandular hairs are present. May smell resinous.

LEAVES: Mostly basal, 7–17 cm; leaflets 15–35 per side, overlapping, with < 5 leaflet lobes; stem leaves 2–4.

FLOWERS: Open inflorescences, flowers 10–15 mm wide; sepals 3–6 mm; petals 3–7 mm, oblanceolate. Stamens 20, with flat filaments, ± forming a tube.

FRUITS: 1.5 mm.

SIERRA DISTRIBUTION: Tulare, Inyo, and Kern Counties.

ELEVATION: 1400–2900 m

HABITAT: Meadow edges; montane zone.

FLOWERING TIME: Jun–Aug

NOTES: Differs from *Horkelia* by having 20 rather than 10 stamens.

Ivesia lycopodioides var. *lycopodioides*

Club Moss Ivesia

Rosaceae (Rose family)

Tufted perennial herb with stems 3–15 cm; horizontal to ascending.

LEAVES: Mostly basal, 1–7 cm, cylindrical, ± hairless; leaflets 10–35 per side, somewhat overlapping, with 4–10 lobes, 1 mm, with no terminal bristle. Stem leaf 1, smaller.

FLOWERS: Dense heads of 3–20 flowers. Flower 6–12 mm wide; petals 2–3 mm. Pistils 5–15, styles 1–2 mm; stamens 5.

FRUITS: 1–1.5 mm, smooth, pale.

SIERRA DISTRIBUTION: Fresno and Mono Counties north to El Dorado County.

ELEVATION: 2740–4000 m

HABITAT: Rocky sites; subalpine and alpine zones.

FLOWERING TIME: Jul–Aug

NOTES: The species name refers to the fact that the leaves of these plants resemble club mosses, which are primitive vascular plants in the genus *Lycopodium*.

Ivesia muirii ✿

Muir's Ivesia

Rosaceae (Rose family)

Tufted perennial herb 5–15 cm; erect or ascending, silvery-hairy.

LEAVES: Mostly basal, 2–5 cm, resembling mouse tails; leaflets 25–40 per side, densely overlapping, with ± 3 lobes, 1–3 mm. Stem leaf 1, much smaller.

FLOWERS: Headlike clusters of 10–20 flowers. Flowers 5–6 mm wide, petals 1–2 mm, yellow, ≤ sepals. Pistils 1–4; stamens 5.

FRUITS: ± 2 mm, smooth, gray with red spots.

SIERRA DISTRIBUTION: Tulare and Inyo Counties north to Tuolumne and Mono Counties.

ELEVATION: 2700–4000 m

HABITAT: Rocky areas; subalpine and alpine zones.

FLOWERING TIME: Jun–Aug

NOTES: Very distinctive! Compare leaves with those of mousetail ivesia, *I. santolinoides*.

Ivesia pygmaea

Pygmy Ivesia

Rosaceae (Rose family)

Matted perennial herb with stems 3–15 cm; decumbent to erect, bright green.

LEAVES: Mostly basal, 1–10 cm; leaflets 10–15 per side, overlapping, with 3–5 lobes, 1–3 mm, oblanceolate. Stem leaf 1.

FLOWERS: Dense to open cymes 8–30 mm wide, with 5–10 flowers. Flowers 9–11 mm wide; petals 2–4 mm, shorter than or the same length as sepals. Pistils 10–30; stamens 10.

FRUITS: 1–1.5 mm, smooth, pale.

SIERRA DISTRIBUTION: High country of Tulare, Fresno, and Inyo Counties (1 record from 1888 from Tuolumne County).

ELEVATION: 2700–4000 m

HABITAT: Rocky sites; subalpine and alpine zones.

FLOWERING TIME: Jul–Sep

NOTES: Other yellow-flowered *Ivesia* spp. have 5 stamens.

Ivesia santolinoides

Mousetail Ivesia

Rosaceae (Rose family)

Erect, tufted perennial herb with stems 15–40 cm; foliage is silvery, and flowering stems are baby's breathlike.

LEAVES: Mostly basal, 4–10 cm, mousetail-like; leaflets 40–80 per side, indistinguishable. Stem leaves 1–3, smaller.

FLOWERS: Open, airy cymes with 30–200 flowers. Flowers 5–8 mm wide; petals 2 mm, > sepals. Pistil 1; stamens 15.

FRUITS: 2 mm, smooth, mottled.

SIERRA DISTRIBUTION: El Dorado County south.

ELEVATION: 1500–3600 m

HABITAT: Open sandy or gravelly areas, rocky outcrops; montane, subalpine, and alpine zones.

FLOWERING TIME: Jun–Sep

NOTES: A most distinctive plant, well camouflaged in its grayish habitat, but quite spectacular upon close examination. The only other plant with leaves like mouse tails is *I. muirii*.

Ivesia sericoleuca

Plumas Ivesia

Rosaceae (Rose family)

Decumbent to ascending, tufted perennial herb with stems 15–45 cm; green to whitish-hairy.

LEAVES: 10–20 cm, sheathing bases densely white-strigose; leaflets 20–35 per side, with 1–4 lobes, 3–15 mm. Stem leaves 3–8.

FLOWERS: Clusters of 5–20 flowers. Flowers 10–15 mm wide, petals 4–7 mm, > sepals. Pistils 2–7; stamens ± 20.

FRUITS: 2–3 mm, smooth, brown.

SIERRA DISTRIBUTION: El Dorado County north.

ELEVATION: 1300–2320 m

HABITAT: Dry, usually volcanic meadows; montane and sagebrush zones.

FLOWERING TIME: May–Sep

NOTES: Rare, CRPR 1B.2; restricted to 4 counties in the northern Sierra. It can be seen in meadows north of Truckee. Yosemite ivesia, *I. unguiculata*, looks similar but their ranges do not overlap. *I. unguiculata* is found on the west slope in dry meadows of montane and subalpine zones in Mariposa, Madera, and Fresno Counties, 1500–2000 m; it has 15–20 leaflets per leaf side.

Ivesia shockleyi var. *shockleyi*

Shockley's Ivesia

Rosaceae (Rose family)

Mat-forming perennial herb with stems < 15 cm; spreading, and glandular hairy.

LEAVES: Mostly basal, 2–8 cm; leaflets 5–10 per side, with 3–5 lobes, 1–5 mm. Stem leaf 1, much smaller.

FLOWERS: Open cyme of 2–10 flowers. Flowers 5–10 mm wide; petals < 2 mm, < sepals. Pistils 2–6 (usually 3); stamens 5.

FRUITS: Pedicels S-shaped in fruit. Achenes 2 mm, smooth, light brown.

SIERRA DISTRIBUTION: Inyo, Fresno, Mono, Alpine, and Placer Counties.

ELEVATION: 2700–4000 m

HABITAT: Dry rocky or gravelly areas; alpine zone.

FLOWERING TIME: Jun–Aug

NOTES: The relatively few leaflets, tiny and narrow petals, few fruits, and S-shaped fruiting pedicels are distinguishing features of this alpine *Ivesia*.

Oemleria cerasiformis

Oso-Berry

Rosaceae (Rose family)

Deciduous shrub or small tree 1–6 dm; dioecious.

LEAVES: Alternate, simple, 5–13 cm x 2–5 cm wide; leaves paler below, hairless above, fine-hairy or hairless below, with margins entire, rolled under, and stipules early deciduous.

FLOWERS: Dioecious; racemes 3–10 cm of fragrant staminate (male) or pistillate (female) flowers on short, lateral branches; pendent.

FRUITS: Drupe 5–15 mm, bean-shaped, blue-black.

SIERRA DISTRIBUTION: West slope, Tulare County north (no records for Madera or Tuolumne Counties).

ELEVATION: < 1850 m

HABITAT: Dry to moist sites; foothill and montane zones.

FLOWERING TIME: Feb–Apr

NOTES: The berries are eaten by a variety of native mammals and birds, who also help disperse the seeds. Native American uses are many; the bark is used to treat tuberculosis and as a laxative, and the fruits are dried or cooked and eaten.

Physocarpus capitatus

Pacific Ninebark

Rosaceae (Rose family)

Deciduous shrub 1–2.5 m; with stellate hairs.

LEAVES: Alternate, with stipules; blades palmately 3- to 5-lobed, and toothed; hairless or hairy.

FLOWERS: Clusters of > 10 flowers; sepals 2.5–3.5 mm; petals 2.5–3 mm. Stamens 25–30; filaments as long as 4–5 mm.

FRUITS: Follicles 8–10 mm, hairless or hairy, splitting along both sides. Seeds 1–4 per follicle.

SIERRA DISTRIBUTION: Primarily west slope, Tulare County north.

ELEVATION: < 1400 m

HABITAT: Streambanks, north-facing slopes; foothill and montane zones.

FLOWERING TIME: May–Jul

NOTES: The common name "ninebark" refers to the fact that the bark peels away, revealing several layers. Available for sale as a garden plant, it can be somewhat drought tolerant, though generally it prefers moist conditions.

Potentilla breweri
Brewer's Cinquefoil
Rosaceae (Rose family)

Silvery-hairy, ascending or decumbent perennial herb 10–40 cm; from branched caudexes, often ± tufted.

LEAVES: Pinnate, with petioles shorter than blades; leaflets 3–7 per side, ± overlapping, deeply palmately toothed. Leaves silvery cottony-hairy.

FLOWERS: Cymes of 3–15 flowers; petals 5–9 mm, heart-shaped. Pistils 15–25; styles 2–3 mm, slender.

FRUITS: 1.5–2 mm, brown, smooth.

SIERRA DISTRIBUTION: Tulare and Inyo Counties north.

ELEVATION: 2000–3800 m

HABITAT: Moist areas, meadows, fellfields; montane, subalpine, and alpine zones.

FLOWERING TIME: Jun–Sep

NOTES: Formerly *P. drummondii* ssp. *breweri*. Compare with *P. drummondii*.

Potentilla drummondii
Drummond's Cinquefoil
Rosaceae (Rose family)

Ascending or decumbent perennial herb 10–60 cm; from branched caudexes, often ± tufted.

LEAVES: Pinnate; petioles larger than blades, and largest blades 5–25 cm; leaflets 2–4 per side, usually not overlapping, deeply irregularly 7- to 15-toothed. Leaves green, hairless, or have hairs straight (not cottony-hairy).

FLOWERS: Cymes of 3–15 flowers; petals 7–10 mm, heart-shaped. Pistils many, styles 2–3 mm.

FRUITS: 1.5–2 mm, brown, smooth.

SIERRA DISTRIBUTION: Tulare and Inyo Counties to Butte County.

ELEVATION: 1110–3500 m

HABITAT: Meadows; subalpine and alpine zones.

FLOWERING TIME: Jun–Sep

NOTES: Formerly *P. drummondii* ssp. *drummondii*. Occurring in the same range is *P. bruceae* (formerly *P. drummondii* ssp. *bruceae*), which also has 2–4 leaflets per side, but they overlap and are densely cottony-hairy or shaggy.

Potentilla flabellifolia

Fanfoil, Fan-Leaf Cinquefoil

Rosaceae (Rose family)

Low-growing perennial herb; forming loose clusters from scaly, branched rootstocks. Stems are 10–30 cm, ascending to erect; plant is hairless or sparsely long-hairy, nonglandular.

LEAVES: Mostly basal, 3–12 cm, hairless; leaflets 3, up to 30 mm, fan-shaped, with teeth 7–15 and uneven, reaching ± ¼ to midvein.

FLOWERS: Cymes of 1–5 yellow flowers with petals 6–10 mm; pedicels 10–20 mm. Pistils > 20, styles > 1 mm and attached near tip of ovary; stamens ± 20.

FRUITS: 1.2 mm, smooth, reddish or brownish.

SIERRA DISTRIBUTION: Tulare and Inyo Counties north.

ELEVATION: 1830–3500 m

HABITAT: Moist meadows; upper montane, subalpine, and alpine zones.

FLOWERING TIME: Jul–Sep

NOTES: This is a common meadow-dwelling *Potentilla* at higher elevations and is distinguished fairly easily by having only 3 leaflets. *P. grayi* also has 3 leaflets but has < 20 pistils, fewer (± 7) leaflet teeth on central leaflet (these divided ¼–½ to midrib), pedicels 10–40 mm, and sparsely strigose stems.

Potentilla glaucophylla var. *glaucophylla*

Blueleaf Cinquefoil

Rosaceae (Rose family)

Tufted perennial herb 10–40 cm; from caudexes, with stems ascending, hairless at bases.

LEAVES: Palmately compound; basal leaves 4–13 cm, with 5 leaflets, the central one 15–40 mm, oblanceolate, 3- to 7-toothed distally ± ¼ to midvein; leaves hairless except on margins and along veins, and glaucous when fresh (bluish).

FLOWERS: Cymes of < 20 flowers; sepals 5, bractlets 5, shorter than sepals; petals 6–8 mm, yellow. Styles ± 2 mm; filaments 1–2.5 mm, anthers 0.4–0.9 mm.

FRUITS: Achenes 1.5 mm, smooth, pale brown.

SIERRA DISTRIBUTION: Both slopes, Tulare and Inyo Counties north to Placer County.

ELEVATION: 2600–3500 m

HABITAT: Moist rocky sites, meadows; subalpine and alpine zones.

FLOWERING TIME: Jul–Sep

NOTES: Intergrades with *P. gracilis* and *P. drummondii*.

Potentilla gracilis var. *fastigiata*

Slender Cinquefoil

Rosaceae (Rose family)

Ascending to erect perennial herb 20–50 cm; from scaly, horizontal rhizomes, stems hairy (stem hairs all the same length), mostly nonglandular.

LEAVES: Palmately compound; leaflets 5–9, with the basal central leaflet 2–6 cm, and each leaflet with 13–23 teeth (teeth shallow, < ½ to midvein); upper and lower leaf surfaces ± similarly hairy.

FLOWERS: Cymes many-flowered; petals 4–7 mm, bright yellow. Styles < 50, 1.5–2 mm.

FRUITS: 1–1.5 mm, brown, smooth.

SIERRA DISTRIBUTION: All.

ELEVATION: 1000–3500 m

HABITAT: Meadows, streamsides, moist soil in forests; montane, subalpine, and alpine zones.

FLOWERING TIME: Jun–Sep

NOTES: The most common *P. gracilis* variety in the SN. Another variety of the higher-elevation west slope, crest, and east slope is var. *elmeri* (inset). It has deeply toothed leaflets (> ¾ to midvein) with woolly undersides, and petals 5–8 mm.

Potentilla millefolia

Feather Cinquefoil

Rosaceae (Rose family)

Prostrate to decumbent perennial herb 5–20 cm; from taproots, hairy, sometimes glandular.

LEAVES: Basal in rosettes, 2–15 cm, with blades longer than petioles, and leaflets 5–13 per side, overlapping, deeply 3- to 10-toothed.

FLOWERS: Inflorescences of < 10 flowers; petals 4–8 mm, yellow. Style 2–3 mm, slender.

FRUITS: Pedicels recurved in fruit; fruit 1.5–2 mm, tan, smooth.

SIERRA DISTRIBUTION: West slope, Nevada County north; east slope, Mono County.

ELEVATION: 900–2000 m

HABITAT: Wet or formerly wet meadows; montane, subalpine, and sagebrush zones.

FLOWERING TIME: Apr–Aug

NOTES: A distinctive *Potentilla*, with its leaves resembling yarrow, *Achillea millefolium*.

Potentilla recta (non-native, invasive)

Sulphur Cinquefoil

Rosaceae (Rose family)

Ascending to erect perennial herb 20–60 cm; from clustered taproots or branched caudexes, hairy (hairs at base of stem are of 2 lengths, easily seen at 10x and sometimes even with the naked eye).

LEAVES: Palmately compound; leaflets usually 6–7, the basal ones withered by flowering time; stem leaves 8–15 cm, with the central leaflet 30–80 mm, and each leaflet 15- to 35-toothed ± halfway to midvein, sparsely hairy.

FLOWERS: Cymes many-flowered; sepals 4–10 mm, petals 7–10 mm, light yellow. Styles ± 1 mm, columnar; stamens about 25, filaments 1–2 mm, anthers ± 1 mm.

FRUITS: Seeds 1–1.5 mm, brown, prominently veined.

SIERRA DISTRIBUTION: West slope, Madera County north (to date, this invasive plant is spreading).

ELEVATION: 150–1900 m, likely will be found at higher elevations

HABITAT: Often in disturbed areas like roadsides and old roadbeds, but also found in dry to moist meadows; montane zone, expected in more zones in the future.

FLOWERING TIME: May–Aug

NOTES: Sulphur cinquefoil is an aggressive CDFA noxious weed that displaces native plants in meadows and grasslands. It is easy to confuse with native slender cinquefoil, *P. gracilis*, and the 2 often grow together. Hints on distinguishing them: Sulphur cinquefoil has both long and short hairs at the stem base, basal leaves that wither by blooming time, distinctive seeds with raised veins (appearing sculpted, in contrast to the smooth seeds of *P. gracilis*), and the flowers tend to be a lighter yellow.

Poteridium annuum

Western Burnet

Rosaceae (Rose family)

Erect annual or biennial herb 10–70 cm; from taproots.
LEAVES: Basal and alternate, pinnately compound, 3–12 cm, leaflets deeply linear-lobed that give a feathery, fernlike appearance. Basal leaves turn brown once blooming begins.
FLOWERS: Spikes 5–35 mm, oblong, headlike, of 10–50 flowers. Sepals 4, green, and no petals.
FRUITS: Achenes 2–4 mm, 4-angled, enveloped in hypanthium.
SIERRA DISTRIBUTION: West slope, Tuolomne County north.
ELEVATION: 225–1890 m
HABITAT: Open, often disturbed areas; foothill and montane zones.
FLOWERING TIME: May–Aug
NOTES: Formerly *Sanguisorba occidentalis*. Similar to a non-native species that appears to be spreading in the SN, garden burnet, *Poterium sanguisorba* (formerly *Sanguisorba minor* ssp. *muricata*), a perennial with shallowly lobed leaflets and basal leaves remaining green at bloom time. Currently it is recorded from Madera, Mariposa, El Dorado, Placer, Nevada, and Butte Counties.

Prunus andersonii

Desert Peach

Rosaceae (Rose family)

Thorny, highly branched, deciduous shrub < 3 m.
LEAVES: Blades 9–30 mm, toothed.
FLOWERS: Inflorescences of 1–2 flowers, positioned close to each other and forming showy flowering branches; sepals 5, hairless with gland-toothed margins; petals 5, 5–18 mm, dark pink (sometimes lighter).
FRUITS: 10–18 mm, round to obovoid, fuzzy, yellowish-greenish to orange.
SIERRA DISTRIBUTION: East slope, Kern County north.
ELEVATION: 900–2600 m
HABITAT: Open areas, rocky slopes; montane and sagebrush zones.
FLOWERING TIME: Mar–May
NOTES: The presence of thorns and the large, fuzzy round fruit, along with the mostly east slope distribution of this species, help to distinguish it from other Sierra *Prunus*.

Prunus emarginata

Bitter Cherry

Rosaceae (Rose family)

Deciduous shrub to small tree < 15 m; forming dense thickets, not spiny; young twigs are shiny red.

LEAVES: Blades 1.5–6 cm, elliptic to obovate, scalloped-toothed, with short petioles 3–12 mm.

FLOWERS: Inflorescence is a raceme of 3–12 flowers; pedicels 3–12 mm; sepals 5, hairless, margins entire; petals 5, 3–8 mm, white.

FRUITS: 7–14 mm, round, hairless, bright red, with fleshy pulp.

SIERRA DISTRIBUTION: All.

ELEVATION: < 3000 m

HABITAT: Dry slopes and rocky sites; montane zone.

FLOWERING TIME: Apr–Jun

NOTES: The narrowly oblanceolate leaves (length > 2x width), prominent veins, small, white flowers in roundish racemes, and bright red berries distinguish this *Prunus* from other west slope species. The plant is a common component of montane chaparral, forming dense, impenetrable stands after fire and logging.

Prunus subcordata

Sierra Plum

Rosaceae (Rose family)

Deciduous shrub < 3 m (up to 6 m); often forming large patches or thickets, with stiff branches (± thorny).

LEAVES: Blades 20–50 mm, roundish, toothed, with rounded or subcordate (heart-shaped) bases.

FLOWERS: Inflorescences of 2–5 white flowers, sparsely spaced on branches; pedicels 5–8 mm; sepals 5, hairless or puberulent, with gland-toothed margins; petals 5, 5–10 mm, white.

FRUITS: 15–25 mm, round, hairless, yellow to dark red, with fleshy pulp.

SIERRA DISTRIBUTION: West slope, all.

ELEVATION: < 1800 m

HABITAT: Dry slopes and rocky sites; foothill and montane zones.

FLOWERING TIME: Mar–May

NOTES: Of the *Prunus* spp. you're likely to confuse on the west slope, Sierra plum has the largest flowers in the most sparing clusters, the roundest leaves, and the petals that fall off the most rapidly.

Prunus virginiana var. *demissa*
Western Choke Cherry
Rosaceae (Rose family)

Deciduous shrub < 6 m (occasionally up to 10 m); can form large patches, not thorny; young twigs are grayish-brown.

LEAVES: Blades 30–100 mm, elliptic to oblanceolate, finely toothed, with pointed tips.

FLOWERS: Inflorescence is an elongate raceme of 18+ white flowers, leafy at base; pedicels 5–8 mm; sepals 5, hairless, with gland-toothed margins; petals 5, 4–7 mm, white.

FRUITS: 6–14 mm, round, hairless, red to black, with fleshy pulp.

SIERRA DISTRIBUTION: All.

ELEVATION: < 3000 m

HABITAT: Moist sites; foothill and montane zones.

FLOWERING TIME: May–Jun

NOTES: Choke cherry has longer inflorescences with more numerous and more densely arranged flowers, and much larger leaves, and it frequents moister sites than the other *Prunus* spp.

Purshia tridentata var. *tridentata*
Bitterbrush
Rosaceae (Rose family)

Shrub 0.2–2.5 m; twig hairs are without glands.

LEAVES: In clusters on short branchlets, evergreen or drought-deciduous, clearly 3-toothed, and densely nonglandular hairy beneath.

FLOWERS: Hypanthium 2.5–5 mm, sepals 2–4 mm, petals 4–8 mm. Pistils 1–2.

FRUITS: 5–7 mm, woolly.

SIERRA DISTRIBUTION: Primarily east slope, but also on west slope.

ELEVATION: 900–3400 m

HABITAT: Open areas, rocky slopes; montane and sagebrush zones.

FLOWERING TIME: Mar–Jul

NOTES: Because bitterbrush was planted in erosion control seed mixes in the past, it is hard to know how many of the populations on the west slope were introduced in this manner. Sometimes it is fairly clear as the plants are strictly along roadsides.

Rosa bridgesii

Bridges' Rose, Sierra Ground Rose

Rosaceae (Rose family)

Low-growing shrub < 1 m; rhizomatous, with dark brown stems. Prickles are sparse and usually in pairs, 3–10 mm, straight, with thickened bases.

LEAVES: Leaflets 5–7, 10–30 mm, elliptic to round, usually with squared-off tips, sparsely glandular-hairy. Leaf margins double-toothed, and teeth glandular; stipules ciliate with gland-tipped hairs.

FLOWERS: 1–2 per inflorescence; axis hairless to hairy, with or without glands; pedicels 10–15 mm, hairless or glandular-hairy but not shaggy. Sepals glandular, lanceolate, with long, tapering tips shorter than the sepal body, and margins entire; petals 10–20 mm, pink to reddish.

FRUITS: 5–10 mm, erect, hairless, with sepals still attached at maturity.

SIERRA DISTRIBUTION: Mostly west slope, Butte and Plumas Counties south.

ELEVATION: 900–2500 m

HABITAT: Rocky areas and forest floors; mostly montane zone.

FLOWERING TIME: May–Aug

NOTES: Distinguished from other native roses by the low-growing habit, square leaflet tips, the straight, often paired prickles, and the erect, persistent fruiting sepals.

Rosa woodsii var. *gratissima*

Mojave Rose

Rosaceae (Rose family)

Densely branched or thicket-forming, erect shrub 0.5–3 m; gray-brown to red-brown stems. Prickles are straight, slender, often dense.

LEAVES: Leaflets 5–7, 10–40 mm, hairless or slightly hairy; margins single-toothed without glands.

FLOWERS: 1–3 per inflorescence; pedicels 10–20 mm, hairless. Hypanthium 3–5 mm wide; sepals have tips ± equal to body, with margins entire; petals 15–20 mm. Pistils 20–35.

FRUITS: 9–12 mm wide, hairless, with sepals present at maturity, erect.

SIERRA DISTRIBUTION: East slope, mostly Mono and Inyo Counties; west slope, limited distribution.

ELEVATION: 800–3400 m

HABITAT: Moist areas; montane and subalpine zones.

FLOWERING TIME: May–Aug

NOTES: Distinguished from other native roses by the lack of glands on leaflets and sepals, straight and slender prickles, and the presence of intermodal prickles in the inflorescence, which the related *R. woodsii* var. *ultramontana*, interior rose, does not have. Interior rose is a common high-elevation species of the central and southern High Sierra and the east slope.

Rubus armeniacus (non-native, invasive)

Himalayan Blackberry

Rosaceae (Rose family)

Woody, prickled vine with mounded to arched stems; stems < 3 m, 20–25 mm diameter, 5-angled, with many stout, wide-based prickles; stems root at tips.

LEAVES: Palmately compound, with 5 (sometimes 3) double-toothed leaflets; stipules < 1 mm wide, threadlike.

FLOWERS: Terminal, paniclelike cymes; sepals hairy, nonglandular; petals 10–15 mm, obovate, white (or pink). Ovary hairless, pistils > 30; filaments threadlike.

FRUITS: Blackberry, hairless.

SIERRA DISTRIBUTION: West slope, Tulare to Butte Counties.

ELEVATION: < 1600 m

HABITAT: Disturbed moist areas, streamsides; foothill and montane zones.

FLOWERING TIME: Mar–Jun

NOTES: Native to Eurasia, and exceedingly invasive. Overtakes riparian vegetation and changes ecosystems quite dramatically. Cutleaf blackberry, *R. laciniatus*, also from Eurasia, is similar in overall growth form, but the 5-angled stems are 3–10 mm, and the 5 leaflets are each deeply divided ("cut"); petals are white and, in contrast to *R. armeniacus*, 3-lobed. Found in similar habitats as Himalayan blackberry and often growing with it, cutleaf blackberry is less abundant.

Rubus glaucifolius

Wax-Leafed Raspberry

Rosaceae (Rose family)

Prostrate to decumbent shrub 1–3 m; hairless, ± glaucous; stems are round in x.s. (not angled), and slender (2–3 mm diameter); stems do not root at nodes or tips; prickles are few, weak (not stout), slender, and straight or ± curved.

LEAVES: Compound; leaflets 3, with the terminal leaflet ovate, sometimes shallowly 3-lobed, coarse-toothed and densely white-tomentose on lower surfaces; stipules ≤ 1 mm wide, threadlike.

FLOWERS: Inflorescence is terminal or in axillary cymes, of 1–4 flowers; sepals hairy, also with stalked glands; petals 4–8 mm, oblong to oblanceolate, and white. Ovary white-hairy, pistils < 10; stamens many, filaments straplike.

FRUITS: Raspberry type, red-purple.

SIERRA DISTRIBUTION: West slope.

ELEVATION: 800–2100 m

HABITAT: Openings in forests, rocky sites; foothill and montane zones.

FLOWERING TIME: Apr–Jul

NOTES: Whitebark raspberry, *R. leucodermis*, is actually more common, found on the west slope < 2400 m; young stems are thicker (4–10 mm diameter), glaucous, white-looking, arched, rooting at tips; prickles are stout, wide-based; lower leaf surface is white-woolly; leaflets are 3–5.

Rubus parviflorus

Thimbleberry

Rosaceae (Rose family)

Erect shrub 0.6–2 m; stems are round with no prickles.

LEAVES: Simple, palmately 5-lobed, 5–15 cm, soft to the touch, deciduous.

FLOWERS: Sepals have long, narrow tips; petals 2–5 cm, white.

FRUITS: Raspberrylike, red; when ripe, fruits are easily removed from receptacle and appear thimblelike.

SIERRA DISTRIBUTION: All.

ELEVATION: < 2600 m

HABITAT: Moist shady areas in conifer forest; montane zone.

FLOWERING TIME: Apr–Aug

NOTES: The soft, pliable leaves of thimbleberry are sometimes referred to as the "Charmin of the Forest" (emergency toilet paper supply!).

Rubus ursinus

California Blackberry

Rosaceae (Rose family)

Prostrate to decumbent shrub with white-glaucous stems 2–10 mm diameter, round in x.s. (not angled). Stems are branched, rooting at tips to form impenetrable masses along streamsides. Prickles are many, slender, straight.

LEAVES: Leaflets 1–3 (rarely 5), triangular-ovate, and grayish and hairy underneath.

FLOWERS: Inflorescence of 1–5 unisexual flowers; sepals hairy, glandular, prickly; petals white, 6–11 mm in female flowers, and 10–15 mm in male; sepals reflexed at flowering time.

FRUITS: Blackberry, hairless.

SIERRA DISTRIBUTION: All.

ELEVATION: < 1500 m

HABITAT: Moist areas in valleys and along streams; foothill and montane zones.

FLOWERING TIME: Mar–Jul

NOTES: The glaucous (whitish) stems, dense fine prickles, and leaves with 3 leaflets that are grayish rather than stark white below distinguish this species.

Sibbaldia procumbens

Creeping Sibbaldia

Rosaceae (Rose family)

Mat-forming perennial herb 2–15 cm.

LEAVES: Basal, with leaflets 3, 5–25 mm, wedge-shaped, tips 3-toothed.

FLOWERS: Inflorescence is a cyme of yellow flowers, with pedicels 3–10 mm; bractlets 5; sepals 5, 2–4 mm; petals 1 mm. Stamens 5.

FRUITS: Achenes 1 mm, brown, smooth.

SIERRA DISTRIBUTION: Tulare and Inyo Counties north.

ELEVATION: 1820–3810 m

HABITAT: Moist forest floor and meadow edges, rocky areas; montane, subalpine, and alpine zones.

FLOWERING TIME: Jun–Aug

NOTES: This is a distinctive species; if it looks just like the picture, it's probably *Sibbaldia*!

Sorbus scopulina

Rocky Mountain Ash

Rosaceae (Rose family)

Deciduous shrub 1–5 m; stem-buds 8–14 mm, shiny, sticky, with sparse white hairs.

LEAVES: Alternate, pinnately compound; leaflets 7–13, 3–8 cm x 1.5–3 cm wide, shiny on the upper surface, with white hairs in axils.

FLOWERS: Inflorescence is a many-flowered panicle, with no pedicel bractlets; pedicels hairy; sepals 5; petals 5, 4–6 mm.

FRUITS: 8–12 mm, orange to reddish-orange, sometimes glaucous. Seeds 4–5 mm, ovate, brown, and flat.

SIERRA DISTRIBUTION: Both slopes, Tulare County north.

ELEVATION: 1000–2800 m

HABITAT: Canyons, moist places, conifer forest; montane zone.

FLOWERING TIME: Jun–Jul

NOTES: The other SN mountain ash is *S. californica*, California mountain ash; it's a shorter shrub at 1–2 m and occurs from Tulare and Inyo Counties north on both slopes, 1200–4300 m. Leaves have 7–9 leaflets that are 2–4 cm x 1–2 cm wide, with red-brown rather than white hairs in axils. Inflorescences differ by having hairless pedicels, petals 3–4 mm, and fruits 6–9 mm, red and glaucous. Flowering time is May–June. In the photo, the fruit and leaves are *S. scopulina*; the flower is *S. californica*.

Spiraea douglasii

Douglas's Spiraea

Rosaceae (Rose family)

Deciduous shrub 10–20 dm; herbage ± tomentose.

LEAVES: Simple, alternate, with petiole < 10 mm; blades 3–9 cm, oblong to obovate, and serrate; leaves tomentose below, ± hairless and darker green above.

FLOWERS: Inflorescence is longer than wide, and consists of a panicle of densely crowded, rose-colored flowers. Hypanthium ± 1 mm; sepals 0.5–1 mm; petals 1.5 mm, pink to rose. Ovary superior; styles terminal; pistils 5, opposite petals and surrounded by the hypanthium; stamens 15 to many, exserted.

FRUITS: Follicles 5, with styles beaklike in fruits; fruit dehiscent on the inner side and top of outer side. Seeds fusiform.

SIERRA DISTRIBUTION: West slope, El Dorado County north.

ELEVATION: < 2060 m

HABITAT: Moist areas, conifer forest; montane zone.

FLOWERING TIME: Jun–Sep

NOTES: A shrub of the northern SN, easily told from *S. splendens* where they overlap by the elongated inflorescence.

Spiraea splendens

Rose Meadowsweet

Rosaceae (Rose family)

Rounded, multistemmed, deciduous shrub 2–9 dm.

LEAVES: Alternate, 1–7 cm, ovate, toothed.

FLOWERS: Panicles of densely crowded, rose-colored flowers. Hypanthium 2–2.5 mm, tubular; sepals 1 mm; petals 1.5 mm. Stamens many, exserted.

FRUITS: Follicles 5, shiny.

SIERRA DISTRIBUTION: Tulare and Inyo Counties north.

ELEVATION: 1820–3810 m

HABITAT: Moist rocky areas, cliffs; montane, subalpine, and alpine zones.

FLOWERING TIME: Jun–Sep

NOTES: Formerly called *Spiraea densiflora*. A popular native shrub for gardening, it is available widely through native plant nurseries.

Cephalanthus occidentalis
California Button Willow
Rubiaceae (Madder family)

Deciduous shrub or small tree 2–10 m.
LEAVES: Opposite or in whorls of 3, 7–20 cm, with short petioles.
FLOWERS: Inflorescence is spheric, of many small flowers; calyx 4-lobed; corollas funnel-shaped with 4 lobes, white or yellowish. Styles protruding; stamens 4, within the tube.
FRUITS: Nutlets 2–4, brown.
SIERRA DISTRIBUTION: West slope, all.
ELEVATION: < 1000 m
HABITAT: Moist to wet sites; foothill and montane zones.
FLOWERING TIME: May–Sep
NOTES: This charming shrub is always fun to find in flower as summer yields to fall. Butterflies and moths are frequent visitors.

Galium aparine
Goose Grass
Rubiaceae (Madder family)

Perennial herb with stems 30–90 dm; climbing, sprawling, erect, or prostrate; stems have small prickles.
LEAVES: In whorls of 6–8, usually held parallel to the ground, ovate, with acute tips, sometimes with terminal bristles.
FLOWERS: In clusters of 2–3 in leaf axils. Flowers bisexual; corollas rotate, with 4 lobes, 2–3 mm wide, greenish, cream, or pink; no sepals present.
FRUITS: Nutlets soft-hairy, with hairs white or brown, hooked.
SIERRA DISTRIBUTION: All.
ELEVATION: < 2300 m
HABITAT: Forest floor, grassy areas; foothill, montane, and sagebrush zones.
FLOWERING TIME: Mar–Jul
NOTES: This is an extremely common *Galium,* found throughout CA, the eastern US, and Europe. The small, downward-pointing prickles on the stem are surprisingly strong, allowing the plant to climb on other plants or to stick to clothing. Low mountain bedstraw, *Galium bifolium*, is another native annual with hooked hairs on the nutlets, occurring up to 3700 m. It is a shorter (5–15 cm) plant, with flowers single in axils, 3 rather than 4 petals, and leaves in whorls of 4 (2 unequal pairs).

Galium bolanderi

Bolander's Bedstraw

Rubiaceae (Madder family)

Perennial herb with stems 15–36 cm; with woody bases, stems tufted or climbing, stout, hairless or scabrous, sometimes gray-hairy.

LEAVES: In whorls of 4 (2 ± unequal pairs), 6–27 mm, elliptic to narrow, acute or with soft, pointed tips, obscurely 3-veined.

FLOWERS: Male flowers in clusters in leaf axils. Female flowers single in axils; corollas rotate with 4 lobes, 1–2 mm wide, greenish or purplish; no sepals present.

FRUITS: Berries 2, hairless or hairy.

SIERRA DISTRIBUTION: All.

ELEVATION: < 2600 m

HABITAT: Open conifer forest, chaparral, rocky sites; foothill, montane, and sagebrush zones.

FLOWERING TIME: May–Aug

NOTES: Berries go from light green to whitish to dark purple-black when fully mature.

Galium porrigens var. *tenue*

Climbing Bedstraw

Rubiaceae (Madder family)

Sprawling or climbing perennial herb with stems 1–1.5 m; woody bases, scabrous, with recurved hairs; plants are dioecious.

LEAVES: In whorls of 4, 3–8 mm, ± linear, with tips acute or obtuse, and terminal hair not persistent.

FLOWERS: Male flowers in clusters in leaf axils. Female flowers single in axils; corollas rotate with 4 lobes, pale yellow to reddish-green; no sepals present.

FRUITS: Berries 1–2, hairless or slightly hairy, white or tinged purple.

SIERRA DISTRIBUTION: West slope, most counties.

ELEVATION: < 1050 m

HABITAT: Chaparral, rocky sites, foothill woodland; foothill and lower montane zones.

FLOWERING TIME: Feb–Jul

NOTES: This is one of the longest-stemmed bedstraws, notable for its habit of climbing on shrubs. Another variety, var. *porrigens*, occurs in El Dorado and Nevada Counties (and possibly others); it has longer (2–18 mm) leaves, which are also wider (widely ovate to oblong) and are obscurely 3-veined; it occurs up to 2130 m, and blooms later, May–Aug.

Galium trifidum ssp. *columbianum*

Three-Petaled Bedstraw

Rubiaceae (Madder family)

Perennial (occasionally annual) herb with stems 10–50 cm; slender, sprawling, and scabrous.
LEAVES: In whorls of 4–6, petioled, 4–19 mm, linear-oblanceolate, with tips rounded, and rough-hairy on margins and midrib.
FLOWERS: In clusters in leaf axils. Flowers bisexual; corollas rotate, usually with 3 (sometimes 4) petals, 0.5 mm, white to pinkish; no sepals present.
FRUITS: Nutlets 2, round, hairless, and black when mature.
SIERRA DISTRIBUTION: All counties, except Inyo County.
ELEVATION: < 3030 m
HABITAT: Moist to wet shaded areas; montane zone.
FLOWERING TIME: Jul–Aug
NOTES: Formerly *G. trifidum* var. *pacificum*. Common in moist areas, and distinctive with its 3 white petals. *G. trifidum* ssp. *subbiflorum* (formerly *G. trifidum* var. *pusillum*) is a dwarfed, matted form found in meadows and lake margins of the subalpine and alpine zones; stems are ≤ 15 cm. It ranges from Tulare County north up to 3200 m.

Galium triflorum

Sweet-Scented Bedstraw

Rubiaceae (Madder family)

Perennial herb with stems 20–76 cm; usually decumbent to erect, hairless or slightly rough-hairy.
LEAVES: In whorls of 6, 18–31 mm, oblong-oblanceolate, rough-hairy on margins, and bristle-tipped.
FLOWERS: In clusters of 3 in leaf axils. Flowers bisexual; corollas rotate, 2–3 mm wide, with 4 lobes, whitish or pinkish lobes; no sepals present.
FRUITS: Nutlets 2, densely hairy, with hairs barbed.
SIERRA DISTRIBUTION: All counties, except Inyo County.
ELEVATION: < 3000 m
HABITAT: Moist to wet shaded areas; montane zone.
FLOWERING TIME: May–Jul
NOTES: This common bedstraw of wet areas in the montane zone is distinguished by the fact that the flowers are bisexual and are held in clusters of 3, along with having 6 ± wide leaves per whorl.

Kelloggia galioides

Mountain Madder

Rubiaceae (Madder family)

Perennial herb with stems 15–40 cm; resembling *Galium*, with stems ascending or sometimes erect, often sprawling, 4-angled.
LEAVES: Opposite, 19–38 mm, lanceoloate, with leaflike stipules.
FLOWERS: Inflorescences terminal and few-flowered; calyx has 4–5 small lobes; pedicels long, slender; corollas 3–4 mm, funnel-shaped, 4- to 5-lobed, and white or pink. Ovary inferior, hairy.
FRUITS: Nutlets 2, 3–4 mm, with hairs hooked.
SIERRA DISTRIBUTION: Tulare and Inyo Counties north.
ELEVATION: 700–3110 m
HABITAT: Shady forest floor, openings in conifer forest; montane and subalpine zones.
FLOWERING TIME: May–Aug
NOTES: An extremely common understory plant of conifer forests. Differs from *Galium* by having opposite rather than whorled leaves.

Salix petrophila

Rocky Mountain Willow

Salicaceae (Willow family)

Low-growing shrub < 10 cm; twigs are yellow-green or yellow-brown, and soft shaggy-hairy. Dioecious.

LEAVES: Alternate; stipules absent; petioles 2–3 mm; mature blades 19–44 mm, elliptic to obovate, and entire, with hairs on undersurface of leaf long, soft, shaggy, wavy (or not hairy).

FLOWERS: Blooming when leaves appear (rather than prior to the plant leafing out), 18–60 mm, borne on leafy shoots 2–40 mm; flower bracts tan to brown. Staminate flowers have 2 stamens; pistillate flowers have silky-hairy ovaries, style 0.4–1.6 mm.

FRUITS: 2-valved, opening to disperse seeds with silky hairs.

SIERRA DISTRIBUTION: Tulare and Inyo Counties north.

ELEVATION: 1670–4000 m

HABITAT: Alpine tundra, meadows; subalpine and alpine zones.

FLOWERING TIME: Jul–Aug

NOTES: We only include this charming, high-elevation willow because it is reasonable to identify to species—there are numerous willows in the SN, and they are challenging to identify and, being wind-pollinated, are not especially showy. An entire book would be required to present most willows adequately.

Acer glabrum var. *glabrum*

Mountain Maple

Sapindaceae (Soapberry family)

Shrub (or small tree) 2–6 m; mostly dioecious.

LEAVES: Opposite; petioles 2.8–9.7 cm; blades 2.9–5.6 cm, 3–5 cm wide, palmately 3-lobed, and the outer lobes have 5–22 acute teeth on the outer side. Leaves generally broader than long, and the central lobe usually longest.

FLOWERS: Inflorescence is ascending, of < 10 flowers; petals and sepals similar, about 2–3 mm. Some male plants have bisexual flowers as well as staminate flowers.

FRUITS: Hairless, with wings of samara spreading 70º–120º.

SIERRA DISTRIBUTION: All, but no records for Calaveras or Madera Counties.

ELEVATION: 1200–3000 m

HABITAT: Rocky sites ranging from moist to dry; montane and subalpine zones.

FLOWERING TIME: May–Jun

NOTES: The hairless base of the fruit is clearly different from the densely hairy fruit base of big-leaf maple, *Acer macrophyllum*, a tree of riparian areas. Another variety of *A. glabrum*, var. *diffusum*, is found sporadically throughout the SN. It differs from var. *glabrum* in having smaller leaves and shorter petioles, and fewer (3–7), usually obtuse teeth on the outer side of the lateral leaf lobes.

Aesculus californica

California Buckeye

Sapindaceae (Soapberry family)

Large shrub or small tree 4–12 m; canopy rounded in outline, deciduous.

LEAVES: Opposite; petioles 1–12 cm; blades palmately compound, with 5–7 leaflets, 6–12 cm, lance-oblong.

FLOWERS: Inflorescences 1–2+ dm, conical or cylindric, ± erect, with numerous flowers. Pedicels 3–10 mm; calyx 5–8 mm, 2-lobed, and much smaller than the petals; petals 4 (sometimes 5), 12–18 mm, white (or pale pink). Stamens 5–7, 18–30 mm (longer than petals); anthers bright orange.

FRUITS: Each many-flowered inflorescence typically produces 1 large fruit 5–8 cm wide. Some years, more (2–9) fruits are produced per inflorescence. Inside the grayish or tan covering of each fruit is a glossy brown seed 2–5 cm.

SIERRA DISTRIBUTION: All west slope counties.

ELEVATION: < 1700 m

HABITAT: Various; foothill and montane zones.

FLOWERING TIME: May–Jun

NOTES: Formerly in Hippocastanaceae (Horse-Chestnut family). Horse-chestnuts and buckeyes are closely related. All parts of California buckeye are toxic, though California Indians were able to process the seeds extensively to remove the poisonous tannins when acorns were in short supply (do not try this at home!). Native bees and other pollinators have no trouble using the pollen and nectar of buckeye, but both are toxic to honeybees (non-native species). Buckeye is distinctive for leafing out much earlier than other deciduous shrubs; new, bright green foliage is often fully open by Feb–Mar; and in addition, this shrub loses its leaves earlier than its associates, usually turning brown in midsummer.

Darlingtonia californica
California Pitcher-Plant
Sarraceniaceae (Pitcher-Plant family)

Perennial herb 1–6 dm (sometimes up to 1 m); with stolons up to 1 m, carnivorous.
LEAVES: Basal in rosettes, each leaf enlarging into a tubular pitcher with fluid containing enzymes, bacteria, etc. that digests insects captured in the fluid. Stiff, downward-pointing hairs on the inner wall of the pitcher prevent easy escape by prey. Each pitcher is domed at the top, with 2 fanglike appendages, ± 5 cm.
FLOWERS: Scape < 1 m with yellow bracts the length of the scape. Flower 1, bisexual, and nodding; sepals 5, 5–6 cm; petals 5, 2–4 cm, with wide, dark purple veins. Ovary superior; style 2–3 mm, 5-lobed; stamens 12–15.
FRUITS: Capsule 2.5–4.5 cm, loculicidal, 5-valved. Seed ± 2 mm, light red, papillate.
SIERRA DISTRIBUTION: Nevada County north.
ELEVATION: 800–2000 m
HABITAT: Bogs, seeps, areas with running water, especially serpentine; montane zone.
FLOWERING TIME: Apr–Jun
NOTES: A great place to see this plant is the Butterfly Valley Botanical Area in the Plumas National Forest about 20 minutes NW of Quincy.

Bolandra californica
Sierra Bolandra
Saxifragaceae (Saxifrage family)

Perennial herb with slender stems 15–25 cm, with rootstocks bulbiferous.
LEAVES: Basal and cauline (the latter with stipules), palmately veined, thin and light green. Petioles 2–10 cm; blades 1–5 cm wide, roundish with heart- or kidney-shaped bases, and 3–7 deep-toothed lobes.
FLOWERS: Inflorescences paniclelike, with glandular hairs; flowers horizontal or nodding. Hypanthium (flower tube) deep (± 5 mm), bell- to urn-shaped, and free from the ovary; sepals 5, with long, slender tips; petals 5, 4–7 mm, awl-shaped (slender and pointed), purplish to green, with purple tips. Stamens 5.
FRUITS: Capsule.
SIERRA DISTRIBUTION: West slope, Mariposa to El Dorado Counties.
ELEVATION: 1000–3000 m
HABITAT: Moist to wet rocky areas; montane and subalpine zones.
FLOWERING TIME: May–Aug
NOTES: CRPR 4.3 (watch list).

Boykinia major

Mountain Boykinia

Saxifragaceae (Saxifrage family)

Perennial herb 5–13 dm; rhizomatous, with unbranched stems; herbage is glandular-hairy.

LEAVES: Basal and alternate, 9–50 cm; blades 5- to 7-palmately cleft, divided more than halfway to midrib, and the leaf margin has gland-tipped teeth; upper leaves have stipules ± 1 cm, leaflike, toothed.

FLOWERS: Inflorescences dense, > 3-branched, of > 15 small, white flowers, and petals 5–7 mm. Ovary ½ inferior, styles 2; stamens 5, opposite the sepals.

FRUITS: Capsule.

SIERRA DISTRIBUTION: West slope, Madera County north.

ELEVATION: 900–2100 m

HABITAT: Moist to wet shady areas, mostly along streams; montane zone.

FLOWERING TIME: Jun–Aug

NOTES: The similar *Boykinia occidentalis* is a shorter (< 6 dm) plant found from Amador County north, with small, brown, bristly stipules and smaller (< 4 mm) petals.

Darmera peltata

Indian Rhubarb, Umbrella Plant

Saxifragaceae (Saxifrage family)

Enormous, tropical-looking perennial herb 0.3–1.5 m, thick, scaly, gnarled rhizomes < 5 cm diameter.

LEAVES: Basal, with petioles 3–15 dm; blades ≤ 1 m, peltate, roundish, with 5–9 lobes, mostly hairless.

FLOWERS: Inflorescences emerge before leaves and are covered with dense, brown hairs; stalk reddish, longer than leaves. Sepals 5, 3–4 mm, becoming reflexed; petals 5–7 mm, pink or white. Stamens 10.

FRUITS: Follicles 8–11 mm.

SIERRA DISTRIBUTION: Fresno County north.

ELEVATION: < 2050 m

HABITAT: Streams in coniferous forest; montane zone.

FLOWERING TIME: Apr–Jul

NOTES: This plant has the biggest leaves in the SN, hands down. Flowering begins in late spring and early summer when streams are at high flow; it is not uncommon to see reddish flowering stalks with buds or flowers underwater.

Heuchera micrantha

Small-Flowered Alum-Root

Saxifragaceae (Saxifrage family)

Perennial herb 1–10 dm; rhizomatous.
LEAVES: Blades 2–12 cm, large, 5- to 7-lobed, hairy, and borne on glandular-hairy petioles 2–12 cm.
FLOWERS: Panicles glandular, of radially symmetrical flowers; calyx lobes 0.5–1.5 mm, green to red-tipped; petals 2–3 mm, white (to pinkish).
FRUITS: Capsule 4–5 mm.
SIERRA DISTRIBUTION: Tulare and Mono Counties north to Plumas County.
ELEVATION: < 2500 m
HABITAT: Moist sites on cliffs and rocks; foothill through upper montane zones.
FLOWERING TIME: Apr–Jul
NOTES: Extremely variable. Where they overlap, *H. micrantha* can be distinguished from *H. rubescens* by its greater height, having radially rather than bilaterally symmetrical flowers, and smaller petals. Also compare habitats and elevation ranges when attempting to distinguish *Heucheras*.

Heuchera rubescens

Pink Heuchera

Saxifragaceae (Saxifrage family)

Perennial herb < 5.5 dm; rhizomatous.
LEAVES: Blades 8–60 mm, 5- to 9-lobed, hairy.
FLOWERS: Inflorescences frequently 1-sided, with flowers bilaterally symmetrical; calyx lobes 1–2 mm, unequal, white to pinkish red; petals 3–6 mm, pinkish.
FRUITS: Capsule 4–5 mm.
SIERRA DISTRIBUTION: All.
ELEVATION: 1000–4000 m
HABITAT: Dry cliffs and rocky sites: montane through alpine zones.
FLOWERING TIME: May–Sep
NOTES: Compare with *H. micrantha*.

Lithophragma affine

Woodland Star

Saxifragaceae (Saxifrage family)

Erect, finely hairy perennial herb 10–60 cm; from slender rhizomes with bulblets.

LEAVES: Basal leaves round in outline, and lobed < ½ to base; lower stem leaves mostly 3-lobed; stem leaves alternate and increasingly deeply lobed toward top of stem; petioles 3–15 mm.

FLOWERS: Racemes of 7–10 flowers. Hypanthium obconic, glandular; calyx lobes 5; petals 5, 5–13 mm, white, narrowed at base into a claw, with tips 3-lobed. Ovary mostly inferior, pistil 1; stamens 10.

FRUITS: Capsule 3-beaked.

SIERRA DISTRIBUTION: All (no formal records for Fresno County, though it likely occurs there).

ELEVATION: < 2000 m

HABITAT: Open, grassy areas, often partially shaded; foothill and montane zones.

FLOWERING TIME: Mar–Apr

NOTES: One of the earliest bloomers in the western foothills.

Lithophragma bolanderi

Bolander's Woodland Star

Saxifragaceae (Saxifrage family)

Erect, finely hairy perennial herb 20–80 cm; from slender rhizomes with bulblets.

LEAVES: Basal leaves round in outline, shallowly 3- to 5-lobed; stem leaves alternate, deeply lobed almost to base.

FLOWERS: Racemes of 5–25 flowers. Hypanthium hemispheric; calyx lobes 5; petals 5, 4–7 mm, white, usually entire but sometimes with > 3 toothlike lobes. Ovary < ½ inferior, pistil 1, styles 3; stamens 10.

FRUITS: Capsule 3-beaked. Seeds spiny.

SIERRA DISTRIBUTION: All west slope counties.

ELEVATION: < 2900 m

HABITAT: Open, grassy areas, often partially shaded; foothill and montane zones.

FLOWERING TIME: Feb–Jul

NOTES: One of the earliest bloomers in the western foothills but also found in upper coniferous forest blooming later. Distinctive because of its (usually) entire petals and the very shallow lobing of the basal leaves.

Lithophragma glabrum

Bulbous Woodland Star

Saxifragaceae (Saxifrage family)

Erect, finely hairy perennial herb 8–25 cm; from slender rhizome with bulblets.

LEAVES: Basal leaves round in outline, deeply lobed; stem leaves alternate, deeply 3-lobed with sharp-tipped teeth.

FLOWERS: Racemes of 1–7 flowers, often with bulblets in axils of lower bracts. Hypanthium bell-shaped, glandular; calyx lobes 5; petals 5, 3–7 mm, deeply 3- to 5-lobed, white or pink. Ovary mostly superior, pistil 1; stamens 10.

FRUITS: Capsule 3-beaked. Seed spiny.

SIERRA DISTRIBUTION: Tulare and Inyo Counties north.

ELEVATION: 1000–3700 m

HABITAT: Dry sandy and gravelly sites; montane, subalpine, and alpine zones.

FLOWERING TIME: Mar–Jul

NOTES: This mid–high elevation woodland star with its unusual bulblets in the inflorescence and dramatically slashed petals is fairly easy to distinguish from other members of its genus.

Lithophragma parviflorum var. *parviflorum*

Smallflower Woodland Star

Saxifragaceae (Saxifrage family)

Erect, finely hairy perennial herb 10–50 cm; from slender rhizome with bulblets.

LEAVES: Basal leaves round in outline, lobed < ¾ to base; lowest stem leaf 3-lobed.

FLOWERS: Racemes of 4–14 flowers. Hypanthium long-obconic (2x longer than wide), glandular; calyx lobes 5; petals 5, 7–16 mm, ± obovate, 3-lobed, white. Ovary mostly inferior, pistil 1; stamens 10.

FRUITS: Capsule 3-beaked. Seeds smooth.

SIERRA DISTRIBUTION: Fresno County north, sporadic.

ELEVATION: < 3000 m

HABITAT: Open gravelly and rocky areas; foothill and montane zones.

FLOWERING TIME: Mar–Jul

NOTES: Compare with *L. glabrum*.

Micranthes aprica

Sierra Saxifrage

Saxifragaceae (Saxifrage family)

Perennial herb < 13 cm; from fibrous-rooted caudexes (usually with bulblets). Entire plant is often purplish or reddish.

LEAVES: Basal in rosettes; blades 10–45 mm, oval, entire or minutely toothed, and margins sometimes wavy. Undersurface of leaves often purple.

FLOWERS: Aggregated into a terminal head (sometimes 1–2 additional heads present); the inflorescence stalk has tiny hairs, which may be glandular toward base of stem. Sepals 1–1.5 mm, ovate; petals 1.5–3 mm, obovate, white, without a spot, base narrowed into a claw.

FRUITS: Follicles 2, 2–4 mm, purple, diverging from each other at full maturity.

SIERRA DISTRIBUTION: Tulare and Inyo Counties north.

ELEVATION: 1000–3700 m

HABITAT: Wet or moist sandy or gravelly soil, meadows; subalpine and alpine zones.

FLOWERING TIME: May–Aug

NOTES: Compare with *M. oregana*, which is a larger, more coarse-looking plant with numerous glandular-hairy inflorescences (rather than 1 inflorescence, as in *M. aprica*). The ripe capsules of *M. aprica* are more elongate than in *oregana*, which has more rounded capsules that don't diverge as much from each other.

Micranthes bryophora

Bud Saxifrage

Saxifragaceae (Saxifrage family)

Delicate, glandular-pubescent annual herb < 2 dm; with threadlike inflorescence stalks.

LEAVES: Basal, sessile, linear-elliptic, entire or toothed, fleshy; blades 1–4 cm.

FLOWERS: Racemes, with axis glandular-hairy; pedicels very slender and threadlike, and lower are deflexed. Sepals 1–2 mm, reflexed; petals 3–5 mm, white with 2 yellow spots. Ovary superior; stamens have filaments widened at base.

FRUITS: Capsule 3–4 mm.

SIERRA DISTRIBUTION: Tulare and Inyo Counties north.

ELEVATION: 1000–3500 m

HABITAT: Moist sandy to gravelly areas; up through alpine zone.

FLOWERING TIME: Jul–Aug

NOTES: Distinctive with its slender, reddish inflorescence stalks held at odd angles, from horizontal to deflexed.

Micranthes californica

California Saxifrage

Saxifragaceae (Saxifrage family)

Perennial herb; from slender rootstocks with bulblets; flowering stem is 15–35 cm, glandular-hairy.

LEAVES: All basal, 4–10 cm, ovate with a tapered base, shallowly toothed, and soft-hairy.

FLOWERS: Branched (panicles); each branch has loose clusters of white flowers. Sepals 1.5–2 mm, ovate to triangular, reflexed; petals 2.5–4 mm, obovate or oval. Ovary ½ inferior.

FRUITS: Follicles (because carpels are only united at the very base) 2.5–3.5 mm.

SIERRA DISTRIBUTION: West slope, all.

ELEVATION: < 1300 m

HABITAT: Moist grassy slopes and rocky areas, usually in at least partial shade; foothill zone.

FLOWERING TIME: Feb–Jun

NOTES: One of the earliest foothill wildflowers to emerge each spring. The leaves of a given patch are often noticeably different sizes, giving this plant a characteristic look and allowing recognition prior to blooming once you get to know them. It often grows near *Primula hendersonii*, which has a similar leaf phenomenon, though its leaves are hairless and generally smaller.

Micranthes odontoloma

Brook Saxifrage

Saxifragaceae (Saxifrage family)

Perennial herb to 50 cm; from rhizomes with no bulblets.

LEAVES: All basal, roundish with regularly arranged, gland-tipped teeth; pale green, hairless. Petioles 2–20 cm, blades 2–7 cm wide.

FLOWERS: Loose panicles, with reddish, glandular-hairy stems. Sepals 1.5–2 mm, reflexed; petals 2.5–4 mm, rounded, white with 2 yellow dots at base. Ovary superior; filaments club-shaped, white.

FRUITS: Capsule 4–8 mm, purple.

SIERRA DISTRIBUTION: Tulare and Inyo Counties north.

ELEVATION: 1500–3400 m

HABITAT: Moist, shady areas along streams, in meadows, and on seepy rock ledges; montane, subalpine, and alpine zones.

FLOWERING TIME: Jun–Aug

NOTES: Of the round-leaved saxifrages in the SN, this is the only one with yellow dots on the petals. And check out those filaments!

Micranthes oregana

Bog Saxifrage

Saxifragaceae (Saxifrage family)

Perennial herb; from creeping rootstocks with no bulblets; flowering plant is 30–80 cm.

LEAVES: All basal; blades 5–15 cm elliptic to oblanceolate, tapering to a winged petiole; leaves entire or toothed, hairless or sparsely hairy.

FLOWERS: Branched, each branch with dense clusters of white flowers. Sepals 2 mm, ovate to triangular, reflexed; petals 2–4 mm, ovate.

FRUITS: Capsule 3.5–5 mm, with beaks spreading, purplish.

SIERRA DISTRIBUTION: Tulare and Inyo Counties north.

ELEVATION: 1000–3200 m

HABITAT: Wet, boggy areas, especially meadows; montane through alpine zones.

FLOWERING TIME: May–Aug

NOTES: Formerly *Saxifraga oregana*. Compare with *M. aprica*.

Pectiantia breweri

Brewer's Miterwort

Saxifragaceae (Saxifrage family)

Perennial herb 10–30 cm.

LEAVES: Basal, petioles 3–8 cm, with long, tan-colored hairs; blades 3–8 cm wide, ± round, with 7–11 indistinct lobes.

FLOWERS: Racemes, flowers light green, spiderlike, with pedicels 2–7 mm. Hypanthium ± 3 mm wide, saucer-shaped; sepals 5, rounded, and shorter than petals; petals 5, pinnately divided into 5–7 green, threadlike segments. Ovary mostly inferior, pistil 1; stamens 5, alternate the petals.

FRUITS: Capsule 2-beaked.

SIERRA DISTRIBUTION: Both slopes, Tulare and Inyo Counties north.

ELEVATION: 1500–3500 m

HABITAT: Moist to wet, shady areas along streams and in meadows; montane zone.

FLOWERING TIME: Jun–Aug

NOTES: Formerly *Mitella breweri*. If the stamens are opposite the petals rather than alternate as shown in the photo, you're looking at *P. pentandra*, which is found in the same range and habitat as *P. breweri.* Both species are pollinated by fungus gnats (Mycetophilidae).

Tellima grandiflora

Fringe Cups

Saxifragaceae (Saxifrage family)

Perennial herb; rhizomatous, spreading hairy, with unbranched stems 4–10 dm.

LEAVES: Basal and cauline. Basal leaves 5–10 cm wide, with 3–7 lobes, bases heart-shaped, hairy; stem leaves smaller than basal, with stipules ± 5 mm.

FLOWERS: Borne on long, glandular-pubescent racemes; flowers have hypanthium (flower tube) 3–5 mm, cuplike, urnlike, inflated; sepals ± 2 mm; petals 4–6 mm, whitish, becoming red or pink, with 5–7 linear lobes (forming the "fringe"). Ovary mostly inferior, styles 2; stamens 10.

FRUITS: Capsule ovate, opening at top.

SIERRA DISTRIBUTION: West slope; El Dorado County north.

ELEVATION: < 2000 m

HABITAT: Moist sites in forests; montane zone.

FLOWERING TIME: Apr–Jun

NOTES: *Tellima* is an anagram of *Mitella*, which has similar, linear-lobed petals. Fringe cups is used horticulturally and is said to be deer-resistant in SN gardens.

Scrophularia californica

California Bee Plant

Scrophulariaceae (Figwort family)

Erect perennial herb 80–120 cm; stem ± square in x.s.

LEAVES: Opposite, petioled, and pairs set at right angles to each other; larger blades toward base of plant are 8–17 cm, cordate or truncate, sometimes wedge-shaped, and bright or dark green.

FLOWERS: Panicle of cymes, glandular, with pedicels glandular and straight. Calyx lobes 5, 2–4 mm, triangular-ovate; corollas 8–12 mm, bilateral, inflated at base, with mouth constricted; upper lip 2-lobed, maroon or red, and lower lip of 3 shorter lobes lighter or greenish. Ovary superior, stigma headlike; fertile stamens 4, staminode club-shaped.

FRUITS: Capsule 6–8 mm. Seeds oblong-ovoid, ridged.

SIERRA DISTRIBUTION: Both slopes, all.

ELEVATION: < 2500 m

HABITAT: Rocky areas, dry to damp places; most zones.

FLOWERING TIME: Mar–July

NOTES: Bees are frequent visitors—while they cannot see the color red, they likely see the UV light reflected by the flowers.

Scrophularia desertorum

Desert Figwort

Scrophulariaceae (Figwort family)

Erect perennial herb 70–120 cm; stem ± square in x.s.
LEAVES: Opposite, petioled, and pairs set at right angles to each other; larger blades toward base of plant are 4–8 cm, with bases wedge-shaped and tapering to petioles, not truncate or cordate, dull green, yellow- to gray-green.
FLOWERS: Panicles of cymes; axes of inflorescence and pedicels glandular-puberulent, and pedicels curved upward. Calyx lobes 5, 2–3 mm, triangular-ovate; corollas 7–9 mm, bilateral, inflated at base, with mouth constricted, upper lip 2-lobed, maroon or red, and lower lip cream-colored with pinkish edges. Ovary superior, stigma headlike; fertile stamens 4, staminode club-shaped.
FRUITS: Capsule. Seeds oblong-ovoid, ridged.
SIERRA DISTRIBUTION: Mostly east slope.
ELEVATION: 850–3000 m
HABITAT: Rocky areas, often in crevices or among boulders, gravelly washes; montane, subalpine, and sagebrush/pinyon-juniper zones.
FLOWERING TIME: Apr–Aug
NOTES: Leaf bases taper to petioles; note how the bases differ from the truncate leaf bases of *S. californica*.

Verbascum blattaria (non-native, invasive)

Moth Mullein

Scrophulariaceae (Figwort family)

Erect biennial herb 30–120+ cm; occasionally branched below inflorescence, hairless below the glandular inflorescence.
LEAVES: Hairless; basal in rosettes, short-petioled, 4–25 cm, oblanceolate, and crenate; cauline leaves sessile, 2–12 cm, elliptic to ovate, cordate, and dentate.
FLOWERS: Racemes long, wandlike, glandular toward top, with lowermost pedicels 10–15 mm (up to 25 mm); inflorescence bracts linear to lanceolate, < pedicels; flowers 1 per node. Calyx 5–8 mm; corollas 25–30 mm wide, yellow or white. Filaments dense and purple-hairy.
FRUITS: Capsule 6–8 mm, spheric. Seeds numerous, small.
SIERRA DISTRIBUTION: West slope, Tulare County north.
ELEVATION: < 1700 m
HABITAT: Disturbed areas, roadsides; foothill and montane zones.
FLOWERING TIME: May–Sep
NOTES: Native to Eurasia, noticeably increasing in abundance in the central Sierra since about 2015. The species name (genus name for cockroaches) derives from the belief that moth mullein repels cockroaches. Seeds can remain viable for over 120 years in the soil under the right conditions. In the same range is the very similar-looking wand mullein, *V. virgatum*, which differs in having hairy leaves, shorter (< 10 mm) proximal pedicels, and 1–4 flowers per node; wand mullein is also spreading and more common than 10–20 years ago.

Verbascum thapsus (non-native, invasive)

Woolly Mullein

Scrophulariaceae (Figwort family)

Erect biennial herb 30–200 cm; occasionally branched, and densely tomentose with stellate hairs (giving herbage a gray appearance).

LEAVES: Basal in rosettes, short-petioled, 8–50 cm, and oblanceolate; stem leaves smaller toward top of plant, and decurrent.

FLOWERS: Dense racemes; bracts 12–18 mm; pedicels < 2 mm, fused to stem. Calyx 7–9 mm, ± radial, deeply 5-lobed; corollas 15–25 mm wide, also ± radial (slightly bilateral), rotate, and yellow glandular. Stigma spheric; stamens 5; upper 3 filaments white or yellow-hairy, and the lower 2 hairless or sparsely hairy.

FRUITS: Capsule 7–10 mm, ovoid. Seeds numerous, small, and can remain viable for at least 35 years in the soil.

SIERRA DISTRIBUTION: Both slopes, all.

ELEVATION: < 2200 m

HABITAT: Disturbed areas, moist or dry; proliferates after fire, or mechanical disturbance; all zones.

FLOWERING TIME: May–Sep

NOTES: Ubiquitous, invasive; can be damaging to recovering ecosystems when growing abundantly and competing with the variety of native seedlings present.

Datura wrightii

Jimson Weed

Solanaceae (Nightshade family)

Bushy annual to perennial herb 5–15 dm. Foul-smelling.

LEAVES: Alternate, 7–20 cm, ovate, coarsely lobed or entire.

FLOWERS: Calyx 8–12 cm, ribbed, with lobes ± 2 cm; corollas 15–20 cm, funnel-shaped, 5-lobed, white, variously tinged purple.

FRUITS: Nodding, densely prickly capsule 2.5–3 cm wide, plus prickles 5–12 mm; opening along irregularly shaped valves. Remnant base of calyx forms a "skirt" at capsule base. Seeds 5 mm, flat, light brown.

SIERRA DISTRIBUTION: West slope, all; east slope, Kern County.

ELEVATION: < 2200 m

HABITAT: Open areas, roadsides; foothill, montane, and sagebrush/pinyon-juniper zones.

FLOWERING TIME: Apr–Oct

NOTES: Used for ceremonial and healing purposes by Native Americans with millennia of knowledge on how to use the plant safely; note that all parts of the plant are poisonous. Each year misguided people end up in emergency rooms and some die because they ingest seeds of *D. wrightii* or related species, seeking a hallucinogenic high.

Leucophysalis nana

Dwarf Chamaesaracha

Solanaceae (Nightshade family)

Perennial herb with decumbent, branched stems 5–25 cm.
LEAVES: Alternate, 15–50 mm, ovate, with margins entire and sometimes wavy.
FLOWERS: Inflorescences are axillary clusters of 1–5 flowers; pedicels 8–18 mm (up to 3 cm in fruit, and recurved); calyx 5 mm; corollas rotate, white. Style 1; filaments 5 mm; anthers shorter.
FRUITS: Berry 1–1.2 cm wide, partially enclosed by calyx lobes elongated.
SIERRA DISTRIBUTION: East slope, Inyo and Mono Counties; both slopes, Alpine County north.
ELEVATION: 1100–2900 m
HABITAT: Sandy soils in coniferous forest; montane and subalpine zones.
FLOWERING TIME: May–Jul
NOTES: Formerly *Chamaesaracha nana*. Since this mouthful of a word is still part of the common name, it helps to know that *chamae-* is the Latin root for "low, close to the ground," and *Saracha* is a related genus from South America named for Spanish monk and botanist Isidoro Saracha.

Solanum xanti

Nightshade

Solanaceae (Nightshade family)

Many-branched perennial herb to subshrub 4–9 dm; herbage hairy, sometimes with glands.
LEAVES: Alternate, 2–7 cm, entire or with 2 lobes at base.
FLOWERS: Inflorescence is an open, umbel-like cyme. Calyx 4–5 cm; corollas 1.5–3 cm wide, wheel-shaped, purple or dark blue, and with 5 lobes, green spots at base. Stamens 5; anthers 4.5–5 mm, closely hugging the style.
FRUITS: Berry 10–15 mm, green.
SIERRA DISTRIBUTION: All.
ELEVATION: < 2900 m
HABITAT: Dry sites; foothill and montane zones.
FLOWERING TIME: Mar–Aug
NOTES: Related to deadly nightshade; all parts of the plant contain the toxin solanine.

Staphylea bolanderi
Sierra Bladdernut
Staphyleaceae (Bladdernut family)

Deciduous shrub to small tree 2–6 m; hairless.
LEAVES: Opposite, pinnately compound; leaflets 3, 2.5–6 cm, with margins toothed.
FLOWERS: Inflorescences drooping, appearing with or before leaves; flowers 5-merous, sepals petal-like, 8–10 mm; petals white, 10–12 mm; stamens exserted; style equal to or longer than stamens, stigmas 3-lobed.
FRUITS: Capsule 2.5–5 cm, inflated, each of the 3 bladderlike chambers with a long, pointed tip. Seeds 1–2 per chamber, 5–7 mm, obovoid, smooth.
SIERRA DISTRIBUTION: West slope, Tulare to Calaveras Counties.
ELEVATION: < 1600 m
HABITAT: Slopes in chaparral, foothill woodland, or lower coniferous forest; foothill and montane zones.
FLOWERING TIME: Mar–May
NOTES: This is a spectacular shrub, equally handsome in full flower or when adorned with the creamy-colored, inflated fruits. It is relatively uncommon but can form stands of hundreds of plants in some areas, such as the South Fork Merced River drainage. The only record for Calaveras County is from 1895, at "McCormick Bridge" on the Stanislaus River. More recent sightings would be valuable!

Styrax redivivus
Snowdrop Bush
Styracaceae (Storax family)

Deciduous shrub 1–4 m.
LEAVES: Alternate, simple, without stipules; petioles 3–14 mm, blades 2–8 cm, ovate to round, entire.
FLOWERS: Drooping racemes of radial, fragrant, white flowers. Calyx teeth 6–9 (minute, not visible in photo); corollas 12–26 mm, bell-shaped, 5- to 10-lobed; stamens 10, filaments white, anthers yellow, style 1, longer. Ovary ± superior.
FRUITS: Spherical, 11–15 mm, hard, indehiscent. Seeds 1–2.
SIERRA DISTRIBUTION: West slope, Tulare County north.
ELEVATION: < 1600 m
HABITAT: Dry slopes and rocky areas in chaparral, foothill woodland; foothill zone.
FLOWERING TIME: Apr–Jun
NOTES: This uncommon shrub is spectacular in bloom and can be grown ornamentally, available for purchase at nurseries specializing in CA native plants. During full bloom, clouds of tiny moths visit the flowers at dusk. The name "storax" refers to a resin that is actually produced by trees in the genus *Liquidambar,* in a different family. Some plants in Styracaceae produce a similar compound called benzoin, which has been called storax, hence the family common name. Confused yet? Storax and benzoin are used in perfumery and medicinally for their antiseptic properties, to name a few uses.

Urtica dioica ssp. *holosericea*

Stinging Nettle, Hoary Nettle

Urticaceae (Nettle family)

Erect perennial herb 5–30 dm; rhizomatous, stinging hairs are mixed with nonstinging hairs.

LEAVES: Opposite, lanceolate; blades 6–20 cm, coarsely toothed.

FLOWERS: Inflorescences are either all male or all female flowers (plant is monoecious), wind-pollinated, with petals absent. Male flower clusters are longer than female (nearly as long as leaves); male flowers have 4 green, bristly sepals and 4 stamens; female flowers have 4 unequal sepals (outer 2 smaller than the inner), ovary superior, and pistil 1.

FRUITS: Achenes < 1 mm, ovate, enclosed by calyx.

SIERRA DISTRIBUTION: All.

ELEVATION: < 3370 m

HABITAT: Moist and wet areas; foothill through subalpine zones.

FLOWERING TIME: Jun–Sep

NOTES: Touching this giant nettle is surprisingly painful!

Plectritis ciliosa

Ciliose Plectritis

Valerianaceae (Valerian family)

Erect annual herb 5–80 cm, with stem angled, single-stemmed, or with few branches.

LEAVES: Basal and opposite; stem leaves sessile.

FLOWERS: Inflorescences headlike; sepals absent, bracts present, often reddish; corollas 1.5–8.5 mm, 2-lipped, and pink; the lower lip has red spots at base, the spur slender, longer than corolla tube.

FRUITS: Convex side has grooved, longitudinal keel, and wings with longitudinal bands of hairs (ciliate).

SIERRA DISTRIBUTION: West slope.

ELEVATION: < 2100 m

HABITAT: Open grassy slopes with partial shade; foothill and lower montane zones.

FLOWERING TIME: Mar–Jun

NOTES: This species has a markedly 2-lipped flower with a long, slender spur relative to other *Plectritis* spp. in the SN. A 10x hand lens is required to see fully the beauty of individual *Plectritis* flowers, and the use of the online key and species descriptions in the Jepson eFlora is helpful for a serious botanist wanting to distinguish species of this genus.

Plectritis congesta ssp. *brachystemon*

Shortspur Seablush

Valerianaceae (Valerian family)

Erect annual herb 5–80 cm, with stem angled, single-stemmed, or with few branches.

LEAVES: Basal and opposite; stem leaves sessile.

FLOWERS: Inflorescences headlike; sepals absent, bracts present; corollas 1.5–3.5 mm, ± radial to ± 2-lipped, uniformly white to pale pink; spur may be a slight swelling or more developed and slenderer with tip enlarged, spur < ½ length of corolla tube.

FRUITS: Convex side keeled lengthwise, with a few small grooves, winged or not; if winged, the wings are hairy near tips, margins, or throughout, and margins are as thick as wing body in winged specimens.

SIERRA DISTRIBUTION: West slope, sporadic.

ELEVATION: < 1900 m

HABITAT: Open grassy slopes with partial shade; foothill and lower montane zones.

FLOWERING TIME: Mar–Jun

NOTES: Compare with *P. ciliosa*, which has a very long spur, and *P. macrocera*, with a short, blunt spur.

Plectritis macrocera

Longhorn Plectritis

Valerianaceae (Valerian family)

Erect annual herb 5–80 cm, with stem angled, single-stemmed, or with few branches.

LEAVES: Basal and opposite; stem leaves sessile.

FLOWERS: Inflorescences headlike; sepals absent, bracts present; corollas 2–3.5 mm, ± radial with lobes ± equal, white to pale pink; the lower lip lacks red spots at base, and spur is stout and blunt rather than pointed as in *P. ciliosa*.

FRUITS: Winged; convex side has longitudinal groove along the keel, and wing margins of fruit are not rolled toward fruit body.

SIERRA DISTRIBUTION: West slope, all.

ELEVATION: < 2000 m

HABITAT: Open grassy slopes with partial shade; foothill and lower montane zones.

FLOWERING TIME: Mar–Jun

NOTES: Compare with *P. congesta* ssp. *brachystemon*, which is similar-looking, but the spur is very short with an enlarged tip (sometimes just a swelling).

Valeriana californica

California Valerian

Valerianaceae (Valerian family)

Perennial herb 2.5–5 dm; hairless or short-hairy, rhizomatous.
LEAVES: Basal 5–15 cm, oblanceolate, entire; stem leaves smaller, pinnate or pinnatifid, with terminal lobe larger than the others.
FLOWERS: Cymes; flowers 3–4.5 mm, slightly asymmetric, funnel-shaped, creamy-white.
FRUITS: Achenes 4–6 mm, ovoid.
SIERRA DISTRIBUTION: All.
ELEVATION: 1500–3700 m
HABITAT: Moist to dry shaded sites in conifer forest; montane and subalpine zones.
FLOWERING TIME: Jun–Aug
NOTES: Related to the European *V. officinalis*, which has many herbal uses.

Verbena lasiostachys var. *lasiostachys*

Western Vervain, Western Verbena

Verbenaceae (Vervain family)

Perennial herb 35–80 cm; with stems erect or decumbent, 4-angled, short-hairy.
LEAVES: Opposite, 4–10 cm, with petioles < 2 cm; blades ovate, deeply lobed, with margins toothed; upper leaf surface gray-green and soft-hairy, sometimes a dull green.
FLOWERS: Spikes 1–3 per stem; calyx 2.5–4 mm, with 5 ribs; corollas 2.5–5 mm, 5-lobed, blue or purple, dense early on and spike elongating with age.
FRUITS: Nutlets 4, 1–2 mm.
SIERRA DISTRIBUTION: West slope, patchy.
ELEVATION: < 2500 m
HABITAT: Open, often disturbed places in dry or moist soil; foothill and montane zones.
FLOWERING TIME: May–Sep
NOTES: Var. *scabrida* has more pronounced green upper leaf surfaces, which are stiff-hairy, and nutlet scars that are bumpy-white (need a 20x lens). This variety is restricted to the far southern Sierra in Kern County and is much less common.

Sambucus mexicana

Blue Elderberry

Viburnaceae (Muskroot family)

Deciduous shrub 2–4 m (up to 8 m); no main trunk (bushy); young stems are pithy; older bark is grayish to brown, irregularly ridged/furrowed.

LEAVES: Pinnately compound, 15–35 cm; 3–9 leaflets, 3–20 cm, toothed, tips acute, often with bases asymmetric.

FLOWERS: Inflorescences up to 33 cm wide, ± flat-topped, with no obvious central axis; flowers cream to yellow.

FRUITS: 5–6 mm wide, black with dense, white "bloom" (appearing blue).

SIERRA DISTRIBUTION: All.

ELEVATION: < 3000 m

HABITAT: Streamsides, valleys, forest openings, roadsides; foothill through subalpine zones.

FLOWERING TIME: Mar–Sep

NOTES: Previously classified as *Sambucus nigra* ssp. *caerulea*, now changed back to *S. mexicana*. By far the most common elderberry in the SN. Sprouts vigorously after fire as well as produces seedlings. Various parts of the shrub are used for tools, medicine, dye, food, and musical instruments by Native Americans; e.g., flutes and clappers are made by hollowing out the pithy stems. Flowers and ripe fruits are edible but must be cooked to deactivate toxic alkaloids. Fruits support a variety of wildlife, including black bears and native songbirds.

Sambucus racemosa var. *racemosa*

Red Elderberry

Viburnaceae (Muskroot family)

Shrub < 2 m, usually < 1 m; no main trunk (bushy).

LEAVES: Leaflets 5–7, thin and often hairless, coarsely toothed, sometimes asymmetric at base.

FLOWERS: Inflorescences 6–12 cm wide, dome-shaped, with an obvious central axis; flowers cream-colored or yellow, with (often) reflexed corolla lobes.

FRUITS: 4 mm, red.

SIERRA DISTRIBUTION: Tulare and Inyo Counties north.

ELEVATION: 1800–3350 m

HABITAT: Moist rocky places; upper montane and subalpine zones.

FLOWERING TIME: May–Aug

NOTES: Formerly *S. racemosa* var. *microbotrys*. Leaves have an unpleasant smell, especially when crushed. The red fruits can be poisonous in some regions. Black elderberry, *S. racemosa* var. *melanocarpa*, occurs at the same elevations as var. *racemosa* but has narrower (4–7 cm wide) inflorescences and purple-black fruits. Currently documented from the following SN counties: Tulare, Mariposa, Tuolumne, Mono, El Dorado, Placer, and Nevada.

Viburnum ellipticum

Common Viburnum

Viburnaceae (Muskroot family)

Deciduous shrub, usually hairy.

LEAVES: Opposite, simple, with coarse teeth above middle, and 3-veined from base. Petioles 6–12 mm; blades 2–6 cm, elliptical to round.

FLOWERS: Compound cymes approximately 3 cm wide, with numerous small, white flowers. Stamens long, white, conspicuous.

FRUITS: Drupe 10–12 mm, reddish to black when mature (resembling an olive).

SIERRA DISTRIBUTION: El Dorado, Placer, and Fresno Counties.

ELEVATION: 300–1400 m

HABITAT: Generally found on north-facing slopes; foothill and montane zones.

FLOWERING TIME: Jun–Aug

NOTES: A lovely shrub with a sparse distribution, it is worth a special trip to see in bloom. Any new county occurrences in the SN would be significant to document.

Viola adunca ssp. *adunca*

Western Dog Violet

Violaceae (Violet family)

Prostrate to erect, many-stemmed perennial herb 3.5–3 dm; rhizomatous.

LEAVES: Basal and cauline, simple, roundish to ovate; blades 0.5–6.6 cm x 3.6–4.8 cm wide, and cauline leaves smaller.

FLOWERS: Inflorescences in leaf axils, with peduncles 1–10.3 cm. Sepals ± 5 mm, linear or lanceolate; petals 6–15 mm, spur 5–12 mm, and flowers are violet (sometimes lower 3 are white with purple veins); lateral petals have long, cylindrical hairs at base. Cleistogamous flowers frequently present.

FRUITS: 6–11 mm, hairless. Seeds 1.5–2 mm, brown-black.

SIERRA DISTRIBUTION: Tulare and Inyo Counties north.

ELEVATION: 1200–3400 m

HABITAT: Streambanks, meadows, moist sites; montane through alpine zones.

FLOWERING TIME: Apr–Aug

NOTES: The only all-purple violet in the SN.

Viola douglasii

Douglas' Violet

Violaceae (Violet family)

Erect or decumbent perennial herb 0.3–2.0 dm, usually with several clustered stems.

LEAVES: Basal blades bipinnate, longer than wide (3.5–5 cm x 2.4–3.5 cm wide); segments 1–2.5 mm wide, densely ciliate.

FLOWERS: Peduncle 2–12.5 cm; sepals lanceolate, ciliate; lower petals 8–12 mm, gold-yellow; upper 2 petals dark reddish-brown on back, and lower 3 veined dark reddish-brown. Cleistogamous flowers absent.

FRUITS: 5–12 mm, round or oblong, hairless. Seeds ± 2.5 mm, pale brown.

SIERRA DISTRIBUTION: West slope, all.

ELEVATION: < 1525 m

HABITAT: Rocky or grassy flats and slopes, vernally moist sites, frequently on serpentine; foothill zone.

FLOWERING TIME: Feb–Jul

NOTES: The only violet with pinnately compound leaves.

Viola glabella

Stream Violet

Violaceae (Violet family)

Erect, bright green, mostly hairless perennial herb < 2.2 dm; from caudexes.

LEAVES: Basal leaves 2–3, ± heart-shaped, with margins scalloped and long petioles; blades 3.3–8.5 cm x 2–9.3 cm wide; stem leaves similar but smaller.

FLOWERS: In upper leaf axils, with peduncles 2–8 cm. Sepals 3–10 mm, lanceolate-linear; petals 6–16 mm, with spur 1–2 mm, yellow on both surfaces; the lower 3 have purple veins. Cleistogamous flowers frequently present.

FRUITS: 7–13 mm, hairless. Seeds 2 mm, pale brown, shiny.

SIERRA DISTRIBUTION: West slope, all.

ELEVATION: < 2600 m

HABITAT: Shaded, moist to wet sites on forest floor and along streams; montane zone.

FLOWERING TIME: Mar–Aug

NOTES: Could be confused with the entire-leafed form of *V. lobata* ssp. *integrifolia*; see notes on *V. lobata* ssp. *lobata*.

Viola lobata ssp. *lobata*

Delta-Leaved Forest Violet, Pine Violet

Violaceae (Violet family)

Erect perennial herb; rhizomatous, hairless or finely hairy, with 1–3 stems 0.5–4.6 dm.

LEAVES: Mostly toward top of stems, divided into 3–12 linear lobes, wider than long, with stipules 10–15 mm, toothed.

FLOWERS: In upper leaf axils, with peduncles 2–13 cm. Sepals 6–10 mm, ciliate; petals 8–19 mm, upper ones yellow with purple to brownish backs. Cleistogamous flowers frequently present.

FRUITS: 6–16 mm, hairless. Seeds 2.1–2.7 mm, shiny, pale, spotted or streaked.

SIERRA DISTRIBUTION: All.

ELEVATION: 43–2300 m

HABITAT: Dry sites in chaparral or oak woodland, foothill zone; floor of coniferous forest, montane zone.

FLOWERING TIME: Apr–Aug

NOTES: Also occurring in the Sierra is ssp. *integrifolia*, which has entire or coarsely toothed leaves, and superficially resembles *V. glabella*. *V. lobata* ssp. *integrifolia* is a plant of drier habitats than *V. glabella* and has hairy, dull green herbage, in contrast to the hairless, bright green herbage of *V. glabella*.

Viola macloskeyi

Macloskey's Violet

Violaceae (Violet family)

Patch-forming perennial herb < 1 dm; stoloniferous, hairless or hairy, lacking true stems (peduncle appears to be a stem).

LEAVES: Basal only, simple, kidney-shaped to ovate, and entire to slightly scalloped; blades 1–6.5 cm x 1–5.5 cm wide.

FLOWERS: Peduncles 2.5–21 cm. Sepals 3–4 mm, ovate or lanceolate; petals 6–12 mm, spur 2–3 mm, white, and lower 3 have purple veins. Cleistogamous flowers frequently present.

FRUITS: 5–7 mm, hairless. Seeds 1–1.5 mm, beige or bronze.

SIERRA DISTRIBUTION: All.

ELEVATION: 600–3400 m

HABITAT: Moist to wet areas; montane through alpine zones.

FLOWERING TIME: Mar–Sep

NOTES: An extremely common species in Sierra meadows, and easy to learn as it is the only truly white violet in the SN.

Viola pinetorum ssp. *pinetorum*

Pine Violet

Violaceae (Violet family)

Perennial herb with stem 0.6–2.2 dm; from woody rhizomes; stem erect or spreading, hairless to finely hairy, sometimes canescent.
LEAVES: Basal leaves have petioles 3.2–9.5 cm; blades 1.3–5 cm x 0.7–2.5 cm wide, linear to lance-elliptic, entire or irregularly toothed, and sometimes purple underneath; cauline leaves smaller.
FLOWERS: Peduncles 3.4–11.5 cm; petals 6–17 mm, yellow, the upper 2 with brownish-purple backs, and spur 1–2 mm.
FRUITS: 3.5–7 mm, ovoid, finely hairy.
SIERRA DISTRIBUTION: West slope.
ELEVATION: 1400–3200 m
HABITAT: Dry to vernally moist forest floor and meadow borders; montane and subalpine zones.
FLOWERING TIME: May–Jul
NOTES: Leaves are longer in proportion to width than the similar *V. purpurea*.

Viola purpurea ssp. *quercetorum*

Oak Violet

Violaceae (Violet family)

Grayish-green, spreading to erect perennial herb 0.5–1.5 dm; finely hairy.
LEAVES: Basal leaves have petioles 1.9–9.5 cm; blades 1–5.3 cm x 1–3.5 cm wide, ovate to round, with margins wavy, and variously toothed to nearly entire, but bases usually tapered to petioles; stem leaves smaller.
FLOWERS: Peduncles 2.7–17 cm; petals 12–17 mm, yellow with reddish-brown to purple backs, spur 1–2 mm. Cleistogamous flowers frequently present.
FRUITS: 8–12 mm.
SIERRA DISTRIBUTION: West slope.
ELEVATION: 300–1400 m
HABITAT: Open grassy flats and slopes; foothill and lower montane zones.
FLOWERING TIME: Feb–Jul
NOTES: *Viola purpurea* ssp. *purpurea* has dark green, purple-tinged herbage, and stems 0.3–2.5 dm.

Viola sheltonii
Fan Violet
Violaceae (Violet family)

Perennial herb 0.2–1 dm; from woody rhizomes, with stems clustered, prostrate or erect, hairless to sparsely hairy.
LEAVES: Compound, width > length. Petioles 8.6–21 cm; blades 2–7 cm x 2–11 cm wide, twice dissected into long, narrow divisions.
FLOWERS: Peduncles 5–19 cm; sepals 6–8 mm, lanceolate, and ciliate. Petals 12–17 mm, yellow; the upper 2 petals have brownish-purple backs, and the lateral 2 sometimes have tiny, club-shaped hairs at base; spur 1–2 mm. Cleistogamous flowers usually present.
FRUITS: 6–8 mm, hairless to finely hairy. Seed 2.5 mm, shiny, brown.
SIERRA DISTRIBUTION: West slope.
ELEVATION: 480–2500 m
HABITAT: Dry, shady sites in conifer forest; montane zone.
FLOWERING TIME: Mar–Jul
NOTES: Compare with *V. lobata* ssp. *lobata*, which has deeply lobed but not compound leaves.

Viola tomentosa
Woolly Violet
Violaceae (Violet family)

Prostrate to erect, densely white-woolly perennial herb 0.7–1.0 dm.
LEAVES: Basal leaves have petioles 2–6 cm; blades 1.5–5 cm x 1.4–2.1 cm wide, narrowly ovate, with margin entire.
FLOWERS: Peduncles 1–4 cm; sepals lanceolate; petals dark yellow, the upper 2 usually with reddish-brown backs, and the lowest 6–11 mm. Cleistogamous flowers absent.
FRUITS: ± 5 mm, roundish, white-woolly. Seed 2.5–2.8 mm, mottled brown.
SIERRA DISTRIBUTION: West slope; El Dorado to Plumas Counties.
ELEVATION: 1350–2030 m
HABITAT: Dry gravelly sites in coniferous forest; montane zone.
FLOWERING TIME: May–Aug
NOTES: This locally endemic violet is only known from 5 counties, and is CRPR 4.2.

Arceuthobium americanum

Lodgepole Pine Dwarf Mistletoe

Viscaceae (Mistletoe family)

Perennial shrub with stem < 6–18 cm, angled when young; male stems are yellow, female stems yellow-green.

LEAVES: Simple, opposite, < 1 mm, scalelike.

FLOWERS: Dioecious; individual plants produce either male or female flowers. Male flowers have 3–4 perianth parts, with anthers sessile and directly on perianth parts; female flowers have 2 perianth parts, ovary inferior.

FRUITS: 2–5 mm, fusiform to spheric, 2-toned (green below), with pedicel recurved. Seeds released by being projected to 15 m in distance when fruits "explode."

SIERRA DISTRIBUTION: Tulare County north.

ELEVATION: 1300–2500 m

HABITAT: Parasitic on lodgepole pine, *Pinus contorta* ssp. *murrayana*; montane zone.

FLOWERING TIME: Apr–May

NOTES: Found almost exclusively on lodgepole pine, rarely on other pines.

Phoradendron juniperinum

Juniper Mistletoe, Incense Cedar Mistletoe

Viscaceae (Mistletoe family)

Hairless subshrub with stem 10–25 cm; dioecious, woody at base, hemiparasitic on branches of *Juniperus* and *Calocedrus*. Stem is green to olive green, with internodes cylindrical (not angled), each internode 5–20 x 1.5–2.5 mm diameter.

LEAVES: Scalelike; blades 1–2 mm, triangular, green to olive green.

FLOWERS: Male plants have staminate inflorescences 3–5 mm, with 1–6 flowers sunken into the axis of the fertile internode, flowers arranged 3 per bract. Female plants have pistillate inflorescences 3–5 mm, the fertile internode with 2 sunken flowers, 1 per bract. Flowers of both sexes have 3–4 petals, 0.5–1 mm.

FRUITS: Berry 4–5 mm x 3 mm, globose or sightly ellipsoid, hairless.

SIERRA DISTRIBUTION: Both slopes, all counties.

ELEVATION: 1700–2600 m

HABITAT: On incense cedar and junipers; montane and pinyon-juniper zones.

FLOWERING TIME: Jul–Sep

NOTES: Strictly speaking, *Phoradendron* spp. are hemiparasites; plants have chlorophyll and photosynthesize, producing their own carbohydrates to at least some extent while tapping into host plants' xylem for water. This mistletoe likely benefits its host plant by producing a steady supply of fruits used by a variety of birds, e.g., Townsend's solitaire, western bluebird, American robin, and cedar waxwing. These birds also feed upon juniper berries, which are produced with more variability from year to year, and disperse their seeds, likely increasing germination success of juniper seeds once they've passed through birds' digestive tracts.

subsp. *macrophyllum* (on *Alnus*)
♂
♀
♀
subsp. *tomentosum*
(on *Quercus*)

Phoradendron leucarpum ssp. *tomentosum*

Oak Mistletoe

Viscaceae (Mistletoe family)

Perennial herb with stem ≤ 80 cm, not angled; male stems are yellow, female stems yellow-green.

LEAVES: Simple, opposite, < 3 cm x < 2 cm wide, dull gray-green (not shiny), ± hairless to densely hairy.

FLOWERS: Dioecious (individual plants produce either male or female flowers); flowers sunken into inflorescence axis; perianth parts 3. Male inflorescences have 25–35 flowers, anthers 2-chambered; female inflorescences have 6–15 flowers.

FRUITS: 4–5 mm, spheric, white or pinkish, pedicels absent. Bird-dispersed.

SIERRA DISTRIBUTION: West slope.

ELEVATION: < 2100 m

HABITAT: Parasitic primarily on oaks, *Quercus* spp., sometimes on *Adenostoma*, *Arctostaphylos*, *Cercocarpus ledifolius*, *Rhus*, and *Umbellularia*; foothill and montane zones.

FLOWERING TIME: Jul–Sep

NOTES: Also found in the SN is *P. leucarpum* ssp. *macrophyllum* (formerly *P. macrophyllum*), bigleaf mistletoe, which has shiny, yellow-green leaves ≤ 6 cm x 5 cm wide. It is found on trees other than *Quercus*, especially trees of riparian areas like *Alnus*, *Fraxinus*, *Juglans*, *Platanus*, *Populus*, and *Salix*, as well as the non-native, nonriparian genus *Robinia*, e.g., black locust, *Robinia pseudoacacia*.

Vitis californica

California Wild Grape

Vitaceae (Grape family)

Climbing or trailing, woody vine with peeling bark; young stems woolly.

LEAVES: Simple, palmately lobed, with small stipules < 3.5 mm; tendrils are opposite leaves.

FLOWERS: Panicles of numerous small (about 1.5 mm), unisexual flowers. Sepals 5, green, with lobes absent or tiny; petals 4–5, united, yellow, forming a caplike structure. Male flowers have 3–9 erect stamens; female flowers have stamens reflexed or absent, and nectaries present.

FRUITS: Grape 4–20 mm, small, purple, may or may not be glaucous. Seeds 1–4.

SIERRA DISTRIBUTION: West slope, all.

ELEVATION: < 1370 m

HABITAT: Shady areas along streams or rivers; foothill and lower montane zones.

FLOWERING TIME: May–Jun

NOTES: Grapes are eaten by California Indians, who also use the roots for basketry and the stems in construction of dwellings.

MONOCOTS

Monocots are a large group of approximately 56,000 species worldwide (22 percent of angiosperms). This group is demonstrated through molecular data to be monophyletic (sharing a common evolutionary history, derived from a common ancestor). Monocots are distinguished by having a single cotyledon (seed leaf); flower parts generally in threes or multiples of three (although this is not strictly limited to monocots); most have leaves with parallel venation. Pollen grains have one aperture (an area with differentiated tissue through which pollen tubes exit to deliver sperm nuclei to eggs). Monocots also differ from eudicots in stem morphology—vascular bundles are scattered throughout the stem rather than in a cambium; and there are more differences in cell types and other nitty-gritty details that can be further investigated, if you're interested. Michael Simpson's *Plant Systematics* textbook is a good place to start.

Camassia leichtlinii ssp. *suksdorfii*

Large Camas

Agavaceae (Century Plant family)

Perennial herb 2–10 dm; from single bulbs.

LEAVES: Basal leaves 4–8, 20–60 cm x 5–25 mm wide, lanceolate, hairless.

FLOWERS: Racemes of 15–40 flowers, usually with only 1–4 open at a time; perianth radially symmetric or nearly so, with tepals 20–40 mm, usually 5-veined, blue-purple.

FRUITS: Capsule 10–30 mm, ovoid.

SIERRA DISTRIBUTION: West slope, Tulare County north.

ELEVATION: 1000–2600 m

HABITAT: Wet meadows; montane zone.

FLOWERING TIME: May–Aug

NOTES: Formerly misclassified as *C. quamash* ssp. *quamash*, which turns out not to occur in CA. To remember how to tell the 2 SN Camas taxa apart, think of *l* for this one; "large" and "*leichtlinii*" both start with an *l*, and large camas plants are taller on average and have tepals up to twice as long as those of small camas. All *Camassia* spp. still are enormously important food items for Native Americans; the sweet, starchy bulbs are roasted and eaten. Plants must be in flower when bulbs are collected, or bulbs of the lethally poisonous death camas, *Toxicoscordion* spp., could be collected and eaten by mistake.

Camassia quamash ssp. *breviflora*

Small Camas

Agavaceae (Century Plant family)

Perennial herb 2–5 dm; from single bulbs.

LEAVES: Basal leaves 3–5, 10–30 cm x 7–20 mm wide, lanceolate to linear, hairless, with glaucous upper surfaces.

FLOWERS: Dense raceme bearing 3–45 flowers, with 2–5+ open at a time; flowers bilaterally symmetrical to varying degrees, sometimes radial. Tepals 10–25 mm x 3–6 mm wide, pale blue to blue-violet, and usually 3-veined (sometimes 5-veined).

FRUITS: Capsule 8–20 mm, ovoid, 3-chambered. Seeds 6–12 per chamber.

SIERRA DISTRIBUTION: West slope, Tulare to Alpine Counties; both slopes farther north.

ELEVATION: 700–2600 m

HABITAT: Wet meadows; montane zone.

FLOWERING TIME: May–Jul

NOTES: This species can be distinguished from *C. leichtlinii* ssp. *suksdorfii* by its shorter stature and racemes of smaller, more densely arranged flowers that are at least somewhat bilaterally symmetrical (most evident in the longer lower tepal). The ssp. name *breviflora* means "small flower," reflecting on the smaller tepal size compared to that of large camas. Large populations of either *Camas* species can occupy many acres, turning meadows purple as they come into bloom. Both species are pollinated primarily by native bees, especially bumblebees and mason bees.

Chlorogalum pomeridianum var. *pomeridianum*

Soap Plant

Agavaceae (Century Plant family)

Perennial herb with flowering stalks 5–25 dm; from bulbs surrounded by coarse, brown fibers.

LEAVES: Basal 20–70 cm x 6–25 mm wide, wavy-margined.

FLOWERS: Inflorescence is a many-branched panicle reaching overall heights up to 2.5 m. Tepals 15–25 mm, linear, white with a green or purple central vein.

FRUITS: Capsule 4–7 mm, ovoid; old tepals twist over fruit as it matures.

SIERRA DISTRIBUTION: West slope, all.

ELEVATION: < 1900 m

HABITAT: Open, dry to moist areas; foothill and lower montane zones.

FLOWERING TIME: May–Aug

NOTES: Flowers open for less than a day: A given flower opens in late afternoon, produces nectar and pollen, then usually closes by midnight. In late afternoon/early evening, a plant with many open flowers is often a hub of insect activity with diurnal pollinators like bumblebees working flowers alongside crepuscular ones like sphinx moths. Soap plant is important for many reasons to Native Americans. The bulbs lather with water into soap, are used to stun fish, and are roasted for food. Strong fibers surrounding large bulbs are used to make brushes, with glue made from boiled the bulbs binding the fibers into a handle. Similar species: Narrowleaf soap plant, *C. angustifolium*, is found < 500 m from Fresno County north; this plant is < 7 dm and has narrow, flat leaves 2–5 mm wide with flowers 8–12 mm. The rare Red Hills soap plant, *C. grandiflorum*, occurs < 500 m on serpentine; plants < 1 m, pedicels 2–5 mm, much shorter than the 15–30 mm long tepals. The Red Hills Recreation Area in Tuolumne County is a good place to see it flowering in May.

Hastingsia alba

White Rushlily

Agavaceae (Century Plant family)

Hairless perennial herb 40–90 cm; from bulbs.

LEAVES: Mostly basal, 28–53 cm, grasslike, keeled, glaucous.

FLOWERS: Dense racemes 14–27 cm, flowers with 6 tepals, 6–8 mm, white to cream. Fresh anthers purplish-brown.

FRUITS: Capsule 6–9 mm. Seeds black, flattened.

SIERRA DISTRIBUTION: West slope, Placer County north.

ELEVATION: < 500–2300 m

HABITAT: Wet meadows, often in areas of peat (fens); rocky sites with a year-round source of water; montane zone.

FLOWERING TIME: Jun–Jul

NOTES: This plant is an indicator of fens (peatlands) and is documented to contribute to peat accumulation, thus maintaining functionality of fens. A good place to see it in bloom is the fen in Butterfly Valley Botanical Area, in the Plumas National Forest near Quincy.

Alisma triviale

Northern Water-Plantain

Alismataceae (Water-Plantain family)

Aquatic perennial herb ≤ 1 m; often from rhizomes.

LEAVES: Basal 7–45 cm, emersed; blades 5.5–15 cm x 1.5–10 cm wide, linear-lanceolate to broadly oval or elliptic. Bases squared off or somewhat lobed, not tapered to the petioles.

FLOWERS: Inflorescence far exceeds leaves in height. Sepals 3, green; petals 3, 3.5–6 mm, white, with somewhat irregular margins. Pistils 15–20; stamens 6.

FRUITS: Each pistil becomes an ovoid achene, and the combined pistils of a flower form a fruiting head 4.1–7 mm wide, smooth, roundish, brown.

SIERRA DISTRIBUTION: Mostly west slope, patchy distribution, Tulare County north.

ELEVATION: < 1600 m

HABITAT: Shallow ponds, streamsides, ditches; foothill and montane zones.

FLOWERING TIME: May–Sep

NOTES: Formerly included with *Alisma plantago-aquatica*, which is native to Eurasia and may not occur in the US. Medicinal uses of *A. triviale* by the Cherokee include a poultice for wounds and sores. The similar-looking *Alisma lanceolatum* is a non-native species with pink petals and leaf bases that are tapered gradually to the petioles. *A. lanceolatum* occurs from Nevada County north but may spread further south over time.

Hesperoyucca whipplei

Chaparral Yucca

Agavaceae (Century Plant family)

Robust perennial with tall flowering stalks up to 3 m far, exceeding basal rosette of leaves. Sometimes secondary rosettes are produced via rhizomes.

LEAVES: 4–10 dm, rigid, spine-tipped, with toothed edges and wider bases; grayish-green and glaucous.

FLOWERS: 1 tall (9 dm–3 m) inflorescence stalk produced per rosette, with numerous flowers in panicles. Perianths bell-shaped to spheric, with 6 tepals, 3.2–4.5 cm, distinct, creamy-white. Ovary superior, stigma headlike; stamens 6.

FRUITS: Capsule 3–5 cm, obovoid, erect. Seeds many, 6–9 mm diameter, flattened, dull black.

SIERRA DISTRIBUTION: West slope, Fresno County (Kings River drainage) south.

ELEVATION: < 1400 m

HABITAT: Dry slopes in chaparral and desert scrub; foothill and sagebrush/pinyon-juniper zones.

FLOWERING TIME: Apr–May

NOTES: Formerly *Yucca whipplei*. Plants may take 5 years to mature and flower, after which they often die. Pollination is via the yucca moth (*Tegeticula* spp.), which deposits balls of pollen from a different yucca plant onto the stigma while laying eggs in the ovary of the flower. Larvae feed on seeds. Flowers, seeds, roots, and stems all have Native American uses: Roots contain saponin and are pounded in water to make a bubbly all-purpose soap used for bathing, shampooing, washing clothes, hides, etc. Fibers from the leaves are prized for their softness, and once extracted from the leaves and cleaned, they are twisted into strands to make string, baskets, fishing nets, clothing, blankets, and much more. Newly opened flowers are eaten, and the bases of flowering stems are roasted and eaten.

Damasonium californicum

California Damasonium

Alismataceae (Water-Plantain family)

Aquatic perennial herb with stems < 35 cm. Vegetative reproduction occurs via corms.

LEAVES: 5–35 cm, with blades 3–9 cm x 0.5–3 cm wide. Leaf shape variable, from linear to narrowly ovate.

FLOWERS: Petals 3, 6–10 mm, white with yellow at base, margins irregularly jagged. Pistils 6–15, fused at base; stamens 6.

FRUITS: When mature, each flower produces a burlike head of beaked achenes.

SIERRA DISTRIBUTION: Calaveras County north.

ELEVATION: < 1700 m

HABITAT: In water and at edges of ponds, vernal pools, streams; many zones.

FLOWERING TIME: Jun–Aug

NOTES: Uncommon in the SN. Reported from the vicinity of Calaveras Big Trees State Park and several sites in Sierra Valley (Plumas County) southeast of Portola along Highway 70 and along roads leading to Marble Hot Springs.

Allium abramsii

Abrams' Onion

Alliaceae (Onion family)

Perennial herb with stems < 15 cm; bulbs 1–3.

LEAVES: 1 per stem, 12–30 cm, cylindrical, withered from tip at flowering time.

FLOWERS: Spherical umbels of 10–45 flowers; flowers 8–15 mm, rose-purple/purple and white, with a tubular or narrowly bell-shaped appearance; pedicels 6–15 mm. Inner 3 tepals have minutely toothed to jagged margins; outer tepals slightly longer and minutely toothed on upper ¼. This character is variable and can be hard to see (best seen with a 10x lens).

FRUITS: Ovary crests 6, 2 per ovary chamber; prominent, triangular.

SIERRA DISTRIBUTION: West slope, Tulare to Madera Counties.

ELEVATION: 900–2200 m

HABITAT: Gravelly or sandy openings, cracks of rocky outcrops; foothill and montane zones.

FLOWERING TIME: May–July

NOTES: Formerly *A. fimbriatum* var. *abramsii*. CRPR 1B.2 as of 2024, but much more common within its 3-county range than formerly thought. The similar-looking *A. fimbriatum* var. *fimbriatum* occurs in Tulare and Kern Counties. It has heads of 6–35 flowers, smooth-margined inner tepals, and fingerlike projections on the 6 ovary crests.

Sagittaria latifolia

Duck Potato

Alismataceae (Water-Plantain family)

Aquatic perennial herb to 45 cm; from oblong tubers (corms). Stolons present; monoecious.

LEAVES: Emergent or floating leaves have 3-angled petioles 7–51 cm; blades 1.5–30 cm, sagittate (arrow-shaped). Basal lobes about the same length as the terminal lobe.

FLOWERS: Flowers either male or female (unisexual), with both sexes on the same plant. Sepals 3–10 mm; petals 3, white, margins smooth. Male flowers have 7–30 hairless filaments; female flowers are at base of inflorescences, with numerous pistils spiraled on convex receptacles.

FRUITS: Achenes 2–3.5 mm, beak 1–2 mm.

SIERRA DISTRIBUTION: West slope, Mariposa County north.

ELEVATION: < 1500 m

HABITAT: Shallow waters of ponds, streams, rivers, lakes; muddy shores; sometimes in deeper water; many zones.

FLOWERING TIME: Jul–Aug

NOTES: In the SN, arrowhead-shaped emergent leaves distinguish *Sagittaria* spp. from other genera in the family (note that underwater leaves have blades tapered to bases rather than being arrowhead-shaped). *S. latifolia* can be seen in Yosemite National Park at Siesta and Laurel Lakes, and near Butterfly Valley in Plumas County. The similar Tule potato, *S. cuneata*, occurs on both slopes up to 2650 m from Mariposa County north. Plants are taller (up to 110 cm); emergent leaf blades are 2.5–17 cm, with basal lobes shorter than the main lobe. The achene beak is shorter (0.1–0.4 mm). *S. cuneata* can be found in Yosemite in Siesta Lake next to Tioga Road and at a pond 0.4 mi due north of the Cathedral Lakes Trailhead in Tuolumne Meadows.

Allium campanulatum

Sierra Onion

Alliaceae (Onion family)

Perennial herb with stems 10–30 cm; stems 1–3, 1–3 bulbs; bulblets are produced below roots and at ends of slender rhizomes ≤ 10 cm.

LEAVES: 2 per plant, 8–40 mm x 1.5 mm wide, upper surface channeled, withered by flowering time.

FLOWERS: Open umbels of 10–50 flowers, shiny, light to dark pink; pedicels 1–2 cm. Tepals 5–8 mm, held open widely, pink with a purple crescent near base, midrib darker pink.

FRUITS: Ovary crests 6, short, triangular with tiny teeth at top.

SIERRA DISTRIBUTION: All.

ELEVATION: 600–2900 m

HABITAT: Dry to moist, often sandy sites; all zones except alpine.

FLOWERING TIME: May–Aug

NOTES: One of the most common onions in the SN. The species name means "bell-like," and the flower does pass through a bell-shaped form as it opens. Twincrest onion, *A. bisceptrum*, could be mistaken for Sierra onion; it has 2–5 leaves, 8–30 cm x 1–13 mm wide, remaining green at flowering; and white to pink tepals without a purple crescent. *A. bisceptrum* occurs in somewhat moister sites on both slopes from Tulare County north, 2000–2900 m.

Allium hyalinum

Glassy Onion

Alliaceae (Onion family)

Erect perennial herb with stem 15–45 cm; from clustered bulbs.

LEAVES: 2–3 per stem, 0.7–1.5x length of stem, cylindrical or rounded with wide channels.

FLOWERS: Open inflorescences of 5–25 flowers; pedicels 10–35 mm. Flowers white to light pink, with or without a dark pink midrib stripe; tepals 6–10 mm, papery in fruit.

FRUITS: No ovary crests; dried tepals hyaline (glassy, translucent) and converge over the ripe fruit.

SIERRA DISTRIBUTION: West slope.

ELEVATION: 50–1500 m

HABITAT: Wet areas in cracks and at edges of rocky outcrops, as well as grassy slopes or even rocky soils where moisture is abundant in early spring; foothill and montane zones.

FLOWERING TIME: Mar–May

NOTES: This onion was and still is an important food for Native Americans; it is one of the easiest onions to gather as the bulbs can be "scooped" off of the shallow bedrock where masses of these plants tend to cluster. A similar-looking species, narrowleaf onion, *A. amplectens*, has shorter (4–15 mm) pedicels, more (10–50) flowers per head, and mature fruits have 6 crests. It grows in open or wooded areas < 1800 m, often on serpentine or clay, and has 2–4 leaves, which are shorter than the stem.

Allium lemmonii

Lemmon's Onion

Alliaceae (Onion family)

Perennial herb with stem 15–20 cm; stems flattened with narrow wings, breaking off at ground level with leaves once fruits are mature.

LEAVES: 2 per stem, 8–30 cm x 3–5 mm, flat, sickle-shaped, withering from the tips at flowering time.

FLOWERS: Inflorescences of 10–40 flowers; pedicels 8–16 mm. Tepals 6–9 mm, ovate, white or pink, with smooth margins.

FRUITS: Ovary crests 6, 2 per ovary lobe; ridgelike, inconspicuous.

SIERRA DISTRIBUTION: Both slopes; Mono and Amador Counties north.

ELEVATION: 1200–1900 m

HABITAT: Open, sunny areas with rocky or clay soils; montane zone.

FLOWERING TIME: May–Jul

NOTES: A good place to see *A. lemmonii* in bloom is Martis Creek Lake National Recreation Area, about 3 mi southeast of Truckee, around June.

Allium obtusum var. *obtusum*

Red Sierra Onion

Alliaceae (Onion family)

Perennial herb with stem 1.5–17 cm; bulbs 1–5, ovoid.

LEAVES: 1–2 per stem, up to 4x the length of the stem and 1–5 mm wide, sickle-shaped, withering at tip by flowering time.

FLOWERS: Inflorescences of 6–30 flowers; pedicels 2–14 mm. Flowers 4–12 mm, bell-shaped; tepals white or light pink, with dark midveins, tips more or less obtuse. Stamens included; style ≥ stamens.

FRUITS: Ovary crests 3, variable in prominence, each with 2 central lobes. Seeds dull.

SIERRA DISTRIBUTION: West slope to crest, Fresno County north.

ELEVATION: 1500–3500 m

HABITAT: Open gravelly or sandy sites, rocky outcrops; montane through alpine zones.

FLOWERING TIME: Apr–Jun

NOTES: This small plant blooms earlier than most of its rocky outcrop/gravel associates and is well camouflaged against the speckled background. Showy red Sierra onion, var. *conspicuum*, grows in the same habitat on the west slope from Fresno County north, 800–3000 m, and differs in always having only 1 leaf per stem, with blades 2–14 mm wide, and heads of 10–65 flowers with lanceolate, acute, pink tepals. *A. burlewii* is a similar-looking plant occurring from Fresno County south, with heads of 8–20 shorter (7–10 mm), conical or narrowly bell-shaped flowers, stamens and styles equal to or longer than tepals, and ovaries crested or not; seeds are shiny.

Allium platycaule

Broad-Stemmed Onion

Alliaceae (Onion family)

Perennial herb with stems 7–25 cm; widely flattened stem, winged at top.

LEAVES: 2 per stem, 10–30 cm, sickle-shaped, withered at tip at flowering time.

FLOWERS: Inflorescences of 30–90 flowers; flowers 8–15 mm, rose-pink; pedicels 10–25 mm. Tepals long, slender, rose to bright pink. Style and stamens about the same length; stamens exserted.

FRUITS: No ovary crests.

SIERRA DISTRIBUTION: El Dorado County north.

ELEVATION: 1500–2500 m

HABITAT: Rocky slopes, sandy soils; montane zone.

FLOWERING TIME: May–Jun

NOTES: This onion is unique and puts on quite a show with its long, pink filaments adding to the pink of the unusually narrow tepals. The species name *platycaule* means "wide stem." The only other onions in the SN with wide, flattened stems are *A. parvum* (not pictured), a small plant 2–12 cm tall with 2 flattened, sickle-shaped leaves, and *A. lemmonii*.

Allium shevockii

Spanish Needle Onion

Alliaceae (Onion family)

Perennial herb with stem up to 20 cm; stem holds clusters of 12–30 shiny, maroon flowers next to a curved, cylindrical, gray-green leaf that noticeably exceeds the length of the stem.

LEAVES: 1 per stem, up to 2.5x stem length.

FLOWERS: Outer tepals maroon near tips, reflexed and curled; inner tepals flaring and much smaller.

FRUITS: Ovary crests 6, prominent, with irregular margins, each with a notched tip.

SIERRA DISTRIBUTION: Only known from northeastern Kern County in the vicinity of Spanish Needle.

ELEVATION: 2000–2500 m

HABITAT: Talus of granitic or metamorphic outcrops; montane zone.

FLOWERING TIME: May–Jun

NOTES: An unmistakable rare onion of the Kern Plateau in the southern SN. Protected by the federal and state governments from collection or disturbance.

Allium validum

Pacific Onion, Swamp Onion

Alliaceae (Onion family)

Perennial herb with stem 50–100 cm; stem angled, topped by clusters of white to pink flowers with protruding stamens.

LEAVES: 3–6, roughly equal in length to stem, flat or somewhat keeled.

FLOWERS: Heads of 15–40 flowers; tepals 6–10 mm, narrowly lanceolate and erect. Stamens longer than tepals and topped by blue anthers.

FRUITS: No ovary crests.

SIERRA DISTRIBUTION: All.

ELEVATION: 1200–3400 m

HABITAT: Wet meadows and streamsides; montane through alpine zones.

FLOWERING TIME: Jul–Sep

NOTES: Distinguished from other onions by tall stature, wet habitat, and strong onion smell.

Iris hartwegii ssp. *hartwegii*

Hartweg's Iris

Iridaceae (Iris family)

Perennial herb < 30 cm; stem unbranched, rhizomatous, rhizomes 7–10 mm in diameter.

LEAVES: 2-ranked, basal arranged in a fanlike manner, 5–10 mm wide. Cauline leaves 1–4, like basal or bractlike.

FLOWERS: Inflorescence of 2 flowers, subtended by 2 bracts arranged alternately, lower (outer) bract 6–11 cm, bracts not concealing perianth tubes. Sepals 4–7 cm x 11–27 mm wide, oblanceolate; petals 5–7 cm x 5–14 mm wide, narrowly elliptic. Ovary inferior; perianth tube above ovary, 5–10 mm, barrel-shaped; style branches 31–45 mm, petal-like, crests 8–15 mm, stigma triangular.

FRUITS: Capsule 2–3 cm, oblong-oval, 3-chambered, somewhat triangular in x.s., and tapering at each end. Seeds brown, D-shaped.

SIERRA DISTRIBUTION: West slope, all.

ELEVATION: 400–2300 m

HABITAT: Dry, often partially shaded sites in conifer forest; montane zone.

FLOWERING TIME: May–Jul

NOTES: In the central Sierra, lavender to purple flowers are by far the most common. Further north, yellow flowers are most common. This is the most common iris of the mixed conifer forests in the SN.

Iris missouriensis
Western Blue Flag
Iridaceae (Iris family)

Perennial herb 20–60 cm; forming large, continuous colonies by freely branching rhizomes; stem usually unbranched; rhizomes 20–30 mm in diameter.
LEAVES: 2-ranked, basal in a fanlike manner, 45–60 cm x 3–9 mm wide; cauline 1–2, bractlike.
FLOWERS: 1–2 per inflorescence; lower 2 inflorescence bracts opposite with scarious edges, mostly enclosing the perianth tubes. Sepals 4–7 cm x 18–22 mm wide, obovate, white to light lavender with prominent purple veins; petals 4–6 cm x 9–12 mm wide. Ovary inferior; perianth tube above ovary, 4–12 mm; style branches 25–35 mm, petal-like, crests 7–14 mm, stigmas 2-lobed.
FRUITS: Capsule 4–5 cm x 1.5 cm wide, roughly round in x.s., 3-chambered, with 6 ridges. Seeds 4–4.5 mm, spheric to pear-shaped, light brown, and wrinkled.
SIERRA DISTRIBUTION: Both slopes, all.
ELEVATION: 400–3200 m
HABITAT: Moist to wet sites; below alpine zone.
FLOWERING TIME: May–Jul
NOTES: Populations can occupy hundreds of acres where conditions are optimal. Undesirable for domestic livestock due to bitter, unpalatable taste and toxicity, so this plant tends to dominate where livestock grazing occurs. Fine, strong fibers from the outside edges of leaves make excellent cordage and are gathered by Native Americans and worked into nets, string, and rope.

Iris pseudacorus (non-native, invasive)
Yellowflag Iris
Iridaceae (Iris family)

Erect, branched perennial herb 50–150 cm, forming enormous colonies in and along water edges; rhizomes 30–40 mm in diameter, pink.
LEAVES: 2-ranked, 4–10 cm x 2–3 cm wide; cauline 3–6, like basal.
FLOWERS: 3–6 per branch; lower 2 inflorescence bracts opposite, 5–7 cm. Sepals 5–8 cm, bright yellow with brown marks and veins; petals 2–3 cm. Perianth tube 12–13 mm; style branches < 25 mm, petal-like, crests 7–10 mm.
FRUITS: Capsule 5–8 cm, 3-chambered.
SIERRA DISTRIBUTION: Both slopes, sporadic, spreading.
ELEVATION: < 1400 m
HABITAT: In and at edges of ponds, streams, lakes, rivers; below subalpine zone.
FLOWERING TIME: Apr–Jun
NOTES: On the CDFA as a noxious weed. This plant is native to Eurasia, extremely invasive, and can become impossible to control, especially where aquatic herbicides are not an option. Early detection of any new locations followed by rapid control is crucial.

Sisyrinchium bellum

Western Blue-Eyed Grass

Iridaceae (Iris family)

Tufted perennial herb < 6.4 dm; rhizomatous, stems 1.5–5.3 mm wide, branched, narrowly winged, not glaucous.

LEAVES: Basal and cauline, 2–4 mm wide, grasslike.

FLOWERS: Outer tepals 10–17 mm, often broadly wedge-shaped, light blue to dark purple or violet, sometimes white. Bases yellow, tips truncate or notched, with short bristles at tips. Filaments fused or free, not stalked-glandular at base.

FRUITS: Capsule 3–5 mm, spheric to obovoid, brownish.

SIERRA DISTRIBUTION: All.

ELEVATION: < 2400 m

HABITAT: Moist, open grassy areas, meadows; most zones.

FLOWERING TIME: Mar–May

NOTES: Highly variable; compare with *S. idahoense*, which has only basal leaves. Find locations to view this common and beautiful plant on Calflora.

Sisyrinchium californicum

Golden-Eyed Grass

Iridaceae (Iris family)

Tufted perennial herb ≤ 62 cm; rhizomatous; stems 2–7 mm wide, flattened and broadly winged; herbage a dull green; drying blackish.

LEAVES: Basal, grasslike.

FLOWERS: Yellow; perianth of 6 tepals, outer tepals 12–18 mm.

FRUITS: Capsule 6–13 mm, dark brown to black.

SIERRA DISTRIBUTION: West slope, sporadic.

ELEVATION: ≤ 1800 m (possibly higher)

HABITAT: Wet meadows, fens; montane zone.

FLOWERING TIME: Apr–Aug

NOTES: The similar Elmer's blue-eyed grass, *S. elmeri*, is the more common of the 2 yellow SN *Sisyrinchium* spp., occurring in wet meadows up to 2700 m on the west slope from Tulare County northward. It is a shorter (≤ 25 cm) plant, with stems < 2 mm wide, and smaller flowers (outer tepals 7.5–8.5 mm). Herbage is a medium green, drying to a dark olive, but not black. The Butterfly Valley Botanical Area in Plumas County is a good place to find both species. Posting good diagnostic photos of any yellow-flowered *Sisyrinchium* encountered on iNaturalist or Calflora would be helpful to better understand the distribution, habitat, and elevation limits of these 2 taxa.

Sisyrinchium idahoense var. *occidentale*

Blue-Eyed Grass

Iridaceae (Iris family)

Tufted perennial < 4.5 dm; rhizomatous, stems unbranched, 1–2 mm wide, narrowly winged; herbage green or glaucous.
LEAVES: Usually all basal, 2–4 mm wide, grasslike.
FLOWERS: Outer tepals 8–13 mm, elliptic or oblanceolate, tips rounded or with a tiny notch, sometimes bristle-tipped, as in photo. Filaments bluish-violet, fused into a tube that is stalked-glandular at base.
FRUITS: Capsule 3–6 mm, brown, glandular-hairy.
SIERRA DISTRIBUTION: All.
ELEVATION: < 2400 m
HABITAT: Moist, open grassy areas, meadows; montane and subalpine zones.
FLOWERING TIME: Mar–May
NOTES: Compare with *S. bellum*.

Triglochin maritima

Common Arrow-Grass

Juncaginaceae (Arrow-Grass family)

Perennial herb 4–11 dm; rhizomatous, rhizomes stout.
LEAVES: Basal 1–8 dm x 2–5 mm wide, elliptic rather than round in x.s., grasslike, erect from sheaths. Sheaths 0.7–2.5 cm; ligule where sheath meets blades 1.5–5 mm, entire or shallowly 2-lobed.
FLOWERS: Racemes elongated, exceeding leaves in length, with pedicels 1–4 mm. Flowers bisexual, with 6 elliptic tepals, 1.3–1.7 x 0.6–1.4 mm, yellow-green; adaxially concave, with acute tips. Stamens 6, pistils 6, all fertile and plumose (as in photo).
FRUITS: Achenes 3–8 mm, linear-obovate, indistinctly ridged with an erect or recurved beak, 0.2 mm.
SIERRA DISTRIBUTION: Both slopes, sporadic, from Kern to Alpine Counties.
ELEVATION: < 2800 m
HABITAT: Alkaline or saline marshes and meadows, often near hot springs; montane and sagebrush zones.
FLOWERING TIME: May–Aug
NOTES: This species is found throughout the northern hemisphere (circumboreal) and in South America. Toxic to livestock due to the presence of cyanide in leaves, fruits, and seeds, and cyanide is in highest concentrations in young, early season foliage. Death of livestock animals can occur within minutes of symptom onset, and although an antidote is available, it must be administered immediately.

Calochortus albus
White Globe-Lily, Fairy Lantern
Liliaceae (Lily family)

Perennial herb; glaucous, from bulbs. Stems are usually branched, 20–80 cm, and herbage is glaucous.
LEAVES: Linear, flat. Basal 30–70 cm, 1–4 cm wide, present at flowering. Stem leaves 2–6.
FLOWERS: Inflorescences of 2 to many, nodding, white to light pinkish flowers. Sepals 10–15 mm, appressed to petals, at least in early flower; petals 20–25 mm, ciliate. Nectary crescent-shaped, with several membranes, each fringed and with yellow or white glandular hairs, the proximal membrane ⅓–⅔ width of petal. Filaments dilated (wide, flattened). Gently pull petals apart to see nectary and filaments.
FRUITS: Capsule 2–4 cm, nodding, winged.
SIERRA DISTRIBUTION: West slope, Madera County north.
ELEVATION: < 1300 m
HABITAT: Moist, shady to partially shady slopes; foothill and montane zones.
FLOWERING TIME: May–Jun
NOTES: An area of overlap with *C. amoenus* occurs in Madera County from North Fork to Oakhurst including the area around Bass Lake. The 2 spp. sometimes grow in mixed patches in this region, and each is also found without the other. A study is underway to elucidate the evolutionary relationship between the 2 spp., including whether they interbreed, share pollinators, or have different microsite requirements.

Calochortus amoenus ✿
Rose Globe-Lily, Purple Fairy Lantern
Liliaceae (Lily family)

Perennial herb with stems 10–50 cm; from bulbs. Stems are usually branched, and herbage is glaucous.
LEAVES: Basal 20–50 cm, present at flowering. Stem leaves 2–5.
FLOWERS: Inflorescences of 2 to many, nodding, rose-pink flowers. Sepals 10–15 mm, held away from petals; petals 16–25 mm, ciliate and sparsely hairy on inner surface. Nectary elliptic, dark maroon or purple, with 4–5 ciliate membranes; bottom membrane extends full width of inner petal surface. Filaments slender, not dilated. Gently pull petals apart to see nectary and filaments.
FRUITS: Capsule 2–3 cm, nodding, winged.
SIERRA DISTRIBUTION: West slope, Madera County south.
ELEVATION: 500–1500 m
HABITAT: Grassy slopes in foothill woodland and chaparral; foothill zone, sometimes extending into lower montane zone.
FLOWERING TIME: Apr–Jun
NOTES: These pretty, rose-colored flowers often appear in extraordinary abundance after wildfires. In April 2021, after the 2020 Creek Fire in the Sierra National Forest, there were masses of many thousands of flowering rose globe-lilies in Jose Basin (Fresno County) in burned chaparral.

Calochortus coeruleus

Beavertail-Grass, Cat's-Ear

Liliaceae (Lily family)

Low-growing perennial herb with stem 3–15 cm; from bulbs, stem erect or leaning, glaucous, unbranched.

LEAVES: Basal 10–20 cm; upper stem leaves with margins rolled in.

FLOWERS: Inflorescences of 1–10 flowers, erect, bell-shaped. Sepals 10 mm; petals 15–25 mm, with inner surfaces densely hairy from the nectary outward. Nectary depressed, crescent-shaped or semicircular with a hairy border. Petal surface below nectary is smooth, with blue or purple markings.

FRUITS: 10–20 mm, nodding.

SIERRA DISTRIBUTION: West slope, Amador County north.

ELEVATION: < 2500 m

HABITAT: Open, gravelly areas and grassy meadows within coniferous forest; montane zone.

FLOWERING TIME: May–Jul

NOTES: Distinctive, unlike any other SN *Calochortus*.

Calochortus invenustus

Plain Mariposa Lily

Liliaceae (Lily family)

Perennial herb with stem 20–50 cm; from bulbs (bulblets also present), glaucous.

LEAVES: Basal 10–20 cm, channeled, withered by flowering time; stem leaves smaller, rolled in.

FLOWERS: Inflorescences ± umbel-like, with 1–6 flowers; bracts 2–5 cm. Flowers erect, open, bell-shaped; sepals 2–3 cm, lance-ovate, adaxial surface yellow-green and lacking a purple blotch at base; petals 2–4 cm, obovate to wedge-shaped with short claw at base (not visible in photo), white or tinged light pink or lilac. Petals sometimes purple below the nectary, but not above (distal to) it. Nectary small, round, ± depressed, covered with dense, short, branched hairs, and encircled by a fringed membrane with longer, often yellow hairs. Filaments 6–7 mm, not dilated at base; anthers 7–8 mm, purple or yellow.

FRUITS: Capsule 5–7 cm, erect, angled, and linear with long-tapered tip. Seeds flat, with netlike coat.

SIERRA DISTRIBUTION: Both slopes, sporadic; most records are in Yosemite National Park and Tulare and Kern Counties.

ELEVATION: 1500–3000 m

HABITAT: Dry soil, often decomposed granite; montane zone.

FLOWERING TIME: May–Aug

NOTES: Most similar to *C. invenustus* is pinyon mariposa lily, *Calochortus bruneaunis*, which is primarily a plant of the eastern Sierra in shrublands or pinyon-juniper woodland, found from Tulare and Inyo Counties north. *C. bruneaunis* nectaries are surrounded by a yellow spot, petals have a dark red or purple arch distal to the nectary, and filament bases are dilated.

Calochortus leichtlinii

Leichtlin's Mariposa Lily

Liliaceae (Lily family)

Erect perennial herb with stems 20–60 cm; unbranched, from bulbs.

LEAVES: Basal 10–15 cm, linear, withered by flowering time; cauline leaves 2–3.

FLOWERS: Inflorescence of 1–5 erect, open, bell-shaped flowers. Sepals 10–20 mm; petals 10–40 mm, white with yellow centers, and maroon to black wedge-shaped or roundish blotch above the nectary. Nectary ovate and densely short-hairy, often depressed below the petal surface. Filaments 5 mm; anthers 5–7 mm, sagittate with distinct tails at base, as in photo, but often not as clearly sagittate.

FRUITS: Capsule 30–60 mm, erect.

SIERRA DISTRIBUTION: Both slopes, Tulare and Inyo Counties north.

ELEVATION: 500–3200 m

HABITAT: Open to partially shaded rocky or gravelly areas; montane and subalpine zones.

FLOWERING TIME: Jun–Aug

NOTES: This delicate, high-elevation mariposa lily is distinguished by the ovate nectary and the arrowhead-shaped anthers (when present, sometimes the anthers don't have tails). Compare with cream-colored variants of *C. venustus* or *C. superbus*, each with its own distinctly different nectary shape).

Calochortus luteus

Yellow Mariposa Lily

Liliaceae (Lily family)

Perennial herb with stems 20–50 cm; from bulbs with bulblets.

LEAVES: Basal 10–20 cm, linear, withering by flowering time.

FLOWERS: Inflorescences of 1–7 erect, open, bell-shaped flowers. Sepals 2–3 cm, long-tapered, straight or recurved, often with a basal blotch; petals 2–4 cm, bright yellow, upper surface with reddish-brown stippling or lines, with or without a central red-brown spot. The patterning of brown on the petals varies enormously from nearly nonexistent to a variety of bold patterns. Nectary crescent-shaped, varies in curvature from nearly linear to somewhat chevron-shaped; matted short-hairy, with longer, sparse, yellow hairs arching above the gland. Filaments 7–9 mm; anthers 4–6 mm, linear-oblong.

FRUITS: Capsule 3–6 cm, erect, angled.

SIERRA DISTRIBUTION: West slope, all.

ELEVATION: < 700 m

HABITAT: Full sun in heavy clay or serpentine soils; foothill zone.

FLOWERING TIME: Apr–Jun

NOTES: The crescent-shaped nectary distinguishes *C. luteus* from yellow forms of *C. superbus* or *C. venustus*. The *C. superbus* nectary is a pronounced upside-down V, and the *C. venustus* nectary is square. Yellow *C. superbus* flowers are usually a paler yellow than the deep, bright yellow of *C. luteus.* Where these 2 grow together, they often hybridize, producing plants with intermediate characters that preclude the satisfaction of a firm ID. It's not their fault; evolution is an ongoing process and does not pause for us to easily assign names!

Calochortus minimus

Lesser Star-Tulip

Liliaceae (Lily family)

Low-growing perennial herb with stem < 10 cm; from bulbs, stem erect or leaning.

LEAVES: Basal 10–20 cm (longer than stems), up to 1 cm wide; lacks stem leaves.

FLOWERS: Inflorescence of 1–5 white flowers; sepals 8–10 mm; petals 10–14 mm, hairless except ciliate membrane near nectary.

FRUITS: Capsule 15–20 mm, nodding, winged.

SIERRA DISTRIBUTION: West slope, Tulare County north.

ELEVATION: 1000–3300 m

HABITAT: Moist areas; often meadow and lake edges; montane zone.

FLOWERING TIME: May–Aug

NOTES: One of the smallest *Calochortus*. This plant is common but sometimes overlooked if growing among a mix of taller plant species. It hybridizes with nude mariposa lily, *C. nudus*, which looks similar but has stems 10–25 cm, longer (14–16 mm) petals, and erect rather than nodding fruits.

Calochortus monophyllus

Yellow Cat's-Ear, Yellow Star-Tulip

Liliaceae (Lily family)

Perennial herb with stems 8–20 cm; from bulb, sometimes branched.

LEAVES: Basal 10–30 cm (often > stems), up to 2.5 cm wide; stem leaves up to 3.

FLOWERS: Inflorescences of 1–6 golden-yellow flowers; sepals 16–20 mm; petals about the same length, all yellow or with brownish-red between nectary and base of petal, and densely hairy from nectary outward.

FRUITS: Capsule 1–2 cm, nodding, winged.

SIERRA DISTRIBUTION: West slope, Mariposa County north.

ELEVATION: < 1400 m

HABITAT: Full sun in heavy clay or serpentine soils; foothill zone.

FLOWERING TIME: Apr–May

NOTES: The only yellow cat's-ear.

Calochortus superbus

Grand Mariposa Lily

Liliaceae (Lily family)

Erect perennial herb with stems 30–60 cm, often branched.
LEAVES: Basal 15–25 cm, linear, withered by flowering time.
FLOWERS: Inflorescences of 1–3 erect, open, bell-shaped flowers. Sepals 2–4 cm, recurved, usually with a basal blotch visible between petal bases; petals 2–4 cm, usually with margins rounded, minutely toothed, and color generally cream, but sometimes pale yellow, pink, or purple with a central dark blotch (blotch often edged in yellow) above the nectary. Nectary crescent- or chevron-shaped, densely short-hairy. Filaments 7–9 mm, slender to dilated at base; anthers 8–10 mm, linear-oblong, white or yellow.
FRUITS: Capsule 5–6 cm, erect, angled.
SIERRA DISTRIBUTION: West slope.
ELEVATION: < 1700 m
HABITAT: Sandy soil in foothills, chaparral, and lower montane coniferous forest; foothill and montane zones.
FLOWERING TIME: May–Jul
NOTES: Similar to *C. venustus,* but the nectary shape distinguishes them; yellow forms can be difficult to distinguish from *C. luteus,* especially where the 2 spp. hybridize, and both can have crescent-shaped to oblong nectaries. *C. luteus* differs by its consistently deep yellow flowers.

Calochortus venustus

Butterfly Mariposa Lily

Liliaceae (Lily family)

Erect perennial herb with stems 10–60 cm, often branched.
LEAVES: Basal 10–20 cm, linear, withered by flowering time.
FLOWERS: Inflorescences of 1–6 (sometimes up to 10) erect, open, bell-shaped flowers. Sepals 3–5 cm, tips recurved; petal color remarkably variable: white, cream, pink, magenta, orange, red, maroon, copper, or yellow with at least 1 dark red blotch above the nectary (not visible in dark red or maroon flowers), and sometimes with a second blotch. Nectary square, not crescent-shaped or linear, covered with short orange to purple hairs. Filaments 7–10 mm, wider at base; anthers 7–10 mm, linear-oblong, cream, white, or pinkish.
FRUITS: Capsule 5–6 cm, erect, angled.
SIERRA DISTRIBUTION: All.
ELEVATION: 300–2000 m
HABITAT: Sandy soil; foothill and montane zones.
FLOWERING TIME: Jun–Aug
NOTES: This is one of the most common and the most variable of the mariposa lilies; flower color and markings vary infinitely. Look for the square nectary covered with short hairs, and the sepals with reflexed tips.

Clintonia uniflora

Bead Lily

Liliaceae (Lily family)

Mat-forming perennial herb with stems 1.5–2.5 dm; from elongated rhizomes.

LEAVES: 2–3 per rhizome, 7–15 cm, 2–7 cm wide, and somewhat hairy.

FLOWERS: Peduncle 4–8 cm, hairy, 1-flowered. Flowers 18–22 mm, bell-shaped. Stigma slightly 2- to 3-lobed; stamens 6.

FRUITS: Berry 6–10 mm, blue.

SIERRA DISTRIBUTION: West slope, Tulare County north.

ELEVATION: 1000–1900 m

HABITAT: Moist shady sites in mixed conifer forest; montane zone.

FLOWERING TIME: May–Jul

NOTES: The only *Clintonia* found in the SN; a common understory plant of giant sequoia groves. Another common name is "bride's bonnet."

Erythronium multiscapideum

Sierra Fawn Lily

Liliaceae (Lily family)

Perennial herb 8–23 cm; from bulbs, as well as bulblets and rhizomes.

LEAVES: 4–16 cm, elliptic to lanceolate, mottled brown or white, margin sometimes wavy.

FLOWERS: Racemes of 1–4 nodding flowers with 6 tepals, 16–35 mm, white with yellow bases. Style 10–13 mm, white, with 3 stigma lobes, 1–4 mm, recurved; stamens have filaments < 0.8 mm wide, slender, white, and anthers white to cream.

FRUITS: Capsule 2–4 cm, obovoid.

SIERRA DISTRIBUTION: West slope, Mariposa County north.

ELEVATION: < 1200 m

HABITAT: Openings in woodland and chaparral; foothill and lower montane zones.

FLOWERING TIME: Mar–May

NOTES: The only CA fawn lily that spreads underground via rhizomes. A great place to see thousands of plants in bloom in April is along Hunter's Valley Access Road on BLM land northeast of Bear Valley (Mariposa County).

Erythronium purpurascens

Purple Fawn Lily

Liliaceae (Lily family)

Perennial herb 7–20 cm; from elongated, slender bulbs.
LEAVES: 6–15 cm, narrowly ovate or lanceolate, usually wavy-margined, solid green (not mottled).
FLOWERS: Racemes of 1–6 nodding flowers with 6 tepals, 10–20 mm, white (or light pink) with a yellow base. Flowers fade pink-purple as they age. Style 4–5 mm, yellow; stigma unlobed; stamens have slender yellow filaments and yellow anthers.
FRUITS: Capsule 2–4 cm, obovoid.
SIERRA DISTRIBUTION: West slope, Placer County north (likely further south as well).
ELEVATION: 1100–2700 m
HABITAT: Streamsides, meadow edges, margins of melting snowfields in conifer forest; montane zone.
FLOWERING TIME: May–Aug
NOTES: Compare with *E. multiscapideum,* which has mottled leaves and occurs < 1200 m.

Erythronium tuolumnense ✿

Tuolumne Fawn Lily

Liliaceae (Lily family)

Perennial herb 15–35 cm; scapes often reddish, from bulbs (which form many offsets).
LEAVES: 15–35 cm, ovate or elliptic, sometimes with wavy margins, solid green.
FLOWERS: Racemes of 1–5 nodding flowers with 6 tepals, 20–35 mm, solid yellow. Style 8–10 mm, white; stigma entire or with tiny lobes (< 1 mm). Stamens have filaments < 0.6 mm wide, slender, white.
FRUITS: Capsule 2–5 cm, obovoid.
SIERRA DISTRIBUTION: Tuolumne County, between the Tuolumne and Middle Fork Stanislaus Rivers.
ELEVATION: 600–1000 m
HABITAT: Openings or shaded sites in woodlands; foothill and montane zones.
FLOWERING TIME: Mar–Jun
NOTES: Restricted to Tuolumne County, CRPR 1B.2. Find locally dense patches blooming in April along Italian Bar Road (Stanislaus National Forest Road 4N26) where it parallels the South Fork Stanislaus River, about a half-hour drive from Sonora. Another rare, yellow-flowered fawn lily with solid green leaves is the Shuteye Peak fawn lily, *E. pluriflorum*, found only in Madera County (Sierra National Forest). Racemes have up to 10 flowers, with tepals 15–28 mm, and yellow styles and filaments. It grows in open, rocky sites of the subalpine zone as well as the understory of red fir and lodgepole pine forest in the montane zone.

Fritillaria micrantha

Brown Bells

Liliaceae (Lily family)

Perennial herb 40–90 cm; from scaly bulbs, flowering stems erect.

LEAVES: Cauline 4.5–15 cm, linear to lanceolate. Lower leaves in 1–3 whorls of 4–6, upper leaves alternate.

FLOWERS: Inflorescence of 4–10 nodding flowers; tepals 1–2 cm, faintly mottled (if at all), greenish to purplish. Style divided ⅓–⅔ its length into 3 strongly recurved branches.

FRUITS: Capsule have 6 distinct wings.

SIERRA DISTRIBUTION: West slope, Tulare to Butte and Sierra Counties.

ELEVATION: 300–2200 m

HABITAT: Forest floor, dry slopes, rocky areas; coniferous forest; montane zone.

FLOWERING TIME: Apr–Jun

NOTES: The long, distinctly recurved styles and nodding flowers with faint mottling distinguish this *Fritillaria* in the SN. Flowering plants are well camouflaged and easy to miss, even when in full bloom. Nonflowering bulbs produce ovate to lanceolate, waxy, green bulb leaves, usually visible among flowering stems (inset).

Fritillaria pinetorum

Pinewoods Fritillary

Liliaceae (Lily family)

Perennial herb 10–40 cm; from scaly bulbs; flowering stems erect, glaucous.

LEAVES: Cauline 5–15 cm, linear, glaucous. Lower leaves in whorls of 2–3 per node, upper leaves alternate and reaching or exceeding the height of the inflorescence.

FLOWERS: Inflorescence of 1–3 broadly bowl-shaped flowers that are erect or face horizontally (rarely nodding); tepals 1.4–1.9 cm, broadly ovate, purplish, mottled with greenish-yellow. Nectary ⅔ the length of the perianth, difficult to see; style divided nearly to base.

FRUITS: Capsule 12–15 mm, angled, obovoid, with hornlike structures at base and top of each wing.

SIERRA DISTRIBUTION: Both slopes, Alpine County south.

ELEVATION: 1500–3200 m

HABITAT: Open coniferous forest and rocky slopes; subalpine zone.

FLOWERING TIME: May–Jul

NOTES: Often confused with mountain fritillary, *F. atropurpurea*, which usually has nodding flowers with tepals held widely apart, and a wider range, occurring sporadically in all SN counties. Taxonomic study of *Fritillaria* in the SN and beyond is sorely needed; these can be very tricky to identify, even with a dissecting microscope and good material.

Fritillaria recurva

Scarlet Fritillary

Liliaceae (Lily family)

Perennial herb 30–90 cm; from scaly bulbs, flowering stems erect.
LEAVES: Cauline 3–15 cm, linear to lanceolate. Lower leaves in 1–3 whorls of 2–5 per node, upper leaves alternate.
FLOWERS: Inflorescences of 1–9 bell- or funnel-shaped, nodding flowers; tepals 1.5–3.7 cm, orange to scarlet with yellow checkerboarding on the inner surface, and tepal tips strongly recurved. Nectary yellow, ¼ perianth length. Style branched ¼–½ of its length.
FRUITS: Capsule winged.
SIERRA DISTRIBUTION: West slope, El Dorado County north.
ELEVATION: < 1700 m
HABITAT: Open to partially shady sites, rocky outcrops, gravelly or sandy soil; foothill and montane zones.
FLOWERING TIME: Apr–Jun
NOTES: These unmistakable, red, trumpetlike flowers are pollinated by hummingbirds. The bulbs (as well as other *Fritillaria* spp.) were used as food by the Shasta and Yani tribes.

Lilium kelleyanum

Kelley's Lily

Liliaceae (Lily family)

Tall, erect perennial herb with stems up to 2.2 m; from bulbs.
LEAVES: 7.5–17.5 cm, in whorls 1–4 per stem, tips drooping, margins not wavy.
FLOWERS: Inflorescences of 1–15 flowers, pendent; tepals 4.2–6 cm, reflexed, uniformly yellow or yellow-orange with reddish-purple dots. Pistil 2.5–3.5 cm; anthers 3–6 mm, magenta or reddish; pollen tan-orange, orange, or reddish brown.
FRUITS: Capsule 1.5–3 cm, erect.
SIERRA DISTRIBUTION: Both slopes; Tulare and Inyo Counties north to Plumas County.
ELEVATION: 2200–3300 m
HABITAT: Streamsides, meadow edges, seeps; montane, subalpine, and alpine zones.
FLOWERING TIME: Jul–Aug
NOTES: Kelley's lily has a sweet fragrance, which is not generally the case for the similar-looking leopard lily, *L. pardalinum* ssp. *pardalinum*. The 2 are also distinguished by tepal color pattern (red-tipped in leopard lily) and the much shorter anthers of Kelley's lily.

Lilium pardalinum ssp. *pardalinum*

Leopard Lily

Liliaceae (Lily family)

Tall, erect perennial herb with stems up to 2.8 m; from elongate bulbs, forming dense colonies via rhizomes.
LEAVES: 4–27 cm, in whorls 1–8 per stem, tips drooping.
FLOWERS: Inflorescences of 1–28 flowers, pendent, with reflexed tepals 5.9–10.4 cm, yellowish-orange at base, red on outer half, speckled with reddish-purple dots. Pistil 5–8 cm; anthers 11–22 mm, magenta, turning purplish or somewhat yellow; pollen reddish-brown to brownish-orange.
FRUITS: Capsule 3–6 cm, erect.
SIERRA DISTRIBUTION: West slope, Tulare County north.
ELEVATION: < 1700 m
HABITAT: Streamsides, moist to wet areas; montane zone.
FLOWERING TIME: May–Aug
NOTES: Compare with Kelley's lily, *L. kelleyanum*, which is smaller, does not have red-tipped tepals, and is fragrant whereas leopard lily is rarely fragrant.

Lilium parvum

Alpine Lily, Sierra Tiger Lily

Liliaceae (Lily family)

Tall, erect perennial herb < 1.7 m; from bulblike rhizomes.
LEAVES: 4–15 cm, in whorls 2–5 per stem, not wavy-margined.
FLOWERS: Inflorescence of 1–26 flowers, erect or horizontal, funnel-shaped, orange, sometimes darker (or red) at tips, with reddish-purple dots; tepals 3.2–4.2 cm, tips usually recurved. Pistil 2.7–3.8 cm; anthers 4–8 mm, yellow; pollen yellow-orange.
FRUITS: Capsule 1.5–2.7 cm, erect, roundish.
SIERRA DISTRIBUTION: Both slopes, northern Kern County north.
ELEVATION: 1400–2900 m
HABITAT: Streamsides, moist to wet areas; montane and subalpine zones.
FLOWERING TIME: Jun–Aug
NOTES: Occurring below the actual alpine zone, the alpine lily's flower is smaller, upright or horizontal, and funnel-shaped, distinguishing this orange lily from *L. kelleyanum* and *L. pardalinum* ssp. *pardalinum*, which have pendent flowers. Alpine lily is pollinated by hummingbirds, swallowtail butterflies, and a variety of native bees.

Lilium washingtonianum ssp. *washingtonianum*

Washington Lily

Liliaceae (Lily family)

Tall, erect perennial herb < 2.6 m; tall, erect, sometimes glaucous.

LEAVES: 3–13 cm, often wavy-margined, in 2–5 whorls, leaves 3–16 per whorl.

FLOWERS: Inflorescence of 1–33 flowers, nodding to ascending funnel-shaped, white. Perianth parts 7.9–11.3 cm, with recurved tips, white with a yellow stripe on lower half of inner surfaces, and with tiny, magenta dots; inner perianth parts wider than outer, oblanceolate. Pistil 7.5–10.4 cm; anthers 8–15 mm, cream-colored (turning pinkish or yellow); pollen cream or light yellow.

FRUITS: Capsule 2.7–5.8 cm, approximately twice as long as wide, erect.

SIERRA DISTRIBUTION: West slope, Fresno County north.

ELEVATION: 1050–2000 m

HABITAT: Dry openings and roadsides in conifer forest; montane zone.

FLOWERING TIME: Jun–Aug

NOTES: Unmistakable! There is no other tall, large-flowered white lily in the SN. Flowers are strongly fragrant, and the scent and production of nectar are timed to peak at dusk to attract sphinx moths, the primary pollinators.

Prosartes hookeri

Fairy Bells

Liliaceae (Lily family)

Perennial herb to 30–80 cm; rhizomatous, branched sparingly.

LEAVES: Alternate, sessile or bases clasping stem, 3–15 cm, prominently veined, sparsely to densely hairy, with dense, short, and sharp hairs on leaf margin.

FLOWERS: In clusters of 1–3, hidden under leaves, bell-shaped, creamy white; tepals 6, 8–18 mm, similar creamy to greenish color. Anthers exserted.

FRUITS: Berry 7–12 mm, spheric, orange to red.

SIERRA DISTRIBUTION: West slope, Tulare County north.

ELEVATION: < 1600 m

HABITAT: Dry to moist floor of shaded mixed conifer forest; montane zone.

FLOWERING TIME: Apr–Jul

NOTES: Formerly *Disporum hookeri*. This is a frequent understory plant of giant sequoia groves. Without stopping to look under the leaves, one could completely miss the charming flowers of "fairy bells."

Toxicoscordion exaltatum

Giant Death Camas

Melanthiaceae (False-Hellebore family)

Erect perennial herb with stem 3–10 dm; hairless; bulb 2–6 cm, ovoid. Bulbs are often clumped, resulting in clumps of plants.
LEAVES: Mostly basal, 20–60 cm x 5–30 mm wide, linear, somewhat folded lengthwise.
FLOWERS: Open panicles 20–40 cm, of 10–80 flowers; inflorescence bracts 3–25 mm, green or white, membranous. Flowers mostly bisexual, except some on lower branches that are staminate (without ovaries or styles, or with stunted ovaries). Tepals 5–12 mm, white with yellow nectary at base, tips rounded; outer 3 tepals often remaining attached, reflexed, below fruits. Ovary superior, styles 3; stamens ⅔–¾ the length of the tepals.
FRUITS: Capsule 10–25 mm, oblong. Seeds numerous. Fruiting pedicels 10–40 mm, spreading.
SIERRA DISTRIBUTION: West slope, Kern to Plumas Counties.
ELEVATION: < 2500 m
HABITAT: Vernally moist areas, meadows, dry ridges and slopes; foothill and montane zones.
FLOWERING TIME: Apr–Jun
NOTES: Formerly *Zigadenus exaltatus*. The tallest and most robust death camas in the SN, with tepals longer than stamens. All parts of any *Toxicoscordion* species, fresh or dried, are extremely toxic to humans and livestock due to alkaloids found throughout the plant, but it is especially concentrated in bulbs. The genus name comes from *Toxico* ("poison") + *scordion* ("garlic").

Toxicoscordion paniculatum

Foothill Death Camas

Melanthiaceae (False-Hellebore family)

Erect perennial herb with stem 2–7 dm; erect, hairless; bulb 3–5 cm, ovoid, bulbs not clumped.
LEAVES: Mostly basal, 20–50 cm x 6–16 mm wide, linear, somewhat folded lengthwise.
FLOWERS: Panicles 10–30 cm, of 10–80 flowers, inflorescence bracts 5–15 mm, green, lanceolate. Most flowers bisexual, with some sterile or staminate. Tepals 2–6 mm; inner 3 narrowed at base and clawed, outer 3 not or barely clawed. Ovary superior, styles 3; stamens the same length as tepals or longer (as in photo).
FRUITS: Capsule 8–20 mm, cylindric. Seeds numerous. Fruiting pedicels 3–25 mm, spreading to ascending.
SIERRA DISTRIBUTION: Mostly east slope, from Kern and Inyo Counties north to Calaveras and Alpine Counties; both slopes northward from there.
ELEVATION: 1200–2300 m
HABITAT: Dry sites; foothill, montane, and sagebrush/pinyon-juniper zones.
FLOWERING TIME: Apr–Jul
NOTES: Formerly *Zigadenus paniculatus*. Compare with *T. venenosum* var. *venenosum*, which has flowers in racemes, not panicles.

Toxicoscordion venenosum var. *venenosum*

Meadow Death Camas

Melanthiaceae (False-Hellebore family)

Erect perennial herb with stem 1.5–7 dm; hairless; bulb 1.2–2.5 cm wide, ovate, bulbs not clumped.

LEAVES: Mostly basal, 10–40 cm x 4–10 mm wide, somewhat folded lengthwise.

FLOWERS: Racemes 5–25 cm (occasionally a panicle branch present at base of inflorescence), of 10–50 flowers, all bisexual. Inflorescence bracts 5–25 mm, usually green, sometimes white-membranous. Tepals 6, 4–6 mm, all clawed. Stamens the same length as or longer than tepals.

FRUITS: Capsule 8–14 mm, cylindrical. Seeds numerous. Fruiting pedicels 3–25 mm, spreading to ascending.

SIERRA DISTRIBUTION: Kern and Inyo Counties north.

ELEVATION: < 3000 m

HABITAT: Dry slopes, rocky areas, moist meadows, burned areas; foothill, montane, and subalpine, and sagebrush/pinyon-juniper zones.

FLOWERING TIME: May–Jul

NOTES: Formerly *Zigadenus venenosus* var. *venenosus*. Frequently among the first flowers to bloom after wildfires, exhibiting fire-induced flowering typical of many bulb-forming monocots, whereby flowers are produced in massive quantities years 1–2 post-fire.

Trillium albidum

Giant White Wakerobin, White Toadshade

Melanthiaceae (False-Hellebore family)

Erect perennial herb 2–6 dm; rhizomatous.

LEAVES: The "leaves" are actually floral bracts that function as leaves, in a whorl of 3 just below the flower, 7–20 cm x 12–15 cm wide, ovate, and usually mottled.

FLOWERS: 1 per stem; sepals 3–6.5 cm, light green; petals 4–8 cm, white or pinkish, fragrant.

FRUITS: Berry green to purplish, fleshy-pulpy.

SIERRA DISTRIBUTION: West slope, El Dorado County north.

ELEVATION: < 2000 m

HABITAT: Moist, shady sites; foothill and montane zones.

FLOWERING TIME: Feb–Jun

NOTES: *Trillium* seeds have a large, white or yellow structure called an eliaosome, which provides nutrition for ants. Ants, in turn, disperse seeds of *Trillium*. The sweet fragrance emitted by the flowers attracts insect pollinators—various bees, butterflies, and moths.

Trillium angustipetalum
Narrow-Petaled Wakerobin
Melanthiaceae (False-Hellebore family)

Erect perennial herb 2–7 dm; rhizomatous.
LEAVES: Floral bracts 3, 9–25 cm x 8–15 cm wide, ovate, and usually sparsely mottled, sometimes solid green, leaflike.
FLOWERS: 1 per stem; sepals 3, 3.5–5.5 cm, green or maroon to dark-purple like the petals; petals 5–11 cm, erect or inward-leaning, obscuring the stamens. Ovary purple; stamens 6, 11–23 mm, with purple connective tissue between anther sacs.
FRUITS: Capsule 6-angled in x.s., winged, dark purple.
SIERRA DISTRIBUTION: West slope, Fresno to Butte Counties.
ELEVATION: < 2000 m
HABITAT: Moist to dry sites in coniferous forest; montane zone.
FLOWERING TIME: Apr–Jun
NOTES: The tropical-looking leafy bracts and unusual purple color of the flowers makes this plant a treat to find. Although not endangered, it is not common, and care should be taken not to damage plants if you're lucky enough to encounter them. The flower emits a musty, sometimes fetid smell.

Veratrum californicum var. *californicum*
Corn Lily
Melanthiaceae (False-Hellebore family)

Erect perennial herb 1–2 m; hairy; spreads by thick rhizomes.
LEAVES: Alternate, 20–40 cm, lanceolate to ovate, curly hairy on lower surface; hairless or veins with hairs on upper surface.
FLOWERS: Inflorescence 30–60 cm, branched, woolly-hairy, with numerous flowers. Flowers 10–15 mm, tepals 6, ovate, white to greenish with green nectar glands at base.
FRUITS: Capsule 2–3 cm. Seeds 10–12 mm, winged.
SIERRA DISTRIBUTION: All.
ELEVATION: < 3500 m
HABITAT: Meadows, streamsides; montane and subalpine zones.
FLOWERING TIME: May–Aug
NOTES: One of the most recognizable plants of mountain meadows. Larger leaves are easily adequate to use as a small umbrella or sun hat. Plants contain toxic steroidal alkaloids in all parts but they're concentrated in the roots and rhizomes; this was first recognized in the 1950s when birth defects in lambs were connected to feeding on corn lily. Ingestion by humans or livestock causes vomiting and abdominal pain, followed by lowered heart rate, hypotension, conduction abnormalities, and sometimes death. Several of these compounds also show promise for medicinal use, including cancer treatment.

Xerophyllum tenax

Beargrass

Melanthiaceae (False-Hellebore family)

Tall, erect perennial herb 1.5–15 dm; from woody, tuberlike, fibrous rhizomes.

LEAVES: Persistent after 1 year. Basal leaves densely clustered, 30–100 cm, stiff, wiry, and grasslike. Stem leaves smaller, replaced by similar-appearing inflorescence bracts toward top of stalk.

FLOWERS: Racemes of numerous small, whitish flowers; tepals 6, 5–10 mm.

FRUITS: 5–7 mm, 3-angled. Seeds 3-angled, black.

SIERRA DISTRIBUTION: West slope, El Dorado to Plumas Counties.

ELEVATION: < 1650 m (higher outside the SN)

HABITAT: Open areas in coniferous forest, often abundant after wildfires; montane zone.

FLOWERING TIME: May–Aug

NOTES: Beargrass is a culturally important plant that is managed by Native American tribes to this day for the leaf fibers, which are used for clothing and to make watertight baskets. The bulbous rhizomes are eaten after being roasted for days. Beargrass is increasingly popular for floral arrangements and overharvesting of entire plants is a problem. On public lands, permits for collection are required for the general public and commercial providers (tribal harvesting and tending requires no permit).

Narthecium californicum

California Bog Asphodel

Nartheciaceae (Asphodel family)

Erect perennial herb with stems 20–60 cm; rhizomatous.

LEAVES: Mostly basal, 10–30 cm, linear and grasslike. Stem leaves 3–6, smaller than basal leaves.

FLOWERS: Racemes of yellow flowers, with pedicels 5–12 mm; tepals 6, 5–10 mm. Filaments fat and woolly with dense, yellow hairs; anthers 3–4 mm, brick red.

FRUITS: Capsule with slender, tapering beak about twice the length of the perianth.

SIERRA DISTRIBUTION: West slope, Tulare to Plumas Counties.

ELEVATION: 900–2500 m

HABITAT: Wet fens, meadows, seeps, and streamsides; montane and subalpine zones.

FLOWERING TIME: Jul–Aug

NOTES: Infrequently encountered, *N. californicum* usually indicates the presence of a fen. The yellow, woolly filaments are unique and diagnostic, and may serve to facilitate self-pollination by allowing raindrops to float pollen up to the stigma as in European *Narthecium* spp. No nectaries are present, thus any pollinator visiting *N. californicum* would be obtaining pollen.

Cephalanthera austiniae

Phantom Orchid

Orchidaceae (Orchid family)

Erect perennial herb 20–65 cm; rhizomatous, completely white, mycoheterotrophic.

LEAVES: Reduced to bracts, absent at flowering.

FLOWERS: Racemes; sepals 3, 12–20 mm, white. Petals 3; the lower petal is a lip with yellow markings, smaller than the lateral 2 petals. Ovary inferior; single stamen fused into a semicylindrical column with stigma and style.

FRUITS: Capsule 15 x 10 mm, erect. Seeds many, tiny.

SIERRA DISTRIBUTION: West slope, Tulare County north.

ELEVATION: < 2200 m

HABITAT: Shady areas in conifer forest; montane zone.

FLOWERING TIME: Apr–Jul

NOTES: The only all-white orchid, though it fades yellowish or brownish. This nonphotosynthetic plant obtains all its nutrition from mycorrhizal fungi that are also connected to a photosynthetic plant (see *mycoheterotroph* in the glossary).

Corallorhiza maculata
Spotted Coralroot
Orchidaceae (Orchid family)

Erect, leafless perennial herb with stem 17–55 cm; red to yellow-brown, lacking chlorophyll, mycoheterotrophic.

LEAVES: Scalelike, sheathing.

FLOWERS: Sepals the same color as the stems; lateral petals pink to red, and the lower petal (lip) 5–7 mm. The lip is white with purple or red spots (rarely plain white), with 2 roundish lobes at base, and parallel sides (lip not becoming wider toward tip). Stamen 1, fused with the style and stigma into a column 3–5 mm, and purple-spotted. Note the mentum, a short, orange projection at the base of each lateral sepal.

FRUITS: Capsule 15–20 mm, twisted, reflexed downward in age.

SIERRA DISTRIBUTION: West slope, Fresno County north.

ELEVATION: < 2800 m

HABITAT: Shady areas in conifer forest; montane zone.

FLOWERING TIME: Jun–Aug

NOTES: This coralroot is easy to recognize by its spotted lip, presence of mentum, and lobes at base of lip. *C. maculata* var. *occidentalis* also occurs in the SN; it has a lip that widens toward the tip, and where the 2 grow together, var. *occidentalis* blooms about a month earlier. The roots do resemble coral, but no need to dig them up for ID.

Corallorhiza striata
Striped Coralroot
Orchidaceae (Orchid family)

Erect, leafless perennial herb with stems 15–50 cm; nonphotosynthetic, varying in color from reddish-purplish to yellowish, mycoheterotrophic.

LEAVES: Scalelike, no leaves present when plants are blooming.

FLOWERS: Sepals 6–17 mm, striped, the lower 2 curved forward. Lateral petals 2 are similar to the sepals; lower petal (lip) 8–15 mm, has bolder striping. Stamen 1, fused with the style and stigma into a column 4–7 mm, and yellow with purple spots near the base.

FRUITS: Capsule 15–20 mm, twisted, reflexed downward in age.

SIERRA DISTRIBUTION: Mostly west slope, Tulare County north.

ELEVATION: 900–2200 m

HABITAT: Shady areas in conifer forest; montane and subalpine zones.

FLOWERING TIME: May–Jul

NOTES: *C. striata* is specialized to a specific group of mycorrhizal fungi in the genus *Tomentella*, which could mean it is more vulnerable to climate change than species with a wider array of fungal partners.

Cypripedium californicum

California Lady's Slipper

Orchidaceae (Orchid family)

Perennial herb with stems 8–130 cm; rhizomatous, puberulent, growing in sometimes large patches.

LEAVES: 5–12 per stem, alternate, 5–15 cm. Lower leaves elliptic, and upper ± lanceolate.

FLOWERS: 4–20 per inflorescence. Upper sepal 15–20 mm, lateral petals 3, 12–18 mm, greenish to yellowish-green; lower 2 sepals fused, pendent behind lip, lip (pouch) 15–20 mm. Functional stamens 2 and stigma united into a column; staminode (nonfunctional stamen) round, reflexed, white with green stripe, covering column.

FRUITS: Capsule. Seeds numerous, tiny.

SIERRA DISTRIBUTION: West slope, Nevada County north.

ELEVATION: 50–2200 m

HABITAT: Moist slopes, streamsides, meadows and fens in conifer forest, in partial shade or sun; montane zone.

FLOWERING TIME: May–Jun

NOTES: *C. californicum* is the tallest of the 3 lady's slippers in the SN, forms the largest patches, and is more confined to wet and moist habitats. All *Cypripedium* take many years to mature from seedlings and produce flowers. All orchids require mycorrhizal fungi for germination of the minute seeds as well as for seedling nutrition, and likely throughout their lifespans. Of the 3 SN spp., *C. californicum* has the least specificity for mycorrhizal fungi partners and associates readily with at least 3 fungal groups, while *C. montanum* and *C. fasciculatum* have much more specific requirements and are limited to a few fungal taxa. Interestingly, *C. californicum* is the most substrate-specific of the 3, found almost exclusively on serpentine geology.

Cypripedium fasciculatum

Clustered Lady's Slipper

Orchidaceae (Orchid family)

Perennial herb 12–20 cm; rhizomatous, with 2 opposite leaves halfway up stem.

LEAVES: Opposite, 5–12 cm, widely elliptic, hairy underneath, pleated.

FLOWERS: In clusters of 1–14 per inflorescence, nodding. Entire flower < 4.5 cm; upper sepal 1.5–2.5 cm; lateral petals similar-looking, all downcurved. Lip (pouch) 8–15 mm, greenish-yellow below, purple above, often partially hidden by lateral petals. Staminode < 5 mm, pale, yellow to light green.

FRUITS: Capsule 2–3 cm. Seeds countless, tiny.

SIERRA DISTRIBUTION: West slope, El Dorado County north.

ELEVATION: 700–1750 m

HABITAT: Shady, usually moist areas; mostly on streambanks; in fir forest; montane zone.

FLOWERING TIME: May–Jun

NOTES: Rare in CA, CRPR 4.2. Populations are vulnerable to extirpation from high severity wildfire and are also threatened by unethical collectors and ground disturbance from heavy equipment. It's also on Forest Service and BLM sensitive plant lists, thus afforded protection via field surveys and requirements for avoidance during ground-disturbing projects.

Cypripedium montanum

Mountain Lady's Slipper

Orchidaceae (Orchid family)

Perennial herb with stems 30–70 cm; rhizomatous, glandular-hairy, often in clumps.

LEAVES: 4–6 per stem, alternate, sessile, 5–15 cm x 4–8 cm wide, ovate, with wavy margins, often pleated, with obvious veins.

FLOWERS: 1–3 per stem; sepals 4–6 cm, the lower 2 united almost entirely. Lateral petals narrower than sepals but similar-looking. Lip (pouch) 2–3 cm, white with purple veins. Staminode 8–12 mm, yellow with red or purple spots.

FRUITS: Capsule 2–3 cm. Seeds numerous, tiny.

SIERRA DISTRIBUTION: West slope, Madera County north.

ELEVATION: < 2000 m

HABITAT: Shady, usually moist areas in conifer forest; montane zone.

FLOWERING TIME: May–Jun

NOTES: Rare in CA, CRPR 4.2. Many populations have been extirpated during recent unusually large wildfires, when the lady's slippers experienced high fire intensity. In the Sierra National Forest, at least 5 populations did not come back after the 2017 Railroad Fire.

Epipactis gigantea

Stream Orchid, Chatterbox

Orchidaceae (Orchid family)

Hairless perennial herb 30–100 cm; from rhizomes.

LEAVES: Alternate, 5–15 cm, lanceolate to elliptic; smaller toward top of stem.

FLOWERS: Racemes 1-sided, of 4–20+ flowers, each flower subtended by a leaflike bract. Sepals 3, alike, lateral longer than upper, variable in color from greenish to reddish, usually purple-veined. Lateral petals 13–15 mm, semierect, ovate. Lip 3-lobed, with the 2 outer basal lobes forming somewhat triangular wings; lip constricted (hinged) at the middle, allowing the distal (outer) section to vibrate in the breeze—the basis for the common name "chatterbox." Petals yellow, orange, light pink, or rose, strongly veined with red or purple.

FRUITS: Capsule 20–28 mm, pendent.

SIERRA DISTRIBUTION: All.

ELEVATION: < 2600 m

HABITAT: Wet areas, streamsides, springs; foothill to subalpine zones.

FLOWERING TIME: Mar–Oct

NOTES: Commonly pollinated by several species of syrphid flies, which resemble bees but have only 2 wings. Syrphid flies normally lay their eggs on plants with aphids, and the larvae hatch and feed on the aphids. This orchid deceives the flies by emitting a fragrance that mimics the sweet honeydew liquid excreted by aphids. The fragrance combined with the bumpy surface of the lip are mistaken by the fly as a nest of aphids, so it enters the flower to lay eggs, depositing any pollen on its back on the stigma. As it flies out, a new load of pollen is deposited, and if the bee proceeds to another plant, cross-pollination is achieved.

Goodyera oblongifolia

Rattlesnake Plantain

Orchidaceae (Orchid family)

Perennial herb 18–35 cm; from rhizomes.
LEAVES: Basal in rosettes; midvein white, sometimes also with white mottling. Recognition by leaves alone is possible once you become familiar with this plant.
FLOWERS: Inflorescences 1-sided below and spiraled above. Sepals green to brown, upper 6–11 mm, adherent to lateral petals, forming a hood that mostly encloses the lip and column; petals white.
FRUITS: Capsule, ascending to erect.
SIERRA DISTRIBUTION: West slope, Tulare County north.
ELEVATION: 700–2200 m
HABITAT: Shaded floor of conifer forest; montane zone.
FLOWERING TIME: May–Sep
NOTES: Flowers of rattlesnake plantain produce nectar and are pollinated by a variety of bumblebee species.

Listera convallarioides

Broadlipped Twayblade

Orchidaceae (Orchid family)

Glandular-hairy perennial herb 10–35 cm; rhizomatous.
LEAVES: 1 pair of opposite leaves, blades 2.5–7 cm.
FLOWERS: Open racemes, flowers green, with lower lip 8–13 mm, much larger than other flower parts.
FRUITS: Capsule 6–8 mm.
SIERRA DISTRIBUTION: West slope, northern Kern County north; east slope, Tahoe region.
ELEVATION: 800–2500 m
HABITAT: Moist, shady areas along streams and meadow edges in conifer forest; montane zone.
FLOWERING TIME: May–Aug
NOTES: Blends into background; it is easily distinguished from other greenish orchids by the long, wedge-shaped lower lip.

Piperia transversa

Royal Rein Orchid

Orchidaceae (Orchid family)

Erect perennial herb 12–57 cm; from roundish tubers, stem < 3 mm wide.

LEAVES: Basal, not rosetted, 6–19 cm x 10–45 mm wide, linear to oblanceolate.

FLOWERS: Spike 7–26 cm, flowers not spirally arranged. Sepals and lateral petals white to cream with green midveins; upper sepal points forward, and lateral petals spread or are curved backward; lip 2.2–5.3 mm. Spur 6–12 mm, straight, usually perpendicular to inflorescence axis. Flowers have a strong, clovelike fragrance at night.

FRUITS: Capsule 4–13 mm, ascending to erect. Seeds cinnamon-brown.

SIERRA DISTRIBUTION: West slope, Tulare County north.

ELEVATION: 300–2450 m

HABITAT: Dry, shady sites in conifer forest; montane zone.

FLOWERING TIME: May–Aug

NOTES: *Piperia transversa* is the only species in the SN with a long spur held horizontally (transverse to the inflorescence axis). *P. michaelii*, an uncommon, all-green species of the foothills < 700 m, also has a long (8–12 mm) spur, but it is usually pointed downward.

Piperia unalascensis

Slender Spire Orchid

Orchidaceae (Orchid family)

Erect perennial herb 9–70 cm; from roundish tubers.

LEAVES: Basal, not rosetted, 5–20 cm x 5–40 mm wide, linear to oblanceolate.

FLOWERS: Open spikes 3–44 cm, flowers not spirally arranged. Perianth green, lip 2–5 mm, pointed down with an upcurved tip. Spur 2–5.5 mm, short, horizontal or curved downward.

FRUITS: Capsule 3.5–10.5 mm, ascending to erect. Seeds tan to cinnamon.

SIERRA DISTRIBUTION: Both slopes, Kern County north.

ELEVATION: 600–2600 m

HABITAT: Dry, shady sites in conifer forest; montane and subalpine zones.

FLOWERING TIME: May–Aug

NOTES: Flowers produce a musky, soapy, or sometimes honeylike fragrance at night that lingers into the day. Pollinators are moths in families Geometridae and Pterophoridae.

Platanthera dilatata var. *leucostachys*

White-Flowered Bog Orchid

Orchidaceae (Orchid family)

Erect perennial herb 15–150 cm; from tuberlike rhizomes.
LEAVES: Cauline and alternate, 5–35 cm x 9–30 cm wide, with 1 creaselike fold.
FLOWERS: Dense spikes of 5–35 white flowers with leaflike bracts at base of each flower, lower bracts larger. Perianth white or cream; lower petal lip 5–10 mm, spur 5–15 mm (spur ≥ 1.5x the lip).
FRUITS: Capsule ascending to erect.
SIERRA DISTRIBUTION: Both slopes, Tulare and Inyo Counties north.
ELEVATION: < 3400 m
HABITAT: Seeps, meadows, wet and moist areas; montane and subalpine zones.
FLOWERING TIME: May–Sep
NOTES: Although the other members of this genus can be tricky to distinguish, this species stands out: It is *the* tall white-flowered orchid with the long spur common in moist to wet areas.

Platanthera sparsiflora

Sparse-Flowered Bog Orchid

Orchidaceae (Orchid family)

Erect perennial herb 25–55 cm; from tuberlike rhizomes.
LEAVES: Cauline and alternate, 4–15 cm x 5–30 cm wide, with 1 creaselike fold.
FLOWERS: Open (sometimes dense) spike 15–40 cm, flowers greenish to greenish-yellow, with leaflike bracts at base. Lower petal (lip) lance-linear and about equal to the long (6–10 mm) spur. Column (fused stamen, style, and stigma) 2.5–4 mm, approximately ⅔ length of the surrounding hood formed by lateral petals and upper sepal.
FRUITS: Capsule ascending to erect.
SIERRA DISTRIBUTION: Both slopes, Tulare and Inyo Counties north.
ELEVATION: 100–3400 m
HABITAT: Seeps, meadows, wet areas; montane and subalpine zones.
FLOWERING TIME: May–Sep
NOTES: Yosemite bog orchid, *P. yosemitensis*, is a rare species only known from fens in montane meadows of the west slope in Fresno, Madera, and Mariposa Counties, 2100–2285 m. The yellowish flowers emit a foul smell, and have a saclike spur that is much shorter than the sepals and petals, very different from the long spur in *P. sparsiflora*. The 2 often grow in the same meadow, and at a given site, *P. sparsiflora* starts flowering a few weeks before *P. yosemitensis*.

Spiranthes porrifolia

Creamy Lady's Tresses

Orchidaceae (Orchid family)

Erect perennial herb 15–60 cm (sometimes taller); roots are swollen and tuberlike.

LEAVES: Mostly basal, 8–14 cm, lanceolate.

FLOWERS: Dense spikes 5–14 cm, flowers tightly spiraled, creamy, with green bracts below. Upper sepal and lateral petals ± fused, with tips flared upward, and the lower 2 sepals free. Lip linear to ovate, not constricted at middle, and puberulent on upper surface.

FRUITS: Capsule ovoid.

SIERRA DISTRIBUTION: Both slopes, Tulare and Inyo Counties north.

ELEVATION: < 2500 m

HABITAT: Seeps, meadows, streamsides, moist areas; foothill and montane zones.

FLOWERING TIME: Jun–Sep

NOTES: Pollinated by insects foraging for nectar at the base of the lip. Compare with *S. romanzoffiana*, which is generally a shorter plant, with white flowers and a constricted lower lip. These 2 spp. can hybridize, making ID to species level difficult.

Spiranthes romanzoffiana

Hooded Lady's Tresses

Orchidaceae (Orchid family)

Erect perennial herb 7–30 cm; roots fleshy.

LEAVES: Mostly basal, 3–13 cm.

FLOWERS: Dense spikes 2–14 cm, flowers spirally arranged, white to light cream. Upper sepal and lateral petals form a hood; lower lip constricted ("violin-shaped"), tip hairless on upper surface.

FRUITS: Capsule.

SIERRA DISTRIBUTION: Tulare and Inyo Counties north.

ELEVATION: < 3300 m

HABITAT: Meadows, wet areas; montane and subalpine zones.

FLOWERING TIME: May–Sep

NOTES: Occurs higher in elevation than *S. porrifolia*. Distinguished from other white-flowered orchids by the lack of a spur on the lower lip. Plants in SN are likely derived from hybrids between *S. porrifolia* and *S. romanzoffiana*, for which *Spiranthes stellata* will be applied in the future. The photo shows the lack of a clearly defined hood and the lower lip less violin-shaped than true *S. romanzoffiana*.

Maianthemum racemosum

False Solomon's Seal, False Lily of the Valley

Ruscaceae (Butcher's Broom family)

Perennial herb with stems ranging in height 50–90 cm; rhizomatous (rhizomes 5–10 mm diameter), stems erect to ascending; hairless or finely hairy above.

LEAVES: Alternate, mostly sessile, 7–20 cm, ovate to oblong-elliptic, with acute to acuminate tips, hairless to puberulent above. Leaves often twisted at the base and arrayed in a plane with the edges facing the stem (inset), but can also be held more symmetrically with the central axis facing the stem. Margins may be straight as in the photo, or wavy-edged.

FLOWERS: > 20 per panicle, white; tepals 6, 1–2 mm, much shorter than the stamens, which have filaments showy, widened, white.

FRUITS: Berry 5–7 mm, red, sometimes with purple spots.

SIERRA DISTRIBUTION: West slope.

ELEVATION: < 3250 m

HABITAT: Variable; dry to moist forest floor, along streams and in meadows; montane through subalpine zones.

FLOWERING TIME: May–Aug

NOTES: Formerly *Smilacina racemosa*. The species name is misleading as the inflorescence is not a raceme but a classic panicle. The wide, white anther filaments contribute more to the showiness of the flowers than do the shorter, white petals. Can be tricky to distinguish leaves from the mountain lady's slipper, *Cypripedium montanum*, if neither flowers nor fruits are present.

Maianthemum stellatum

Starry False Lily of the Valley

Ruscaceae (Butcher's Broom Family)

Perennial herb with stems 30–70 cm; rhizomatous, herbage hairless or puberulent. Stems are erect, either straight or somewhat zigzag above.

LEAVES: Alternate, sessile, 5–17 cm; shape variable; lanceolate, oblanceolate, or elliptic; acuminate; puberulent below.

FLOWERS: Racemes 2–8 cm, of 5–15 white flowers. Tepals 6, 4–7 mm, oblong to lanceolate, longer than the stamens.

FRUITS: Berry 7–10 mm, reddish-purple to black.

SIERRA DISTRIBUTION: All.

ELEVATION: 1000–3350 m

HABITAT: Variable; dry to moist forest floor, along streams, and in meadows; montane, and subalpine, and sagebrush/pinyon-juniper zones.

FLOWERING TIME: Jul–Sep

NOTES: Formerly *Smilacina stellata*. *M. stellatum* has fewer and larger flowers than *M. racemosum*, and tends to have narrower leaves; and ironically, this species does have a racemose inflorescence, while *M. racemosum* does not (its inflorescence is a panicle).

Smilax californica

California Greenbriar

Smilacaceae (Smilax family)

Vine from a short, knotty caudex, with stems 2–5 m, sometimes climbing to 12 m; prickly.

LEAVES: Evergreen, alternate; with tendril-like stipules. Petioles 0.5–2 cm, blades 5–10 cm, ovate, dull green, with conspicuous veins, cordate bases, and entire margins.

FLOWERS: Dioecious (plants either with only female or only male flowers); inflorescences of axillary umbels. Both types of flowers have 6 tepals, green (as in photo), white, or yellowish, in 2 petal-like whorls; reflexed or curled back from middle. Staminate inflorescences (as in photo) have 15–30 flowers, each flower with pistils reduced or absent, stamens 6, anthers white. Pistillate inflorescences have 15–35 flowers; ovary superior, stigmas 3, and staminodes 6.

FRUITS: Berry ± 6 mm, black. Seeds 3.

SIERRA DISTRIBUTION: West slope, Nevada county north.

ELEVATION: < 1300 m

HABITAT: Streamsides; foothill and montane zones.

FLOWERING TIME: May–Jun

NOTES: *S. californica* is only found in CA and southern Oregon. The leaves are unusual for a monocot in that between the major parallel veins, there are netlike veins (reticulate). The photo shows staminate flowers (male plant).

Odontostomum hartwegii

Hartweg's Odontostomum

Tecophilaeaceae (Tecophilaea family)

Erect perennial herb with stem 12–50 cm; from corms, stems sometimes zigzagged or curved, branched above.

LEAVES: Mostly basal, cauline reduced above, 10–30 cm, linear.

FLOWERS: Panicles or racemes, flowers white to yellowish; tepals 6, fused into a 4–5 mm tube below, with tepal lobes also 4–5 mm, reflexed in maturity. Stamens 6, alternating with 6 white staminodes.

FRUITS: Capsule 3-chambered. Seeds brown.

SIERRA DISTRIBUTION: West slope, Fresno County north.

ELEVATION: < 950 m

HABITAT: Clay soils, often serpentine, sometimes at vernal pool edges; foothill zone.

FLOWERING TIME: Apr–May

NOTES: *Odonto* is the Latin root for "tooth," and *stomum* for "mouth," referring to the staminodes (sterile stamens) at the opening of the perianth tube. Sometimes called "inside-out lily."

Brodiaea elegans ssp. *elegans*

Harvest Brodiaea

Themidaceae (Brodiaea family)

Erect perennial herb 10–50 cm; from corms.

LEAVES: Basal 3–5, linear, hairless.

FLOWERS: Blue-purple or violet; pedicels 5–10 cm. Tube 8–19 mm, funnel-shaped and not splitting in fruit. Perianth lobes 15–30 mm (total flower length 23–49 mm). Staminodes white, held away from stamens.

FRUITS: Capsules 3-chambered, sessile, ovoid. Seeds 2+ per chamber, oblong, black, ridged.

SIERRA DISTRIBUTION: West slope, all.

ELEVATION: < 2200 m

HABITAT: Open areas with moist to dry soils; foothill and montane zones.

FLOWERING TIME: May–Sep

NOTES: Blooms after most other spring wildflowers and once grasses have dried at a given elevation. Distinguished from the similar *B. coronaria* (inset) by the staminodes, which are positioned near the tepals in *B. elegans* rather than held close to the stamens around the stigma, as in *B. coronaria*.

Dichelostemma multiflorum

Wild Hyacinth, Many Flowered Brodiaea

Themidaceae (Brodiaea family)

Erect or leaning perennial herb 20–90 cm; from corms, scabrous.

LEAVES: Basal 3–4, 30–85 cm, glaucous, not keeled.

FLOWERS: Inflorescences umbel-like or headlike (often appearing spherical), of 6–35 purple to pink flowers; subtending bracts 7–12 mm, ovate; pedicels 3–15 mm. Each flower has perianth tube 8–10 mm, cylindric and constricted above ovary; lobes 8–10 mm, spreading. Appendages 5–6 mm, white or purple, arising from base of outer perianth lobes, forming a crown; appendages move apart as flowering progresses. Style 5–8 mm; stamens 3, anthers 4–5 mm.

FRUITS: 8 mm, ovoid.

SIERRA DISTRIBUTION: West slope, Mariposa County north.

ELEVATION: < 2000 m

HABITAT: Open areas in foothill grassland, montane forest; foothill and montane zones.

FLOWERING TIME: May–Jun

NOTES: Sometimes mistaken for blue dicks, *Dipterostemon capitatus* ssp. *capitatus*, which blooms earlier; *D. multiflorum* has 3 anthers, while blue dicks has 6. The heads of wild hyacinth are usually spherical, unlike blue dicks, and blue dicks does not have constricted flower tubes. Often seen in great abundance (and beauty) after wildfire.

Dichelostemma volubile

Snake Lily, Twining Brodiaea

Themidaceae (Brodiaea family)

Twining perennial herb; from corms, with scapes 40–150 cm, scabrous, often reddish.

LEAVES: Basal 3–4, 30–70 cm, keeled.

FLOWERS: Inflorescences umbel-like or headlike (often spherical), of 6–30, pink to white flowers; bracts at inflorescence base are 12–15 mm, ovate, pinkish; pedicels 10–40 mm. Each flower has perianth tube 5–7 mm, spherical, constricted above ovary; lobes 6, 5–7 mm, spreading (ascending in fruit). Style 3–4 mm; stamens 3, each with 2 white appendages folding inward, forming a crown; staminodes 3, on outer 3 perianth lobes, white; anthers 3–4 mm.

FRUITS: Ovoid.

SIERRA DISTRIBUTION: West slope, all.

ELEVATION: < 2800 m

HABITAT: Open areas; foothill and montane zones.

FLOWERING TIME: Apr–Jun

NOTES: The long, twining "stem" is actually a scape (flowering stalk). This plant is unique and unmistakable. After wildfires, it can be found in extreme abundance, turning entire hillsides pink, twining and climbing on other plants as well as on itself (as in photo).

Dipterostemon capitatus ssp. *capitatus*

Blue Dicks

Themidaceae (Brodiaea family)

Erect perennial herb; from corms, with scapes 9–60 cm.

LEAVES: Basal 2–3, 10–70 cm linear, hairless, usually keeled.

FLOWERS: Each scape topped by a head of 2–16 flowers, subtended by bracts dark purple, wide-lanceolate, sometimes paler with dark purple stripes. Perianth 12–20 mm, tube 4–8 mm; perianth lobes 7.4–13 mm, oblong, with the outer 3 narrower than the inner 3. Stamens 6, unequal in size; inner filaments have white, winglike appendages.

FRUITS: Capsule 4–6 mm, ovoid. Seeds black, sharp-angled.

SIERRA DISTRIBUTION: All.

ELEVATION: < 2500 m

HABITAT: Open areas in a variety of soil types; most zones below subalpine.

FLOWERING TIME: Feb–Jun

NOTES: Formerly *Dichelostemma capitatum* ssp. *capitatum*. One of the most common early blooming wildflowers each spring, appearing in the foothills by February along with popcorn flower and several fiddlenecks. Flowers of *Dichelostemma* spp. have 3 anthers and 3 staminodes, while there are 6 anthers and no staminodes in *Dipterostemon*. The corms have been a staple food for Native Americans for millennia.

Triteleia bridgesii

Bridges' Brodiaea

Themidaceae (Brodiaea family)

Erect perennial herb; from spheric corms, with scapes 10–60 cm.

LEAVES: Basal 1–3, 20–55 cm x 3–10 mm wide, narrowly lanceolate, hairless, keeled.

FLOWERS: Inflorescence open, umbel-like, with pedicels 20–90 mm. Perianth 27–45 mm, blue-purple to lilac; tube 17–25 mm, funnel-shaped; throat shiny. Perianth lobes 10–20 mm, abruptly spreading. Stamens 6, attached at the same level and of equal length; filaments 3–4 mm, triangular; anthers 3.5–4.5 mm, bluish.

FRUITS: Capsule ovoid, stalked. Seeds spheric, black-crusted.

SIERRA DISTRIBUTION: West slope, Mariposa County north.

ELEVATION: < 1200 m

HABITAT: Open rocky areas, often serpentine; foothill zone.

FLOWERING TIME: Apr–Jun

NOTES: This brodiaea is distinguished from similar-looking taxa in having the 6 stamens attached at the same level, of equal lengths, and with blue or bluish anthers.

Triteleia hyacinthina

White Brodiaea, Fool's Onion

Themidaceae (Brodiaea family)

Erect perennial herb; from spheric corms, with scapes 30–60 cm.

LEAVES: Basal 1–3, 10–40 cm x 4–22 mm wide, narrowly lanceolate, hairless, keeled.

FLOWERS: Inflorescences open and umbel-like, with pedicels 5–50 mm. Perianth 9–16 mm, white; tube 2–4 mm; lobes 7–12 mm, spreading to ascending, with green midvein. Stamens 6, attached at the same level and of equal length; filaments 2–4 mm, triangular; anthers 1–2 mm, white to yellow.

FRUITS: Capsule 4–6 mm, ovoid, stalked.

SIERRA DISTRIBUTION: All, except Inyo and Mono Counties.

ELEVATION: < 2000 m

HABITAT: Most often in moist sites, sometimes found on drier slopes; foothill and montane zones.

FLOWERING TIME: Mar–Jul

NOTES: Compare with similar-looking glassy onion, *Allium hyalinum*. All onions (*Allium* spp.) smell like onions, and white brodiaea does not! Often called wild hyacinth, although this is a common name for *Dichelostemma multiflorum* as well.

Triteleia ixioides ssp. *anilina*

Mountain Pretty Face

Themidaceae (Brodiaea family)

Erect perennial herb; from spheric corms, with scapes 10–30 cm.

LEAVES: Basal 1–2, 10–25 cm x 3–15 mm wide, linear, keeled, hairless, often withered at full bloom.

FLOWERS: Inflorescences open and umbel-like, subtending bracts lanceolate and scarious. Pedicels 10–70 mm; perianth yellow with a dark midvein; tube 4–7 mm, funnel-shaped; lobes 7–12 mm, spreading. Stamens 6, attached at the same level, unequal; short filaments 1–2 mm, long filaments 3–3.5 mm. Appendages at tip of filaments short; straight or incurved; anthers 1–2 mm, blue (or cream).

FRUITS: Capsule 3-angled.

SIERRA DISTRIBUTION: All.

ELEVATION: < 3000 m

HABITAT: Moist sandy soil or gravel, edges of conifer forest, meadows; montane zone.

FLOWERING TIME: May–Jul

NOTES: Foothill prettyface, *T. ixioides* ssp. *scabra*, is the most common ssp. in the foothills and is found up to 2200 m in the montane zone. It is taller (scape 20–50 cm) and scabrous. Perianths are pale yellow with a dark midvein; perianth tube 3–7 mm, perianth lobes 10–12 mm, ascending to reflexed; short filaments 4–5 mm, long filaments 5–7 mm. Appendages at tip of filaments are long and slender, straight or recurved; anthers are yellow or cream, occasionally blue.

Triteleia laxa

Ithuriel's Spear

Themidaceae (Brodiaea family)

Erect perennial herb; from spheric corms, with scapes 10–70 cm.

LEAVES: Basal 1–3, 20–40 cm x 4–25 mm wide, linear, keeled, hairless, often withered at full bloom.

FLOWERS: Inflorescences open and umbel-like, subtending bracts lanceolate and scarious. Pedicels 10–100 mm; perianth 18–47 mm, blue-purple to lavender to white; tube 12–25 mm, funnel-shaped; lobes 8–20 mm. Stamens 6, attached at 2 levels to perianth tube, with filaments 3–6 mm, linear, and anthers 2–5 mm, bluish to white.

FRUITS: Capsule ovoid.

SIERRA DISTRIBUTION: West slope, all.

ELEVATION: < 1500 m

HABITAT: Open areas; foothill and montane zones.

FLOWERING TIME: Apr–Jun

NOTES: Ithuriel's spear has the largest flowers of the SN *Triteleia* taxa, and in a good rain year these tall, graceful plants are abundant and eye-catching.

Triantha occidentalis ssp. *occidentalis*

Western False Asphodel

Tofieldiaceae (False Asphodel family)

Erect perennial herb with stems 30–80 cm; rhizomatous, hairless below, densely glandular above.

LEAVES: Mostly basal, 5–20 cm, 3–8 mm wide, narrow and grasslike.

FLOWERS: Inflorescences of white flowers densely arranged in heads, elongating as fruits develop. Tepals 6, white, about equal in length to the stamens 6; filaments flat, attached at base.

FRUITS: Capsule 5–9, 3-chambered, often reddish or maroon toward top.

SIERRA DISTRIBUTION: Both slopes along crest, Tulare and Inyo Counties north.

ELEVATION: < 3300 m

HABITAT: Wet meadows, especially in fens; montane zone.

FLOWERING TIME: Jun–Sep

NOTES: Formerly *Tofieldia occidentalis* ssp. *occidentalis*. Recent research showed that this sticky, glandular plant is actually carnivorous—sticky hairs on the inflorescence catch and digest small insects, transferring nitrogen to other parts of the plant, especially the leaves.

Glossary

abaxial. Facing away from the central axis; lower surface, dorsal.

above. The top surface of a plant part, usually a leaf.

achene (also spelled **akene**). Dry, indehiscent one-seeded fruit, with seed attached at only one point on the ovary wall, e.g., sunflower "seeds."

acuminate. Tapering to a sharp tip with concave sides.

acute. Tapering to a sharp tip with straight or nearly straight sides, converging at an angle less than 90°.

adaxial. Facing toward the central axis; upper surface, ventral.

adnate. Fusion of unlike parts—e.g., filaments fused to corolla (compare with **connate**).

aerenchyma. Spongy, porous tissue with large air chambers, which provides buoyancy and gas transfer in aquatic plants.

alternate. (1) Attached singly and on alternating sides of a stem or other axis; (2) positioned between other organs, e.g., stamens attached between petals are alternate to petals (see *Pectiantia breweri*).

angiosperm. Plants producing flowers with ovules contained in an ovary and sperm contained in pollen grains (flowering plants).

annual. A plant that completes its life cycle within a year—germinates from seed, flowers and produces seed, then dies, leaving seeds to carry on the population in subsequent year(s).

anther. The pollen-bearing part of a stamen, at the apex of a filament.

anther sac. Pollen-bearing chamber of an anther; there may be one, two, or four chambers.

apex. The tip of a plant or plant part, furthest from base.

appendage. Any protuberance or projection attached to a plant part, e.g., appendages on petals may be small flaps of tissue or elaborate, fringed structures (see *Hackelia nervosa*).

appressed. Pressed closely against another organ but still separate; e.g., mature fruits held tightly against inflorescence stalk (compare with **reflexed**, **spreading**).

ascending. A stem with the lower section growing at a 30°–60° angle, then curving to become erect.

awn. A slender bristle, usually at tip of a leaf or other organ.

awl-like. Narrowly triangular, broad at the base, tapering to a slender sharp point.

axil. The angle formed by a stem and another organ, usually a leaf; buds arise from axils.

axillary. Arising from an axil, positioned in an axil.

beak. A narrow, elongated tip or projection on some seeds or fruits; also, the beaklike structure formed by the fused upper corolla lobes in *Castilleja*, elongated styles on Geraniaceae fruits.

below. The bottom surface of a plant part, usually a leaf (underneath).

bract. A modified leaf associated with inflorescences, often small, green, and scalelike; sometimes large and petaloid, e.g., *Cornus nuttallii*.

bractlet. A small bract, often secondary to primary bracts—e.g., in compound umbellate inflorescences in Apiaceae, bracts subtend the rays of the umbrellalike inflorescence, and bractlets subtend the outer flower clusters.

bulb. An underground structure composed of a small stem segment topped by fleshy leaves modified for food storage, as in onions.

bulblet. A small bulb or bulblike structure; produced as offsets of larger bulbs or borne in leaf axils, in an inflorescence, or stolons.

burl. A swollen area at the base of a woody plant containing buds that sprout after fire.

calyx. All the sepals of a flower.

calyx lobe. Used rather than "sepals" when referring to flowers with fused sepals; each calyx lobe corresponds to a sepal, and lobes are positioned above calyx tube.

canescent. Covered with short, white or gray hairs, causing the surface to look white or gray.

capsule. A dry, dehiscent fruit composed of more than one carpel.

carpel. A modified leaf (evolutionarily) consisting of a single pistil (ovary, style, stigma) or one unit of a compound pistil corresponding to one modified leaf of several that are fused together. The segments of an orange or the chambers of a bell pepper are carpels.

catkin. A spike or raceme of unisexual flowers with no petals (or inconspicuous petals), often with a bract below each flower, often pendent; wind-pollinated (see *Corylus cornuta* ssp. *californica*).

caudex. Persistent, expanded, semiwoody root crown in some herbaceous perennials.

cauline. Pertaining to or on the stem.

ciliate. Fringed with hairs; eyelashlike.

circumboreal. With a distribution throughout the northern hemisphere, but south of the arctic.

circumscissile. Opening along a line around the circumference—e.g., the top of a capsule opens and falls as a lid.

clasping. Base of leaf enclosing stem (partially or entirely).

claw. The narrowed basal portion of some petals, sepals, or tepals.

cleistogamous. Flowers that remain closed, self-pollinate, and set fertile seed.

connate. Fusion of like parts—e.g., fusion of petals into a corolla tube, of filaments into a tube (compare with **adnate**).

corm. A short, swollen, underground stem, topped with thin, papery scales or leaves (e.g., *Calochortus* spp.).

cormlet. Small offshoot of a corm.

corolla. All the petals of a flower.

corolla lobe. Used rather than "petals" when referring to flowers with fused petals; e.g., *Plagiobothrys nothofulvus* has five corolla lobes above the corolla tube.

cotyledon. The leaf (monocotyledons) or leaves (dicotyledons) of the embryo—"seed leaves."

crenate. Edged with rounded teeth, scalloped.

cyathium (plural: **cyathia**). The unique inflorescence type of the genus *Euphorbia*, consisting of a cuplike involucre containing one to many reduced staminate flowers and one pistil, which matures into an ovary that protrudes from the involucre.

cyme. A flat-topped or round-topped paniculate inflorescence in which the terminal (central) flower blooms first, and flowering progresses toward the outer flowers.

cymose. Having flowers in a cyme.

decumbent. Stems lying flat on the ground with tips turned upward.

decurrent. Leaf tissue extending down the stem below the node (e.g., *Centaurea solstitialis*, *Helenium bigelovii*).

dehiscent. Breaking apart along predetermined lines to release contents (as in seeds from a fruit, pollen from an anther). Includes release through pores.

dentate. Toothed, as in a leaf margin.

denticulate. With tiny teeth.

dichotomous. Branched or forked into two more or less equal divisions, each forking again, etc.; also, a dichotomous key—a series of two consecutive choices based on characters of plants (see Jepson eFlora).

dioecious. Male and female flowers borne on different plants, flowers all unisexual (compare with **monoecious**).

disciform. A flower head in Asteraceae comprised primarily, but not entirely, of disk flowers; with marginal sterile or pistillate (female) flowers without rays or with miniscule rays.

discoid. A flower head in Asteraceae composed entirely of disk flowers.

distal. Toward the tip rather than near the base (e.g., of a stem, a leaf).

dorsal. The side away from the stem or central axis, as in the lower surface of a leaf, the outer surface of a nutlet in relation to the attachment point; abaxial (compare with **ventral**).

drupe. A fleshy fruit with thin outer skin or "peel" (exocarp) usually surrounding a single seed within a hardened protective coating, e.g., plum, peach, apricot, cherry, olive, date, mango, walnut. Also referred to as stone fruit.

elaiosome. A small, usually starchy part of a seed with chemicals that attract ants and stimulate them to transport seeds back to the nest where the elaiosome is consumed and the seed is left to germinate.

elliptic. A narrow oval, widest at center and narrowing equally to the ends.

endemic. Restricted to a particular geographic area or substrate (e.g., Sierra Nevada endemic, serpentine endemic).

entire. Smooth-edged (not toothed, fringed, lobed, or divided). Said of a margin (e.g., of a leaf).

epithet. In the binomial system of nomenclature, organisms have a genus and a species name. The species name is its epithet. E.g., for *Rosa californica*, *californica* is the epithet.

exserted. Protruding beyond enclosing or surrounding parts (e.g., stamens or stigmas sticking out of a corolla tube).

family. A group of genera with similar characteristics and a common evolutionary history (sometimes a family consists of only one genus).

filament. The stalk supporting an anther.

flower. The reproductive structure of angiosperms (flowering plants) typically (but not always) comprised of sepals, petals, carpels (female), and stamens (male).

follicle. A dry, dehiscent fruit formed from one carpel that opens along one side, like a milkweed "pod."

fusiform. Elongate, widest near middle and tapering to both ends.

genus. The first of the two names in the binomial system of nomenclature, the genus name is applied to a group of species with similar characteristics that are closely related, or sometimes only one species.

gland. A small, round structure on or embedded in the surface of a plant organ, or at the tip of a hair, exuding a sticky, oily, and often aromatic substance. Look for hairs with ball-like structures at the ends with a 10x hand lens.

glandular. With glands.

glaucous. Covered with a whitish or bluish coating ("bloom"), as on some grapes, blue elderberry fruits, and stems of California blackberry, and easily removed by light rubbing. The coat is usually due to a wax layer on the cuticle of plant tissues. Benefits to plants for being glaucous include water conservation, light reflectance, making surfaces too slippery for insects to climb.

hairless. Lacking hairs on a surface; word used as it's a more familiar word to nonbotanists than "glabrous."

head. In Asteraceae, a dense inflorescence of sessile flowers closely arranged to form a round or spheric shape. Also used for other groups when flowers are densely arranged into a sphere; see *Cephalanthus occidentalis* (Rubiaceae), *Monardella breweri* ssp. *lanceolata* (Lamiaceae).

herbage. The leaves and nonwoody stems of a plant, not including the flowers and fruits.

hypanthium. A cup or bowl-shaped structure formed by the fusion of basal portions of perianth and filaments (resulting in sepals, petals, and stamens appearing to arise from the hypanthium). When the hypanthium is also fused to the ovary wall and the ovary appears positioned below the attachment points for sepals, petals, pistils, and stamens, the ovary is said to be inferior.

included. Enclosed in, not protruding, as in stamens and stigmas in a corolla tube (see **exserted**).

indehiscent. Not opening at maturity (e.g., fruits that do not open along predetermined lines of weakness or pores to release seeds).

inferior. An ovary positioned below the point of attachment of sepals and petals in a flower (compare with **superior**; see **hypanthium**).

inflorescence. Assemblage of flowers, starting from the uppermost node on a stem; can be difficult to interpret.

inflorescence bract. Bracts associated with an inflorescence, different from but may resembling leaves; subtending or surrounding the inflorescence and/or flowers.

intergrade. When plants of a species, subspecies, or variety interbreed/ hybridize and some characteristics of each parent taxon manifest in offspring.

internode. The region on a stem between nodes.

involucel. A smaller, secondary involucre; especially common in inflorescences of Apiaceae where the rays of the umbel are subtended by an involucre, and each group of flowers borne on the rays has yet another set of bracts subtending them.

involucre. A whorl of bracts, free or fused, subtending a flower or a flower cluster; in Asteraceae, the combined phyllaries of a flower head.

keel. A distinct, longitudinal ridge like the keel of a boat, said of leaves, phyllaries, sepals, petals, bracts, etc. Also, the two lower petals in many Pea family flowers that are joined to form a small canoelike structure (e.g., *Lupinus*).

l.s. Longitudinal section, whereby a structure such as a stem, cylindrical fruit, or flower head is sliced from top to bottom to examine its internal structure (compare with **x.s.**).

lanceolate. Shaped like a lance (a medieval weapon), longer than wide, widest below middle.

ligulate flower. In Asteraceae, a bisexual, bilateral flower with the strap-shaped, five-lobed extension of the corolla appearing as one petal, along with other ligulate flowers making up a liguliflorous head.

ligule. From Latin for "tongue." (1) In Asteraceae, the five-lobed part of a ligulate flower that extends beyond the corolla tube; (2) in grasslike monocots, a flap or band of tissue between the leaf blade and the sheath.

liguliflorous. A flower head in Asteraceae composed of only ligulate flowers (e.g., chicory, dandelions).

limb. The expanded upper/outer part of a corolla with fused petals. Measurements refer to the diameter of the entire outer, expanded area of the corolla lobes when looking straight down at the corolla (not the radius from the center to the outer edge, as might be inferred from the term "limb").

linear. Long and narrow with parallel sides, wider than threadlike and narrower than elliptic.

lip. In bilaterally symmetric flowers, there is often an upper and a lower projection, reminiscent of lips; e.g., monkeyflowers.

lobe. A projection from a margin, as in round or jagged projections on a leaf margin, or the lobes of a calyx or corolla with fused sepal or petal bases.

locule. A cavity within a plant organ, usually referring to chambers within an ovary or a fruit (locules may equal the number of carpels or may be fewer). Also refers to chambers (sacs) that produce pollen within an anther.

meadow. An open, flat or gently sloped wetland where water from snowmelt and runoff loses momentum and drops its load of sediment, rocks, and organic matter carried off the surrounding slopes. Meadows generally have few to no trees and maintain a high water table for all or part of the summer. The acreage and water content of meadows depends on the slope and size of the water catchment basin feeding it. Many meadows are also spring-fed.

membranous. Thin, translucent or nearly so, dry or moist, sometimes green.

-merous (suffix)**.** The number of sepals and when they're the same as the number of petals (and sometimes stamens) in a flower. E.g., Brassicaceae flowers are 4-merous because there are four sepals and four petals.

monoecious. A plant with unisexual—male (staminate) and female (pistillate)—flowers on the same plant, and usually no bisexual flowers (compare with **dioecious**).

mucronate. With a sharp, abrupt, apical point.

mycoheterotroph. A plant that is (usually) nonphotosynthetic, and obtains carbon from fungal associates rather than through photosynthesis (*myco* is "fungi," *hetero* is "other," and *troph* is "food"). An autotroph is a typical green plant that photosynthesizes to produce its own carbohydrates (*auto* is "self," and *trophic* is "food").

nectary. A multicellular, glandular structure that produces nectar; often within the flower, and if so, associated with stomata and a supply of phloem.

ob- (prefix)**.** Inversely or reversely, in the reverse direction or inverted; basically, upside down—obovate means an ovate shape with the widest part at the top and narrowest at the base (like an upside-down egg).

obconic. Cone-shaped (like a snow cone cup, not a pine cone), with an attachment at the narrow end.

obcordate. Cordate (heart-shaped), turned upside down so that the lobes of the heart are toward the tip.

oblanceolate. With a lanceolate shape, turned upside down, so that the narrowest point is at the base, widest toward the tip.

oblong. Like an oval, but two to four times longer than broad and with more or less parallel sides.

obovate. Upside-down egg in outline shape, attached at the narrow end (applied to a plane or two-dimensional parts).

obtuse. Blunt- or round-tipped with sides converging at an angle more than 90°.

ocrea (also spelled **ochrea**, plural: **ocreae**). A sheath around the stem formed by fused stipules, characteristic of many species in Polygonaceae.

offset. A short, often prostrate shoot originating near the ground at the base of the main shoot.

organ. A part of an organism (a plant, in this case) with a specific function, e.g., leaf, stem, flower, anther.

opposite. (1) Leaves directly across from each other on a stem; (2) positioned in line with other organs, e.g., stamens attached at base of petals rather than between them.

ovary. The expanded basal part of a carpel (or group of carpels) containing the ovules, below the style and stigma.

ovate. Egg-shaped in outline, attached at the broad end (applied to a plane or two-dimensional parts)

ovoid. Egg-shaped—refers to three-dimensional structures.

ovule. An unfertilized immature seed.

palea. In some Asteraceae, a scalelike bract accompanying each individual flower on the receptacle, except in tarweeds, where a ring of paleae occurs between disk flowers and ray flowers but not elsewhere on the receptacle.

palmate. Like fingers on a hand; lobed, veined, or divided from a central point.

panicle. A branched inflorescence, with pedicels of individual flowers attached to stalks that branch from the central inflorescence axis, and flowers opening from the bottom upward, paniculate (compare with **raceme**).

papillate. Covered with short, rounded projections, usually protruding from an epidermal cell.

papilionaceous. Butterflylike, referring to the bilateral corolla of Pea family flowers; the top petal being the banner, side petals wings, and lower two petals fused into a keel.

pappus. In Asteraceae, the modified calyx found at the top of the inferior ovary (achene), varying from awns to bristles to scales.

pedicel. The stalk bearing an individual flower.

peduncle. The stalk bearing an inflorescence, or a single flower if borne singly (used interchangeably with pedicel for one-flowered stalks), technically starts above leaves on stem/branch.

peltate. Shieldlike, with stalk attached at center.

perianth. The combined sepals and petals of a flower.

perianth part. An individual component of a perianth used when sepals and petals look alike. E.g., see *Calycanthus*. Often used interchangeably with **tepal**.

petaloid. Sepals or bracts that are brightly colored or marked, resembling petals.

petiole. Stalk of a leaf.

petiolule. Stalk of a leaflet.

phloem. The food-conducting tissue of the vascular system in plants (food being sugars manufactured during photosynthesis)

phyllary. An inflorescence bract of the involucre in Asteraceae.

pilose. With long, soft, straight hairs.

pinnate. Veined, lobed, or divided like a feather.

pistil. The combined female reproductive parts of a flower; ovary, style, stigma.

pistillate. A flower with pistils only, a female flower.

plumose. Featherlike, with fine hairs or bristles extending from both sides of a main axis.

pod. Vernacular for fruits of the Pea family, often meant to describe any single-chambered, dry fruit that dehisces (splits open) along preexisting lines to release seeds. "Pod" does not have a single technical meaning and is also used to describe indehiscent dry fruits.

pollen. Mature microspores (haploid male reproductive structures), exists as pollen grains, each grain with two sperm nuclei.

pollen aperture. Put simply, the hole on the wall of a pollen grain through which a pollen tube protrudes and grows down from a stigmatic surface through the style to fertilize an egg within an ovule.

pollinarium. Complex structure in *Asclepias* and Orchidaceae composed of pollinium, translators, and corpusculum. Two or more pollinia are attached by translators to a central corpusculum.

pollinium. A waxy mass of pollen grains that are transported by pollinators as a unit in Asclepias and Orchidaceae.

proximal. Toward the base or point of attachment of an organ.

pubescent. Covered with short, soft, erect hairs.

puberulent. Minutely pubescent—covered with fine, very short, usually dense hairs.

raceme. A simple inflorescence with all flowers directly attached by their pedicels to the central axis, racemose (compare with **panicle**).

radiate. Asteraceae—refers to flower heads with ray flowers surrounding disk flowers. Rays are the outer flowers of the flower head with five corolla lobes fused together to one side and elongated, appearing as petals if the entire head is viewed as one flower.

radiant. Asteraceae—a flower head with all corollas radially symmetrical, and peripheral corollas enlarged.

ray. (1) In Asteraceae, the expanded, petal-like part of a ray flower's corolla, usually three-lobed (compare with **ligule**); (2) the radiating peduncles of an umbel in Apiaceae, all originating from the same point on the inflorescence stalk; in a compound umbel the secondary umbels are found at the end of the rays.

receptacle. The surface upon which flowers are attached, at the tip of the pedicel or peduncle.

recurved. Curved backward or downward.

reflexed. Bent sharply backward or downward (compare with **appressed**, **spreading**).

rhizome. Underground stems, usually growing horizontally, and unlike roots having leaves, leaf scars, axillary buds, produces new shoots that emerge above ground.

rhizome leaf. In *Cardamine*, a leaf emerging above ground from a rhizome, with no stem or flowers but connected to them.

rhizomatous. Having rhizomes; this is often evidenced by plants forming patches or mats, resulting from underground growth of rhizomes that give rise to new stalks at various intervals along their lengths.

rotate. Referring to a corolla that has no corolla tube; the corolla lobes are all in one plane, like a dinner plate.

scabrous. Rough to the touch, like sandpaper; due to short, stiff hairs.

scape. The stalk of a flower or an inflorescence that looks like a leafless stem.

scapose. A plant with inflorescences or flowers borne on a scape.

scarious. Dry, translucent, and membranous in texture.

secund. With all parts (usually flowers or fruits) grouped on one side of a stalk.

series. The arrangement of flower parts in the same row or whorl, most often used to describe arrangement of phyllaries in Asteraceae.

serrate. Resembling a saw; a margin toothed with sharp teeth pointing forward.

sessile. Attached without a stalk, e.g., a leaf without a petiole, a flower without a pedicel.

sickle-shaped. Shaped like a sickle (a handheld harvesting tool with a curved blade).

silicle. In Brassicaceae, a short, dry, fruit, its length less than two times its width, that dehisces by the two valves separating from the central placenta/ septum where seeds are attached.

silique. In Brassicaceae, a long dry, fruit, length much longer than twice the width; that dehisces by the two valves separating from the central placenta/septum where seeds are attached.

species. A taxon composed of plants/ populations capable of interbreeding and producing fertile offspring; the second name (epithet) of a binomial scientific name.

spike. An unbranched inflorescence with sessile or subsessile flowers, matures from the bottom upward.

spreading. Extending from point of attachment to a nearly horizontal position; said of leaves, pedicels, bracts, fruits, etc (compare with **appressed**, **reflexed**).

stalk. General term for a supporting structure of a plant part, e.g., the pedicel of a flower, peduncle of an inflorescence, petiole of a leaf.

stamen. Male flower part, consists of filament and anther.

staminate. A flower with stamens only (a male flower).

staminode. A sterile stamen.

stigma. The part of the pistil that receives pollen.

stipule. One of a pair of appendages at the base of petioles in eudicot leaves, may be leaflike, may be fused with each other.

stomate (plural: **stomata**). A pore surrounded by two guard cells, through which gas exchange is regulated (e.g., carbon dioxide in, oxygen out), usually on epidermis.

strigillose. Minutely strigose.

strigose. With straight, stiff, appressed hairs.

style. The (usually) elongated part of the pistil between the ovary and stigma through which pollen tubes grow.

sub- (prefix)**.** Meaning somewhat, almost, slightly, or almost. E.g., subequal phyllaries are more or less equal in length to each other, but there is slight variation (usually in comparison to another species with notably unequal phyllaries).

subshrub. A perennial with a woody base.

subspecies. A taxonomic category below the species level, technically a higher rank than variety.

subtend. Positioned just below a given part, e.g., a bract subtending a flower.

superior. An ovary that appears positioned above the point of attachment of sepals and petals in a flower (compare with **inferior**).

tepal. A perianth part of a corolla with similar-looking sepals and petals.

terminal. At apex or tip.

ternate. In groups of three; divided three ways, a compound leaf with three leaflets.

tomentose. Covered with dense, matted, woolly hairs.

translator. Threadlike part of a pollinarium derived from stigmatic secretions that connects the corpusculum with the pollinia, as in *Asclepias*.

truncate. With the base or tip squared off abruptly as if cut off.

tubercle. A small, wartlike bump or projection.

turion. A condensed, short shoot or bud that overwinters in the soil, can look like a small, clustered rosette (e.g., in some *Epilobium*).

umbel. An inflorescence with pedicels or peduncles radiating like the ribs of an umbrella.

valve. One segment of a dehiscent fruit, separated by lines of dehiscence.

variety. A taxonomic category below the species level, technically a lower rank than subspecies.

ventral. The side facing the stem or central axis, as in the upper surface of a leaf, the inner surface of a nutlet in relation to the center point where nutlets are attached; adaxial (compare with **dorsal**).

villous. With long, soft, unmatted hairs.

whorl. Three or more leaves or flowers arising from the same node, as in leaves of *Galium* spp. or flowers of *Lupinus microcarpus* var. *densiflorus*.

x.s. Cross section—a slice through an organ from side to side, e.g., to see the shape of a stem, a leaf, or a fruit and its inner morphology (compare with **l.s.**).

xylem. Conducts water and minerals from the roots to the rest of the plant; along with phloem, makes up the vascular system of the plant.

Bibliography

Abrams, Leroy, and Roxanna S. Ferris. *Illustrated Flora of the Pacific States in Four Volumes*. Stanford University Press, 1965.

Anderson, M. Kat, and Michael J. Moratto. "Native American land-use practices and ecological impacts." In *Sierra Nevada ecosystem project: final report to Congress*, vol. 2, pp. 187–206. Davis, CA: University of California, Centers for Water and Wildland Resources Davis, 1996.

Anderson, M. Kat. *Tending the Wild: Native American Knowledge and the Management of California's Natural Resources*. University of California Press, 2005.

Baldwin, Bruce G., Douglas H. Goldman, David J. Keil, Robert Patterson, Thomas J. Rosatti, and Dieter H. Wilken, eds. *The Jepson Manual: Vascular Plants of California*. University of California Press, 2012.

Bishop, Jim. "Using Alpine Flora to Measure Ecological Effects of Climate Change." *Fremontia* 38, no. 4 / 39, no. 1 (2012): 2–9.

Bliss, M. "Plant Guide for Great Valley Gumweed (*Grindelia camporum*)." USDA Natural Resources Conservation Service, Plant Materials Center, 2012. plants.usda.gov/DocumentLibrary/plantguide/pdf/pg_grca.pdf.

Botti, Stephen J. *Illustrated Flora of Yosemite National Park*. Yosemite Association, 2001.

Brink, Don. "Tuberous Aconitum (Ranunculaceae) of the continental United States: morphological variation, taxonomy and disjunction." Bulletin of the Torrey Botanical Club (1982): 13–23.

California Consortium of Herbaria 2. Biodiversity occurrence data published by CCH2 Portal. Accessed November 5, 2023. www.cch2.org/portal.

Coleman, Ronald A. *The Wild Orchids of California*. Cornell University Press, 2002.

Flora of North America Editorial Committee, eds. *Flora of North America North of Mexico*. Accessed July 21, 2024. beta.floranorthamerica.org.

Gerritsen, Mary E., and Ron Parsons. *Calochortus: Mariposa Lilies & Their Relatives*. Timber Press, 2007.

Gonzalves, Pete, and Dale Darris. "Indian Plum Fact Sheet." USDA Natural Resources Conservation Service, Plant Materials Center, 2009. plants.usda.gov/DocumentLibrary/factsheet/pdf/fs_oece.pdf.

Grant, Verne. "The pollination of *Calycanthus occidentalis*." *American Journal of Botany* 37, no. 4 (1950): 294–297.

Harris, James G., and Melinda Woolf Harris. *Plant Identification Terminology: An Illustrated Glossary*. 2nd ed. Spring Lake Publishing, 2001.

Hill, Mary. *Geology of the Sierra Nevada*. Rev. ed. Vol. 80. University of California Press, 2006.

Hynson, Nicole A., and Thomas D. Bruns. "Evidence of a Myco-heterotroph in the Plant Family Ericaceae That Lacks Mycorrhizal Specificity." *Proceedings of the Royal Society B: Biological Sciences* 276, no. 1675 (2009): 4053–4059.

Jepson Flora Project. *Jepson eFlora*. Accessed July 21, 2024. ucjeps.berkeley.edu/eflora/.

Kauffmann, Michael, Tom Parker, and Michael Vasey. *Field Guide to Manzanitas: California, North America, and Mexico*. 2nd ed. Backcountry Press, 2021.

Keeley, Jon E., and C. J. Fotheringham. "Smoke-Induced Seed Germination in California Chaparral." *Ecology* 79, no. 7 (1998): 2320–2336. doi.org/10.1890/0012-9658(1998)079[2320:SISGIC]2.0.CO;2.

Keeley, Jon E., Thomas W. McGinnis, and Kim A. Bollens. "Seed Germination of Sierra Nevada Postfire Chaparral Species." *Madroño* 52, no. 3 (2005): 175–181. doi.org/10.3120/0024-9637(2005)52[175:SGOSNP]2.0.CO;2.

Layberry, Ross A., Peter W. Hall, and J. Donald Lafontaine. *The Butterflies of Canada*. University of Toronto Press, 2018.

Lee, Gaylen D. *Walking Where We Lived: Memoirs of a Mono Indian Family*. University of Oklahoma Press, 1999.

Long, Jonathan W., Frank K. Lake, and Ron W. Goode. "The Importance of Indigenous Cultural Burning in Forested Regions of the Pacific West, USA." *Forest Ecology and Management* 500 (2021): 119597. doi.org/10.1016/j.foreco.2021.119597.

Long, Jonathan W., Frank K. Lake, Ron W. Goode, and Benrita Mae Burnette. "How Traditional Tribal Perspectives Influence Ecosystem Restoration." *Ecopsychology* 12, no. 2 (2020): 71–82. doi.org/10.1089/eco.2019.0055.

Marsh, C. Dwight, A. B. Clawson, and Hadleigh Marsh. "Zygadenus, or Death Camas." *US Department of Agriculture* 125 (1915).

Miles, Scott R., and Charles B. Goudey. "Ecological Subregions of California." USDA, Forest Service, Pacific Southwest Region, R5-EM-TP-005 (1997).

Nesom, Guy. "Blue Elderberry." *Plant Guide*. USDA, NRCS, 2000. Contributed by USDA, NRCS, National Plant Data Center, and the Biota of North America Program. plants.usda.gov/DocumentLibrary/plantguide/pdf/pg_sanic5.pdf.

North American Orchid Conservation Center. 2024. goorchids.northamericanorchidcenter.org/

Padgett, Donald J. "A Monograph of *Nuphar* (Nymphaeaceae)." *Rhodora* 109, no. 937 (2007): 1–95. www.jstor.org/stable/23314744.

Sawyer, John O., Todd Keeler-Wolf, and Julie M. Evens. *A Manual of California Vegetation*. 2nd ed. California Native Plant Society Press, 2009.

Schmidt, Marjorie G. *Growing California Native Plants*. University of California Press, 1980.

Shevock, James R. "Status of Rare and Endemic Plants." *Status of the Sierra Nevada*. Sierra Nevada Ecosystem Project: Final Report to Congress, vol. 2. Wildland Resources Center Report, no. 37. Centers for Water and Wildland Resources, University of California, David, 1996. pubs.usgs.gov/dds/dds-43/VOL_II/VII_TOC.PDF.

Spira, Timothy P., and Oren D. Pollak. "Comparative Reproductive Biology of Alpine Biennial and Perennial Gentians (Gentiana: Gentianaceae) in California." *American Journal of Botany* 73, no. 1 (1986): 39–47.doi.org/10.1002/j.1537-2197. 1986.tb09678.x.

Stevens, Michelle. "Western Blue Flag." USDA, NRCS, 2000. Contributed by USDA, NRCS, National Plant Data Team. plants.usda.gov/Document Library/plantguide/pdf/pg_irmi.pdf.

Stevens, P. F. Angiosperm Phylogeny. Version 14, July 2017. www.mobot.org/mobot/research/apweb/welcome.html

Taylor, Dean William. *Flora of the Yosemite Sierra: Being a Transect Flora of the Central Sierra Nevada, Including All of Tuolumne, Mariposa and Madera Counties, the Mono Basin, and Adjacent Areas of Mono County*. Lulu Press, 2010.

Telewski, Frank W. and Jan A. D. Zeevaart. "The 120-Yr Period for Dr. Beal's Seed Viability Experiment." *American Journal of Botany* 89, no. 8 (2002): 1285–1288. www.jstor.org/stable/4131429.

Van Ee, Josh, Matthew Westbrook, Sunita Yadav, Jens Roland, and Stephen F. Matter. "Predicting the Effects of Forest Encroachment for *Sedum lanceolatum* and *Rhodiola integrifolia* at a Subalpine Ecotone." *Alpine Botany* 125 (2015): 125–135. dx.doi.org/10.1007/s00035-015-0148-0.

Photo Credits

Unless otherwise indicated, credit is for the main photo including insets, if any. All photos not included in the following lists were taken by Stephen Sharnoff.

Acmispon americanus var. *americanus* (Joanna Clines; inset); *Actaea rubra* (Aaron Schusteff; inset); *Agoseris parviflora* (Steve Matson); *Agoseris aurantiaca* var. *aurantiaca* (Mike Russler, inset); *Alisma triviale* (Aaron Schusteff; flowers inset); *Alisma triviale* (Neal Kramer; main); *Alisma triviale* (Steve Matson; leaf inset); *Amsinckia eastwoodiae* (Michael Simpson; insets); *Antirrhinum leptaleum* (Barry Breckling; inset of *A. vexillocalyculatum* ssp. *intermedium*); *Aphyllon corymbosum* (Joanna Clines); *Apocynum cannabinum* (Joanna Clines); *Arctostaphylos manzanita* ssp. *manzanita* (Julie Kierstead; main); *Arctostaphylos manzanita* ssp. *manzanita* (Neal Kramer; inset); *Arctostaphylos mewukka* ssp. *mewukka* (Joanna Clines; inset); *Arnica parryi* (Barry Breckling); *Artemisia douglasiana* (Joanna Clines; insets); *Athysanus pusillus* (Keir Morse; flower inset); *Athysanus pusillus* (Joanna Clines; leaf inset); *Bidens frondosa* (Joanna Clines); *Boechera lemmonii* (Aaron Schusteff; inset); *Boechera retrofracta* (Steve Matson); *Bolandra californica* (Steve Matson); *Brassica nigra* (Keir Morse); *Brickellia californica* (Neal Kramer; main); *Brickellia californica* (Barry Breckling; inset); *Brodiaea elegans* ssp. *elegans* (Steve Matson; inset of *B. coronaria*); *Calochortus coeruleus* (Sylvia Sharnoff); *Calochortus luteus* (Sylvia Sharnoff, main); *Calyptridium umbellatum* (Joanna Clines); *Cardamine breweri* (Steve Matson); *Cardamine californica* (Joanna Clines); *Cardamine hirsuta* (Joanna Clines); *Centaurea solstitialis* (Joanna Clines); *Centaurea stoebe* ssp. *australis* (Joanna Clines); *Chaenactis alpigena* (Neal Kramer; main); *Chaenactis alpigena* (Steve Matson; inset); *Chlorogalum pomeridianum* var. *pomeridianum* (Joanna Clines; main); *Claytonia perfoliata* ssp. *perfoliata* (Joanna Clines; inset); *Clintonia uniflora* (Joanna Clines); *Collomia linearis* (Keir Morse); *Collomia tinctoria* (Steve Matson); *Cornus nuttallii* (Sylvia Sharnoff; main); *Cornus nuttallii* (Joanna Clines; fruit inset); *Corylus cornuta* ssp. *californica* (Joanna Clines; inset); *Crassula connata* (Keir Morse); *Cryptantha* (all taxa) (Michael Simpson; insets); *Cuscuta subinclusa* (Keir Morse; flower inset); *Cymopterus terebinthinus* var. *californicus* (Steve Matson); *Dicentra pauciflora* (Sylvia Sharnoff); *Dicentra uniflora* (Sylvia Sharnoff); *Dichelostemma multiflorum* (Joanna Clines; inset); *Diplacus bicolor* (Neal Kramer; inset); *Diplacus congdonii* (Joanna Clines); *Diplacus leptaleus* (Debra Cook); *Draba densifolia* (Steve Matson; main, leaves inset); *Draba densifolia* (Barry Breckling; fruits inset); *Draba longisquamosa* (Aaron Schusteff); *Drymocallis pseudorupestris* var. *crumiana* (Steve Matson); *Ehrendorferia chrysantha* (Joanna Clines; main); *Enemion occidentale* (Joanna Clines; main); *Epilobium brachycarpum* (Joanna Clines); *Epilobium ciliatum* ssp. *ciliatum* (Joanna Clines; inset); *Epilobium glaberrimum* ssp. *glaberrimum* (Keir Morse); *Epilobium torreyi* (Joanna Clines); *Ericameria arborescens* (Joanna Clines); *Ericameria nauseosa* var. *speciosa* (Keir Morse); *Eriophyllum lanatum* var. *integrifolium* (Joanna Clines; inset of var. *croceum*); *Eryngium castrense* (Carol Witham; inset); *Erythranthe arenaria* (Joanna Clines); *Erythranthe bicolor* (Joanna Clines; inset); *Erythranthe floribunda* (Joanna Clines); *Euphorbia albomarginata* (Neal Kramer, inset); *Eurybia radulina* (Joanna Clines); *Euthamia occidentalis* (Joanna Clines; main); *Euthamia occidentalis* (Steve Matson; inset); *Fritillaria pinetorum* (Steve Matson; main, inset of *F. atropurpurea*); *Galium aparine* (Joanna Clines); *Galium porrigens* var. *tenue* (Joanna Clines; inset);

Galium trifidum ssp. *columbianum* (Debra Cook; main); *Galium trifidum* ssp. *columbianum* (Steve Matson; inset); *Gaultheria humifusa* (Steve Matson); *Gayophytum diffusum* ssp. *parviflorum* (Keir Morse); *Gentiana calycosa* (Sylvia Sharnoff); *Gentiana newberryi* var. *tiogana* (Sylvia Sharnoff); *Gentianopsis holopetala* (Sylvia Sharnoff); *Greeneocharis circumscissa* var. *circumscissa* (Michael Simpson; insets); *Hazardia whitneyi* var. *whitneyi* (Joanna Clines); *Helianthus californicus* (Debra Cook; main); *Helianthus californicus* (Neal Kramer; inset); *Heterotheca grandiflora* (Joanna Clines; main); *Heterotheca grandiflora* (Neal Kramer; inset); *Heuchera micrantha* (Sylvia Sharnoff); *Hoita orbicularis* (Joanna Clines); *Isatis tinctoria* (Julie Kierstead; inset); *Jamesia americana* var. *rosea* (Steve Matson); *Lamium amplexicaule* (Joanna Clines); *Lasthenia debilis* (Debra Cook); *Lasthenia gracilis* (Joanna Clines); *Layia fremontii* ssp. *alba* (Neal Kramer); *Lessingia nemaclada* (Joanna Clines); *Lewisia disepala* (Joanna Clines; outcrop); *Limnanthes alba* ssp. *alba* (Neal Kramer); *Limnanthes montana* (Joanna Clines); *Linanthus parryae* (Steve Matson; inset); *Linnaea borealis* var. *longiflora* (Sylvia Sharnoff); *Listera convallarioides* (Aaron Schusteff); *Lomatium dissectum* (Joanna Clines; inset); *Lonicera conjugialis* (Joanna Clines; inset); *Lupinus andersonii* (Barry Breckling; inset); *Lupinus andersonii* (Steve Matson; main); *Lupinus covillei* (Joanna Clines); *Lupinus lepidus* var. *confertus* (Joanna Clines); *Lycopus uniflorus* (Steve Matson); *Madia exigua* (Aaron Schusteff; main); *Madia exigua* (Keir Morse; inset); *Mentha canadensis* (Barry Breckling); *Mentha pulegium* (Joanna Clines); *Mentzelia laevicaulis* var. *laevicaulis* (Sylvia Sharnoff); *Micranthes aprica* (Steve Matson; inset); *Mimetanthe pilosa* (Joanna Clines; main); *Mimetanthe pilosa* (Aaron Schusteff; inset); *Nemacladus interior* (Joanna Clines; main); *Nemophila pedunculata* (Keir Morse; inset); *Oenanthe sarmentosa* (Joanna Clines); *Oreocarya confertiflora* (Michael Simpson; insets); *Oreocarya nubigena* (Michael Simpson; insets); *Oreonana clementis* (Steve Matson); *Orochaenactis thysanocarpha* (Aaron Schusteff); *Osmorhiza berteroi* (Steve Matson); *Osmorhiza occidentalis* (Keir Morse; main); *Osmorhiza occidentalis* (Steve Matson; insets); *Oxyria digyna* (Steve Matson; inset); *Pedicularis racemosa* (Steve Matson); *Phacelia vallicola* (Joanna Clines; main); *Phoenicaulis cheiranthoides* (Aaron Schusteff; inset); *Phoradendron juniperinum* (Keir Morse); *Phoradendron leucarpum* (Joanna Clines; insets of ssp. *macrophyllum*); *Physaria occidentalis* ssp. *occidentalis* (Steve Matson; inset); *Phytolacca americana* var. *americana* (Joanna Clines); *Plagiobothrys* (all taxa) (Michael Simpson; insets); *Platystemon californicus* (Sylvia Sharnoff); *Potentilla drummondii* (Barry Breckling); *Potentilla recta* (Joanna Clines; main, upper insets); *Primula fragrans* (Sylvia Sharnoff; main); *Prunus virginiana* var. *demissa* (Joanna Clines); *Pseudognaphalium beneolens* (Joanna Clines); *Pseudognaphalium luteoalbum* (Joanna Clines); *Pycnanthemum californicum* (Neal Kramer); *Quercus chrysolepis* (Neal Kramer; acorns inset); *Ranunculus aquatilis* var. *aquatilis* (Joanna Clines); *Raphanus raphanistrum* (Joanna Clines; inset); *Rhamnus ilicifolia* (Joanna Clines; inset); *Rhododendron occidentale* (Sylvia Sharnoff); *Ribes inerme* var. *inerme* (Steve Matson); *Ribes roezlii* var. *roezlii* (Joanna Clines; inset); *Rubus armeniacus* (Joanna Clines; inset); *Rubus ursinus* (Joanna Clines; inset); *Rumex acetosella* (Joanna Clines; left inset); *Rumex paucifolius* (Joanna Clines; inset); *Sagittaria latifolia* (Keir Morse); *Sanicula crassicaulis* (Steve Matson; inset); *Sidalcea reptans* (Debra Cook); *Silene bridgesii* (Joanna Clines); *Smilax californica* (Neal Kramer; inset); *Solidago velutina* ssp. *californica* (Joanna Clines; inset); *Sorbus scopulina* (Steve Matson; inset of *S. californica*); *Spiranthes porrifolia* (Joanna Clines); *Stachys rigida* var. *rigida* (Joanna Clines); *Stellaria media* (Joanna Clines); *Trifolium depauperatum* var. *truncatum* (Joanna Clines); *Trifolium wormskioldii* (Joanna Clines); *Vaccinium uliginosum* ssp. *occidentale* (Joanna Clines; inset); *Viola sheltonii* (Steve Matson); *Viola tomentosa* (Steve Matson); *Vitis californica* (Barry Breckling; flowers inset); *Vitis californica* (Neal Kramer; purple grapes inset); *Wyethia helenioides* (Joanna Clines)

Resources

Plant Identification Apps for Smartphones

A word about artificial intelligence apps that ID plants from photos (e.g., PictureThis, PlantSnap): In my experience, these apps can be quite useful, but results should always be taken with a large grain of salt. In my informal test drives of a few of the more popular AI apps using photos I'd taken of common native plants, the app got it right less than 50 percent of the time. They do better for ornamental plants, but still get some wrong. They are a lot of fun; just keep in mind that they are far from perfect. Don't base your decision to eat a plant or not on AI apps! Seek is different from the commercially available apps in that it's connected to iNaturalist, a citizen science platform.

In addition to AI apps that require a cellular signal or WiFi to operate fully, there are several good apps that can be downloaded to your device and work without a connection. Examples are Yosemite Wildflowers, which is sold for a reasonable price; and Central California Wildflowers, which is free and covers the central SN to the coast.

Learning from Real People, in Person

A great way to learn about the plants of an area is to go on field trips with your local California Native Plant Society chapter, where you can get to know your local experts. If there's a land trust in your area, they often offer hikes focused on natural history, including native plants. Agency botanists in the US Forest Service, Bureau of Land Management, National Park Service, and California State Parks sometimes lead public wildflower walks and are often willing to help identify plants from emailed photographs and/or specimens. University herbaria (e.g., UC Davis, UC Berkeley/Jepson herbaria) assist the public with plant identification on a regular basis. Calflora and iNaturalist are citizen science apps where you can upload photos and get feedback from others. The Jepson eFlora provides a link to CalPhotos, where you can query by county, elevation, etc.; or you can go directly to CalPhotos. Keep in mind that there are some wrongly identified photos on any site used for citizen science.

Many herbaria offer one-day indoor classes and longer field classes for people who want to learn more about native plants. The Friends of the Jepson Herbarium has some beginner classes each year, and the Friends of the Ahart Herbarium in Chico offers a variety of classes, as does the Davis Botanical Society.

Additional Resources

There are excellent tools online for learning to garden with California native plants. For example, the California Native Plant Society website has tools under Calscape that allow you to enter your address and receive a list of native plants that occur in your area, as well as the Calscape Garden Planner, which guides you through some questions to help you plan the layout of your garden and select appropriate species that you'll enjoy. Cal-IPC, the California Invasive Plant Council, has great resources about invasive plants, including a section on avoiding or eliminating invasive horticultural pests, and PlantRight is a handy site to find non-invasive (often native) alternatives to invasive horticultural plants of major California regions. If you live in a wildland/urban interface area, the more you can plant natives that are known to originate nearby (within your subwatershed, within 5 mi, no more than ~150 m (500 ft) higher or lower in elevation), the better.

Index

All species with a dedicated entry are listed here in bold. Unbolded text denotes names mentioned in passing.

About the Authors

Joanna Clines has been intrigued by the flora (and fauna) of the Sierra Nevada since taking a field biology class at Fresno City College in 1982. This led to obtaining an environmental biology degree and a master's degree, both at California State University, Fresno. Her thesis research focused on reproductive biology and fire ecology of the rare shrub carpenteria (*Carpenteria californica*), a central Sierra Nevada endemic. She has been the Forest Botanist for the Sierra National Forest since 1990. She served on the boards of the California Native Plant Society, California Invasive Plant Council, and is past president of the Sequoia Chapter, CNPS. She lives in the Sierra Nevada, and can be found botanizing on or off the job, often at dusk.

Stephen Sharnoff developed a lifelong love for the Sierra Nevada through numerous backpacking trips beginning in the early 1960s. He and his late wife, Sylvia Sharnoff, did the photographic fieldwork for *Lichens of North America*, and he wrote and illustrated *A Field Guide to California Lichens*. He received a Guggenheim Foundation Fellowship in support of that work. His photographs have appeared in numerous books, exhibitions and magazines, among them *National Geographic*, *Smithsonian*, and *Science News*. He currently lives in Santa Fe, New Mexico. A collection of his photographic work can be seen at www.sharnoffphotos.com.

About Sierra College Press

The mission of the Sierra College Press is to inform and inspire scholars, students, and general readers by disseminating ideas, knowledge, and academic scholarship of value concerning the Sierra Nevada region. The Sierra College Press endeavors to reach beyond the library, laboratory, and classroom to promote and examine the Sierra Nevada. For more information, please visit www.sierracollege.edu/press.

SIERRA COLLEGE PRESS

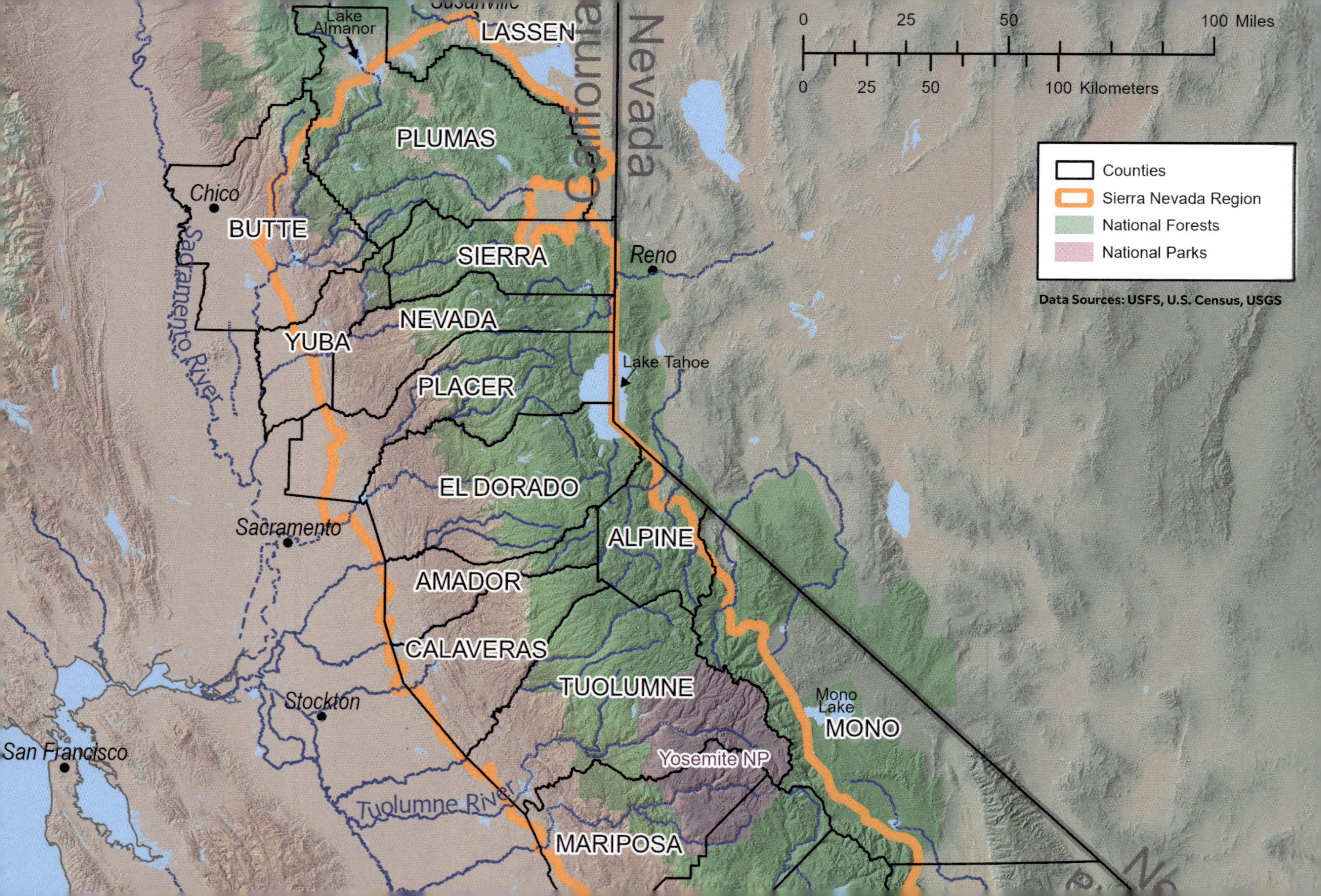

0 25 50 100 Miles
0 25 50 100 Kilometers
Counties
Sierra Nevada Region
National Forests
National Parks
Data Sources: USFS, U.S. Census, USGS
California
Nevada
Lake Almanor
LASSEN
PLUMAS
Chico
BUTTE
Sacramento River
SIERRA
Reno
NEVADA
YUBA
Lake Tahoe
PLACER
EL DORADO
Sacramento
ALPINE
AMADOR
CALAVERAS
TUOLUMNE
Stockton
Mono Lake
MONO
San Francisco
Yosemite NP
Tuolumne River
MARIPOSA